Fluid Phase Behavior for Conventional and Unconventional Oil and Gas Reservoirs

Fluid Phase Behavior for Conventional and Unconventional Oil and Gas Reservoirs

Alireza Bahadori, PhD, CEng, MIChemE, CPEng, MIEAust, NER, RPEQ
School of Environment, Science & Engineering
Southern Cross University, Lismore, NSW, Australia
Managing Director of Australian Oil and Gas Service, Pty, Ltd, Lismore, NSW, Australia

ELSEVIER

Amsterdam • Boston • Heidelberg • London
New York • Oxford • Paris • San Diego
San Francisco • Singapore • Sydney • Tokyo
Gulf Professional Publishing is an imprint of Elsevier

G P
P

Gulf Professional Publishing is an imprint of Elsevier
50 Hampshire Street, 5th Floor, Cambridge, MA 02139, United States
The Boulevard, Langford Lane, Kidlington, Oxford, OX5 1GB, United Kingdom

Library of Congress Cataloging-in-Publication Data
A catalog record for this book is available from the Library of Congress

British Library Cataloguing-in-Publication Data
A catalogue record for this book is available from the British Library

ISBN: 978-0-12-803437-8

For information on all Gulf Professional Publishing publications visit our website at https://www.elsevier.com/

Publisher: Joe Hayton
Acquisition Editor: Katie Hammon
Editorial Project Manager: Kattie Washington
Production Project Manager: Mohana Natarajan
Cover Designer: Vicky Pearson

Typeset by TNQ Books and Journals

Dedicated to the loving memory of my parents and grandparents, and to all who contributed so much to my work over the years.

CONTENTS

LIST OF CONTRIBUTORS

M.A. Ahmadi
Petroleum University of Technology (PUT), Ahwaz, Iran

A. Bahadori
Southern Cross University, Lismore, NSW, Australia
Australian Oil and Gas Services Pty Ltd, Lismore, NSW, Australia

E. Mahdavi
Sharif University of Technology, Tehran, Iran

M. Mesbah
Sharif University of Technology, Tehran, Iran

N. Rahmanian
University of Bradford, Bradford, United Kingdom

E. Soroush
Sahand University of Technology, Tabriz, Iran

M. Suleymani
Sharif University of Technology, Tehran, Iran

BIOGRAPHY

Alireza Bahadori, PhD, CEng, MIChemE, CPEng, MIEAust, NER, RPEQ is an academic staff member in the School of Environment, Science and Engineering at Southern Cross University, Lismore, New South Wales (NSW), Australia, and the managing director of Australian Oil and Gas Services, Pty Ltd (www.australianoilgas.com.au). He received his PhD from Curtin University, Perth, Western Australia. During the past 20 years, Dr. Bahadori has held various process and petroleum engineering positions and was involved in many large-scale projects at National Iranian Oil Co. (NIOC), Petroleum Development Oman (PDO), and Clough AMEC Pty Ltd.

He is the author of around 300 articles and 14 books. His books have been published by multiple major publishers, including Elsevier, John Wiley & Sons, Springer, and Taylor & Francis Group.

Dr. Bahadori is the recipient of the highly competitive and prestigious Australian Government's Endeavor International Postgraduate Research Award as part of his research in the oil and gas area. He also received a Top-Up Award from the State Government of Western Australia through Western Australia Energy Research Alliance (WA:ERA) in 2009. Dr. Bahadori serves as a member of the editorial board and reviewer for a large number of journals. He is a Chartered Engineer (CEng) and Chartered Member of Institution of Chemical Engineers, London, UK (MIChemE), Chartered Professional Engineer (CPEng) and Chartered Member of Institution of Engineers Australia (MIEAust). Registered Professional Engineer of Queensland (RPEQ). Registered Chartered Engineer of Engineering Council of United Kingdom, London, UK and Engineers Australia's National Engineering Register (NER).

PREFACE

The demand for primary energy is ever growing. As the world struggles to find new sources of energy, it is clear that the fossil fuels will continue to play a dominant role in the foreseeable future. Within the hydrocarbon family, the fastest-growing hydrocarbon is natural gas. Most estimates put the average rate of growth at 1.5–2.0%.

Unconventional oil and natural gas activity is revolutionizing the world's energy future and generating enormous economic benefits. As oil and gas production from resource plays continues to expand, substantial growth is expected in capital expenditures and industry employment to support this activity, generating millions of jobs and billions in government receipts. Even for the United States, the world's biggest gas market, this represents almost 100 years of supply. Many perceive the discovery of unconventional gas and, in particular, "Shale Gas" to be a game changer.

Growing production from unconventional sources of oil—tight oil, oil sands—is expected to provide all of the net growth in global oil supply to 2020, and over 70% of growth to 2030. By 2030, increasing production and moderating demand will result in the United States being 99% self-sufficient in net energy. A number of books are available on the market.

Fluid-phase behavior represents the behavior of hydrocarbon reservoir fluids (i.e., oil, gas, and water) during the life of a reservoir as well as the effect of changes in temperature and pressure during fluid transfer from reservoir to surface/processing facilities.

In this book, we discuss the role of pressure–volume–temperature (PVT) tests/data in various aspects of Petroleum Engineering for both conventional and unconventional reservoirs.

After introducing various laboratory facilities, PVT tests, and reports for various hydrocarbon systems are discussed in detail. This book provides the following information for both conventional and unconventional reservoirs in detail.

- Provide professionals with the knowledge on thermodynamic aspects of reservoir fluids.
- Learn the importance of PVT test design and results.
- Evaluate the quality of PVT data.
- Identify the relevant PVT data for various tasks, best practices, and avoidance of common mistakes.

- Understand the reasons behind various PVT tests.
- Understand natural gas hydrate phase behavior.
- Develop an effective knowledge of shale gas and shale oil characterization.
- Awareness of various equations of state, their strengths, and weaknesses.
- Gain perspective of fluid characterization.
- Awareness of various techniques for characterizing the heavy end.
- Appreciate the need for equation of state (EOS) tuning, the role of experimental data and parameters used for tuning.
- Gain knowledge of generating the necessary PVT input data for reservoir simulation using industry standard software.

Dr. Alireza Bahadori
School of Environment, Science & Engineering
Southern Cross University, P.O. Box 157, Lismore, New South Wales (NSW), 2480, Australia
Australian Oil and Gas Services, Pty Ltd, Lismore, NSW, Australia
www.AustralianOilGas.com.au
July 7, 2016

ACKNOWLEDGMENTS

I would like to thank the Elsevier editorial and production team, and Ms Katie Hammon and Ms Katie Washington of Gulf Professional Publishing, for their editorial assistance.

CHAPTER ONE

Oil and Gas Properties and Correlations

E. Mahdavi[1], M. Suleymani[1], N. Rahmanian[2]
[1]Sharif University of Technology, Tehran, Iran
[2]University of Bradford, Bradford, United Kingdom

1.1 INTRODUCTION

Crude oil and gas are naturally occurring mixtures composed of mainly hydrocarbons and small amounts of nonhydrocarbon compounds such as sulfur, oxygen, and nitrogen. Crude oil and gas samples are characterized in petroleum engineering by their different physical properties. The composition of reservoir fluid is known as the most significant factor, which affects its pressure–volume–temperature (PVT) behavior. The phase behavior of the reservoir fluid and reservoir temperature are two important factors; the type of reservoir fluid is determined based on them. Crude oil and gas properties are used in various steps of petroleum engineering in order to evaluate oil and gas reserves, recovery efficiency, production optimization, etc. More particularly, the phase behavior of natural gas should be addressed precisely not only for gas reservoirs but also because of its substantial role in oil production mechanisms of saturated oil reservoirs. Therefore an accurate evaluation of reservoir fluid properties is required for the modeling and simulation of oil and gas production during the lifetime of a hydrocarbon reservoir. The best source for a description of properties is laboratory experiments of actual reservoir fluid samples. However, there are many correlations that can be used in lieu of experimental data for the prediction of oil and gas properties.

In the following sections, the most physical and thermodynamic properties of crude oil (oil density, oil gravity, compressibility, bubble point pressure, solution gas ratio, oil formation volume factor, and viscosity) and gas (gas density, gas compressibility, gas formation volume factor, and viscosity) are defined, and the corresponding correlations are presented.

Fluid Phase Behavior for Conventional and Unconventional Oil and Gas Reservoirs
http://dx.doi.org/10.1016/B978-0-12-803437-8.00001-4

1.2 CRUDE OIL PROPERTIES

1.2.1 Oil Density

Density is defined as the mass of a unit volume at a specified pressure and temperature. There are several theoretical and empirical expressions for estimating oil density. The predictive capability of theoretical approaches, such as the all cubic Equation of State (EOS), for liquid density is poor, so a correction must be applied. Also, it is needed to perform a tough calculation to find the density of liquid phases. Although empirical correlations are much easier to use, they are developed from experimental data points. Therefore the correlations are valid for limited ranges of pressure, temperature, and composition. Some theoretical and empirical correlations are provided in the following sections.

1.2.1.1 Equation of State Method

Gas and liquid densities can be determined from molar volumes predicted by cubic EOS. Generally, the results of EOSs like Soave–Redlich–Kwong (SRK) give a reliable value for gas density. However, for the liquid phase it leads to an underestimation (Pedersen et al., 1984b). Péneloux et al. (1982) introduced a correction for molar volume obtained from SRK EOS. A modified form of SRK EOS proposed by Péneloux is presented:

$$P = \frac{RT}{(V-b)} - \frac{a}{[(V+c)(V+b+2c)]} \tag{1.1}$$

where c is a measure of deviation from true density. For a mixture, this parameter is obtained from

$$c = \sum_i c_i z_i \tag{1.2}$$

where z_i and c_i are the mole fraction and a constant for component i.

For nonhydrocarbon components and hydrocarbon components with a carbon number less than 7, the c_i is computed as follows:

$$c_i = 0.40768\left(\frac{RT_{c_i}}{P_{c_i}}\right)(0.29441 - (Z_{\mathrm{RA}})_i) \tag{1.3}$$

where Z_{RA} is the Racket compressibility factor defined as (Spencer and Danner, 1972):

$$(Z_{\mathrm{RA}})_i = 0.29056 - 0.08775\,\omega_i \tag{1.4}$$

Péneloux et al. (1982) suggested a correlation for the approximation of the c parameter for the C_{7+} fraction, but it only works well for the gas phase or the gas condensate phase.

1.2.1.2 Alani–Kennedy Equation

Alani and Kennedy (1960) presented an equation for the prediction of fluid density. The experimental results revealed that this equation is an effective and highly accurate method for estimating liquid phase density, but it is not reliable for vapor density calculation. Alani and Kennedy's equation is as follows:

$$V^3 - \left(\frac{RT}{P} + b\right)V^2 + a\frac{V}{P} - \frac{ab}{P} = 0 \tag{1.5}$$

where P is the pressure, psia; T is the temperature, °R; V is the specific volume, ft^3/lb mol; $R = 10.7335$ lb ft^3/in.^{2} °R lb mol; and a and b for pure components are obtained by

$$a = K\exp\left(\frac{n}{T}\right) \tag{1.6}$$

$$b = mT + C \tag{1.7}$$

K, m, n, and C are constants that are tabulated in Table 1.1.

For the C_{7+} fraction, a and b are obtained by the following equations:

$$\ln a_{C_{7+}} = 3.8405985 \times 10^{-3}\mathrm{MW} - 9.5638281 \times 10^{-4}\frac{\mathrm{MW}}{\rho} + 2.6180818 \times \frac{10^2}{T} + 7.3104464 \times 10^{-6}\mathrm{MW}^2 + 10.753517 \tag{1.8}$$

Table 1.1 Values of Constants Utilized in Eqs. (1.6) and (1.7) for Different Pure Hydrocarbons

Component	*K*	*N*	*m* × 10^4	*C*
C_1 (70–300°F)	9160.6413	61.893223	3.3162472	0.50874303
C_1 (301–460°F)	147.47333	3247.4533	−14.072637	1.8326695
C_2 (100–249°F)	46,709.573	−404.48844	5.1520981	0.52239654
C_2 (250–460°F)	17,495.343	34.163551	2.8201736	0.62309877
C_3	20,247.757	190.24420	2.1586448	0.90832519
i-C_4	32,204.420	131.63171	3.3862284	1.1013834
n-C_4	33,016.212	146.15445	2.9021257	1.1168144
n-C_5	37,046.234	299.62630	2.1954785	1.4364289
n-C_6	52,093.006	254.56097	3.6961858	1.5929406

$$b_{C_{7+}} = 3.4992740 \times 10^{-2}\text{MW} - 7.2725403\rho + 2.2323950 \times 10^{-4}T - 1.6322572 \times 10^{-2}\frac{\text{MW}}{\rho} + 6.2256545 \tag{1.9}$$

where MW is the molecular weight of the C_{7+} fraction, lb_m/lb mol; ρ is the density of C_{7+} at 14.7 psi and 520°R, g/cm^3; and T is the temperature, °R.

For a mixture, the following simple mixing rule is used for the calculation of a and b constants:

$$a = \sum_i a_i z_i \tag{1.10}$$

$$b = \sum_i b_i z_i \tag{1.11}$$

Example 1.1

Estimate oil density with the following composition at 650°R and 2000 psi.

Component	Composition
C_1	36.30
C_2	7.90
C_3	4.15
i-C_4	0.71
n-C_4	1.44
n-C_5	1.97
n-C_6	0.81
N_2	0.60
CO_2	3.34
H_2S	0.00
C_{7+}	42.78
Total	100.00

$\text{MW}_{C_{7+}} = 180$ lb/lb mol,
Specific gravity of $C_{7+} = 0.9$

Solution

a and b must be specified first for each component by using Table 1.1 and Eqs. (1.6) and (1.7), except for the C_{7+} fraction. These two parameters for the C_{7+} fraction can be computed easily by 1.8 and 1.9 equations. Therefore a, b, and MW for mixture are obtained. The results of the calculations are presented in the next table.

Composition	a_i	b_i	$a_i\,z$	$b_i\,z$	MW	MW z
C_1	10,075.8002	0.72	3657.52	0.2629	16.04	5.82
C_2	25,069.6819	0.86	1980.50	0.0677	30.07	2.38
C_3	27,132.3680	1.05	1125.99	0.0435	44.10	1.83
$i\text{-}C_4$	39,433.4339	1.32	279.98	0.0094	58.12	0.41
$n\text{-}C_4$	41,340.8747	1.31	595.31	0.0188	58.12	0.84
$n\text{-}C_5$	58,740.5470	1.58	1157.19	0.0311	72.15	1.42
$n\text{-}C_6$	77,066.0587	1.83	624.24	0.0148	86.18	0.70
N_2	4315.1959	0.68	25.89	0.0041	44.01	0.26
CO_2	9912.7851	0.51	331.09	0.0169	28.01	0.94
C_{7+}	146,274.0676	2.86	62,576.05	1.2230	180.00	77.00
Total			72,353.75	1.6922		91.602

MW, molecular weight.

The specific molar volume is the root of the following cubic equation:

$$V^3 - 5.18V^2 + 36.18V - 61.22 = 0$$

The above equation has a unique real root:

$$V = 2.0578 \ \text{ft}^3/\text{lb mol}$$

The density is given by

$$\rho = \frac{MW}{V} = \frac{91.602}{2.0578} = 44.51 \ \text{lb/ft}^3$$

1.2.1.3 Standing–Katz Method

Standing and Katz (1942a,b) originally suggested a correlation for density in a graphical form. Later, it was converted to the following set of equations by Pedersen et al. (1984b), which is used for the determination of fluid density. The results of this method are more acceptable for the liquid phase. It is important to note that all of the densities are in g/cm^3.

For the determination of density, initially the density of the ($H_2S + C_{3+}$) fraction is computed by

$$\rho_{(H_2S+C_{3+})} = \frac{\sum_i \text{MW}_i x_i}{\sum_i \frac{\text{MW}_i x_i}{\rho_i}} \tag{1.12}$$

where $\rho_{(H_2S+C_{3+})}$ is the density of the ($H_2S + C_{3+}$) fraction, g/cm^3; MW_i is the molecular weight of component i; and ρ_i is the pure component density at the standard condition, g/cm^3.

The i index includes H_2S, C_3, and heavier components with a carbon number more than 3. The densities for some of the pure components at the standard condition (at 1 atm and 15.6°C) are listed in the table below.

Component	**Density, g/cm^3**
C_3	0.5072
i-C_4	0.5625
n-C_4	0.5836
i-C_5	0.6241
n-C_5	0.6305
n-C_6	0.6850
H_2S	0.7970

The effect of the C_2 component was introduced into the model. So, the density of the ($H_2S + C_{2+}$) fraction is determined by

$$\rho_{(H_2S+C_{2+})} = \rho_{(H_2S+C_{3+})} - A_0 - A_1a_1 - A_2a_2 \tag{1.13}$$

where A_0, A_1, A_2, a_1, and a_2 are given by

$$A_0 = 0.3158\, w_2 \tag{1.14}$$

$$A_1 = -0.2583\, w_2 \tag{1.15}$$

$$A_2 = 0.01457 w_2 \tag{1.16}$$

$$a_1 = 3.3 - 5.0\, \rho_{(H_2S+C_{3+})} \tag{1.17}$$

$$a_2 = 1 + 15\left(\rho_{H_2S+C_{3+}} - 0.46\right)\left(2.5\, \rho_{(H_2S+C_{3+})} - 2.15\right) \tag{1.18}$$

and w_2 is the weight fraction of the C_2 component.

The effect of the CO_2 fraction has also been considered. Therefore the density of ($CO_2 + H_2S + C_{2+}$) is obtained by an additive volume basis using the density of the ($H_2S + C_{2+}$) fraction and the CO_2 density at standard conditions, i.e., $P = 14.7$ psi and $T = 520°R$.

Next, the density of the ($H_2S + C_{2+}$) fraction plus the C_1 and N_2 components is calculated at a standard condition as follows:

$$\rho_0 = \rho_{(CO_2+H_2S+C_{2+})} - B_0 - B_1b_1 \tag{1.19}$$

where B_0, B_1, B_2, b_1, b_2, and b_3 are obtained by

$$B_0 = 0.088255 - 0.095509b_2 + 0.007403b_3 - 0.00603b_4 \tag{1.20}$$

$$B_1 = 0.142079 - 0.150175b_2 + 0.006679b_3 + 0.001163b_4 \quad (1.21)$$

$$b_1 = \rho_{(CO_2+H_2S+C_{2+})} - 0.65 \quad (1.22)$$

$$b_2 = 1 - 10w_1 \quad (1.23)$$

$$b_3 = 1 + 30w_1(5w_1 - 1) \quad (1.24)$$

$$b_4 = 1 - 60w_1 + 750w_1^2 - 2500w_1^3 \quad (1.25)$$

and w_1 is the mole fraction of $(C_1 + N_2)$.

Afterward, the density at the standard conditions should be adjusted to the desired temperature and pressure conditions. First, the density at the prevailing pressure and standard temperature is calculated:

$$\rho_p = \rho_0 - C_0 - C_1c_1 - C_2c_2 - C_3c_3 \quad (1.26)$$

C_0, C_1, C_2, C_3, c_1, c_2, and c_3 are presented here:

$$C_0 = -0.034674 + 0.026806c_4 + 0.003705c_5 + 0.000465c_6 \quad (1.27)$$

$$C_1 = -0.022712 + 0.015148c_4 + 0.004263c_5 + 0.000218c_6 \quad (1.28)$$

$$C_2 = -0.007692 + 0.0035218c_4 + 0.002482c_5 + 0.000397c_6 \quad (1.29)$$

$$C_3 = -0.001261 - 0.0002948c_4 + 0.000941c_5 + 0.000313c_6 \quad (1.30)$$

$$c_1 = 1 - \frac{2(P - 500)}{10000} \quad (1.31)$$

$$c_2 = 1 + 6(P - 500)\left[\frac{(P - 500)}{10000} - 1\right] \Big/ 10000 \quad (1.32)$$

$$c_3 = 1 - \frac{12(P - 500)}{10000} + 30\left[\frac{(P - 500)}{10000}\right]^2 - 20\left[\frac{(P - 500)}{10000}\right]^3 \quad (1.33)$$

$$c_4 = 3.4 - 5\rho_0 \quad (1.34)$$

$$c_4 = 3.4 - 5\rho_0 \quad (1.35)$$

$$c_5 = 1 + 15(\rho_0 - 0.48)(2.5\rho_0 - 2.2) \quad (1.36)$$

$$c_6 = 1 - 30(\rho_0 - 0.48) + 187.5(\rho_0 - 0.48)^2 - 312.5(\rho_0 - 0.48)^3 \quad (1.37)$$

where P is in psi.

Finally, ρ_p must be corrected for temperature.

$$\rho = \rho_p - E_0 - E_1 e_1 - E_2 e_2 - E_3 e_3 \quad (1.38)$$

where

$$E_0 = 0.055846 - 0.060601 e_4 + 0.005275 e_5 - 0.000750 e_6 \quad (1.39)$$

$$E_1 = 0.037809 - 0.060601 e_4 + 0.012043 e_5 + 0.000455 e_6 \quad (1.40)$$

$$E_2 = 0.021769 - 0.032396 e_4 + 0.011015 e_5 + 0.000247 e_6 \quad (1.41)$$

$$E_3 = 0.009675 - 0.015500 e_4 + 0.006520 e_5 - 0.000653 e_6 \quad (1.42)$$

$$e_1 = 1 - 2\left[\frac{(T-520)}{200}\right] \quad (1.43)$$

$$e_2 = 1 + 6\left[\frac{(T-520)}{200}\right]\left[\frac{(T-520)}{200} - 1\right] \quad (1.44)$$

$$e_3 = 1 - 12\left[\frac{(T-520)}{200}\right] + 30\left[\frac{(T-520)}{200}\right]^2 - 20\left[\frac{(T-520)}{200}\right]^3 \quad (1.45)$$

$$e_4 = 3.6 - 5\rho_r \quad (1.46)$$

$$e_5 = 1 + 15(\rho_p - 0.52)(2.5\rho_p - 2.3) \quad (1.47)$$

$$e_6 = 1 - 30(\rho_p - 0.52) + 187.5(\rho_p - 0.52)^2 - 312.5(\rho_p - 0.52)^3 \quad (1.48)$$

Note that T is in °R in the above equations.

Example 1.2

Calculate the oil density for oil with the following composition at 650°R and 2000 psi by the Standing–Katz Method.

Component	Composition
C_1	6.24
C_2	3.10
C_3	3.27
i-C_4	0.89
n-C_4	2.44
n-C_5	2.20
n-C_6	3.97
N_2	0.05
CO_2	0.00
H_2S	0.68
C_{7+}	77.23

$MW_{C7+} = 180$ lb/lb mol,
specific gravity of $C_{7+} = 0.9$

The following table summarizes the procedure that was used to calculate the ($H_2S + C_{3+}$) fraction density.

Component	Mol%	% Weight	$wt_i\ MW_i$	$\rho_{sc\ i}$, g/cm^3	$wt_i\ MW_i/\rho_{sc\ i}$
C_3	0.033	1.442	0.425	0.5072	0.839
i-C_4	0.009	0.517	0.201	0.5625	0.358
n-C_4	0.024	1.418	0.552	0.5836	0.945
n-C_5	0.022	1.587	0.766	0.6305	1.215
n-C_6	0.040	3.421	1.973	0.6850	2.880
H_2S	0.007	0.0016	0.053	0.7970	0.066
C_{7+}	0.772	0.93	167.267	0.9002	185.804
Total			171.237		192.107

$$\rho(H_2S + C_{3+}) = \frac{\sum x_i MW_i}{\sum \frac{x_i MW_i}{\rho_{sc\ i}}} = \frac{171.237}{192.107} = 0.891\ \mathrm{g/cm^3}$$

In order to correct $\rho(H_2S + C_{3+})$ for the C_2 component, some parameters must first be computed. These parameters are listed in the subsequent table:

Parameter	Value
A_0	0.002
A_1	−0.002
A_2	0.000
a_1	−1.157
a_2	1.507

and then

$$\rho_{(H_2S+C_{2+})} = \rho_{(H_2S+C_{3+})} - A_0 - A_1 a_1 - A_2 a_2 = 0.887$$

Next, the density of the mixture in the presence of C_1 and N_2 (ρ_0) must be obtained. The required parameters are listed in the following table:

$$\rho_0 = \rho_{(CO_2+H_2S+C_{2+})} - B_0 - B_1 b_1 = 0.884\ \mathrm{g/cm^3}$$

(Continued)

Parameter	Value
B_0	0.001
B_1	0.008
b_1	0.237
b_2	0.932
b_3	0.802
b_4	0.624

So far, we have just achieved oil density at the standard condition. The oil density at the desired pressure and standard temperature can be obtained as follows:

$$\rho_p = \rho_0 - C_0 - C_1c_1 - C_2c_2 - C_3c_3 = 0.9864\ \mathrm{g/cm^3}$$

The following table contains all of the parameters incorporated into the above equation.

Parameter	Value
C_0	−0.059
C_1	−0.034
C_2	−0.009
C_3	0.000
c_1	0.700
c_2	0.235
c_3	−0.193
c_4	−1.020
c_5	1.060
c_6	−1.122

Finally, the density at the desired temperature and pressure is attained by

$$\rho = \rho_p - E_0 - E_1e_1 - E_2e_2 - E_3e_3 = 0.8786\ \mathrm{g/cm^3}$$

where E_0, E_1,... are listed in the succeeding table.

Parameter	Value
E_0	0.143
E_1	0.119
E_2	0.081
E_3	0.043
e_1	−0.600
e_2	0.040
e_3	0.360
e_4	−1.242
e_5	1.813
e_6	−2.925

1.2.1.4 American Petroleum Institute Method

The American Petroleum Institute (API) (Daubert and Danner, 1997) proposed the following equation for the density of a mixture at the standard conditions:

$$\rho_{\ell} = \frac{\sum_{i=1}^{n} x_i \mathrm{MW}_i}{\sum_{i=1}^{n} \frac{x_i \mathrm{MW}_i}{\rho_i^{\mathrm{o}}}} \tag{1.49}$$

where ρ^{o} is the pure component density at standard conditions, g/cm^3.

The values of density for some nonhydrocarbons and pure hydrocarbons are given in Table 1.2.

Density at standard conditions has to be corrected by C_1 and C_2, which are the density correlation factors for the standard condition and the actual condition, respectively. Densities at the desired condition and the standard condition are correlated as follows:

$$\rho = \left(\frac{C_2}{C_1}\right)\rho_{\ell} \tag{1.50}$$

Generally, the C parameter is given by

$$C = A_1 + A_2 T_r' + A_3 T_r'^2 + A_4 T_r'^3 \tag{1.51}$$

where A_i can be expressed as

$$A_i = B_1 + B_2 P_r' + B_3 P_r'^2 + B_4 P_r'^3 + B_5 P_r'^4 \tag{1.52}$$

The values of B_i for each A_i are presented in Table 1.3.

T_r' (T/T_c) and P_r' (P/P_c) are the reduced temperature and pressure of mixture, respectively. In order to calculate T_r' and P_r' for a mixture, a

Table 1.2 Density of Some Pure Hydrocarbon and Nonhydrocarbon Components

Component	Density (g/cm^3)	Component	Density (g/cm^3)
N_2	0.804	i-C_4	0.563
CO_2	0.809	n-C_4	0.584
H_2S	0.834	i-C_5	0.625
C_1	0.300	n-C_5	0.631
C_2	0.356	C_6	0.664
C_3	0.508		

Table 1.3 B_i Values of A_1, A_2, A_3, and A_4 Equations

	B_1	B_2	B_3	B_4	B_5
A_1	1.6368	−0.04615	2.1138×10^{-3}	-0.7845×10^{-5}	0.6923×10^{-6}
A_2	−1.9693	0.21874	-8.0028×10^{3}	-8.2328×10^{-5}	5.2604×10^{-6}
A_3	2.4638	−0.36461	12.8763×10^{-3}	14.8059×10^{-5}	-8.6895×10^{-6}
A_4	−1.5841	0.25136	-11.3805×10^{-3}	9.5672×10^{-5}	2.1812×10^{-6}

method for determining T_c and P_c of a mixture is required. For this purpose, molar averaging of critical properties can be applied as a simple mixing rule.

Example 1.3

Estimate the oil density for oil discussed in the previous example at 13 atm and 377K by the API method.

Solution

The calculation of oil density at the standard condition is illustrated in the following table.

Component	Mol%	MW	$\rho_{sc\,i}$ (g/cm³)	× MW	× MW/$\rho_{sc\,i}$
C_1	6.24	16.04	0.30	100.11	333.69
C_2	3.10	30.07	0.36	93.22	261.85
C_3	3.27	44.10	0.51	144.19	283.85
i-C_4	0.89	58.12	0.56	51.73	91.96
n-C_4	2.44	58.12	0.58	141.82	243.01
n-C_5	2.20	72.15	0.63	158.73	251.75
n-C_6	3.97	86.18	0.66	342.12	515.25
N_2	0.05	44.01	0.80	2.20	2.74
CO_2	0.00	28.01	0.81	0.00	0.00
H_2S	0.68	34.08	0.80	23.18	29.08
C_{7+}	77.16	180.00	0.90	13,888.80	15,432.00

MW, molecular weight.

The oil density at the standard condition is

$$\rho_\ell = \frac{\sum x_i MW_i}{\sum \frac{x_i MW_i}{\rho_{sc\,i}}} = \frac{13888.80}{15432.00} = 0.86$$

The calculation of C for condition T_c and P_c of mixture is roughly given by molar averaging critical properties, as shown in the following table. For heptane plus fraction, the empirical correlations of Riazi and Daubert (1980) are applied to determine the critical properties. Slightly more detail is given at the bottom of the table.

Component	Mol%	T_c, K	P_c, MPa	$\times T_c/100$	$\times P_c/100$
C_1	6.24	190.56	4.60	11.89	0.29
C_2	3.10	305.32	4.87	9.46	0.15
C_3	3.27	369.83	4.25	12.09	0.14
i-C_4	0.89	408.18	3.65	3.63	0.03
n-C_4	2.44	425.12	3.80	10.37	0.09
n-C_5	2.20	469.50	3.37	10.33	0.07
n-C_6	3.97	507.60	3.03	20.15	0.12
N_2	0.05	126.10	3.39	0.06	0.00
CO_2	0.00	132.92	3.50	0.00	0.00
H_2S	0.68	373.53	8.96	2.54	0.06
C_{7+}	77.16	732.78	2.19	565.41*	1.69*
Total	100.00			**645.95**	**2.65**

$^*T_c = 308\exp(-0.00013478\text{MW} - 0.61641\gamma_o)\text{MW}^{0.2998}\,\gamma_o^{1.0555}$

$^*P_c = 311.66\exp(-0.0018078\text{MW} - 0.3084\gamma_o)\text{MW}^{-0.8063}\,\gamma_o^{1.6015}$

Therefore P_c and T_c are 2.65 MPa and 645.95K, respectively.

The C parameter should be calculated for the actual condition and the standard condition as follows:

Standard Condition		Actual Condition	
A_1	1.635	A_1	1.461
A_2	−1.961	A_2	−1.095
A_3	2.450	A_3	0.997
A_4	−1.575	A_4	−0.612
C_1	1.108	C_2	1.045

So, oil density at the desired condition is

$$\rho = \left(\frac{C_2}{C_1}\right)\rho_\ell = 0.811\ \text{g/cm}^3$$

1.2.1.5 *Other Methods*

Above the bubble point pressure, density can be written using the definition of oil compressibility as

$$\rho_o = \rho_{ob}\exp[C_o(P - P_b)] \tag{1.53}$$

where ρ_{ob} is the density of oil at the bubble point pressure, lb/ft^3, and C_o is the oil compressibility at an average pressure of P and P_b, 1/psi.

Therefore the oil density can be calculated by incorporating the above equation and the empirical correlations of oil compressibility, which will be discussed later.

The following equation describes the oil density below the bubble point pressure using the oil formation volume factor, the solution gas ratio, oil specific gravity, and gas specific gravity, all of which will be defined later:

$$\rho_o = \frac{62.4\,\gamma_o + 0.0136\,R_s\gamma_g}{B_o} \tag{1.54}$$

where γ_o is the oil specific gravity; R_s is the solution gas oil ratio, SCF/STB; γ_g is the gas specific gravity; and B_o is the oil formation volume factor, bbl/STB.

It should be noted that there are several correlations for the oil formation volume factor and the solution gas ratio that can be coupled by Eq. (1.54) for estimating the oil density below the bubble point pressure.

1.2.2 Oil Gravity

Oil specific gravity is defined as the ratio of oil density at a certain pressure and temperature to the density of water at the same P and T. It is usually reported at the standard condition (60°F/60°F), i.e., a temperature of 60°F and 14.7 psi.

$$\gamma_o = \frac{\rho_o}{\rho_w} \tag{1.55}$$

where γ_o is the oil specific gravity; ρ_o is the oil density; and ρ_w is the water density.

In the petroleum engineering field, another parameter, API gravity, is usually used and is expressed as

$$\text{API} = \frac{141.5}{\gamma_o} - 131.5 \tag{1.56}$$

where γ_o is the oil specific gravity at (60°F/60°F).

1.2.3 Oil Compressibility

The pressure dependency of an oil sample is expressed by the isothermal compressibility coefficient of the oil or oil compressibility. Oil compressibility plays the most significant role in oil production as the main mechanism of oil recovery in undersaturated oil reservoirs. Oil compressibility is defined as the ratio of the change in the oil relative volume per unit pressure drop, and it is expressed as follows:

$$C_o = -\frac{1}{V}\left(\frac{dV}{dP}\right)_T \tag{1.57}$$

Above the bubble point pressure, it can be written by the following expression using the formation volume factor:

$$C_o = -\frac{1}{B_o}\left(\frac{dB_o}{dP}\right)_T \tag{1.58}$$

where C_o is the oil compressibility, 1/psi; B_o is the oil formation volume factor, bbl/STB; and P is pressure, *psi*.

There are also some correlations that can be used for the computation of oil compressibility above the bubble point pressure.

1.2.3.1 Vasquez and Beggs Correlation

Vasquez and Beggs (1980) presented a correlation for oil compressibility based on 4036 experimental data points as follows:

$$C_o = \frac{-1433 + 5\,R_{sb} + 17.2\,T - 1180\,\gamma_{gn} + 12.61\,\mathrm{API}}{10^5 P} \tag{1.59}$$

where R_{sb} is the solution gas ratio at the bubble point pressure, SCF/STB; T is the temperature, °R; and P is the pressure, *psi*.

In this correlation, it was postulated that the gas gravity depends on the separator operating condition. Gas specific gravity at 100 psig separator can be taken as a reference because most separators operate near 100 psig working pressure in oil fields. The normalized gas specific gravity is defined as follows:

$$\gamma_{gn} = \gamma_g\left[1 + 5.912\times10^{-5}\mathrm{API}\cdot T_{sep}\log\left(\frac{P_{sep}}{114.7}\right)\right] \tag{1.60}$$

where γ_{gn} is the normalized gas specific gravity at the reference separator pressure; γ_g is the gas specific gravity at the separator condition (P_{sep} and T_{sep}); T_{sep} is the separator temperature, °F; and P_{sep} is the separator pressure, *psi*.

1.2.3.2 Petrosky Correlation

Petrosky (Petrosky and Farshad, 1993) correlated the oil compressibility of oil samples above the bubble point pressure with R_{sb}, γ_g, API, T, and P by the following expression:

$$C_o = 1.705\times10^{-7}R_{sb}^{0.69357}\gamma_g^{0.1885}\mathrm{API}^{0.3272}T^{0.6729}P^{-0.5906} \tag{1.61}$$

where γ_g is gas specific gravity and T is temperature, °R.

At pressure below the bubble point pressure, the oil compressibility is defined as

$$C_o = -\frac{1}{B_o}\frac{dB_o}{dP} + \frac{B_g}{B_o}\frac{dR_s}{dP} \tag{1.62}$$

Note that in the above equation, B_g should be used in bbl/SCF.

Example 1.4

Calculate the oil compressibility for a crude oil sample with the PVT properties given below using the Vasquez and Beggs and the Petrosky correlations.

$$P = 1800\text{ psi} \quad T = 80°\text{F} \quad \text{API} = 28 \quad R_{sb} = 850\text{SCF/STB} \quad \gamma_g = 0.8$$

$$\text{Separator condition:} \quad P_{sep} = 100 \quad T_{sep} = 65\ °\text{F}$$

Solution

Vasquez and Beggs:

As the first step, the normalized gas specific gravity has to be calculated:

$$\gamma_{gn} = \gamma_g\left[1 + 5.912 \times 10^{-5}\text{API}\cdot T_{sep}\log\left(\frac{P_{sep}}{114.7}\right)\right]$$

$$\gamma_{gn} = 0.8\left[1 + 5.912 \times 10^{-5} \times 28 \times 65 \times \log\left(\frac{100}{114.7}\right)\right] = 0.795$$

$$C_o = \frac{-1433 + 5\,R_{sb} + 17.2\,T - 1180\,\gamma_{gn} + 12.61\text{ API}}{10^5 P}$$

$$C_o = \frac{-1433 + 5 \times 850 + 17.2(80 + 460) - 1180 \times 0.795 + 12.61 \times 28}{10^5 \times 1800}$$

$$= 6.4 \times 10^{-5}\text{psi}^{-1}$$

Petrosky correlation:

$$C_o = 1.705 \times 10^{-7} R_{sb}^{0.69357}\gamma_g^{0.1885}\text{API}^{0.3272}T^{0.6729}P^{-0.5906}$$

$$C_o = 1.705 \times 10^{-7} \times 850^{0.69357} \times 0.8^{0.1885} \times 28^{0.3272} \times (80 + 460)^{0.6729} \times 1800^{-0.5906}$$

$$C_o = 4.3 \times 10^{-5}\text{psi}^{-1}$$

1.2.4 Oil Bubble Point Pressure

Bubble point pressure is a crucial characteristic of the reservoir fluid that is used for forecasting reservoir performance. The bubble point pressure is defined as the highest pressure at which gas bubbles coexist with oil. Several correlations have been reported in the literature to estimate the bubble point pressure of crude oil samples. The bubble point pressure is handled as a function of solution gas oil ratio, gas gravity, oil gravity, and temperature.

1.2.4.1 Standing Correlation

Standing (1947) proposed an empirical correlation for bubble point pressure with 105 experimental data points from California oil fields. He designed a two-step flash liberation test to collect experimental data. The reported average error in this method is about 4.8%. The Standing correlation was first presented in graphical form, and later a mathematical formalism was introduced as follows:

$$P_b = 18.2\left[\left(\frac{R_s}{\gamma_g}\right)^{0.83} 10^a - 1.4\right] \tag{1.63}$$

where P_b is the bubble point pressure, in psi; R_s is the solution gas and ratio, in SCF/STB; and γ_g is gas specific gravity.

$$a = 0.00091(T - 460) - 0.0125\ \text{API} \tag{1.64}$$

where T is temperature, °R.

It should be noted that the above correlation might result in big errors in the presence of nonhydrocarbon components.

1.2.4.2 Vasquez and Beggs Correlation

Vasquez and Beggs (1980) used an extensive set of data from different oil fields for the derivation of their correlation. The subsequent formula is the result of regression over more than 5000 data points. The authors proposed the following expression:

$$P_b = \left[\left(C_1\frac{R_s}{\gamma_{gn}}\right)10^a\right]^{C_2} \tag{1.65}$$

where P_b is the bubble point pressure, psi; R_s is the solution gas and ratio, SCF/STB; and γ_{gn} is the normalized gas specific gravity at the reference separator condition (Eq. (1.60)).

Table 1.4 Coefficients of Vasquez and Beggs Correlation

Coefficient	API ≤ 30	API > 30
C_1	27.62	56.18
C_2	0.914328	0.84246
C_3	11.127	10.393

a is defined as

$$a = \frac{-C_3 \mathrm{API}}{T} \tag{1.66}$$

and the temperature unit is °R.

The constants are presented in Table 1.4.

1.2.4.3 Al-Marhoun Correlation

Based on experimental PVT data from Middle East oil mixtures, Al-Marhoun (1988) established the following equation. The author reported an average absolute relative error of 3.66%. The following correlation has been proposed based on nonlinear regression:

$$P_b = aR_s^b \gamma_g^c \gamma_o^d T^e \tag{1.67}$$

where P_b is the bubble point pressure, psi; T is the temperature, °R; R_s is the gas oil ratio, SCF/STB; γ_g is gas specific gravity; and γ_o is oil specific gravity.

The constants are as follows:

$$a = 5.38088 \times 10^{-3}, \quad b = 0.715082, \quad c = -1.87784, \quad d = 3.1437,$$
$$e = 1.32657$$

1.2.4.4 Glaso Correlation

Glaso (1980) developed a correlation for bubble point prediction based on experimental data mostly from North Sea reservoirs. The average error and standard deviation with respect to the experimental data are 1.28% and 6.98%, respectively. The Glaso correlation is more accurate than the Standing correlation for the North Sea. Glaso introduced the effect of oil paraffinicity in the presence of methane on the prediction of the gas/oil equilibrium condition. The correlation is as follows:

$$\log(P_b) = 1.7669 + 1.7447 \log(A) - 0.30218[\log(A)]^2 \tag{1.68}$$

A is given by

$$A = \left(\frac{R_s}{\gamma_g}\right)^{0.816} \frac{(T - 460)^{0.172}}{\text{API}^{0.989}} \tag{1.69}$$

where P_b is the bubble point pressure, psi; R_s is the solution gas and ratio, SCF/STB; T is the temperature, °R; and γ_g is the gas specific gravity.

1.2.4.5 Petrosky Correlation

Petrosky (Petrosky and Farshad, 1993) developed a correlation for Gulf of Mexico oil. An analysis of a set of 128 PVT data of oil mixtures has been utilized to develop a nonlinear regression model. The authors claimed that the forecasting results provided an average error of 3.28% relative to the database. Their correlation is as follows:

$$P_b = \left[\frac{112.727\ R_s^{0.577421}}{\gamma_g^{0.8439} 10^a}\right] - 1391.051 \tag{1.70}$$

where a is

$$a = 7.916 \times 10^{-4} \times (\text{API})^{1.5410} - 4.561 \times 10^{-5} \times (T - 460)^{1.3911} \tag{1.71}$$

and P_b is the bubble point pressure, psi; R_s is the solution gas and ratio, SCF/STB; T is the temperature, °R; and γ_g is the gas specific gravity.

Example 1.5

Calculate the bubble point pressure for a crude oil with the following properties by using the methods of Standing, Vasquez and Beggs, Al-Marhoun, Glaso, and Petrosky.

Property	Value
API	45
γ_{gas}	0.8
T, R	680
R_s, SCF/STB	600

Solution

Standing correlation:

$$a = 0.00091\ (T - 460) - 0.0125\ \text{API}$$
$$= 0.00091\ (600 - 460) - 0.0125 \times 45 = 0.196603917$$

(Continued)

$$P_b = 18.2\left[\left(\frac{R_s}{\gamma_g}\right)^{0.83} 10^a - 1.4\right] = 18.2\left[\left(\frac{R_s}{\gamma_g}\right)^{0.83} 10^{0.196603917} - 1.4\right]$$
$$= 2087.26$$

Vasquez and Beggs correlation:

$$a = \frac{-C_3 \text{API}}{T} = \frac{-10.393 \times 45}{680} = -0.6878$$

$$P_b = \left[\left(C_1 \frac{R_s}{\gamma_{gn}}\right) 10^a\right]^{C_2} = \left[\left(56.18 \frac{600}{.8}\right) 10^{-0.6878}\right]^{0.84246} = 2073.27$$

Al-Marhoun correlation:

$$P_b = aR_s^b \gamma_g^c \gamma_o^d T^e$$
$$= 5.38088 \times 10^{-3} \times 600^{0.715082} \times .8^{-1.87784} \times 680^{1.32657} = 2265.70$$

Glaso correlation:

$$A = \left(\frac{R_s}{\gamma_g}\right)^{0.816} \frac{(T-460)^{0.172}}{\text{API}^{0.989}} = \left(\frac{600}{.8}\right)^{0.816} \frac{(680-460)^{0.172}}{45^{0.989}} = 12.99$$

$$P_b = 10^{1.7669+1.7447 \log(A) - 0.30218[\log(A)]^2} = 2164.74$$

Petrosky correlation:

$$a = 7.916 \times 10^{-4} \times (\text{API})^{1.5410} - 4.561 \times 10^{-5} \times (T-460)^{1.3911}$$
$$= 7.916 \times 10^{-4} \times (45)^{1.5410} - 4.561 \times 10^{-5} \times (680-460)^{1.3911}$$
$$= 0.1966$$

$$P_b = \left[\frac{112.727 R_s^{0.577421}}{\gamma_g^{0.8439} 10^a}\right] - 1391.051$$
$$= \left[\frac{112.727 \times 600^{0.577421}}{0.8^{0.8439} \times 10^{0.1966}}\right] - 1391.051 = 2087.2668$$

1.2.5 Solution Gas Oil Ratio

The amount of gas dissolved in a unit volume of oil at a specified temperature and pressure is defined as the solution gas oil ratio. When the reservoir pressure is above the bubble point, all of the available gases are dissolved in

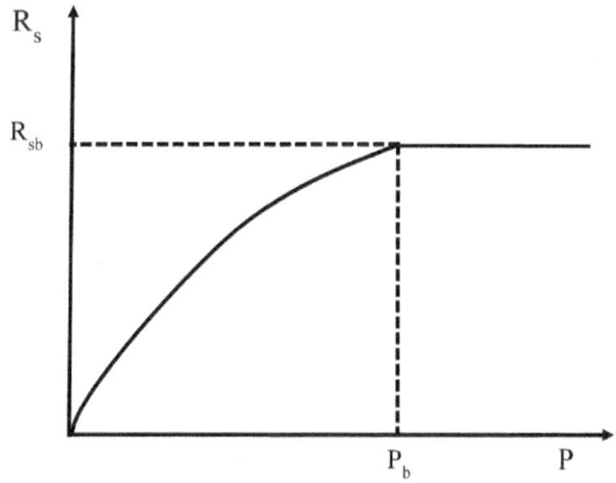

Figure 1.1 Solution gas oil ratio versus pressure.

oil, leading to a maximum and constant solution gas oil ratio. Below the bubble point pressure, the liberation of gas bubbles from crude oil reduces the solution gas oil ratio (see Fig. 1.1). It is well known that the solution gas oil ratio strongly depends on the reservoir pressure, reservoir temperature, oil density, and gas density. Some of the most popular correlations for predicting the solution gas oil ratio are presented below.

1.2.5.1 Standing Correlation

Standing's correlation (Standing, 1947) for bubble point pressure can be rearranged and written for the solution gas oil ratio. They suggested the following correlation and reported a relative average error of 4.8%:

$$R_s = \gamma_g \left[\left(\frac{P}{18.2} + 1.4 \right) 10^a \right]^{1.2048} \tag{1.72}$$

where a is defined by Eq. (1.64).

1.2.5.2 Vasquez–Beggs Correlation

The bubble point estimating correlation proposed by Vasquez and Beggs (1980) can be solved for the gas oil ratio. An analysis of 5008 measured data points has been used for constructing the following correlation of the gas oil ratio:

$$R_s = C_1 \gamma_{gn} P^{C_2} \exp\left[C_3 \left(\frac{API}{T} \right) \right] \tag{1.73}$$

where R_s is the solution gas oil ratio, SCF/STB; T is the temperature, °R; P is the pressure, psi; and γ_{gn} is calculated using Eq. (1.60). The coefficients are presented in Table 1.5.

Table 1.5 Coefficients of the Vasquez and Beggs Correlation for the Solution Gas Oil Ratio

Coefficient	API ≤ 30	API > 30
C_1	0.0362	0.0178
C_2	1.0937	1.1870
C_3	25.7240	23.931

1.2.5.3 Al-Marhoun Correlation

The Al-Marhoun (1988) bubble point pressure correlation can be solved for the gas oil ratio determination. Several oil samples from Middle East reservoirs have been subjected to research. Results of this correlation can be reliable for fluids with similar bulk properties to original oil samples used for derivation. This correlation is as follows:

$$R_s = \left[a\gamma_g^b \gamma_o^c T^d P\right]^e \tag{1.74}$$

where R_s is the solution gas oil ratio, SCF/STB; T is the temperature, °R; P is the pressure, psi; and γ_g is the gas specific gravity.

The constants are defined as

$$a = 185.843208, \quad b = 1.877840, \quad c = -3.1437, \quad d = -1.32657,$$
$$e = 1.39844$$

1.2.5.4 Glaso Correlation

A correlation for the solution gas oil ratio was derived by Glaso (1980) based on 45 North Sea crude oil samples. Glaso suggested the following correlation with an average error of 1.28%:

$$R_s = \gamma_g \left[\frac{\text{API}^{0.989}}{T^{0.172}} 10^a\right]^{1.2255} \tag{1.75}$$

where T is the temperature, °R, and γ_g is the dissolved gas specific gravity.

a is defined as

$$a = 2.8869 - [14.1811 - 3.3093 \log P]^{0.5} \tag{1.76}$$

1.2.5.5 Petrosky Correlation

As explained before, the Petrosky correlation (Petrosky and Farshad, 1993) has been developed for Gulf of Mexico reservoirs. They correlated the gas oil ratio with the temperature, pressure, gas specific gravity, and API of stock tank oil by nonlinear regression as follows:

$$R_s = \left[\left(\frac{P}{112.727} + 12.34\right)\gamma_g^{0.8439}10^a\right]^{1.73184} \tag{1.77}$$

a is defined by

$$a = 7.916 \times 10^{-4} \times \text{API}^{1.541} - 4.561 \times 10^{-5} \times (T - 460)^{1.3911} \tag{1.78}$$

where R_s is the solution gas oil ratio, SCF/STB; T is the temperature, °R; P is the pressure, psi; and γ_g is the gas specific gravity.

Example 1.6

The fluid properties of an oil reservoir are provided in the next table. The separator condition is 60°F and 100 psi. Calculate the gas oil ratio at its bubble point pressure by applying the methods of Standing, Vasquez and Beggs, Al-Marhoun, Glaso, and Petrosky.

Property	Value
API	37.9
Gamma gas	0.804
T, R	580
P_b, psi	2480

Solution

Standing correlation:

$$a = 0.00091(T - 460) - 0.0125\ \text{API}$$
$$= 0.00091(580 - 460) - 0.0125 \times 37.9 = -0.36455$$

$$R_s = \gamma_g\left[\left(\frac{P}{18.2} + 1.4\right)10^{-a}\right]^{1.2048}$$
$$= 0.804\left[\left(\frac{2480}{18.2} + 1.4\right)10^{0.36455}\right]^{1.2048} = 834.29$$

Vasquez and Beggs correlation:

$$\gamma_{gn} = \gamma_g\left[1 + 5.912 \times 10^{-5}\text{API}\cdot T_{sep}\log\left(\frac{P_{sep}}{114.7}\right)\right]$$
$$= 0.804\left[1 + 5.912 \times 10^{-5} \times 37.9 \times 60 \times \log\left(\frac{100}{114.7}\right)\right] = 0.79$$

(Continued)

$$R_s = C_1 \gamma_{gn} P^{C_2} \exp\left[C_3 \left(\frac{\text{API}}{T}\right)\right]$$

$$= 0.0178 \times 0.79 \times 2480^{1.187} \exp\left[23.931 \left(\frac{A37.9}{580}\right)\right] = 725.32$$

Al-Marhoun correlation:

$$R_s = \left[a \gamma_g^b \gamma_o^c T^d P\right]^e$$

$$= \left[185.843208 \times 0.804^{1.87784} \times 0.8353^{-3.1437} \times 580^{-1.32657} \times 2480\right]^{1.39844}$$

$$= 773.54$$

Glaso correlation:

$$a = 2.8869 - [14.1811 - 3.3093 \log P]^{0.5}$$

$$= 2.8869 - [14.1811 - 3.3093 \log 2480]^{0.5} = 1.17$$

$$R_s = \gamma_g \left[\frac{API^{0.989}}{(T-460)^{0.172}} 10^a\right]^{1.2255} = 0.804 \left[\frac{37.9^{0.989}}{(580-460)^{0.172}} 10^{1.17}\right]$$

$$= 651.83$$

Petrosky correlation:

$$a = 7.916 \times 10^{-4} \times \text{API}^{1.541} - 4.561 \times 10^{-5} \times (T - 460)^{1.3911}$$

$$= 7.916 \times 10^{-4} \times 37.9^{1.541} - 4.561 \times 10^{-5} \times (580 - 460)^{1.3911}$$

$$= 0.179$$

$$R_s = \left[\left(\frac{2480}{112.727} + 12.34\right) 0.804^{0.8439} \times 10^{0.179}\right]^{1.73184} = 677.52$$

1.2.6 Oil Formation Volume Factor

During oil production, as the oil pressure reduces at surface conditions, dissolved gas is evolved from oil; therefore oil shrinkage occurs. The relationship between the oil volume at the reservoir condition and at the surface condition is defined as the oil formation volume factor (B_o). The B_o is the ratio of oil volume at the reservoir condition to the volume of the produced oil at the standard condition. Because there is always an amount of expelled gas from crude oil at the surface condition, the oil formation volume factor is higher than unity.

$$B_o = \frac{V_{\text{res condition}}}{V_{\text{st condition}}} \tag{1.79}$$

where B_o is the oil formation volume factor, bbl/STB; $V_{\text{res condition}}$ is the oil volume at the reservoir condition, bbl; and $V_{\text{st condition}}$ is the oil volume at the standard condition, STB.

When the reservoir pressure is above the bubble point pressure, there is no free gas in the reservoir, so all of the dissolved gas is evolved at the surface. In other words, as the reservoir pressure decreases, crude oil expands. As a result, the oil formation volume factor increases slightly. On the other hand, below the bubble point pressure, as the reservoir pressure reduces during production, gas is liberated from oil in the reservoir, so the oil volume and oil formation volume factor decrease (see Fig. 1.2).

As discussed before for the calculation of oil density, using the definition of the oil formation volume factor and the material balance equation, the B_o formula can be written as

$$B_o = \frac{62.4\,\gamma_o + 0.0136\,R_s\gamma_g}{\rho_o} \tag{1.80}$$

where R_s is the solution gas oil ratio, SCF/STB, and γ_g is the gas specific gravity.

As previously mentioned, above the bubble point pressure, the oil formation volume factor increases due to the oil expansion; therefore using oil compressibility, the formation volume factor can be calculated as follows:

$$C_o = -\frac{1}{B_o}\left(\frac{dB_o}{dP}\right)_T$$

$$-\int_{P_b}^{P} C_o dP = \int_{B_{ob}}^{B_o} \frac{1}{B_o} dB_o$$

$$B_o = B_{ob}\exp[-C_o(P - P_b)] \tag{1.81}$$

where B_{ob} is the oil formation volume factor at the bubble point pressure, bbl/STB, and P_b is the bubble point pressure, psi.

The oil formation volume factor is a function of different parameters such as temperature and the solution gas oil ratio. There are several

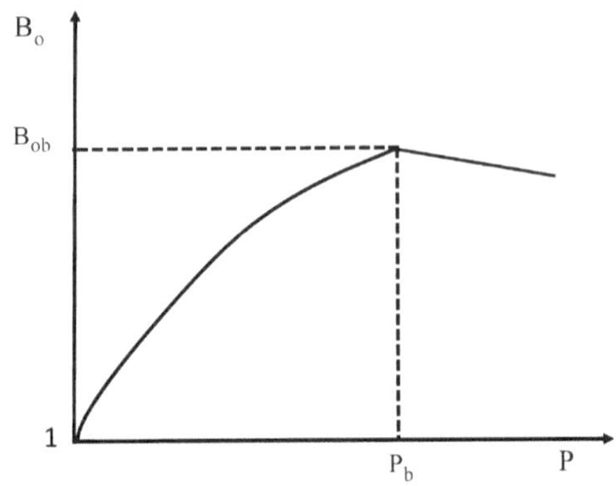

Figure 1.2 Oil formation volume factor versus pressure.

correlations that have been presented based on experimental data of oil samples from reservoirs all over the world. At the pressure equal to or below the bubble point pressure, most of the correlations use the following parameters for calculating B_o:

$$B_o = f(T, R_s, \gamma_o, \gamma_g)$$

In the following section, several correlations for the calculation of B_o are explained.

1.2.6.1 Standing Correlation

Standing (1947) showed the oil formation volume factor as a function of the solution gas oil ratio, gas specific gravity, oil specific gravity, and reservoir temperature in a graphical form using 105 experimental data points from US reservoirs. In 1981, he presented the correlation in a mathematical form as follows:

$$B_o = 0.9759 + 0.000120\left[R_s\left(\frac{\gamma_g}{\gamma_o}\right)^{0.5} + 1.25(T - 460)\right]^{1.2} \tag{1.82}$$

where R_s is the gas oil ratio, SCF/STB, and T is the temperature, °R.

An average standard error of 1.17% was reported for the correlation by the author.

1.2.6.2 Vasquez and Beggs Correlation

Vasquez and Beggs (1980) used 6000 experimental data points and developed a correlation for the oil formation volume factor. The experimental

Table 1.6 Coefficients of the Vasquez and Beggs Correlation for the Oil Formation Volume Factor

Coefficient	API ≤ 30	API > 30
C_1	4.677×10^{-4}	4.670×10^{-4}
C_2	1.751×10^{-5}	1.100×10^{-5}
C_3	-1.811×10^{-8}	1.337×10^{-9}

data covers a wide range of crude oils with API between 15.3 and 59.5. Their correlation is given by

$$B_o = 1.0 + C_1 R_s + (T - 520)\left(\frac{\text{API}}{\gamma_{gn}}\right)(C_2 + C_3 R_s) \quad (1.83)$$

γ_{gn} is the normalized gas specific gravity, as presented in Eq. (1.60). The coefficients are reported in Table 1.6.

It is worth noting that the authors reported an average relative error of 4.7% for the correlation.

1.2.6.3 Kartoatmodjo and Schmidt Correlation

Kartoatmodjo and Schmidt (1994) suggested a new correlation based on 5392 data points from oil reservoirs all over the world:

$$B_o = 0.98496 + 0.0001\ F^{1.5} \quad (1.84)$$

F is a correlating parameter that is expressed by the following equation:

$$F = R_{sb}^{0.755} \gamma_g^{0.25} \gamma_o^{-1.5} + 0.45(T - 460) \quad (1.85)$$

where T is the temperature, °R. R_{sb} is the solution gas oil ratio at the bubble point pressure in SCF/STB, presented by the following correlations:

$$R_{sb} = 0.05958\ \gamma_g^{0.7972} p^{1.0014} 10^{13.1405\ \text{API} \times T} \quad \text{API} \le 30 \quad (1.86)$$

$$R_{sb} = 0.03150\ \gamma_g^{0.7587} p^{1.0937} 10^{11.2895\ \text{API} \times T} \quad \text{API} > 30 \quad (1.87)$$

The authors reported average relative errors of −0.104% for the oil formation volume factor.

1.2.6.4 Al-Marhoun Correlation

Al-Marhoun (1988) presented a correlation using 160 experimental data points obtained from 69 bottom hole fluid samples.

$$B_o = 0.497069 + 8.62963 \times 10^{-4} T + 1.82594 \times 10^{-3} F + 3.18099 \times 10^{-6} F^2 \tag{1.88}$$

$$F = R_s^{0.742390} \gamma_g^{0.323294} \gamma_o^{-1.202040} \tag{1.89}$$

where T is the temperature, °R.

Al-Marhoun reported an average relative error of −0.01%.

1.2.6.5 Glaso Correlation

Glaso (1980) presented a correlation for estimating the oil formation volume factor. The correlation was developed using PVT data of 45 oil samples.

$$B_o = 1 + 10^{\left[-6.58511 + 2.91329 \log\left(B_{ob}^*\right) - 0.27683\left[\log\left(B_{ob}^*\right)\right]^2\right]} \tag{1.90}$$

B_{ob}^* is a correlating parameter that is defined as follows:

$$B_{ob}^* = R_s \left(\frac{\gamma_g}{\gamma_o}\right)^{0.526} + 0.986(T - 460) \tag{1.91}$$

where T is in °R.

Glaso reported an average relative error of −0.43% for the oil formation volume factor.

1.2.6.6 Petrosky Correlation

Petrosky (Petrosky and Farshad, 1993) developed a correlation based on 128 laboratory analysis as follows:

$$B_o = 1.0113 + 7.2046 \times 10^{-5} \left[R_s^{0.3738}\left(\frac{\gamma_g^{0.2914}}{\gamma_o^{0.6265}}\right) + 0.24626\, T^{0.5371}\right]^{3.0936} \tag{1.92}$$

Petrosky reported an average relative error of −0.01%.

T is in °F.

1.2.6.7 Arps Correlation

Arps (Frick, 1962) proposed the following correlation for estimating the oil formation volume factor when there is no extensive data of oil and gas samples:

$$B_o = 1.05 + 0.0005\, R_s \tag{1.93}$$

The correlation is not accurate; however, it can be used as a rough estimation for the oil formation volume factor.

The reciprocal of the oil formation volume factor is called the oil shrinkage factor:

$$b_o = \frac{1}{B_o}\ \text{STB/bbl} \tag{1.94}$$

Example 1.7

Using the following experimental data, estimate the oil formation volume factor for a crude oil sample at a pressure of 2200 psi with the methods of Standing, Vasquez and Beggs, Kartoatmodjo and Schmidt, Al-Marhoun, Glaso, and Petrosky.

$$P_b = 2800\ \text{psi}$$

$$T = 80°F$$

$$T_{sep} = 70°\text{F}$$

$$P_{sep} = 100°\text{F}$$

$$\gamma_o = 0.85$$

$$\gamma_g = 0.8$$

$$P = 2200\ \text{psi},\ \ R_s = 680\ \text{SCF/STB}$$

$$P = 2800\ \text{psi},\ \ R_s = 840\ \text{SCF/STB}$$

Solution

Standing:

$$B_o = 0.9759 + 0.000120\left[R_s\left(\frac{\gamma_g}{\gamma_o}\right)^{0.5} + 1.25(T-460)\right]^{1.2}$$

$$B_o = 0.9759 + 0.000120\left[680\times\left(\frac{0.8}{0.85}\right)^{0.5} + 1.25(540-460)\right]^{1.2}$$

$$= 1.32\ \text{bbl/STB}$$

Vasquez and Beggs:

The API should be calculated based on the values of API suitable coefficients selected from Table 1.6:

$$\text{API} = \frac{141.5}{\gamma_o} - 131.5 = \frac{141.5}{0.85} - 131.5 = 34.97$$

(Continued)

So the following coefficients should be used

Coefficient	API > 30
C_1	4.670×10^{-4}
C_2	1.100×10^{-5}
C_3	1.337×10^{-9}

$$\gamma_{gn} = \gamma_g \left[1 + 5.912 \times 10^{-5} \text{API} \cdot T_{sep} \log\left(\frac{P_{sep}}{114.7}\right)\right]$$

$$\gamma_{gn} = 0.8\left[1 + 5.912 \times 10^{-5} \times 34.97 \times 70 \times \log\left(\frac{100}{114.7}\right)\right] = 0.793$$

$$B_o = 1.0 + C_1 R_s + (T - 520)\left(\frac{\text{API}}{\gamma_{gn}}\right)(C_2 + C_3 R_s)$$

$$B_o = 1.0 + 4.670 \times 10^{-4} \times 680 + (540 - 520)\left(\frac{34.97}{0.793}\right)$$
$$\left(1.1 \times 10^{-5} + 1.337 \times 10^{-9} \times 680\right)$$

$$B_o = 1.33\ \text{bbl/STB}$$

Kartoatmodjo and Schmidt:
The solution gas ratio at the bubble point pressure is provided, so Eq. (1.87) is not required.

$$F = R_{sb}^{0.755} \gamma_g^{0.25} \gamma_o^{-1.5} + 0.45(T - 460)$$

$$F = 840^{0.755} \times 0.8^{0.25} \times 0.85^{-1.5} + 0.45(540 - 460) = 230.75$$

$$B_o = 0.98496 + 0.0001\, F^{1.5} = 0.98496 + 0.0001 \times 230.75^{1.5}$$
$$= 1.34\ \text{bbl/STB}$$

Al-Marhoun:

$$F = R_s^{0.742390} \gamma_g^{0.323294} \gamma_o^{-1.202040}$$

$$F = 680^{0.742390} \times 0.8^{0.323294} \times 0.85^{-1.202040} = 143.33$$

$$B_o = 0.497069 + 8.62963 \times 10^{-4} T + 1.82594 \times 10^{-3} F$$
$$+ 3.18099 \times 10^{-6} F^2$$

$$B_o = 0.497069 + 8.62963 \times 10^{-4} \times (80 + 460) + 1.82594 \times 10^{-3} \times 143.33 + 3.18099 \times 10^{-6} \times 143.33^2$$

$$B_o = 1.29\,\text{bbl/STB}$$

Glaso:

$$B^*_{ob} = R_s \left(\frac{\gamma_g}{\gamma_o}\right)^{0.526} + 0.986(T - 460)$$

$$B^*_{ob} = 680 \times \left(\frac{0.8}{0.85}\right)^{0.526} + 0.986(540 - 460) = 737.54$$

$$B_o = 1 + 10^{\left[-6.58511 + 2.91329 \log\left(B^*_{ob}\right) - 0.27683\left[\log\left(B^*_{ob}\right)\right]^2\right]}$$

$$B_o = 1 + 10^{\left[-6.58511 + 2.91329 \log(737.54) - 0.27683\left[\log(737.54)\right]^2\right]} = 1.31\ \text{bbl/STB}$$

Petrosky:

$$B_o = 1.0113 + 7.2046 \times 10^{-5} \left[R_s^{0.3738} \left(\frac{\gamma_g^{0.2914}}{\gamma_o^{0.6265}}\right) + 0.24626\ T^{0.5371}\right]^{3.0936}$$

$$B_o = 1.0113 + 7.2046 \times 10^{-5} \left[680^{0.3738} \left(\frac{0.8^{0.2914}}{0.85^{0.6265}}\right) + 0.24626 \times 80^{0.5371}\right]^{3.0936}$$

$$B_o = 1.29\ \text{bbl/STB}$$

1.2.7 Oil Viscosity

If an external stress is applied to a part of a fluid, it will cause movement in the direction of the fluid. The affected parts exert a portion of applied stress to nearby portions and make them move with lower velocity. Viscosity is defined as

$$\eta = \frac{\tau}{\frac{dv}{dy}} \tag{1.95}$$

where τ is the external stress and v is the velocity.

Viscosity plays a key role in fluid flow in porous media and pipes. Several correlations for oil viscosity have been developed. Generally, the authors correlated oil viscosity with oil bulk properties such as temperature and API or the composition of the fluid. Some of the more applicable correlations are presented here.

1.2.7.1 Corresponding State Method

The principle of correspondence is an effective tool for determining how a dependent variable is related to independent variables. As an example, the theory of corresponding state illustrates that for all gases the compressibility factors (Z) at the same reduced pressure (P_r) and reduced temperature (T_r) are the same (Standing and Katz, 1942a,b). The same idea has been used for the prediction of fluid viscosity, and reduced viscosity (defined as $\eta_r = \frac{\eta}{\eta_c}$) was correlated with T_r and P_r. The experimental determination of near-critical viscosity is so difficult, and there is no extensive available experimental data; thereby, some models were proposed by researchers. Hirschfelder et al. (1954) proposed the following expression for dilute gases:

$$\eta_c = \frac{P_c^{\frac{2}{3}} MW^{\frac{1}{3}}}{T_c^{\frac{1}{6}}} \tag{1.96}$$

where η_c is the near-critical viscosity; P_c is the critical pressure; T_c is the critical temperature; and MW is the molecular weight.

Therefore a complete set of viscosity data is needed for a dilute gas in order to determine the relation of η_r using T_r and P_r. Such a dilute gas will be selected as a reference component. According to the corresponding state theory, this relationship is the same for all components of a group. So, the viscosity of other components can be calculated based on the reference component. The following relationship is developed for determining the viscosity of each component at a specified temperature and pressure:

$$\eta(P, T) = \frac{\left(\frac{P_c}{P_{C_{ref}}}\right)^{\frac{2}{3}} \left(\frac{MW}{MW_{ref}}\right)^{\frac{1}{3}}}{\left(\frac{T_c}{T_{C_{ref}}}\right)^{\frac{1}{6}}} \eta_{ref}\left(P\frac{P_{C_{ref}}}{P_c}, T\frac{T_{C_{ref}}}{T_c}\right) \tag{1.97}$$

where $P_{C_{ref}}$ is the critical pressure of the reference component; $T_{C_{ref}}$ is the critical temperature of the reference component; and MW_{ref} is the molecular weight of the reference component.

Because of the extensive available set of experimental data for methane viscosity in the literature it has been selected as the reference component. The following correlation was suggested for the prediction of methane viscosity (Hanley et al., 1975):

$$\eta(\rho, T) = \eta_{\mathrm{ref}}(T) + \eta_{\ell}(T)\rho + \Delta\eta'(\rho, T) \tag{1.98}$$

where ρ is the density, mol/L; T is the temperature, K; and η_{ref} is the viscosity of reference gas, 10^{-4} cp.

Also, the correlation that describes η_{ref} is as follows:

$$\eta_{\mathrm{ref}} = \frac{\mathrm{GV}(1)}{T} + \frac{\mathrm{GV}(2)}{T^{\frac{2}{3}}} + \frac{\mathrm{GV}(3)}{T^{\frac{1}{3}}} + \mathrm{GV}(4) + \mathrm{GV}(5)T^{\frac{1}{3}} + \mathrm{GV}(7)T + \mathrm{GV}(8)T^{\frac{4}{3}} + \mathrm{GV}(9)T^{\frac{5}{3}} \tag{1.99}$$

The coefficients are presented in Table 1.7.

η_{ℓ} (in 10^{-4} cp) can be computed by the following empirical correlation. The constants are listed in Table 1.8:

$$\eta_{\ell}(T) = A + B\left(C - \ln\frac{T}{F}\right)^2 \tag{1.100}$$

$\Delta\eta'$ (in 10^{-4} cp) is expressed as

$$\Delta\eta'(\rho, T) = \exp\left(j_1 + \frac{j_4}{T}\right)\left[\exp\left[\rho^{.1}\left(j_2 + \frac{j_3}{T^{1.5}}\right) + \theta\rho^{.5}\left(j_5 + \frac{j_6}{T} + \frac{j_7}{T^2}\right)\right] - 1.0\right] \tag{1.101}$$

For the above equation constants j_1–j_7 can be found in Table 1.9, and θ is defined by

$$\theta = \frac{\rho - \rho_c}{\rho_c} \tag{1.102}$$

Table 1.7 Coefficients of Eq. (1.99)

Constant	**Value**
GV(1)	-2.090975×10^5
GV(2)	2.647269×10^5
GV(3)	-1.472818×10^5
GV(4)	4.716740×10^4

Table 1.8 Coefficients of Eq. (1.100)

Constant	Value
A	1.696985927
B	−0.133372346
C	1.4
F	168.0

Table 1.9 Constants of Eq. (1.101)

Constant	Value
j_1	−10.3506
j_2	17.5716
j_3	−3019.39
j_4	188.730
j_5	0.0429036
j_6	145.290
j_7	6127.68

McCarty (1974) proposed the following equation for methane density based on the Benedict–Webb–Rubin EOS:

$$P = \sum_{n=1}^{9} a_n(T)\rho^n + \sum_{n=10}^{15} a_n(T)\rho^{2n-17}e^{-\gamma\rho^2} \tag{1.103}$$

where P is the pressure, atm; ρ is the density, mol/L; and $R = 0.08205616$ L. atm mol^{-1} K^{-1}

The constants are listed in Table 1.10.

Finally, the methane density and viscosity are calculated at a desired temperature and pressure, and then the viscosity of other components can be obtained using the presented equations. The results of this method are in good agreement with the experimental data for light components. However, this method is not reliable for mixtures containing heavy components. Pedersen et al. (1984a) introduced the α parameter into the classical corresponding state principle, which shows deviation from the theory. They suggested the following expression for the viscosity of a mixture:

$$\eta_{\text{mix}}(P,T) = \frac{\left(\frac{P_{\text{c,mix}}}{P_{\text{c}_{\text{ref}}}}\right)^{\frac{2}{3}}\left(\frac{\text{MW}_{mix}}{\text{MW}_{\text{ref}}}\right)^{\frac{1}{3}}}{\left(\frac{T_{\text{c,mix}}}{T_{\text{c}_{\text{ref}}}}\right)^{\frac{1}{6}}} \frac{\alpha_{\text{mix}}}{\alpha_{\text{ref}}}\eta_{\text{ref}}\left(P\frac{P_{\text{c}_{\text{ref}}}}{P_{\text{c}}}\frac{\alpha_{\text{mix}}}{\alpha_{\text{ref}}}, T\frac{T_{\text{c}_{\text{ref}}}}{T_{\text{c}}}\frac{\alpha_{\text{mix}}}{\alpha_{\text{ref}}}\right) \tag{1.104}$$

Table 1.10 Constants of Eq. (1.103)
Pressures (P) in atm (1 atm = 1.01325 bar), Densities (ρ) in mol/L, and Temperature (T) in K. R = 0.08205616 L atm mol^{-1} K^{-1}

Constant	Value	Constant	Value
a_1	RT	N_{10}	$-3.7521074532 \times 10^{-5}$
a_2	$N_1T + N_2T^5 + N_3 + N_4/T + N_5/T^2$	N_{11}	$2.8616309259 \times 10^{-2}$
a_3	$N_6T + N_7 + N_8/T + N_9/T^2$	N_{12}	-2.8685298973
a_4	$N_{10}T + N_{11} + N_{12}/T$	N_{13}	$1.1906973942 \times 10^{-4}$
a_5	N_{13}	N_{14}	$-8.5315715698 \times 10^{-3}$
a_6	$N_{14}/T + N_{15}/T^2$	N_{15}	3.8365063841
a_7	N_{16}/T	N_{16}	$2.4986828379 \times 10^{-5}$
a_8	$N_{17}/T + N_{18}/T^2$	N_{17}	$5.7974531455 \times 10^{-6}$
a_9	N_{19}/T^2	N_{18}	$-7.1648329297 \times 10^{-3}$
a_{10}	$N_{20}/T^2 + N_{21}/T^2$	N_{19}	$1.2577853784 \times 10^{-4}$
a_{11}	$N_{22}/T^2 + N_{23}/T^4$	N_{20}	$2.2240102466 \times 10^{4}$
a_{12}	$N_{24}/T^2 + N_{25}/T^3$	N_{21}	$-1.4800512328 \times 10^{6}$
a_{13}	$N_{26}/T^2 + N_{27}/T^4$	N_{22}	5.0498054887×10
a_{14}	$N_{28}/T^2 + N_{29}/T^3$	N_{23}	$1.6428375992 \times 10^{6}$
a_{15}	$N_{30}/T^2 + N_{31}/T^3 + N_{32}/T^4$	N_{24}	$2.1325387196 \times 10^{-1}$
N_1	$-1.8439486666 \times 10^{-2}$	N_{25}	3.7791273422×10
N_2	1.0510162064	N_{26}	$-1.1857016815 \times 10^{-5}$
N_3	-1.6057820303×10	N_{27}	-3.1630780767×10
N_4	$8.4844027563 \times 10^{2}$	N_{28}	$-4.1006782941 \times 10^{-6}$
N_5	$-4.2738409106 \times 10^{4}$	N_{29}	$1.4870043284 \times 10^{-3}$
N_6	$7.6565285254 \times 10^{-4}$	N_{30}	$3.1512261532 \times 10^{-9}$
N_7	$-4.8360724197 \times 10^{-1}$	N_{31}	$-2.1670774745 \times 10^{-6}$
N_8	8.5195473835×10	N_{32}	$2.4000551079 \times 10^{-5}$
N_9	$-1.6607434721 \times 10^{4}$	γ	0.0096

Murad and Gubbins (1977) developed a mixing rule for the critical properties of a mixture. According to their suggestion, the critical temperature is calculated as follows:

$$T_{c,\text{mix}} = \frac{\sum_{i=1}^{N}\sum_{j=1}^{N} z_i z_j T_{c_{ij}} V_{c_{ij}}}{\sum_{i=1}^{N}\sum_{j=1}^{N} z_i z_j V_{c_{ij}}} \tag{1.105}$$

where z_i and z_j are mole fractions of the components i and j; N is the number of mixture components; and $T_{c_{ij}}$ is the critical temperature for two different components. It can be expressed as

$$T_{c_{ij}} = \sqrt{T_{c_i} T_{c_j}} \tag{1.106}$$

$V_{c_{ij}}$ is the critical molar volume for two different components and can be computed by the following equation:

$$V_{c_{ij}} = \frac{1}{8}\left(V_{c_i}^{\frac{1}{3}} + V_{c_j}^{\frac{1}{3}}\right)^3 \tag{1.107}$$

For each component V_{c_i} is described by

$$V_{c_i} = \frac{RZ_{c_i}T_{c_i}}{P_{c_i}} \tag{1.108}$$

and for the mixture

$$V_{c,\mathrm{mix}} = \sum_{i=1}^{N}\sum_{j=1}^{N} z_i z_j V_{c_{ij}} \tag{1.109}$$

For the calculation of $P_{c,\mathrm{mix}}$ and $\mathrm{MW}_{\mathrm{mix}}$ the following mixing rules are applied:

$$P_{c,\mathrm{mix}} = \frac{8\sum_{i=1}^{N}\sum_{j=1}^{N} z_i z_j \left[\left(\frac{T_{c_i}}{P_{c_i}}\right)^{\frac{1}{3}} + \left(\frac{T_{c_j}}{P_{c_j}}\right)^{\frac{1}{3}}\right]^3 \sqrt{T_{c_i}T_{c_j}}}{\left(\sum_{i=1}^{N}\sum_{j=1}^{N} z_i z_j \left[\left(\frac{T_{c_i}}{P_{c_i}}\right)^{\frac{1}{3}} + \left(\frac{T_{c_j}}{P_{c_j}}\right)^{\frac{1}{3}}\right]^3\right)^2} \tag{1.110}$$

$$\mathrm{MW}_{\mathrm{mix}} = 1.304 \times 10^{-4}\left(\overline{\mathrm{MW}}_{\mathrm{w}}^{2.303} - \overline{\mathrm{MW}}_{\mathrm{n}}^{2.303}\right) + \overline{\mathrm{MW}}_{\mathrm{n}} \tag{1.111}$$

Based on experimental data points, the following equations were proposed for $\overline{\mathrm{MW}}_{\mathrm{w}}$ and $\overline{\mathrm{MW}}_{\mathrm{n}}$:

$$\overline{\mathrm{MW}}_{\mathrm{w}} = \frac{\sum_{i=1}^{N} z_i \mathrm{MW}_i^2}{\sum_{i=1}^{N} z_i \mathrm{MW}_i} \tag{1.112}$$

$$\overline{\mathrm{MW}}_{\mathrm{n}} = \sum_{i=1}^{N} z_i \mathrm{MW}_i \tag{1.113}$$

Finally, the α parameter is defined as

$$\alpha = 1.000 + 7.378 \times 10^{-3} \rho_{\mathrm{r}}^{1.847} \mathrm{MW}^{0.5173} \tag{1.114}$$

α_{mix} and α_{ref} can be computed by replacing $\mathrm{MW}_{\mathrm{mix}}$ and $\mathrm{MW}_{\mathrm{ref}}$ in this equation, respectively.

ρ_{r} is expressed by the following equation:

$$\rho_{\mathrm{r}} = \frac{\rho_{\mathrm{o}}\left(\frac{TT_{c,\mathrm{ref}}}{T_{c,\mathrm{mix}}}, \frac{PP_{c,\mathrm{ref}}}{P_{c,\mathrm{mix}}}\right)}{\rho_{c,\mathrm{ref}}} \tag{1.115}$$

It is worth mentioning that for methane as the reference point, the critical density ($\rho_{c,ref}$) is equal to 0.16284 g/cm^3.

1.2.7.2 Lohrenz–Bary–Clark Method

Lohrenz et al. (1964) developed a widely used correlation for a mixture of petroleum fluids. They proposed that both gas and oil viscosities can be related to the reduced density (ρ_r) by a fourth order polynomial function as follows:

$$\left[(\eta - \eta^*)\xi + 10^{-4}\right]^{\frac{1}{4}} = a_1 + a_2\rho_r + a_3\rho_r^2 + a_4\rho_r^3 + a_5\rho_r^4 \qquad (1.116)$$

where η^* is the dilute gas mixture viscosity at low pressure, cp, and ξ is the viscosity reducing parameter.

The constants are presented in Table 1.11.

ξ for a mixture with an N component can be computed by the following equation:

$$\xi = \frac{\left[\sum_{i=1}^{N} z_i T_{c_i}\right]^{\frac{1}{6}}}{\left[\sum_{i=1}^{N} z_i M_i\right]^{\frac{1}{2}}\left[\sum_{i=1}^{N} z_i P_{c_i}\right]^{\frac{2}{3}}} \qquad (1.117)$$

where z_i is the mole fraction of component i.

In order to calculate ρ_r, the critical mixture density must be determined using a mixing. Lohrenz et al. (1964) suggested the following equation for the critical density of a mixture:

$$\rho_c = \frac{1}{V_c} = \frac{1}{\sum_{\substack{i=1 \\ i \neq C_{7+}}}^{N} z_i V_{ci} + z_{C_{7+}} V_{cC_{7+}}} \qquad (1.118)$$

where the critical molar volume (in ft^3/lb mol) of the C_{7+} fraction is described by

$$V_{cC_{7+}} = 21.573 + 0.015122M_i - 27.656\rho_i + .070615\rho_i M_i \qquad (1.119)$$

Table 1.11 Constants of Eq. (1.116)

Constant	Value
a_1	0.10230
a_2	0.023364
a_3	0.058533
a_4	−0.040758
a_5	0.0093324

The η^* parameter is given by Herning and Zipperer (1936):

$$\eta^* = \frac{\sum_{i=1}^{N} z_i \eta_i^* \sqrt{M_i}}{\sum_{i=1}^{N} z_i \sqrt{M_i}} \tag{1.120}$$

η_i^* is computed as follows (Stiel and Thodos, 1961):

$$\eta_i^* = 34 \times 10^{-5} \frac{1}{\xi_i} T_{ri}^{0.94} \quad \text{for } T_{ri} < 1.5 \tag{1.121}$$

$$\eta_i^* = 17.78 \times 10^{-5} \frac{1}{\xi_i} (4.58 T_{ri} - 1.67)^{\frac{5}{8}} \quad \text{for } T_{ri} > 1.5 \tag{1.122}$$

where ξ_i is expressed by the following equation:

$$\xi_i = \frac{T_{c_i}^{1/6}}{\mathrm{MW}_i^{1/2} P_{c_i}^{2/3}} \tag{1.123}$$

In the above equations, T_{c_i} and P_{c_i} are in K and atm, respectively, and the unit of computed viscosity is in mPa s.

Example 1.8

Estimate the viscosity for oil studied in Example 1.1 by the Lohrenz–Bary–Clark Method. Assume that the oil density is 0.441 g/cm^3.

Solution

For the calculation of oil viscosity, dilute gas viscosity at a low pressure is required. The method of calculation is shown in the following table. Note that the P_c is in atm.

Component	Mole Fraction	MW	T_c, K	P_c, atm	V_c, m^3/kg mol	T_r	ξ_i	η_i^*	$x_i \eta_i^* \sqrt{MW}$	$x_i \sqrt{MW}$
C_1	0.3630	16.04	190.56	45.39	0.10	1.31	0.047	0.00945	0.0136	1.44
C_2	0.0790	30.07	305.32	48.08	0.15	0.82	0.036	0.00787	0.0035	0.44
C_3	0.0415	44.10	369.83	41.92	0.20	0.68	0.033	0.00704	0.0019	0.27
i-C_4	0.0071	58.12	408.14	36.00	0.26	0.61	0.033	0.00655	0.0005	0.08
n-C_4	0.0144	58.12	425.12	37.46	0.26	0.59	0.032	0.00643	0.0005	0.08
n-C_5	0.0197	72.15	469.70	33.26	0.31	0.53	0.032	0.00592	0.0010	0.17
n-C_6	0.0081	72.15	507.60	29.85	0.37	0.49	0.035	0.00506	0.0004	0.08
C_{7+}	0.4278	180.	565.41	16.68	0.68	0.44	0.033	0.00481	0.0277	5.77
N_2	0.0006	28.00	126.10	33.50	0.09	1.98	0.041	0.01589	0.0008	0.05
CO_2	0.0334	44.01	304.19	72.85	0.09	0.82	0.022	0.01262	0.0025	0.20
H_2S	0.0000	34.08	373.53	88.46	0.10	0.67	0.023	0.01007	0.0000	0.00
Total	1.00								0.05	8.57

*By using Eq. (1.119).

MW, molecular weight.

Therefore

$$\eta* = \frac{0.05}{8.57} = 0.0061 \text{ mPa.s}$$

The critical properties of a mixture can be estimated by molar averaging the individual critical properties of species. The results are

$$\text{MW} = \sum x_i \text{MW}_i = 92.27$$

$$T_c = \sum x_i T_{c_i} = 384.14\,\text{K}$$

$$P_c = \sum x_i P_{c_i} = 33.25 \text{ atm}$$

$$V_c = \sum x_i V_{c_i} = 0.3651 \text{ m}^3/\text{kg mol} = 365.1 \text{ cm}^3/\text{g mol}$$

$$\rho_c = \frac{\text{MW}}{V_c} = \frac{\text{MW}}{\sum_{\substack{i=1 \\ i \neq C_{7+}}}^{N} z_i V_{c_i} + z_{C_{7+}} V_{cC_{7+}}} = \frac{92.27}{365.1} = 0.25 \text{ g/cm}^3$$

$$\xi = \frac{\left[\sum_{i=1}^{N} z_i T_{c_i}\right]^{\frac{1}{6}}}{\left[\sum_{i=1}^{N} z_i M_i\right]^{\frac{1}{2}} \left[\sum_{i=1}^{N} z_i P_{c_i}\right]^{\frac{2}{3}}} = 0.0271$$

By substitution of the computed parameters into Eq. (1.116) the viscosity at the actual condition can be determined as follows:

$$\rho_r = \frac{\rho}{\rho_c} = \frac{0.441}{0.25} = 1.74$$

$$\begin{aligned}\left[(\eta - 0.0061)0.0271 + 10^{-4}\right]^{\frac{1}{4}} = {} & 0.10230 + 0.023364 \times 1.74 \\ & + 0.058533 \times 1.74^2 + (-0.040758) \\ & \times 1.74^3 + 0.0093324 \times 1.74^4\end{aligned}$$

$$\eta = 0.0517 \text{ mPa.s}$$

1.2.7.3 Quiñones-Cisneros et al. Method

Quiñones-Cisneros et al. (2003) suggested a model based on friction theory. They expressed the viscosity of a mixture as the summation of two terms: the dilute gas viscosity (η_0) and the residual friction term (η_f).

$$\eta = \eta_0 + \eta_f \tag{1.124}$$

The correlation for dilute gas viscosity η_0 has already been described, and η_f is determined by the following equation:

$$\eta_f = \kappa_r P_r + \kappa_a P_a + \kappa_{rr} P_r^2 \tag{1.125}$$

where P_r and P_a are the repulsive and attractive parts of well-known equations of state.

If the van der Waals EOS is assumed for the calculation of P_r and P_a, the following equation is obtained:

$$P_r = \frac{RT}{V - b} \tag{1.126}$$

$$P_a = -\frac{a}{V^2} \tag{1.127}$$

κ_r, κ_a, and κ_{rr} are given by

$$\kappa_r = \sum_{i=1}^{N} \zeta_i \kappa_{ri} \tag{1.128}$$

$$\kappa_a = \sum_{i=1}^{N} \zeta_i \kappa_{ai} \tag{1.129}$$

$$\kappa_{rr} = \sum_{i=1}^{N} \zeta_i \kappa_{rri} \tag{1.130}$$

where the parameter ζ_i is defined as

$$\zeta_i = \frac{z_i}{\mathrm{MW}_i^{0.3} \times \mathrm{MM}} \tag{1.131}$$

and z_i is the mole fraction of component i.

MM is calculated by

$$\mathrm{MM} = \sum_{i=1}^{N} \frac{z_i}{\mathrm{MW}_i^{0.3}} \tag{1.132}$$

In order to calculate κ_r, κ_a, and κ_{rr} for a mixture, the values of κ_{ri}, κ_{ai}, and κ_{rri} for component i must first be found:

$$\kappa_{ri} = \frac{\eta_{ci}\widehat{\kappa}_{ri}}{P_{C_i}}, \tag{1.133}$$

$$\kappa_{ai} = \frac{\eta_{ci}\widehat{\kappa}_{ai}}{P_{C_i}} \tag{1.134}$$

$$\kappa_{rri} = \frac{\eta_{ci}\widehat{\kappa}_{rri}}{P_{C_i}^2} \tag{1.135}$$

$\widehat{\kappa}_{ri}$, $\widehat{\kappa}_{ai}$, and $\widehat{\kappa}_{rri}$ are defined as functions of reduced temperature. The critical viscosity for pure component can be adapted from the literature. However, the below equation was suggested for C_{7+}:

$$\eta_{ci} = K_c \frac{\sqrt{MW_i} P_{ci}^{2/3}}{T_{ci}^{1/6}} \tag{1.136}$$

1.2.7.4 Vasquez and Beggs Correlation

Vasquez and Beggs (1980) applied regression analysis to more than 3000 data points for the empirical correlation of oil viscosity at a pressure above the bubble point. Again, oil viscosity computation by this correlation needs the value of oil viscosity at the bubble point pressure. In order to do this, previous correlations can be used.

$$\mu_o = \mu_{ob}\left(\frac{P}{P_b}\right)^D \tag{1.137}$$

where D is defined by

$$D = 2.6P^{1.187}\exp\left(-11.513 - 8.98 \times 10^{-5}P\right) \tag{1.138}$$

and μ_o is the oil viscosity, cp; μ_{ob} is the dead oil viscosity at the bubble point pressure, cp; P is the pressure, psi; and P_b is the bubble point pressure, psi.

1.2.7.5 Glaso Correlation

Regression analysis for the determination of dead oil viscosity was performed on the basis of experimental data from 26 oil mixtures by Glaso (1980). Dead oil viscosity is defined as the viscosity of oil at 14.7 psia and reservoir temperature. For utilized data, the temperature and API of samples were within the range of 50–300°F and 20–48, respectively. Therefore the correlation is applicable for a wide range of crude oil samples. The proposed correlation is as follows:

$$\mu_{od} = 3.141 \times 10^{10} \times (T - 460)^{-3.444} (\log(\text{API}))^{A} \tag{1.139}$$

where A is

$$A = 10.313 \times \log(T - 460) - 36.447 \tag{1.140}$$

and μ_{od} is the dead oil viscosity, cp; R_s is the solution gas oil ratio, SCF/STB; and T is the temperature, °R.

1.2.7.6 Chew and Connally Correlation

Chew and Connally (1959) proposed an empirical correlation for oil viscosity at the bubble point pressure. This correlation takes into account the influence of the gas oil ratio on oil viscosity and corrects the dead oil viscosity for the prediction of oil viscosity at the bubble point pressure. The Chew–Connally model was originally available only as published graphs. Standing (1981) formulated this relationship as follows:

$$\mu_{ob} = 10^{a} \mu_{od}^{b} \tag{1.141}$$

where a and b are calculated by

$$a = R_s\left(2.2 \times 10^{-7} R_s - 7.4 \times 10^{-4}\right) \tag{1.142}$$

$$b = 0.68 \times 10^{c} + 0.25 \times 10^{d} + 0.062 \times 10^{e} \tag{1.143}$$

$$c = -0.0000862 R_s \tag{1.144}$$

$$d = -0.0011 R_s \tag{1.145}$$

$$e = -0.00374 R_s \tag{1.146}$$

μ_{ob} is the oil viscosity at the bubble point pressure, cp; μ_{od} is the dead oil viscosity, cp; R_s is the gas oil ratio, SCF/STB; and T is the temperature, °R.

1.2.7.7 Beggs and Robinson Correlation

Beggs and Robinson (1975) collected a comprehensive set of data on the oil viscosity (at the bubble point pressure and dead oil condition) from different oil fields, covering a wide range of pressure and temperature. In the first step, dead oil viscosity can be calculated as

$$\mu_{od} = 10^{A(T-460)^{-1.163}} - 1 \tag{1.147}$$

A is defined as

$$A = 10^{3.0324 - 0.02023 \text{ API}} \tag{1.148}$$

where μ_{od} is the dead oil viscosity, cp, and T is the temperature, °R.

For the determination of oil viscosity at the bubble point pressure, the dead oil viscosity must be corrected with respect to the influence of dissolved gas. The below correlation describes the oil viscosity at the bubble point pressure:

$$\mu_{ob} = a(\mu_{od})^b \tag{1.149}$$

a and b are as follows:

$$a = 10.715(R_s + 100)^{-0.515} \tag{1.150}$$

$$b = 5.44(R_s + 150)^{-0.338} \tag{1.151}$$

where μ_{ob} is the oil viscosity at the bubble point pressure, cp; μ_{od} is the dead oil viscosity, cp; and R_s is the gas oil ratio, SCF/STB.

1.2.7.8 Beal Correlation

According to the Beal (1946) Method, the dead oil viscosity first needs to be calculated, and then the oil viscosity above the bubble point pressure can be predicted. Originally, this relationship was presented in graphical form, but it has been converted into a mathematical expression by Standing (1981). For the calculation of dead oil viscosity, Beal studied 655 dead oil samples, mostly from US reservoirs. He found that the dead oil viscosity can be related to the API of dead oil and temperature as follows:

$$\mu_{od} = 0.32 + \frac{18 \times 10^7}{\text{API}^{4.53}} \left(\frac{360}{T - 260} \right)^A \tag{1.152}$$

A can be obtained by

$$A = 10^{0.42 + \frac{8.33}{\text{API}}} \tag{1.153}$$

where μ_{od} is the dead oil viscosity, cp, and T is the temperature, °R.

The oil viscosity above the bubble point pressure can be found by the following relationship, where the oil viscosity at the bubble point pressure is determined by predescribed correlations:

$$\mu_o = \mu_{ob} + 0.001(p - p_b)\left[0.024\mu_{ob}^{1.6} + 0.038\mu_{ob}^{0.56}\right] \tag{1.154}$$

where μ_o is the oil viscosity, cp; μ_{od} is the dead oil viscosity, cp; P is the pressure, psi; and P_b is the bubble point pressure, psi.

Example 1.9

Calculate the oil viscosity at the following conditions and a temperature of 680°R:

- dead oil
- oil at the bubble point pressure
- oil at the pressure of 5000 psi

Parameter	Value
P_b, psi	2635
API	40
$\mu_{d,\,measured}$, cp	1.3
$\mu_{b,\,measured}$, cp	0.38
R_s, SCF/STB	770

Solution

Dead oil viscosity:

Glaso method:

$$A = 10.313 \times \log(T - 460) - 36.447$$
$$= 10.313 \times \log(680 - 460) - 36.447 = -12.29$$

$$\mu_{od} = 3.141 \times 10^{10} \times (T - 460)^{-3.444}(\log(\mathrm{API}))^{-12.9}$$
$$= 3.141 \times 10^{10} \times (680 - 460)^{-3.444}(\log(40))^{-12.9} = .77\ \mathrm{cp}$$

Beal method:

$$A = 10^{0.42+\frac{8.33}{\mathrm{API}}} = 10^{0.42+\frac{8.33}{40.7}} = 161.8$$

$$\mu_{od} = \left(0.32 + \frac{1.8 \times 10^7}{40.7^{4.53}}\right)\left(\frac{360}{680 - 260}\right)^{161.8} = 1.02\ \mathrm{cp}$$

Beggs and Robinson method:

$$A = 10^{3.0324-0.02023\times 40.7} = 4.311$$

$$\mu_{od} = 10^{4.311(680-460)^{-1.163}} - 1 = 0.638\ \mathrm{cp}$$

Viscosity at the bubble point pressure:

Chew and Connally method:

$$a = R_s(2.2 \times 10^{-7} R_s - 7.4 \times 10^{-4})$$
$$= 770(2.2 \times 10^{-7} \times 770 - 7.4 \times 10^{-4}) = -0.44$$

$$c = -0.0000862\ R_s = -0.0000862 \times 770 = -0.066$$

$$d = -0.0011\ R_s = -0.0011 \times 770 = -0.847$$

$$e = -0.00374\ R_s = -0.00374 \times 770 = -2.88$$

$$\begin{aligned} b &= 0.68 \times 10^c + 0.25 \times 10^d + 0.062 \times 10^e \\ &= 0.68 \times 10^{-0.066} + 0.25 \times 10^{-0.847} + 0.062 \times 10^{-2.88} = 0.62 \end{aligned}$$

$$\mu_{ob} = 10^a \mu_{od}^b = 10^{-0.44} 1.3^{0.62} = 0.427 \text{ cp}$$

Beggs and Robinson method:

$$a = 10.715(R_s + 100)^{-0.515} = 10.715(770 + 100)^{-0.515} = 0.329$$

$$b = 5.44(R_s + 150)^{-0.338} = 5.44(770 + 150)^{-0.338} = 0.541$$

$$\mu_{ob} = a(\mu_{od})^b = 0.329(1.3)^{0.541} = 0.378 \text{ cp}$$

Viscosity at 5000 psi:

Vasquez and Beggs method:

$$\begin{aligned} D &= 2.6\ P^{1.187} \exp(-11.513 - 8.98 \times 10^{-5} P) \\ &= 2.6 \times 5000^{1.187} \exp(-11.513 - 8.98 \times 10^{-5} \times 5000) = 0.4 \end{aligned}$$

$$\mu_o = \mu_{ob} \left(\frac{P}{P_b}\right)^D = 0.38 \left(\frac{5000}{2635}\right)^{0.4} = 0.49 \text{ cp}$$

Beal method:

$$\begin{aligned} \mu_o &= \mu_{ob} + 0.001(p - p_b)\left[0.024\mu_{ob}^{1.6} + 0.038\mu_{ob}^{0.56}\right] \\ &= 0.38 + 0.001(5000 - 2635)\left[0.0240 \times 0.38^{1.6} + 0.038 \times 0.38^{0.58}\right] \\ &= 0.44 \end{aligned}$$

1.3 GAS PROPERTIES

1.3.1 Gas Density

As mentioned before, density is defined as the ratio of mass per unit volume of a material. Using a real gas law yields the following equation for gas density at a prevailing pressure and temperature. Here, MW is molecular weight, R is the universal gas constant, and Z is the gas compressibility factor.

$$\rho = \frac{P\,\mathrm{MW}}{ZRT} \tag{1.155}$$

In contrast with the liquid phase, the cubic EOSs give reliable gas densities. Therefore EOSs can be used as a suitable method for the determination of gas density.

1.3.1.1 Theoretical Determination of Gas Density

For the determination of gas density at a specified temperature and pressure, the compressibility factor is required. Standing and Katz (1942a,b) charts for the Z factor are useful tools for engineering purposes. They expressed the Z factor as a function of T_r and P_r. Standing and Katz used data from 16 natural gas mixtures over a wide range of compositions. The acceptable accuracy of the Standing–Katz charts encouraged many researchers to convert them to a set of equations. Abou-Kassem (Dranchuk and Kassem, 1975) related the Z factor to T_r and P_r over ranges of 1–3 and 2–30 for T_r and P_r, respectively, as follows:

$$Z = 1 + \left(A_1 + \frac{A_2}{T_r} + \frac{A_3}{T_r^3} + \frac{A_4}{T_r^4} + \frac{A_5}{T_r^5}\right)\rho_r + \left(A_6 + \frac{A_7}{T_r} + \frac{A_8}{T_r^3}\right)\rho_r^2 - A_9\left(\frac{A_7}{T_r} + \frac{A_8}{T_r^2}\right)\rho_r^5 + A_{10}\left(1 + A_{11}\rho_r^2\right)\left(\frac{\rho_r^2}{T_r^3}\right)\exp\left(-A_{11}\rho_r^2\right) \tag{1.156}$$

where the pseudo reduced density (ρ_r) is defined as follows

$$\rho_r = \frac{0.27P_r}{Z \times T_r} \tag{1.157}$$

The constants are presented in Table 1.12:

The critical pressure and critical temperature of a mixture are calculated by

$$P_r = \frac{P}{P_c} \quad , \quad P_c = \sum_i z_i P_{c_i} \tag{1.158}$$

$$T_r = \frac{T}{T_c} \quad , \quad T_c = \sum_i z_i T_{c_i} \tag{1.159}$$

Usually, petroleum fluids contain a nonhydrocarbon component. Wichert and Aziz (1972) presented the following relationships to consider

Table 1.12 Constants of Eq. (1.156)

Constant	Value
A_1	0.3265
A_2	−1.0700
A_3	−0.5339
A_4	0.01569
A_5	−0.05165
A_6	0.5475
A_7	−0.7361
A_8	0.1844
A_{10}	0.1056
A_{11}	0.6134
A_{12}	0.7210

the effects of nonhydrocarbon components in critical temperature and pressure calculations:

$$P_{cp}^{*} = P_{cp}\frac{T_{cp} - \Delta T_{wa}}{T_{cp} + y_{H_2S}(1 - y_{H_2S})\Delta T_{wa}} \tag{1.160}$$

$$T_{cp}^{*} = T_{cp} - \Delta T_{wa} \tag{1.161}$$

where

$$\Delta T_{wa} = a\left[\left(y_{CO_2} + y_{H_2S}\right)^{.9} - \left(y_{CO_2} + y_{H_2S}\right)^{1.6} + 0.125\left(y_{CO_2}{}^{0.5} - y_{H_2S}{}^{4}\right)^{.9}\right] \tag{1.162}$$

Note that in the above equation a is equal to 120°R or 66.666 K.

Example 1.10

Calculate the gas density at 700°R and 5000 psi for the gas given below:

Component	Mole
C_1	0.637
C_2	0.0832
C_3	0.0412
i-C_4	0.0098
n-C_4	0.0197
n-C_5	0.0036
n-C_6	0.0084
C_{7+}	0.033
N_2	0.0027
CO_2	0.155
H_2S	0.0064

$MW_{C7+} = 180$ lb/lb mol, specific gravity of $C_{7+} = 0.9$

(Continued)

Solution

First, determine the critical temperature and pressure by the simple rules of mixing:

Component	Mole	MW	T_c, °R	P_c, psi	$\times T_c$	$\times P_c$	$\times$MW
C_1	0.637	16.04	343.00	667.03	218.50	424.90	10.22
C_2	0.0832	30.07	549.57	706.63	45.72	58.79	2.50
C_3	0.0412	44.10	665.69	616.12	27.43	25.38	1.82
i-C_4	0.0098	58.12	734.65	529.10	7.20	5.19	0.57
n-C_4	0.0197	58.12	765.21	550.56	15.07	10.85	1.15
n-C_5	0.0036	72.15	845.46	488.78	3.04	1.76	0.26
n-C_6	0.0084	86.17	913.68	438.74	7.67	3.69	0.72
C_{7+}	0.033	180.00	1017.73*	245.11*	33.59	8.09	5.94
N_2	0.0027	28.00	226.98	492.26	0.61	1.33	0.08
CO_2	0.155	44.01	547.542	1070.67	84.87	165.95	6.82
H_2S	0.0064	34.08	672.35	1299.98	4.30	8.32	0.22
Total	1				448.01	714.24	30.29

*Critical properties of C_{7+} from Example 1.3
MW, molecular weight.

Suggested modification due to the presence of H_2S and CO_2

$$\Delta T_{wa} = a\left[\left(y_{CO_2} + y_{H_2S}\right)^{.9} - \left(y_{CO_2} + y_{H_2S}\right)^{1.6} + 0.125\left(y_{CO_2}{}^{0.5} - y_{H_2S}{}^{4}\right)^{.9}\right]$$

$$= 120\left[(0.155 + 0.0064)^{.9} - (0.155 + 0.0064)^{1.6} + 0.125\left(0.155^{0.5} - 0.0064^{4}\right)^{.9}\right]$$

$$= 23.24°R$$

$$T^*_{cp} = T_{cp} - \Delta T_{wa} = 448.01 - 23.24 = 424.77°\mathrm{R}$$

$$P^*_{cp} = P_{cp}\frac{T_{cp} - \Delta T_{wa}}{T_{cp} + y_{H_2S}(1 - y_{H_2S})\Delta T_{wa}}$$

$$= 714.24\frac{448.01 - 23.24}{448.01 + 0.0064 \times (1 - 0.0064) \times 23.24} = 676.96\ \mathrm{psi}$$

Abou-Kassem and Dranchuk method for the determination of the *Z* factor:

$$\rho_r = \frac{0.27\,P_r}{Z \times T_r} = \frac{0.27 \times \dfrac{5000}{676.96}}{Z \times \dfrac{770}{424.77}} = \frac{1.21}{Z}$$

$$Z = 1 + \left(0.3265 + \frac{-1.07}{1.65} + \frac{-0.5339}{1.65^3} + \frac{0.01569}{1.65^4} + \frac{-0.05165}{1.65^5}\right)\frac{1.21}{Z}$$

$$+ \left(0.5475 + \frac{-0.7361}{1.65} + \frac{0.1844}{1.65^2}\right)\left(\frac{1.21}{Z}\right)^2$$

$$- 0.1056\left(\frac{-0.7361}{1.65} + \frac{0.1844}{1.65^2}\right)\left(\frac{1.21}{Z}\right)^5$$

$$+ 0.6134\left(1 + 0.721 \times \left(\frac{1.21}{Z}\right)^2\right)\left(\frac{\left(\frac{1.21}{Z}\right)^2}{1.65^3}\right)$$

$$\exp\left(-0.721 \times \left(\frac{1.21}{Z}\right)^2\right)$$

Trial and error method:

$$Z = 0.973$$

Gas density:

$$\rho = \frac{P\,MW}{ZRT} = \frac{5000 \times 30.29}{0.973 \times 10.73 \times 700} = 20.64\ \text{lb/ft}^3$$

1.3.2 Gas Compressibility

The isothermal gas compressibility is defined as the change in relative volume per unit pressure drop at a constant temperature:

$$C_g = -\frac{1}{V}\left(\frac{dV}{dP}\right)_T \tag{1.163}$$

C_g is the gas compressibility, $\frac{1}{\text{psi}}$.

In the case of real gas

$$V = \frac{nRTZ}{P} \rightarrow \frac{dV}{dP} = nRT\frac{P\frac{dZ}{dP} - Z}{P^2} \tag{1.164}$$

$$C_g = \frac{1}{P} - \frac{1}{Z}\left[\frac{dZ}{dP}\right]_T \tag{1.165}$$

where Z is the gas compressibility factor.

For an ideal gas ($Z = 1$)

$$C_g = \frac{1}{P} \tag{1.166}$$

In terms of pseudo reduced pressure and temperature, it can be expressed as

$$P_{pr} = \frac{P}{P_{pc}} \tag{1.167}$$

$$C_g = \frac{1}{P_{pr}P_{pc}} - \frac{1}{Z}\left[\frac{dZ}{d(P_{pr}P_{pc})}\right]_{T_{pr}} \tag{1.168}$$

by using C_{pr}, which is called pseudo reduced compressibility as follows

$$C_g P_{pc} = C_{pr} \tag{1.169}$$

$$C_{pr} = \frac{1}{P_{pr}} - \frac{1}{Z}\left[\frac{dZ}{dP_{pr}}\right]_{T_{pr}} \tag{1.170}$$

P_{pr} is the pseudo reduced pressure.

1.3.3 Gas Formation Volume Factor

The gas formation volume factor is defined as the ratio of volume of a certain weight gas at the reservoir condition to the volume of the same weight of gas at the standard condition.

$$B_g = \frac{V_{res.condition}}{V_{st.conditon}} \tag{1.171}$$

Using the real gas law

$$B_g = \frac{\frac{nZT}{P}}{\frac{nT_{sc}}{P_{sc}}} = \frac{P_{sc}}{T_{sc}}\frac{T}{P} \tag{1.172}$$

in field unit $P_{sc} = 14.7$ psi and $T_{sc} = 520°R$

$$B_g = 0.02829\frac{ZT}{P}\ ft^3/SCF \tag{1.173}$$

or

$$1\ bbl = 5.615\ ft^3 \rightarrow B_g = 0.00504\frac{ZT}{P}\ bbl/SCF \tag{1.174}$$

where B_g is the gas formation volume factor, ft^3/SCF; Z is the gas compressibility factor; T is the temperature, °R; and P is pressure, psi.

The gas expansion factor is the reciprocal of the gas formation volume factor:

$$E_g = \frac{1}{B_g} \tag{1.175}$$

$$E_g = 35.4\frac{P}{ZT}\ \text{SCF/ft}^3 \tag{1.176}$$

$$E_g = 198.6\frac{P}{ZT}\ \text{SCF/bbl} \tag{1.177}$$

Example 1.11

A reservoir with a pore volume of 100 million m^3 and a temperature of 240°F is selected for natural gas storage. Using the gas composition presented in Example 1.10 and the following experimental PVT data, calculate the volume of gas (in SCF) that can be stored in the reservoir at a pressure of 3000 psi.

Solution

$$E_g = 35.4\frac{P}{ZT}\ \text{SCF/ft}^3$$

From Example 1.10, at $T = 700°\text{R}$ and $P = 5000$ psi for this gas sample, Z is equal to 0.973:

$$E_g = 35.4\frac{P}{ZT} = 35.4\frac{5000}{0.973 \times 700} = 259.87\ \text{SCF/ft}^3$$

$$\begin{aligned} V_{g,ST} &= V_{g,res} \times E_g \\ &= \left[100 \times 10^6 \text{m}^3 \times \left(\frac{1\ \text{ft}}{0.3048\ \text{m}}\right)^3\right] \times 259.87\ \text{SCF/ft}^3 \\ &= 917.7 \times 10^9 \text{SCF} \end{aligned}$$

1.3.4 Total Formation Volume Factor

For the purpose of simplifying the material balance equation expressions, the total oil formation volume is defined as the ratio of the total volume of oil and its dissolved gas at the reservoir condition to the one STB produced oil:

$$B_t = B_o + B_g(R_{sb} - R_s) \tag{1.178}$$

where B_t is the total formation volume factor, bbl/STB, and R_{sb} is the solution gas oil ratio at the bubble point pressure, SCF/STB.

Some correlations for estimating B_t are provided below.

1.3.4.1 Al-Marhoun Correlation

Al-Marhoun (1988) proposed the following correlation:

$$B_t = 0.314693 + 1.06253 \times 10^{-5} F + 1.8883 \times 10^{-11} F^2 \tag{1.179}$$

$$F = \frac{R_s^{0.644516} \gamma_o^{0.724874} T^{2.00621}}{\gamma_g^{1.079340} p^{0.761910}} \tag{1.180}$$

where R_s is the original solution gas oil ratio, i.e., the summation of dissolved gas and evolved gas at the prevailing pressure.

1.3.4.2 Glaso Correlation

Glaso (1980) suggested a correlation for the estimation of the total formation volume factor:

$$\log B_t = 0.080135 + 0.47257 \log B_t^* + 0.17351 \left[\log B_t^*\right]^2 \tag{1.181}$$

where B_t^* is a correlating parameter as follows

$$B_t^* = \frac{R_s T^{0.5} \gamma_o^a}{\gamma_g^{0.3} p^{1.1089}} \tag{1.182}$$

$$a = \frac{2.9}{10^{0.00027\, R_s}} \tag{1.183}$$

and R_s is the original solution gas oil ratio, i.e., the summation of dissolved gas and evolved gas at the prevailing pressure.

1.3.5 Gas Viscosity

In this section some of the empirical correlations for gas viscosity are introduced. Generally, gas viscosity can be determined precisely by correlations as a function of temperature, pressure, and composition. For liquids, viscosity increases by increasing the pressure or decreasing the temperature. The behavior of gas viscosity with respect to the temperature is different from liquids, and it decreases by increasing the temperature. Both gas and liquids exhibit the same trend with regard to the impact of the pressure.

1.3.5.1 Carr et al. Method

Carr et al. (1954) proposed a correlation for natural gas viscosity. Their model was originally published in graphical form. Standing (1951) and Dempsey (1965) used their results and generated some correlations based on the proposed graphs. Initially, the viscosity of natural gas at the atmospheric pressure and the desired temperature is predicted by the following equation:

$$\mu_h = \left[1.709 \times 10^{-5} - 2.062 \times 10^{-6}\gamma_g\right](T - 460) + 8.188 \times 10^{-3} - 6.15 \times 10^{-3} \log \gamma_g \tag{1.184}$$

where μ_h is in cp and T is in °R.

Note that the presence of nonhydrocarbon components can significantly affect natural gas viscosity at the atmospheric pressure. Thus the correction of μ_h is crucial:

$$\mu_l = \mu_h + \lambda_{N_2} + \lambda_{CO_2} + \lambda_{H_2S} \tag{1.185}$$

where μ_l is in cp. The equations that describe λ_{N_2}, λ_{CO_2}, λ_{H_2S} and are as follows:

$$\lambda_{N_2} = y_{N_2} \times 10^{-3}[9.59 + 8.48 \log S_g] \tag{1.186}$$

$$\lambda_{CO_2} = y_{CO_2} \times 10^{-3}[6.24 + 9.08 \log S_g] \tag{1.187}$$

$$\lambda_{H_2S} = y_{H_2S} \times 10^{-3}[3.73 + 8.49 \log S_g] \tag{1.188}$$

After a correction for the nonhydrocarbon component, another correction should be applied to account for the pressure change from the atmospheric pressure to the desired pressure. Finally, the viscosity of natural gas at the desired temperature and pressure (μ_g) in cp is obtained as follows:

$$\ln\left(T_r\frac{\mu_g}{\mu_l}\right) = a_0 + a_1P_r + a_2P_r^2 + a_3P_r^3 + T_r\left(a_4 + a_5P_r + a_6P_r^2 + a_7P_r^3\right) + T_r^2\left(a_8 + a_9P_r + a_{10}P_r^2 + a_{11}P_r^3\right) + T_r^3\left(a_{12} + a_{13}P_r + a_{14}P_r^2 + a_{15}P_r^3\right) \tag{1.189}$$

Table 1.13 Constants of Eq. (1.189)

Constant	Value	Constant	Value
a_0	−2.46211820E-00	a_8	−7.93385684E-01
a_1	2.97054714E-00	a_9	1.39643306E-00
a_2	−2.86264054E-01	a_{10}	−1.49144925E-01
a_3	8.05420522E-03	a_{11}	4.41015512E-03
a_4	80860949E-00	a_{12}	8.39387178E-02
a_5	−3.49803305E-00	a_{13}	−1.86408848E-01
a_6	3.60373020E-01	a_{14}	2.03367881E-02
a_7	−1.04432413E-02	a_{15}	−6.09579263E-04

where the dimensionless parameters P_r and T_r are reduced pressure and temperature, respectively.

This correlation can be used in the ranges of 1−3 for gas pseudo reduced temperature and 1−20 for gas reduced pressure. The constants of a_0−a_{15} are listed in Table 1.13.

Example 1.12

Determine the viscosity at 660°R and 5000 psi for the gas given below:

Component	Mole Fraction
C_1	0.8400
C_2	0.0512
C_3	0.0071
i-C_4	0.0241
n-C_4	0.0197
n-C_5	0.0036
n-C_6	0.0017
N_2	0.0027
CO_2	0.0013
H_2S	0.0486

Solution

MW, P_c, and T_c can be calculated as follows:

$$\gamma_g = \frac{20.07}{28.97} = 0.69, \quad T_r = 1.68, \quad P_r = 7.22$$

Component	Mole	MW	×MW	T_c, R	P_c, psi	$\times T_c$	$\times P_c$
C_1	0.8400	16.04	13.48	343.01	667.03	288.13	560.31
C_2	0.0512	30.07	1.54	549.58	706.63	28.14	36.18
C_3	0.0071	44.10	0.31	665.69	616.12	4.73	4.37
i-C_4	0.0241	58.12	1.40	734.65	529.10	17.71	12.75
n-C_4	0.0197	58.12	1.15	765.22	550.56	15.07	10.85
n-C_5	0.0036	72.15	0.26	845.46	488.78	3.04	1.76
n-C_6	0.0017	86.17	0.14	913.68	438.74	1.53	0.73
N_2	0.0027	28.00	0.08	226.98	492.26	0.61	1.33
CO_2	0.0013	44.01	0.06	547.54	1070.67	0.71	1.39
H_2S	0.0486	34.08	1.66	672.35	1299.98	32.70	63.22
Total	1		20.07			392.36	692.89

MW, molecular weight.

$$\begin{aligned}\mu_h &= \left[1.709\times10^{-5} - 2.062\times10^{-6}\gamma_g\right](T-460) + 8.188\times10^{-3} \\ &\quad - 6.15\times10^{-3}\log\gamma_g \\ &= \left[1.709\times10^{-5} - 2.062\times10^{-6}\times0.69\right](660-460) + 8.188 \\ &\quad \times10^{-3} - 6.15\times10^{-3}\log 0.69 \\ &= 0.01\ \text{cp}\end{aligned}$$

$$\begin{aligned}\lambda_{N_2} &= y_{N_2}\times10^{-3}[9.59 + 8.48\log S_g] \\ &= 0.0027\times10^{-3}[9.59 + 8.48\log - 0.69] = 2.2\times10^{-5}\end{aligned}$$

$$\begin{aligned}\lambda_{CO_2} &= y_{CO_2}\times10^{-3}[6.24 + 9.08\log S_g] \\ &= 0.0013\times10^{-3}[6.24 + 9.08\log 0.69] = 0.0062\end{aligned}$$

$$\begin{aligned}\lambda_{H_2S} &= y_{H_2S}\times10^{-3}[3.73 + 8.49\log S_g] \\ &= 0.4863\times10^{-3}[3.73 + 8.49\log 0.69] = 0.00029\end{aligned}$$

$$\begin{aligned}\mu_l &= \mu_h + \lambda_{N_2} + \lambda_{CO_2} + \lambda_{H_2S} = 0.01 + 2.2\times10^{-5} \\ &\quad + 0.0062 + 0.00029 = 0.0166\ \text{cp}\end{aligned}$$

by using Eq. (1.189)

$$\frac{\mu_g}{\mu_l} = 2.08 \rightarrow \mu_g = 2.08\times\mu_l = 2.08\times0.0166 = 0.0346\ \text{cp}$$

1.3.5.2 Lee et al. Method

Lee et al. (1966) suggested the following correlation for the prediction of gas viscosity (in cp).

$$\eta = 10^{-4} k_v \exp\left[x_v \left(\frac{\rho}{62.4}\right)^{y_v}\right] \tag{1.190}$$

where ρ is the gas density (in lb_m/ft^3) at the prevailing pressure and temperature. The following equation can be used to calculate x_v, y_v, and k_v:

$$x_v = 3.448 + \frac{986.4}{T} + 0.01009\ \mathrm{MW} \tag{1.191}$$

$$y_v = 2.4 - 0.2 x_v \tag{1.192}$$

$$k_v = \frac{(9.379 + 0.0160\ \mathrm{MW})T^{1.5}}{209.2 + 19.26\ \mathrm{MW} + T} \tag{1.193}$$

where MW and T are the molecular weight of mixture (in lb_m/lb mol) and temperature (in °R), respectively. This correlation is based on experimental data measured in the range of 560–800°R for temperature and up to 8000 psi for pressure. For most engineering purposes, the Lee et al. model provides results with acceptable accuracy (a standard deviation of ±3%).

Example 1.13

Calculate the viscosity for gas given in Example 1.10 at 700°R and 5000 psi using the Lee et al. method.

Solution

$$x_v = 3.448 + \frac{986.4}{T} + 0.01009\ \mathrm{MW} = 3.448 + \frac{986.4}{700} + 0.01009 \times 30.29$$
$$= 5.16$$

$$y_v = 2.4 - 0.2x_v = 2.4 - 0.2 \times 5.615 = 1.368 = 1.367$$

$$k_v = \frac{(9.379 + 0.0160 \times 30.29)700^{1.5}}{209.2 + 19.26 \times 30.29 + 700} = 122.39$$

$$\eta = 10^{-4} k_v \exp\left[x_v\left(\frac{\rho}{62.4}\right)^{y_v}\right] = 10^{-4} \times 122.39 \exp\left[5.16\left(\frac{20.16}{62.4}\right)^{1.367}\right]$$
$$= 0.036\ \mathrm{cp}$$

1.4 INTERFACIAL TENSION

Interfacial tension (IFT) is known as one of the most important parameters affecting sweep efficiency, particularly during enhanced oil recovery processes such as gas flooding. When two immiscible phases are in contact, the surface layer between the two phases is in tension due to an imbalance of molecular forces at the interface. The forces on the molecules located on the surface of a phase differing from the molecules in the bulk and the surface layer tend to form the smallest area. Generally, the IFT between a liquid and a vapor is called surface tension.

IFT is defined as the energy required to impose an increase in the surface area.

$$\sigma = \left[\frac{\partial G}{\partial A}\right]_{T,V,N} \tag{1.194}$$

There are several experimental methods for IFT measurement such as the pendant drop method, the spinning drop method, the Wilhelmy plate method, and the ring method. Moreover, some models and correlations have been proposed for estimating the IFT of different fluid systems.

1.4.1 Parachor Model

This model was proposed for the IFT prediction of pure compounds using the density of two phases by Macleod (1923) and Sugden (1932) as follows:

$$\sigma = \left[\mathrm{Pa}\left(\rho_{\mathrm{m}}^{\mathrm{L}} - \rho_{\mathrm{m}}^{\mathrm{g}}\right)\right]^4 \tag{1.195}$$

where σ is the IFT, mN/m; $\rho_{\mathrm{m}}^{\mathrm{L}}$ is the molar density of the liquid phase, mol/cm^3; $\rho_{\mathrm{m}}^{\mathrm{g}}$ is the molar density of the gas phase, mol/cm^3; and Pa is the Parachor value for pure components.

The equation was modified for hydrocarbon mixtures using an averaging technique:

$$\sigma = \left[\rho_{\mathrm{m}}^{\mathrm{L}} \sum x_i \mathrm{Pa}_i - \rho_{\mathrm{m}}^{\mathrm{g}} \sum y_i \mathrm{Pa}_i\right]^4 \tag{1.196}$$

where x_i is the mole fraction of the component i in the liquid phase, and y_i is the mole fraction of the component i in the gas phase.

Note that the Parachor value of a component in a mixture and in the pure form is the same. Generally, the model is used for IFT prediction of liquid–vapor systems in the petroleum industry; however, the Parachor

Table 1.14 Parachor Values of Pure Components

Component	Parachor	Component	Parachor
C_1	77.0	n-C_7	312.5
C_2	108.0	n-C_8	351.5
C_3	150.3	n-C_9	393.0
i-C_4	181.5	n-C_{10}	433.5
n-C_4	189.9	N_2	41.0
i-C_5	225.0	CO_2	78.0
n-C_5	233.9	H_2S	80.1
n-C_6	271.0		

model considers each component of a mixture independently, and therefore it does not account for the mass transfers between two phases. The Parachor values of pure components are given in Table 1.14.

There are several correlations for estimating the Parachor value of components using MW. The Parachor of the C_{7+} component can be calculated by

$$\mathrm{Pa}_{C_{7+}} = 59.3 + 2.34\ \mathrm{MW}_{C_{7+}} \tag{1.197}$$

Example 1.14

The equilibrium composition of a crude oil and its associated gas is presented in the following table. Also, some PVT data are available below. Calculate the IFT of the system.

$$\mathrm{MW}_{C_{7+}} = 220$$

$$\rho_L = 50\ \mathrm{lb/ft^3}$$

$$\rho_g = 16\ \mathrm{lb/ft^3}$$

Component	Liquid Composition	Component	Gas Composition
C_1	0.10	C_1	0.66
C_2	0.09	C_2	0.08
C_3	0.04	C_3	0.05
i-C_4	0.08	i-C_4	0.05
n-C_4	0.07	n-C_4	0.02
i-C_5	0.10	i-C_5	0.05
n-C_5	0.09	n-C_5	0.04
n-C_6	0.05	n-C_6	0.03
C_{7+}	0.38	C_{7+}	0.02

Solution

$$MW_L = \sum x_i MW_i = 116.34$$

$$MW_g = \sum y_i MW_i = 32.68$$

$$\rho_m^L = \frac{\rho_L}{62.4 \times MW_L} = \frac{50}{62.4 \times 116.34} = 0.0069$$

$$\rho_m^g = \frac{\rho_g}{62.4 \times MW_g} = \frac{16}{62.4 \times 32.68} = 0.0078$$

$$Pa_{C_{7+}} = 59.3 + 2.34\ MW_{C_{7+}} = 59.3 + 2.34 \times 220 = 574.1$$

$$\sigma = \left[\rho_m^L \sum x_i Pa_i - \rho_m^g \sum y_i Pa_i\right]^4$$
$$= (0.0069 \times 326.5 - 0.0078 \times 120.1)^4 = 2.99 \text{ dyne/cm}$$

Problems

1.1 Estimate the oil density with the following composition at 650°R and 2000 psi using the Alani–Kennedy, Standing–Katz, and API methods.

Component	Composition
C_1	0.145
C_2	0.081
C_3	0.065
i-C_4	0.007
n-C_4	0.032
n-C_5	0.014
n-C_6	0.007
N_2	0.003
H_2S	0.006
C_{7+}	0.640

$MW_{C7+} = 220$ lb/lb mol,
specific gravity of $C_{7+} = 0.92$

1.2 The following experimental data are available for a crude oil sample. Calculate the oil compressibility factor and the total formation volume factor at 3200 and 1700 psi.

$$T = 200\ F \quad API = 32 \quad \gamma_g = 0.75$$

Pressure (psi)	B_o (bbl/STB)	R_s (SCF/STB)
3500	1.30	
3200	1.32	
2900	1.35	
2600	1.38	672
2300	1.31	659
2000	1.26	540
1700	1.21	452
1400	1.13	338
1100	1.08	249

1.3 In the table below, the oil sample properties for a reservoir are shown. Calculate the bubble point pressure using the Standing, Vasquez–Beggs, Al-Marhoun, Glaso, and Petrosky methods.

Property	Value
API	41.2
γ_{gas}	0.83
T, R	650
Rs, SCF/STB	760

1.4 Calculate the oil formation volume factor of the crude oil with the following available data using the Standing, Vasquez and Beggs, and Petrosky correlations. Calculate the absolute average error for each method using the experimental value of 1.529 for the formation volume factor.

$$T = 260\ ^\circ F \quad P = P_b = 2051\ psi \quad R_s = 693$$

$$P_{sep} = 100\ psi \quad T_{sep} = 72\ psi \quad API = 48.6$$

$$\gamma_g = 0.9$$

1.5 Calculate the gas formation volume factor for a gas mixture with a specific gravity of 0.8 and a density of 10 lb/ft^3 at the reservoir pressure and temperature.

1.6 Compute the viscosity using the Lohrenz–Bary–Clark and Quiñones-Cisnero methods for oil given as

Component	Composition
C_1	0.2836
C_2	0.1193
C_3	0.0756
i-C_4	0.0133
n-C_4	0.0373
n-C_5	0.0144
n-C_6	0.0487
N_2	0.0042
H_2S	0.0013
C_{7+}	0.4023

$MW_{C7+} = 205$ lb/lb mol,
specific gravity of $C_{7+} = 0.83$

1.7 The following table provides measured properties for an oil sample. According to the Vasquez–Beggs, Glaso, Chew–Connally, Beggs–Robinson, and Beal methods, calculate

a. dead oil viscosity at 620°R

b. oil viscosity at the bubble point pressure

c. at 4300 psi and 620°R

Parameter	Value
P_b, psi	3000
API	35
$\mu_{d,\ \text{measured}}$, cp	2.58
$\mu_{b,\ \text{measured}}$, cp	0.54
R_s, SCF/STB	793

1.8 The gas composition for a gas reservoir is presented in the following table. Calculate the gas compressibility factor (Z) for 700°R and 2600 psi using the Abou-Kassem and Dranchuk method, and then determine the gas density.

Component	Mole
C_1	0.9246
C_2	0.0318
C_3	0.0101
i-C_4	0.0028
n-C_4	0.0024
n-C_5	0.0013
n-C_6	0.0014
N_2	0.0013
CO_2	0.0051
H_2S	0.0192

1.9 Estimate the viscosity for the gas given in the previous example at the specified condition by applying the Carr et al. and Lee et al. methods.

1.10 Calculate the interfacial tension of an oil mixture with the following equilibrium composition:

$$MW_{C7+} = 200$$

$$\rho_L = 48 \text{lb}/\text{ft}^3$$

$$\rho_g = 15 \text{lb}/\text{ft}^3$$

Component	Liquid Composition	Component	Gas Composition
C_1	0.354	C_1	0.802
C_2	0.071	C_2	0.091
C_3	0.052	C_3	0.041
i-C_4	0.009	i-C_4	0.006
n-C_4	0.024	n-C_4	0.013
i-C_5	0.011	i-C_5	0.004
n-C_5	0.014	n-C_5	0.004
n-C_6	0.019	n-C_6	0.005
C_{7+}	0.446	C_{7+}	0.034

REFERENCES

Alani, G.H., Kennedy, H.T., 1960. Volumes of Liquid Hydrocarbons at High Temperatures and Pressures.

Al-Marhoun, M.A., 1988. PVT correlations for Middle East crude oils. Journal of Petroleum Technology 40 (05), 650–666.

Beal, C., 1946. The viscosity of air, water, natural gas, crude oil and its associated gases at oil field temperatures and pressures. Transactions of the AIME 165 (01), 94–115.

Beggs, H.D., Robinson, J., 1975. Estimating the viscosity of crude oil systems. Journal of Petroleum Technology 27 (09), 1140–1141.

Carr, N.L., Kobayashi, R., Burrows, D.B., 1954. Viscosity of hydrocarbon gases under pressure. Journal of Petroleum Technology 6 (10), 47–55.

Chew, J., Connally, C., 1959. A viscosity correlation for gas-saturated crude oils. Transactions of the AIME 216, 23–25.

Daubert, T.E., Danner, R.P., 1997. API Technical Data Book-Petroleum Refining. American Petroleum Institute (API), Washington, DC.

Dempsey, J.R., 1965. Computer routine treats gas viscosity as a variable. Oil and Gas Journal 63, 141–143.

Dranchuk, P., Kassem, H., 1975. Calculation of Z Factors for Natural Gases Using Equations of State.

Frick, T.C., 1962. Petroleum Production Handbook: Reservoir Engineering. McGraw-Hill.

Glaso, O., 1980. Generalized pressure-volume-temperature correlations. Journal of Petroleum Technology 32 (05), 785–795.

Hanley, H., McCarty, R., Haynes, W., 1975. Equations for the viscosity and thermal conductivity coefficients of methane. Cryogenics 15 (7), 413–417.

Herning, F., Zipperer, L., 1936. Calculation of the viscosity of technical gas mixtures from the viscosity of individual gases. Gas-u. Wasserfach 79, 69.

Hirschfelder, J.O., Curtiss, C.F., Bird, R.B., Mayer, M.G., 1954. Molecular Theory of Gases and Liquids. Wiley, New York.

Kartoatmodjo, T., Schmidt, Z., 1994. Large data bank improves crude physical property correlations. Oil and Gas Journal (United States) 92 (27).

Lee, A.L., Gonzalez, M.H., Eakin, B.E., 1966. The viscosity of natural gases. Journal of Petroleum Technology 18 (08), 997–1000.

Lohrenz, J., Bray, B.G., Clark, C.R., 1964. Calculating viscosities of reservoir fluids from their compositions. Journal of Petroleum Technology 16 (10), 1171–1176.

Macleod, D.B., 1923. On a relation between surface tension and density. Transactions of the Faraday Society 19 (July), 38–41.

McCarty, R., 1974. A modified Benedict-Webb-Rubin equation of state for methane using recent experimental data. Cryogenics 14 (5), 276–280.

Murad, S., Gubbins, K., 1977. Corresponding states correlation for thermal conductivity of dense fluids. Chemical Engineering Science 32 (5), 499–505.

Pedersen, K.S., Fredenslund, A., Christensen, P.L., Thomassen, P., 1984a. Viscosity of crude oils. Chemical Engineering Science 39 (6), 1011–1016.

Pedersen, K.S., Thomassen, P., Fredenslund, A., 1984b. Thermodynamics of petroleum mixtures containing heavy hydrocarbons. 1. Phase envelope calculations by use of the Soave-Redlich-Kwong equation of state. Industrial & Engineering Chemistry Process Design and Development 23 (1), 163–170.

Péneloux, A., Rauzy, E., Fréze, R., 1982. A consistent correction for Redlich-Kwong-Soave volumes. Fluid Phase Equilibria 8 (1), 7–23.

Petrosky Jr., G., Farshad, F., 1993. Pressure-volume-temperature correlations for Gulf of Mexico crude oils. In: SPE Annual Technical Conference and Exhibition. Society of Petroleum Engineers.

Quiñones-Cisneros, S.E., Zéberg-Mikkelsen, C.K., Stenby, E.H., 2003. Friction theory prediction of crude oil viscosity at reservoir conditions based on dead oil properties. Fluid Phase Equilibria 212 (1), 233–243.

Riazi, M.R., Daubert, T.E., 1980. Simplify property predictions. Hydrocarbon Processing 60 (3), 115–116.

Spencer, C.F., Danner, R.P., 1972. Improved equation for prediction of saturated liquid density. Journal of Chemical and Engineering Data 17 (2), 236–241.

Standing, M.B., Katz, D.L., 1942a. Density of crude oils saturated with natural gas. Transactions of the AIME 146 (01), 159–165.

Standing, M.B., Katz, D.L., 1942b. Density of natural gases. Transactions of the AIME 146 (01), 140–149.

Standing, M.B., 1947. A Pressure-Volume-Temperature Correlation for Mixtures of California Oils and Gases. Drilling and Production Practice, American Petroleum Institute.

Standing, M.B., 1951. Volumetric and Phase Behavior of Oil Field Hydrocarbon Systems: PVT for Engineers. California Research Corp.

Standing, M.B., 1981. Volumetric and Phase Behaviour of Oil Field Hydrocarbon Systems, 9th Printing. SPE, Dallas, Texas.

Sugden, S., 1932. The Parachors and Valency (Routledge, 1930); A List of Parachors. Brit. Assoc. Report. Brit. Assoc. Report.

Stiel, L.I., Thodos, G., 1961. The viscosity of nonpolar gases at normal pressures. AIChE Journal 7 (4), 611–615.
Vasquez, M., Beggs, H.D., 1980. Correlations for fluid physical property prediction. Journal of Petroleum Technology 32 (06), 968–970.
Wichert, E., Aziz, K., 1972. Calculate Zs for sour gases. Hydrocarbon Processing 51 (5), 119.

CHAPTER TWO

Equations of State

M. Mesbah[1], A. Bahadori[2,3]
[1]Sharif University of Technology, Tehran, Iran
[2]Southern Cross University, Lismore, NSW, Australia
[3]Australian Oil and Gas Services Pty Ltd, Lismore, NSW, Australia

2.1 INTRODUCTION

An equation of state (EOS) simply refers to any relation that describes the relationship between various macroscopically measurable properties of a system. Usually, the interconnection between pressure, volume, and temperature can be described by an EOS. Although the EOSs have been developed for pure components, they can also be applied for mixtures by employing some mixing rules. The calculation of properties and the phase condition of hydrocarbon mixture are some applications of the EOSs in the oil and gas industry.

The Van der Waals EOS, which is the simplest cubic EOS, originated in 1873. Van der Waals improved the ideal gas equation by introducing the repulsive and attractive intermolecular interactions. This EOS is the first EOS capable of representing vapor—liquid coexistence. Many authors revised and modified the Van der Waals EOS (Redlich and Kwong, 1949; Soave, 1972, 1993; Peng and Robinson, 1976; Boston and Mathias, 1980; Harmens and Knapp, 1980; Mathias, 1983; Mathias and Copeman, 1983; Stryjek and Vera, 1986a,b,c, Yu and Lu, 1987; Carrier et al., 1988; Androulakis et al., 1989; Twu et al., 1991; 1995a,b; Gasem et al., 2001; Farrokh-Niae et al., 2008; Haghtalab et al., 2011; Forero and Velásquez, 2013). These equations are usually called *cubic EOSs*. Liquid density prediction, for saturated liquid and compressed liquid by two constant EOSs, such as Van der Waals EOS, Redlich—Kwong EOS, Soave—Redlich—Kwong EOS, and Peng—Robinson EOS, is poor. However, the predicted liquid density by cubic EOS can be corrected by using a volume-translation parameter. Noncubic equations [such as Benedict—Webb—Rubin (BWR) (Bendict et al., 1940) EOS and its modifications] are more suitable for liquid-density prediction and phase-behavior calculation. However, this advantage is followed by a disadvantage. Noncubic equations have many

Fluid Phase Behavior for Conventional and Unconventional Oil and Gas Reservoirs
http://dx.doi.org/10.1016/B978-0-12-803437-8.00002-6

more adjustable constants than cubic EOSs, which require more computational time for phase-equilibrium calculation.

Cubic EOS, noncubic EOS, and mixing rules will be discussed in this chapter.

2.2 CUBIC EQUATION OF STATE (EOS)

The EOS that is cubic with respect to volume, like the Van der Waals EOS, is usually called a cubic EOS. These EOSs generally are modifications of the Van der Waals EOS. The repulsive term of most of these equations is identical with the repulsive term in the Van der Waals EOS and the attractive term is modified. The most commonly used cubic EOS is reviewed in this section.

The simplest cubic EOS is the Van der Waals EOS that originated in 1873. Van der Waals EOS is composed of contributions of repulsive and attractive terms. The Van der Waals EOS is given by:

$$\left(P+\frac{a}{V^2}\right)(V-b)=RT \tag{2.1}$$

in which a and b are the constants and have different values for each component, but are independent of temperature and pressure. If a and b are set to zero, Eq. (2.1) reduces to an ideal gas EOS. Eq. (2.1) is usually written as:

$$P=\frac{RT}{V-b}-\frac{a}{V^2} \tag{2.2}$$

To find the molar volume from pressure and temperature, this equation may be rearranged in the following form:

$$V^3-\left(b+\frac{RT}{P}\right)V^2+\left(\frac{a}{P}\right)V-\frac{ab}{P}=0 \tag{2.3}$$

Eq. (2.3) is a cubic equation in terms of volume. For this reason, the Van der Waals EOS (and its modifications) is called a cubic EOS. The terms $\frac{a}{V^2}$ and b in Eq. (2.1) are the attractive and repulsive terms, respectively. The constants a and b in Eq. (2.1) have physical meaning. The term $\frac{a}{V^2}$ corrects the pressure due to forces of attraction between molecules; in other words, the amount of pressure exerted by an ideal gas minus $\frac{a}{V^2}$ is equal to the amount of pressure exerted by a Van der Waals gas. If the pressure approaches infinity, the molar volume equals b. Therefore, the b parameter can be considered as the volume of 1 mol of hard-sphere volume and is usually called covolume. Covolume is always less than V, and $(V\text{-}b)$ is a positive term that represents the free space between molecules.

Example 2.1

Determine a and b parameters in Eq. (2.1) in terms of critical temperature and critical pressure. Note that, at the critical point of a pure component, the first and second derivatives of pressure with respect to pressure at constant temperature are zero.

$$\left(\frac{\partial P}{\partial V}\right)_{P_c,V_c,T_c} = \left(\frac{\partial^2 P}{\partial V^2}\right)_{P_c,V_c,T_c} = 0$$

It means the critical isotherm shows a horizontal inflection at the critical point.

Solution

The first and second derivatives of pressure with respect to pressure at constant temperature are calculated from Eq. (2.2).

$$\left(\frac{\partial P}{\partial V}\right)_T = -\frac{RT}{(V-b)^2} + \frac{2a}{V^3}$$

$$\left(\frac{\partial^2 P}{\partial V^2}\right)_T = \frac{2RT}{(V-b)^3} - \frac{6a}{V^4}$$

At the critical point we have:

$$\left(\frac{\partial P}{\partial V}\right)_{T_c} = -\frac{RT_c}{(V_c-b)^2} + \frac{2a}{V_c^3} = 0$$

$$\left(\frac{\partial^2 P}{\partial V^2}\right)_{T_c} = \frac{2RT_c}{(V_c-b)^3} - \frac{6a}{V_c^4} = 0$$

From the last two equations we have:

$$b = \frac{V_c}{3}$$

$$a = \frac{9}{8}T_c V_c$$

At the critical point, the volume is equal to the critical volume. This can be written in the following form:

$$V - V_c = 0 \text{ or } (V - V_c)^3 = 0$$

Expansion of this equation gives:

$$(V - V_c)^3 = V^3 - 3V_cV^2 + 3V_c^2V - V_c^3 = 0$$

Comparing this equation with Eq. (2.3) at T_c and P_c, we can write:

$$-\left(b + \frac{RT_c}{P_c}\right) = -3V_c \quad \text{Coefficients of } V^2$$

$$\frac{a}{P_c} = 3V_c^2 \quad \text{Coefficients of } V$$

$$-\frac{ab}{P_c} = -V_c^3 \quad \text{Coefficients of } V^0$$

hence, a and b can be determined as follows:

$$\frac{-\frac{ab}{P_c}}{\frac{a}{P_c}} = \frac{-V_c^3}{3V_c^2} \rightarrow b = \frac{V_c}{3} \xrightarrow{-\left(b+\frac{RT_c}{P_c}\right)=-3V_c} b = \frac{RT_c}{8P_c}$$

$$a = \frac{9}{8}T_cV_c \xrightarrow{b=\frac{RT_c}{8P_c},\, b=\frac{V_c}{3}} a = \frac{27R^2T_c^2}{64P_c}$$

Eq. (2.3) may be written in terms of compressibility factor:

$$Z^3 - (1 + B)Z^2 + AZ - AB = 0 \tag{2.4}$$

in which dimensionless parameters A and B are defined as follows.

Eqs. (2.3) and (2.4) give three roots for molar volume or compressibility factor at subcritical temperature (at pressure $P1$) as shown in Fig. 2.1. The biggest root for volume ($V1$) or compressibility factor corresponds to saturated vapor, the smallest root for volume ($V2$) or compressibility factor corresponds to saturated liquid, and the intermediate root does

$$A = \frac{aP}{(RT)^2} \tag{2.5}$$

$$B = \frac{bP}{RT} \tag{2.6}$$

not have physical meanings. At this point, the value of $\left(\frac{\partial P}{\partial V}\right)_T$ is positive, that is, not physically possible for a pure component. For a pure component, as the pressure increases the molar volume decreases. Therefore, $\left(\frac{\partial P}{\partial V}\right)_T$ has to be

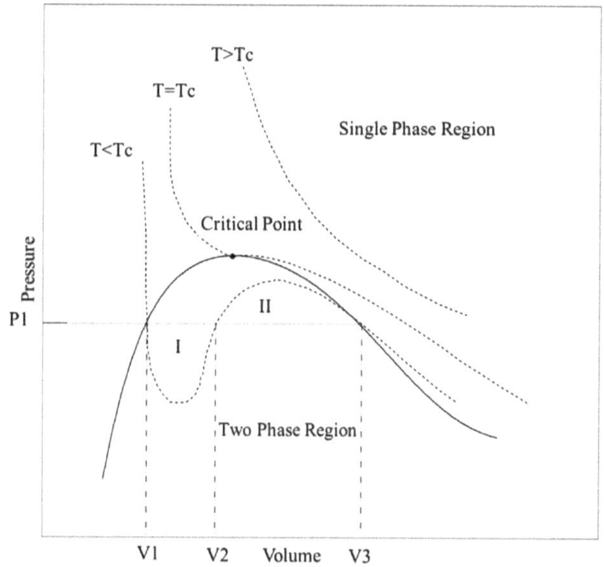

Figure 2.1 Predicted pressure–volume behavior of a pure component at subcritical, critical, and supercritical temperatures by Van der Waals-type equation of state (EOS).

negative. Note that at points $V1$ and $V3$, $\left(\frac{\partial P}{\partial V}\right)_T$ is negative. For a liquid phase for a very large pressure, molar volume change is very small. In other words, $\left(\frac{\partial P}{\partial V}\right)_T$ is relatively high for a liquid phase (which is seen in left-hand side of curve I).

At supercritical temperature, equations give one real root (acceptable root) and two complex roots (not acceptable roots). It should be checked whether the value of the root is near b or RT/P. If the value of root is near b, then the phase is compressed liquid, and if the value of root is near RT/P, then the phase is gas or superheated vapor. At critical temperature, all three roots are equal to critical volume.

Example 2.2

Determine the critical compressibility factor by Van der Waals EOS.

Solution

From previous example, we have:

$$b = \frac{RT_c}{8P_c}$$

$$b = \frac{V_c}{3}$$

(Continued)

Equating these two equations,

$$\frac{V_c}{3}=\frac{RT_c}{8P_c}\rightarrow\frac{P_cV_c}{RT_c}=\frac{3}{8}\xrightarrow{Z_c=\frac{P_cV_c}{RT_c}}Z_c=\frac{3}{8}=0.375$$

So the Van der Waals EOS gives a constant value for all components, whereas very few components such as quantum gas have a critical compressibility factor greater than 0.30 (for most real gas, critical compressibility factor ranges from 0.22 to 0.30).

The a and b parameters in Van der Waal EOS use a boundary condition (as seen in Example 2.1). As mentioned before, this equation cannot accurately predict the behavior of dense fluids. Several modifications have been done to improve the capability of the equation by modifying the attractive and repulsive terms. In modified-version equations, the boundary conditions are also satisfied. In addition, experimental data on pure fluids have been used in the determination of parameters of EOS. Therefore, these equations are semiempirical EOSs.

Example 2.3

Estimate the vapor molar volume and compressibility factor of normal octane at 552.65K and 1.99 MPa from Van der Waals EOS. The experimental value of vapor molar volume at this condition is 0.001216 m^3/mol (Riazi, 2005).

Solution

The critical temperature and critical pressure of normal octane are 568.7K and 2.49 MPa, respectively (Danesh, 1998). The a and b parameters are determined using the results of Example 2.1.

$$a=\frac{27R^2T_c^2}{64P_c}=\frac{27(8.314)^2(568.7)^2}{64(2.49\times10^6)}=3.7877\ \text{Pa}\ \left(\text{m}^3/\text{mol}\right)^2$$

$$b=\frac{RT_c}{8P_c}=\frac{8.314(568.7)}{8(2.49\times10^6)}=2.3736\times10^{-4}\ \left(\text{m}^3/\text{mol}\right)$$

The dimensionless parameter is calculated by Eqs. (2.5) and (2.6) as follows.

$$A=0.3570, B=0.1028$$

Substituting the A and B values in Eq. (2.4) results in the following cubic equation:

$$Z^3-1.1028Z^2+0.3570Z-0.0367=0$$

Solving the previous equation gives two complex roots and one real root equal to 0.6263. Hence, the molar volume is:

$$V = \frac{ZRT}{P} = \frac{0.6263 \times 8.314 \times 552.65}{1.99 \times 10^6} = 0.001446\ (\mathrm{m^3/mol})$$

which is near to *RT/P* (*RT/P* = 0.002308 $\mathrm{m^3}$/mol, b = 2.3736E−4 $\mathrm{m^3}$/mol) and correspond to vapor phase.

Redlich and Kwong (1949) modified the attractive term of Van der Waals EOS. They proposed temperature dependencies of attractive term as follows:

$$P = \frac{RT}{V - b} - \frac{a_c \alpha}{V(V + b)} \tag{2.7}$$

in which

$$\alpha = T_r^{-0.5} \tag{2.8}$$

T_r is the reduced temperature and defined as the ratio of temperature to critical temperature.

The repulsive term in Redlich–Kwong (RK) EOS is identical to the Van der Waals EOS. The form of expressions that described the parameters are similar to Van der Waals EOS, but with a different coefficient.

Eqs. (2.9) and (2.10) describe a_c and b parameters.

$$a_c = 0.42747 \frac{R^2 T_c^2}{P_c} \tag{2.9}$$

$$b = 0.08664 \frac{RT_c}{P_c} \tag{2.10}$$

Zudkevitch and Joffe (Zudkevitch and Joffe, 1970) and Joffe et al. (Joffe et al., 1970) assume the coefficients in Eqs. (2.9) and (2.10) are temperature dependent. The coefficients are obtained for each pure component by matching the calculated liquid density and vapor pressure by experimental values with the help of a generalized fugacity correlation for saturated vapor.

Soave (1972) proposed a more general form of temperature-dependent term in the attractive term in RK EOS, $T_r^{-0.5}$.

$$\alpha = \left[1 + \kappa\left(1 - T_r^{0.5}\right)\right]^2 \tag{2.11}$$

Soave correlated κ against acentric factor by equating the fugacities of saturated liquid and vapor phase at reduced temperature equal to 0.7.

$$\kappa = 0.480 + 1.574\omega - 0.176\omega^2 \tag{2.12}$$

where ω is the acentric factor. Soave calculated the vapor pressure of several pure components and binary mixture system with Soave–Redlich–Kwong (SRK) EOS, and compared with the experimental data SRK EOS showed superior results compared to the RK EOS.

Later in 1993, Soave et al. proposed that dividing the value of κ determined by Eq. (2.12) by 1.18 can improve the accuracy of results.

The SRK and RK equations in terms of the compressibility factor are given by Eq. (2.13).

$$Z^3 - Z^2 + \left(A - B - B^2\right)Z - AB = 0 \tag{2.13}$$

The dimensionless parameters A and B are given by Eqs. (2.5) and (2.6). SRK EOS is well capable to predict the vapor–liquid equilibrium but does not gives reliable results for liquid density.

Peng and Robinson (1976) developed a new EOS mainly to improve the liquid density in comparison with SRK EOS.

$$P = \frac{RT}{V - b} - \frac{a_c\alpha}{V(V + b) + b(V - b)} \tag{2.14}$$

in which:

$$a_c = 0.457235\frac{R^2T_c^2}{P_c} \tag{2.15}$$

$$b = 0.077796\frac{RT_c}{P_c} \tag{2.16}$$

They used a similar form function for α that has been suggested by Soave, Eq. (2.11). Peng and Robinson correlated κ against acentric factor by equating the fugacities of saturated liquid and vapor phases, at temperature ranges from normal boiling point temperature to critical temperature.

$$\kappa = 0.37464 + 1.54226\omega - 0.26992\omega^2 \tag{2.17}$$

Peng–Robinson (PR) EOS in terms of compressibility factor takes the following form:

$$Z^3 - (1 - B)Z^2 + \left(A - 2B - 3B^2\right)Z - \left(AB - B^2 - B^3\right) = 0 \tag{2.18}$$

in which dimensionless parameters A and B are defined similar to previous EOSs.

Example 2.4

Joule–Thomson coefficient is an important property of a given gas. This coefficient is important from two standpoints, intermolecular interaction and liquefaction of gases. Joule–Thomson coefficient is defined as the change in gas temperature due to change in pressure at constant enthalpy (i.e., there is no heat transfer to or from the gas and no external work is done).

$$\mu_{J-T} = \left(\frac{\partial T}{\partial P}\right)_H$$

in which H is the enthalpy. The Joule–Thomson coefficient for an ideal gas is always equal to zero. For real gases, there is a temperature at which the Joule–Thomson coefficient changes sign (or Joule–Thomson coefficient equal to zero), this temperature is called inversion temperature. Below the inversion temperature, the Joule–Thomson coefficient is positive and gas cools due to expansion process, and for temperature above the inversion temperature the Joule–Thomson coefficient is negative and gas warms due to expansion. Determine the Joule–Thomson coefficient for a gas that obeys the Van der Waals EOS.

Solution

Consider that enthalpy is a function of temperature and pressure, $H = H(P,T)$. Then the total differential of H is defined as:

$$dH = \left(\frac{\partial H}{\partial P}\right)_T dP + \left(\frac{\partial H}{\partial T}\right)_P dT \qquad \text{(a)}$$

In Eq. (a), $\left(\frac{\partial H}{\partial T}\right)_P$ is the heat capacity at constant pressure. Hence:

$$dH = \left(\frac{\partial H}{\partial P}\right)_T dP + C_P dT \qquad \text{(b)}$$

At constant enthalpy, $dH = 0$ and we have:

$$\mu_{J-T} = \left(\frac{\partial T}{\partial P}\right)_H = -\left(\frac{\partial H}{\partial P}\right)_T \Big/ C_P \qquad \text{(c)}$$

From fundamental property relation, total differential of enthalpy is defined as:

$$dH = VdP + TdS \qquad \text{(d)}$$

in which S is the entropy. Taking derivative from both sides of Eq. (d) with respect to P at constant T:

$$\left(\frac{\partial H}{\partial P}\right)_T = V + T\left(\frac{\partial S}{\partial P}\right)_T \qquad \text{(e)}$$

(Continued)

From Maxwell's equation, $\left(\frac{\partial S}{\partial P}\right)_T = -\left(\frac{\partial V}{\partial T}\right)_P$. Hence, Eq. (e) can be written as follows:

$$\left(\frac{\partial H}{\partial P}\right)_T = V - T\left(\frac{\partial V}{\partial T}\right)_P \tag{f}$$

Substituting Eq. (f) in Eq. (c) results in

$$\mu_{J-T} = \frac{\left[T\left(\frac{\partial V}{\partial T}\right)_P - V\right]}{C_P} \tag{g}$$

Evaluate the $\left(\frac{\partial T}{\partial V}\right)_P$ (note that $\left(\frac{\partial V}{\partial T}\right)_P = 1/\left(\frac{\partial T}{\partial V}\right)_P$) by taking derivative from Eq. (2.2).

$$0 = \frac{R}{(V-b)}\left(\frac{\partial T}{\partial V}\right)_P - \frac{RT}{(V-b)^2} + 2\frac{a}{V^3} \tag{h}$$

Rearranging Eq. (h) gives:

$$\left(\frac{\partial V}{\partial T}\right)_P = \frac{\frac{R}{(V-b)}}{\frac{RT}{(V-b)^2} - 2\frac{a}{V^3}} \tag{i}$$

Finally, substituting Eq. (i) in Eq. (g) gives the Joule–Thomson coefficient as follows:

$$\mu_{J-T} = \frac{\left[\frac{\frac{RT}{(V-b)}}{\frac{RT}{(V-b)^2} - 2\frac{a}{V^3}} - V\right]}{C_P} \tag{j}$$

Example 2.5

Pressure–volume data for water at temperature 448.15K is reported in the following table.

P (Pa)	V_{mass} (m^3/g)
325,000	0.000622
350,000	0.000577
375,000	0.000537
400,000	0.000503
425,000	0.000472
450,000	0.000445
475,000	0.000421
500,000	0.000399
525,000	0.000380
550,000	0.000362
575,000	0.000345
600,000	0.000330
625,000	0.000316
650,000	0.000304
675,000	0.000292
700,000	0.000281
725,000	0.000270

Estimate the molecular weight of water. The experimental value is 18.

Solution

In the limit of zero pressure, all gases are ideal and obey the following relation.

$$PV = RT$$

in which V is the molar volume and related to V_{mass} by the following equation.

$$V = V_{mass} \times \mathrm{MW}$$

V_{mass} is equal to the inverse of density; hence, we can write:

$$\mathrm{MW} = RT\left(\frac{\rho}{P}\right)_{P=0}$$

The best value for molecular weight is calculated from an extrapolation of ρ/P versus P to zero pressure. In other words, the intercept of line ρ/P versus P is MW/RT. Determine the ρ/P at each pressure.

P (Pa)	V_{mass} (m^3/g)	ρ (g/m^3)	ρ/P (g/(m^3 Pa))
325,000	0.000622	1606.658	0.0049436
350,000	0.000577	1733.403	0.0049526
375,000	0.000537	1860.604	0.0049616
400,000	0.000503	1988.348	0.0049709
425,000	0.000472	2116.536	0.0049801
450,000	0.000445	2245.274	0.0049895
475,000	0.000421	2374.507	0.0049990
500,000	0.000399	2504.32	0.0050086

—cont'd

P (Pa)	V_{mass} (m^3/g)	ρ (g/m^3)	ρ/P (g/(m^3 Pa))
525,000	0.000380	2634.63	0.0050183
550,000	0.000362	2765.487	0.0050282
575,000	0.000345	2896.871	0.0050380
600,000	0.000330	3028.835	0.0050481
625,000	0.000316	3161.456	0.0050583
650,000	0.000304	3294.567	0.0050686
675,000	0.000292	3428.297	0.0050790
700,000	0.000281	3562.649	0.0050895
725,000	0.000270	3697.541	0.0051001

Using least-squares method to fit a straight line gives:

$$\frac{\rho}{P} = (4 \times 10^{-10})P + 0.0048$$

From the previous equation $\left(\frac{\rho}{P}\right)_{P=0}$ equals 0.0048 and the molecular weight is calculated as follows:

$$MW = 8.314 \times 448.15 \times 0.0048 = 17.88$$

That is near to the experimental value.

Example 2.6

The value of the compressibility factor is a description of intermolecular forces. When the compressibility factor is less than 1, attractive intermolecular forces dominate, and when the compressibility factor is greater than 1, repulsive intermolecular forces dominate. The temperature at which the gas behaves ideally and the attractive intermolecular forces are equal to repulsive intermolecular forces is called the Boyle temperature. Below the Boyle temperature, the gas is less compressible than an ideal gas, and above the Boyle temperature, the gas is more compressible than ideal gas.

Boyle temperature is formally defied as:

$$\lim_{V\to\infty}\left(\frac{\partial Z}{\partial(1/V)}\right)_T = 0 \text{ at } T_{Boyle}$$

Determine the Boyle temperature for a gas that obeys the Van der Waals EOS.

Solution

To calculate the Boyle temperature it is convenient to calculate the compressibility factor in terms of pressure and molar volume.

$$Z = \frac{PV}{RT} = \frac{\left(\frac{RT}{V-b} - \frac{a}{V^2}\right)V}{RT} = \frac{V}{V-b} - \frac{a}{VRT}$$

$\frac{\partial Z}{\partial(1/V)}$ can be rewritten as follows:

$$\frac{\partial Z}{\partial(1/V)} = \left(\frac{\partial Z}{\partial V}\right) \times \left(\frac{\partial V}{\partial(1/V)}\right) = -V^2\left(\frac{\partial Z}{\partial V}\right)$$

$\left(\frac{\partial Z}{\partial V}\right)$ is calculated by taking the derivative from the compressibility factor relation with respect to volume.

$$\left(\frac{\partial Z}{\partial V}\right) = \frac{\partial}{\partial V}\left(\frac{V}{V-b} - \frac{a}{VRT}\right) = \left(\frac{-V}{(V-b)^2} + \frac{1}{V-b} + \frac{a}{V^2RT}\right)$$

$$= \frac{-b}{(V-b)^2} + \frac{a}{V^2RT}$$

So the Boyle temperature is calculated with the following relation:

$$\lim_{V\to\infty}\left[-V^2\left(\frac{-b}{(V-b)^2} + \frac{a}{V^2RT}\right)\right] = b - \frac{a}{RT} = 0$$

Van der Waals, RK, SRK, and PR equations give a consistent critical compressibility factor (0.375, 0.333, 0.333, and 0.307 respectively), whereas the critical compressibility factor of components ranges from 0.24 to 0.30 as mentioned earlier. Previous EOSs used critical temperature and critical pressure as input data.

In 1980, Schmidt and Wenzel introduced a Van der Waals-type cubic EOS which uses three input data sets of critical temperature, critical pressure, and acentric factor. This EOS yields a substance-dependent critical compressibility factor. The repulsive term in the Schmidt–Wenzel EOS is similar to Van der Waals EOS and the denominator of attraction term in the original Van der Waals EOS was replaced with a more general second-order polynomial in terms of volume. Schmidt–Wenzel EOS is expressed in the following form:

$$P = \frac{RT}{V-b} - \frac{a_c\alpha}{V^2 + (1+3\omega)bV - 3\omega b^2} \tag{2.19}$$

in which α is a function of temperature and b is temperature independent. PR and SRK equations can be considered as a general form of Schmidt–Wenzel EOS. If the acentric factor in Eq. (2.19) substituted by values zero and 1/3, Schmidt–Wenzel EOS reduces to PR and SRK equations, respectively.

The a_c and b parameters in Schmidt–Wenzel EOS determine by Eqs. (2.20) and (2.21), respectively (these equation obtained by apply the condition at critical point, $\left(\frac{\partial P}{\partial V}\right)_{P_c,V_c,T_c} = \left(\frac{\partial^2 P}{\partial V^2}\right)_{P_c,V_c,T_c} = 0$.

$$a_c = \Omega_{ac}\frac{R^2T_c^2}{P_c} \tag{2.20}$$

$$b = \Omega_b\frac{RT_c}{P_c} \tag{2.21}$$

in which:

$$\Omega_{ac} = [1 - \chi(1 - q)]^3 \tag{2.22}$$

$$\Omega_b = \chi q \tag{2.23}$$

in which χ represents the critical compressibility factor, as predicted by Eq. (2.19). The q parameter related to χ by Eq. (2.24).

$$\chi = \frac{1}{[3(1 + q\omega)]} \tag{2.24}$$

Parameter q is defined as b over critical volume and is the solution of the following equation:

$$(6\omega + 1)q^3 + 3q^2 + 3q - 1 = 0 \tag{2.25}$$

If Eq. (2.25) gives more than one root, the smallest positive root is q. The approximate value of q is given by Eq. (2.26).

This approximation uses an initial guess to solve Eq. (2.25).

The form of α in the Schmidt–Wenzel EOS is the same as that proposed by Soave, Eq. (2.11).

$$q = 0.25989 - 0.0217\omega + 0.00375\omega^2 \tag{2.26}$$

however, here, κ is a function of acentric factor and reduced temperature. Eqs. (2.27) through (2.31) represent κ function for different ranges of acentric factor (Danesh, 1998).

$$\kappa \equiv \kappa_1 = \kappa_0 + 0.01429(5T_r - 3\kappa_0 - 1)^2 \quad \text{for } \omega \le 0.4 \tag{2.27}$$

$$\kappa \equiv \kappa_2 = \kappa_0 + 0.71(T_r - 0.779)^2 \quad \text{for } \omega \geq 0.55 \tag{2.28}$$

$$\kappa \equiv \kappa_1 \left[\frac{0.55 - \omega}{0.15}\right] + \kappa_2 \left[\frac{\omega - 0.4}{0.5}\right] \quad \text{for } 0.4 < \omega < 0.55 \tag{2.29}$$

in which:

$$\kappa_0 = 0.465 + 1.347\omega - 0.528\omega^2 \quad \text{for } \omega \leq 0.3671 \tag{2.30}$$

$$\kappa_0 = 0.5361 + 0.9593\omega \quad \text{for } \omega > 0.3671 \tag{2.31}$$

and for supercritical compounds:

$$\alpha = 1 - (0.4774 + 1.328\omega)\ln T_r \tag{2.32}$$

Example 2.7

Estimate the vapor and liquid molar volume of normal octane at 552.65K and 1.99 MPa from the Schmidt–Wenzel EOS. The experimental value of vapor and liquid molar volume at this condition is 0.001216 m^3/mol and 0.000304 m^3/mol, respectively (Riazi, 2005).

Solution

The critical temperature, critical pressure, and acentric factor of normal octane are 568.7K, 2.49 MPa, and 0.3996 respectively (Danesh, 1998). To calculate a_c and b parameters, we need to know the values of χ and q. The acentric factor is less than 0.3671; hence, κ_0 can be calculated from Eq. (2.31).

$$\kappa_0 = 0.5361 + 0.9593(0.3996) = 0.9194$$

The reduced temperature equal to $T_r = 552.65/568.7 = 0.9718$. κ and α can be determined using Eqs. (2.27) and (2.11):

$$\kappa = 0.9194 + 0.01429[5(0.9718) - 3(0.9194) - 1]^2 = 0.9367$$

$$\alpha = \left[1 + 0.9367(1 - 0.9718^{0.5})\right]^2 = 1.0268$$

Solution of Eq. (2.25) gives one real root, 0.2518, and two complex roots, $-0.5674 \pm 0.9202i$, therefore, q is equal to 0.2518 and χ is determined from Eq. (2.24).

$$\chi = \frac{1}{[3(1 + (0.2518 \times 0.3996))]} = 0.3029$$

(Continued)

Coefficients Ω_{ac} and Ω_b are determined by Eqs. (2.22) and (2.23), respectively.

$$\Omega_{ac} = [1 - 0.3029(1 - 0.2518)]^3 = 0.4626$$

$$\Omega_b = 0.3029 \times 0.2518 = 0.0763$$

Hence, parameters ac and b are 4.1535 Pa $(m^3/mol)^2$ and 1.4482E−04 m^3/mol, respectively [calculated from Eqs. (2.20) and (2.21)]. Substituting pressure, temperature, α, acentric factor, and parameters ac and b in Eq. (2.19) gives the following equation in terms of volume:

$$\left(1.99 \times 10^6\right) - \frac{8.314 \times 552.65}{V - 1.4482 \times 10^{-4}} + \frac{4.1535 \times 1.0268}{V^2 + 1.4482 \times 10^{-4}(2.1988)V - 2.5143 \times 10^{-8}} = 0$$

Solving this equation gives three roots, 0.000348, 0.000604, and 0.001183. The smallest root corresponds to the liquid phase, the largest root corresponds to the vapor phase, and the intermediate root does not have physical meaning. Therefore: $V^{\text{liquid}} = 0.000348\ m^3/mol$ and $V^{\text{vapor}} = 0.001183$.

Patel and Teja (1982) presented an extension of the works of Soave, of Peng and Robinson, and of Schmidt and Wenzel. The Patel−Teja EOS also uses three input data sets of critical temperature, critical pressure, and acentric factor as in the Schmidt−Wenzel EOS. The Patel−Teja EOS gives a substance-dependent critical compressibility factor; in addition, this EOS can be applied for polar fluids such as alcohols, water, and ammonia (Patel and Teja, 1982). The EOS presented by Patel and Teja has the following form:

$$P = \frac{RT}{V - b} - \frac{a_c\alpha}{V(V + b) + c(V - b)} \tag{2.33}$$

Similar to previous equations, the repulsive term is identical to the repulsive term in the Van der Waals EOS. The denominator of the attraction term in the original Van der Waals EOS is replaced by a new second-order polynomial.

The c parameter in Eq. (2.33) is defined as:

$$c = \Omega_c \frac{RT_c}{P_c} \tag{2.34}$$

in which

$$\Omega_c = 1 - 3\chi \tag{2.35}$$

χ is the adjusted critical compressibility factor and is calculated by matching liquid density. It was correlated with the acentric factor for nonpolar substances by Eq. (2.36).

$$\chi = 0.329032 - 0.076799\omega + 0.0211947\omega^2 \tag{2.36}$$

The values of χ for water, ammonia, methanol, and ethanol are 0.269, 0.282, 0.272, and 0.300 respectively.

Note that the prediction of the critical compressibility factor (which in Schmidt–Wenzel and Patel–Teja equations was denoted byχ) is not a significant indicator of overall performance of any equation (Leland and Chappelear, 1968). For this reason, Schmidt and Wenzel and Patel and Teja assume that the critical compressibility factor is an empirical parameter instead of a value equal to the experimental value.

The parameters a_c and b are defined similar to the Schmidt–Wenzel EOS by Eqs. (2.20) and (2.21) respectively, with different values for Ω_{ac} and Ω_b which are determined by applying the condition at critical point (i.e., $\left(\frac{\partial P}{\partial V}\right)_{P_c, V_c, T_c} = \left(\frac{\partial^2 P}{\partial V^2}\right)_{P_c, V_c, T_c} = 0$). Ω_b is the smallest positive root of Eq. (2.37).

$$\Omega_b^3 + (2 - 3\chi)\Omega_b^2 + 3\chi^2\Omega_b - \chi^3 = 0 \tag{2.37}$$

The initial guess for solving Eq. (2.37) is chosen from Eq. (2.38).

$$\Omega_b = 0.32429\chi - 0.022005 \tag{2.38}$$

Ω_{ac} is determined in terms of Ω_b and χ by Eq. (2.39).

$$\Omega_{ac} = 3\chi^2 + 3(1 - 2\chi)\Omega_b + \Omega_b^2 + (1 - 3\chi) \tag{2.39}$$

The form of α in the Patel–Teja EOS is also the same as proposed by Soave, Eq. (2.11). κ correlated with the acentric factor for nonpolar substances by Eq. (2.40).

$$\kappa = 0.452413 + 1.30982\omega - 0.295937\omega^2 \tag{2.40}$$

The values of κ for water, ammonia, methanol, and ethanol are 0.689803, 0.627090, 0.704657, and 1.230395 respectively.

If the values of 0.307 and 0.333 are substituted for χ, Patel–Teja EOS reduces to PR and SRK equations, respectively.

Example 2.8

Repeat Example 2.7 by Patel–Teja EOS.

Solution

χ is calculated by Eq. (2.36) as follows:

$$\chi = 0.329032 - 0.076799(0.3996) + 0.0211947(0.3996)^2 = 0.3017$$

Substitute the χ value in Eq. (2.37). Solving Eq. (2.37) gives one real root, 0.0759, and two complex roots $-0.5853 \pm 0.1390i$. Hence, the Ω_b parameter is equal to 0.0759. Substituting the values 0.3017 and 0.0759 for χ and Ω_b, respectively, in Eq. (2.39) gives 0.4640 for Ω_{ac} value. The values of κ can be calculated by substituting 0.3996 for the acentric value in Eq. (2.40), 0.9286. Hence, α is equal to 1.0266.

The *a*, *b*, and *c* parameters can be calculated as follows:

$$a_c = 0.4640\frac{(8.314 \times 568.7)^2}{2490000} = 4.1656\ \mathrm{Pa}\left(\mathrm{m}^3/\mathrm{mol}\right)^2$$

$$b = 0.0759\frac{8.314 \times 568.7}{2490000} = 1.4410 \times 10^{-4}\left(\mathrm{m}^3/\mathrm{mol}\right)$$

$$c = [1 - (3 \times 0.3017)]\frac{8.314 \times 568.7}{2490000} = 1.8005 \times 10^{-4}\left(\mathrm{m}^3/\mathrm{mol}\right)$$

Solving the following equation gives three roots 0.000346, 0.000602, and 0.001181.

$$\left(1.99 \times 10^6\right) - \frac{8.314 \times 552.65}{V - 1.4410 \times 10^{-4}} + \frac{4.1656 \times 1.0266}{V(V + 1.4410 \times 10^{-4}) + 1.8005 \times 10^{-4}(V - 1.4410 \times 10^{-4})} = 0$$

The smallest root corresponds to the liquid phase, the largest root corresponds to the vapor phase, and the intermediate root does not have physical meaning. Therefore: $V^{\text{liquid}} = 0.000346\ \mathrm{m}^3/\mathrm{mol}$ and $V^{\text{vapor}} = 0.001181\ \mathrm{m}^3/\mathrm{mol}$.

Example 2.9

Consider two EOSs, (A) a first order in volume, and (B) a second order in volume. Is it possible either equation (A) or (B) predicts the liquefaction of a gas? Is it possible either equation (A) or (B) predicts the critical temperature?

Solution

The maximum temperature at which a gas can be liquefied by increasing the pressure is known as critical temperature. In other words, the gas with a temperature greater than critical temperature cannot be liquefied by increasing pressure. Below the critical temperature, the liquid and vapor phases coexist in equilibrium. It means at the same temperature and pressure two molar volumes exist. In fact, a reliable EOS should predict two molar volumes (two real roots) at some temperature and pressure. On the other hand, as the temperature increases above the critical temperature, there is only one phase. It means a reliable EOS should predict one molar volume (one real root).

Consider the polynomial *P* with degree *n* with the real coefficients. The polynomial *P* gives *n* roots. These roots may be real or complex. Based on the complex conjugate root theorem, the complex roots come in conjugate pairs (i.e., if $a + bi$ is a zero of *P*, the $a - bi$ is a zero of *P*) (McGuire and O'Farrell, 2002). In other words, any polynomial with even degree gives an even number of roots and any polynomial with odd degree gives at least one real root (Jeffrey, 2005). Hence, a second-order EOS (B) never has just one real root (it has two real roots or two complex conjugate roots) and cannot predict the critical temperature. The EOS (B) may predict the two-phase condition, because it may have two real roots in some situation, but it cannot predict the liquefaction process. The EOS (A) cannot predict critical temperature, liquefaction process, or two-phase coexistence.

2.3 NONCUBIC EOS

One of the well-known noncubic EOSs is the virial equation. The virial EOS is based on theories of statistical mechanics (Mason and Spurling, 1969). The original version of virial EOS was presented by Onnes in 1901 and it may be written in a power series of molar density (pressure explicit) or pressure (volume explicit) as follows:

$$Z = 1 + \frac{B}{V} + \frac{C}{V^2} + \frac{D}{V^3} + \dots \quad \left(Z = 1 + B\rho_{\mathrm{M}} + C\rho_{\mathrm{M}}^2 + D\rho_{\mathrm{M}}^3 + \dots\right) \tag{2.41}$$

$$Z = 1 + B'P + C'P^2 + D'P^3 + \dots \tag{2.42}$$

in which Z is the compressibility factor, V is the molar volume, ρ_{M} is the molar density, P is the pressure and B, C, D,… are the second, third, fourth, and so on, virial coefficients. The coefficient B corresponds to interaction between two molecules, coefficient C corresponds to interaction between three molecules, and so on. For a given substance, the virial coefficients depend only on the temperature.

The virial series expansion, in theory, is an infinite series, but in practice, terms above the third virial coefficients are rarely used. More data is available for the second virial coefficient, but fewer data are available for the third virial coefficient. Second virial coefficients for several gases at different temperatures are reported in Table 2.1.

It is remarkable to note that the EOS of a real gas coincides with the EOS of a perfect gas as the pressure approaches zero; however, it is not necessary that all properties of real gas coincide with those of a perfect gas at this limit. For example, the slope of a graph of compressibility factor against the pressure (i.e., dZ/dP) for a perfect gas is always equal to zero (for a perfect gas, $Z = PV/RT = 1$ at all pressures); however, for a real gas that obeys the virial EOS according to Eq. (2.42), this can be written:

$$\frac{dZ}{dP} = B' + C'P + 2D'P + \ldots \xrightarrow{\text{as } P \to 0} \frac{dZ}{dP} = B' = \frac{B}{RT} \tag{2.43}$$

So the slope of a graph of compressibility factor against pressure for a real gas that obeys the virial EOS is zero if the second virial coefficient (B or B') is equal to zero. The temperature at which second virial coefficient is equal to zero known as *Boyle temperature*. The Boyle temperature for Ar, CH_4, CO_2, H_2, and N_2 are 411.5K, 510K, 714.8K, 110K, and 327.2K respectively (Atkins and De Paula, 2006).

Table 2.1 Second Virial Coefficient for Several Gases (Dymond and Smith, 1980; Atkins and De Paula, 2006)

Temperature (K)	**100**	**200**	**273**	**300**	**373**	**400**	**500**	**600**
Air	−167.3		−13.5		3.4			19
Ar	−187		−21.7		−4.2			11.9
CH_4		−105	−53.6	−42	−21.2	−15	−0.5	8.1
C_2H_6		−410		−182		−96	−52	
C_3H_8				−382		−208	−124	
CO_2			−142	−122.7	−72.2		−29.8	−12.4
H_2	−2		13.7		15.6			
He	11.4		12		11.3			10.4
Kr			−62.9		−28.7			1.7
N_2	−160	−35.2	−10.5	−4.2	6.2	9	16.9	21.7
Ne	−6		10.4		12.3			13.8
O_2	−197.5		−22		−3.7			12.9
Xe			−153.7		−81.7			−19.6

Values of second virial coefficient given in cm^3/mol.

Example 2.10

Convert the Van der Waals EOS to a virial EOS and obtain the second and third virial coefficients in terms of parameters of the Van der Waals EOS.

Note that according to Taylor series for $-1 < x < 1$, it can be written (Abramowitz and Stegun, 1966; Perry et al., 1997):

$$\frac{1}{1-x} = 1 + x + x^2 + x^3 + \ldots$$

Solution

Multiplying both sides of Van der Waals EOS by V/RT gives:

$$\frac{PV}{RT} = \frac{V}{V-b} - \frac{a}{RTV}$$

$$Z = \frac{V}{V-b} - \frac{a}{RTV}$$

Dividing numerator and denominator of the first term on right-hand side of the previous equation by V:

$$Z = \frac{1}{1-\frac{b}{V}} - \frac{a}{RTV}$$

The (b/V) is less than 1. Hence, the expansion of $\frac{1}{1-\frac{b}{V}}$ is:

$$\frac{1}{1-\frac{b}{V}} = 1 + \frac{b}{V} + \left(\frac{b}{V}\right)^2 + \ldots$$

Substituting the previous expression in "$Z = \frac{1}{1-\frac{b}{V}} - \frac{a}{RTV}$" gives:

$$Z = 1 + \frac{b}{V} + \left(\frac{b}{V}\right)^2 + \ldots - \frac{a}{RTV} = 1 + \left(b - \frac{a}{RT}\right)\frac{1}{V} + (b^2)\frac{1}{V^2} + \ldots$$

Comparing the previous equation with Eq. (2.41), the second and third virial coefficients in terms of the parameters of the Van der Waals EOS are as follows:

$$B = b - \frac{a}{RT}$$

$$C = b^2$$

Example 2.11

Estimate the Boyle temperature for methane using the Van der Waals EOS.

Solution

As mentioned earlier, at Boyle temperature the second virial coefficient is equal to 0. According to Example 2.10, the second virial coefficient for a gas that obeys Van der Waals EOS is:

$$B = b - \frac{a}{RT}$$

Hence, Boyle temperature may be calculated as follows:

$$b - \frac{a}{RT_{\mathrm{Boyle}}} = 0 \rightarrow T_{\mathrm{Boyle}} = \frac{a}{bR}$$

The critical temperature and critical pressure for methane are 190.56K and 4,599,000 Pa, respectively (Danesh, 1998). The a and b parameters are 0.2302 Pa(m^3/mol)^2 and 4.3061E−05 (m^3/mol), respectively. Therefore, the Boyle temperature is:

$$T_{\mathrm{Boyle}} = \frac{0.2302}{4.3061 \times 10^{-5} \times 8.314} = 643.14$$

Many authors proposed correlations to estimate the second virial coefficient (McGlashan and Potter, 1962; McGlashan and Wormald, 1964; Tsonopoulos, 1974; Schreiber and Pitzer, 1989; Prausnitz et al., 1998). Prausnitz et al. (1998) reviewed a number of correlations for estimating the second virial coefficient. McGlashan et al. (McGlashan and Potter, 1962; McGlashan and Wormald, 1964). proposed a correlation to estimate second virial coefficient for normal alkanes and alpha olefin.

$$\frac{B}{V_{\mathrm{c}}} = 0.430 - 0.886 T_{\mathrm{r}}^{-1} - 0.694 T_{\mathrm{r}}^{-2} - 0.0375(n-1) T_{\mathrm{r}}^{-4.5} \quad \text{for } n \leq 8 \tag{2.44}$$

in which T_{r} is the reduced temperature and n is the number of carbon atoms. For example, the second virial coefficient for methane is obtained by substituting $n = 1$ in Eq. (2.44). The critical volume and critical temperature for methane are 98.6 cm^3/mol and 19.56K, respectively. Therefore, the second virial coefficient relation for methane is:

$$B = 98.6 \left[0.430 - 0.886 \left(\frac{T}{190.56} \right)^{-1} - 0.694 \left(\frac{T}{190.56} \right)^{-2} \right] \tag{2.45}$$

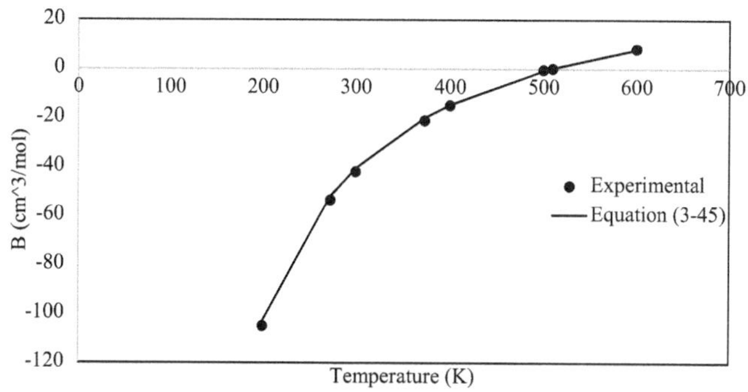

Figure 2.2 Second virial coefficient for methane.

Eq. (2.45) and the experimental value for second virial coefficient of methane are plotted in Fig. 2.2. Good agreement exists between predicted and experimental data.

Schreiber and Pitzer (1989) suggested Eq. (2.46) for estimation of second virial coefficient:

$$\frac{BP_c}{RT_cZ_c} = c_1 + c_2 T_r^{-1} + c_3 T_r^{-2} + c_4 T_r^{-6} \tag{2.46}$$

in which Z_c is the critical compressibility factor and estimated by $Z_c = 0.291 - 0.08\omega$. The coefficients c_1, c_2, c_3, and c_4 related to acentric factor by Eq. (2.47).

$$c_i = c_{i,0} + \omega c_{i,1} \tag{2.47}$$

Coefficients $c_{i,0}$ and $c_{i,1}$ for nonpolar and slightly polar fluid are reported in Table 2.2.

The coefficients reported in Table 2.2 are not valid for highly polar fluids such as H_2O, alcohols, and acids, and for quantum gases such as H_2, He, and Ne.

Table 2.2 Coefficients $c_{i,0}$ and $c_{i,1}$ for Eq. (2.47)

i	$c_{i,0}$	$c_{i,1}$
1	0.442259	0.725650
2	−0.980970	0.218714
3	−0.611142	−1.249760
4	−0.00515624	−0.189187

Table 2.3 Coefficients A_i for Eqs. (2.47), (2.49), and (2.50) (Tsonopoulos, 1974)

i	A_i
0	0.1445
1	−0.330
2	−0.1385
3	−0.0121
4	−0.000607
5	0.0637
6	0.331
7	−0.423
8	−0.008

Tsonopoulos (1974) gives another correlation for prediction of second virial coefficient.

$$\frac{BP_c}{RT_c} = B^{(0)} + \omega B^{(1)} \tag{2.48}$$

$$B^{(0)} = A_0 + \frac{A_1}{T_r} + \frac{A_2}{T_r^2} + \frac{A_3}{T_r^3} + \frac{A_4}{T_r^8} \tag{2.49}$$

$$B^{(1)} = A_5 + \frac{A_6}{T_r^2} + \frac{A_7}{T_r^3} + \frac{A_8}{T_r^8} \tag{2.50}$$

The coefficients for Eqs. (2.49) and (2.50) are reported in Table 2.3.

The second virial coefficient is negative at low and moderate temperature and approaches a positive number as temperature approaches infinity [as can be seen in Fig. 2.2 and from Eqs. (2.44), (2.46), and (2.48)].

As mentioned before, fewer data are available for third virial coefficient; hence, the correlations for third virial coefficient are less accurate and developed based on fewer data. A generalized correlation for third virial coefficient has been proposed by Orbey and Vera (1983).

$$\frac{CP_c^2}{(RT_c)^2} = C^{(0)} + \omega C^{(1)} \tag{2.51}$$

$$C^{(0)} = A_0 + \frac{A_1}{T_r^{2.8}} + \frac{A_2}{T_r^{10.5}} \tag{2.52}$$

$$C^{(1)} = A_3 + \frac{A_4}{T_r^{2.8}} + \frac{A_5}{T_r^3} + \frac{A_6}{T_r^6} + \frac{A_7}{T_r^{10.5}} \tag{2.53}$$

The coefficients for Eqs. (2.49), (2.52), and (2.53) are reported in Table 2.4.

Table 2.4 Coefficients A_i for Eqs. (2.52) and (2.53) (Orbey and Vera, 1983)

i	A_i
0	0.01407
1	0.02432
2	−0.00313
3	−0.02676
4	0.0177
5	0.040
6	−0.003
7	−0.00228

Example 2.12

Estimate the second virial coefficient for methane at temperature 200K, 273K, 300K, 373K, 400K, 500K, and 600K (a) by Eq. (2.48), (b) Eq. (2.46), and compare with experimental value.

Solution

1. The critical temperature, critical pressure, and acentric factor for methane are 190.56K, 4,599,000 Pa, and 0.0115, respectively (Danesh, 1998). At temperature 200K, the reduced temperature is equal to 1.050. Parameters $B^{(0)}$ and $B^{(1)}$ determined by Eqs. (2.49) and (2.50) are as follows:

$$B^{(0)} = 0.1445 - \frac{0.330}{1.050} - \frac{0.1385}{1.050^2} - \frac{0.0121}{1.050^3} - \frac{0.000607}{1.050^8} = -0.3065$$

$$B^{(1)} = 0.0637 + \frac{0.331}{1.050^2} - \frac{0.423}{1.050^3} - \frac{0.008}{1.050^8} = -0.00713$$

Hence, the value of second virial coefficient is:

$$B = \left(B^{(0)} + \omega B^{(1)}\right)\frac{RT_c}{P_c}$$

$$= [-0.3065 - (0.0115 \times 0.0071)]\frac{8.314 \times 190.56}{4599000}$$

$$= -1.0563 \times 10^{-4}\ \mathrm{m^3/mol}$$

$$B = -105.63\ \mathrm{cm^3/mol}$$

(Continued)

The relative deviation is $\frac{-105.63-(-105)}{-105} \times 100 = 0.60\%$. The results for other temperature are reported in the following table.

T (K)	T_r	$B^{(0)}$ Eq. (2.49)	$B^{(1)}$ Eq. (2.50)	$B^{(0)} + \omega B^{(1)}$	B (m^3/mol) Eq. (2.48)	B (cm^3/mol)	$B_{exp.}$ (cm^3/mol)	RD%
200	1.050	−0.3065	−0.0071	−0.3066	−1.0563E−04	−105.63	−105	0.60
273	1.433	−0.1575	0.0807	−0.1566	−5.3930E−05	−53.93	−53.6	0.62
300	1.574	−0.1241	0.0886	−0.1231	−4.2405E−05	−42.41	−42	0.97
373	1.957	−0.0619	0.0937	−0.0608	−2.0938E−05	−20.94	−21.2	−1.23
400	2.099	−0.0455	0.0931	−0.0444	−1.5290E−05	−15.29	−15	1.94
500	2.624	−0.0021	0.0884	−0.0010	−3.5863E−07	−0.36	−0.5	−28.27
600	3.149	0.0253	0.0835	0.0263	9.0582E−06	9.06	8.1	11.83

2. The critical compressibility factor is $Z_c = 0.291 - 0.08(0.0115) = 0.2901$. The values of coefficients c_1, c_2, c_3, and c_4 are 0.450604, −0.978455, −0.625514, and −0.007332 [determined by Eq. (2.47)]. At temperature 200K, the reduced temperature is equal to 1.050.

$$\frac{BP_c}{RT_cZ_c} = 0.450604 - 0.978455(1.050)^{-1} - 0.625514(1.050)^{-2}$$
$$- 0.007332(1.050)^{-6}$$
$$= -1.0550$$

$$B = -1.0550\left[\frac{8.314 \times 190.56 \times 0.2901}{4599000}\right] = -1.0543 \times 10^{-4}\ \text{m}^3/\text{mol}$$

$$B = -105.43\ \text{cm}^3/\text{mol}$$

The results for other temperatures are reported in the following table.

T (K)	T_r	$c_1 + c_2T_r^{-1} + c_3T_r^{-2} + c_4T_r^{-6}$	B (m^3/mol) Eq. (2.48)	B (cm^3/mol)	$B_{exp.}$ (cm^3/mol)	RD%
200	1.050	−1.0550	−1.0543E−04	−105.43	−105	0.41
273	1.433	−0.5380	−5.3762E−05	−53.76	−53.6	0.30
300	1.574	−0.4238	−4.2348E−05	−42.35	−42	0.83
373	1.957	−0.2127	−2.1252E−05	−21.25	−21.2	0.24
400	2.099	−0.1576	−1.5747E−05	−15.75	−15	4.98
500	2.624	−0.0132	−1.3175E−06	−1.32	−0.5	163.51
600	3.149	0.0767	7.6690E−06	7.67	8.1	−5.32

Example 2.13

The pressure–volume data for superheated steam at 250°C are reported in the following table (Abbott et al., 2001).

P (Pa)	V_{mass} (m^3/g)
200,000	1198.9
225,000	1064.7
250,000	957.41
275,000	869.61
300,000	796.44
1,350,000	169.96
1,400,000	163.55
1,450,000	157.57
1,500,000	151.99
1,550,000	146.77
1,600,000	141.87
1,650,000	137.27
1,700,000	132.94
1,750,000	128.85
1,800,000	124.99
1,850,000	121.33
1,900,000	117.87
1,950,000	114.58
2,000,000	111.45
2,100,000	105.64
2,200,000	100.35
2,300,000	95.513

Find the second and third virial coefficients (neglect the higher-order terms). The reported values for second and third virial coefficients are −152.2 cm^3/mol and −5800 cm^6/mol^2 (Riazi, 2005).

Solution

The molar volume may be calculated by multiplying V_{mass} by 18×10^{-6}. The compressibility factor is equal to $Z = PV/RT$ in which R is 8.314 J mol^{-1} K^{-1}. Assuming $f = Z - 1$, $x = 1/V$ and $y = 1/V^2$

P (Pa)	V_{mass} (m^3/g)	V (m^3/mol)	Z	f	x	y
200,000	1198.9	0.02158	0.9923	−0.0077	46.3	2,147.3
225,000	1064.7	0.01916	0.9914	−0.0086	52.2	2,722.7
250,000	957.41	0.01723	0.9905	−0.0095	58.0	3,367.1
275,000	869.61	0.01565	0.9897	−0.0103	63.9	4,081.4
300,000	796.44	0.01434	0.9888	−0.0112	69.8	4,865.7

(Continued)

—cont'd

P (Pa)	V_{mass} (m^3/g)	V (m^3/mol)	Z	f	x	y
1,350,000	169.96	0.00306	0.9495	−0.0505	326.9	106,846.8
1,400,000	163.55	0.00294	0.9476	−0.0524	339.7	115,386.2
1,450,000	157.57	0.00284	0.9455	−0.0545	352.6	124,310.5
1,500,000	151.99	0.00274	0.9435	−0.0565	365.5	133,605.7
1,550,000	146.77	0.00264	0.9415	−0.0585	378.5	143,278.3
1,600,000	141.87	0.00255	0.9394	−0.0606	391.6	153,346.5
1,650,000	137.27	0.00247	0.9373	−0.0627	404.7	163,796.2
1,700,000	132.94	0.00239	0.9353	−0.0647	417.9	174,640.0
1,750,000	128.85	0.00232	0.9332	−0.0668	431.2	185,902.9
1,800,000	124.99	0.00225	0.9311	−0.0689	444.5	197,562.5
1,850,000	121.33	0.00218	0.9289	−0.0711	457.9	209,661.5
1,900,000	117.87	0.00212	0.9268	−0.0732	471.3	222,151.1
1,950,000	114.58	0.00206	0.9247	−0.0753	484.9	235,091.7
2,000,000	111.45	0.00201	0.9225	−0.0775	498.5	248,481.9
2,100,000	105.64	0.00190	0.9181	−0.0819	525.9	276,565.6
2,200,000	100.35	0.00181	0.9136	−0.0864	553.6	306,492.8
2,300,000	95.513	0.00172	0.9091	−0.0909	581.7	338,321.8

Use the least squares method to find the regression coefficients B and C in the following equation:

$$f = Bx + Cy$$

Note that the coefficients B and C are identical with B and C in Eq. (2.41). The results of the least squares method are as follows:

$$f = -0.0001533x - 4.259 \times 10^{-9}y \xrightarrow{x=\frac{1}{V}, y=\frac{1}{V^2}, f=Z-1}$$

$$Z = 1 - \frac{0.0001533}{V} - \frac{4.259 \times 10^{-9}}{V^2}$$

Correlation coefficient = 0.999.

Hence, the second and third virial coefficients −153.3 cm^3/mol and −4259 cm^6/mol^2 are close to reported values.

The best-known and mostly widely used noncubic EOSs are the Benedict–Webb–Rubin (BWR)-type EOS. The BWR EOS is an empirical extension of virial EOS. The BWR EOS can be expressed as follows (Benedict et al., 1940):

$$P = RT\rho_M + \left(B_0RT - A_0 - C_0T^{-2}\right)\rho_M^2 + (bRT - a)\rho_M^6 + cT^{-2}\rho_M^3\left(1 + \gamma\rho_M^2\right)\exp\left(-\gamma\rho_M^2\right) \quad (2.54)$$

in which ρ_M is the molar density. The constants of Eq. (2.54) for several substances are given in Table 2.5 (taken from Novak et al., 1972). Many variations of BWR EOS have been proposed since the introduction of the BWR EOS (Benedict et al., 1940; Starling, 1966, 1973; Nishiumi and Saito, 1975; Nishiumi, 1980; Nishiumi et al., 1991; Soave, 1995; Wang et al., 2001). The most widely used BWR-type EOS is the Benedict–Webb–Rubin–Starling (BWRS) EOS, which has been introduced by Starling (1973).

The BWRS EOS is given by Eq. (2.55).

$$\begin{aligned} P = {} & RT\rho_M + \left(B_0RT - A_0 - C_0T^{-2} + D_0T^{-3} - E_0T^{-4}\right)\rho_M^2 \\ & + \left(bRT - a - dT^{-1}\right)\rho_M^3 + \alpha\left(a + dT^{-1}\right)\rho_M^6 \\ & + cT^{-2}\rho_M^3\left(1 + \gamma\rho_M^2\right)\exp\left(-\gamma\rho_M^2\right) \end{aligned} \tag{2.55}$$

in which $\rho_{Mc} = 1/V_c$. The coefficients of Eq. (2.55) can be determined by the following equations.

$$\rho_{Mc}B_0 = A_1 + A_2\omega \tag{2.56}$$

$$\frac{\rho_{Mc}A_0}{RT_c} = A_3 + A_4\omega \tag{2.57}$$

$$\frac{\rho_{Mc}C_0}{RT_c^3} = A_5 + A_6\omega \tag{2.58}$$

$$\frac{\rho_{Mc}D_0}{RT_c^4} = A_7 + A_8\omega \tag{2.59}$$

$$\frac{\rho_{Mc}E_0}{RT_c^5} = A_9 + A_{10}\exp(A_{11}\omega) \tag{2.60}$$

$$\rho_{Mc}^2 b = A_{12} + A_{13}\omega \tag{2.61}$$

$$\frac{\rho_{Mc}^2 a}{RT_c} = A_{14} + A_{15}\omega \tag{2.62}$$

$$\frac{\rho_{Mc}^2 d}{RT_c^2} = A_{16} + A_{17}\omega \tag{2.63}$$

$$\rho_{Mc}^3\alpha = A_{18} + A_{19}\omega \tag{2.64}$$

Table 2.5 Values of Constants of BWR EOS, Taken From Novak et al. (1972)

Substance	A_0	B_0	C_0	a	b
Hydrogen[a]	9.73E−02	1.8041E−02	3.8914E+02	−9.2211E−03	1.7976E−04
Nitrogen	1.1925	0.0458	5.8891E+03	0.0149	1.98154E−03
Nitrogen	0.872086	2.81066E−02	7.81375E+03	3.12319E−02	3.2351E−03
Oxygen	1.4988	4.6524E−02	3.8617E+03	−4.0507E−02	−2.7963E−04
CO	1.34122	5.45425E−02	8.562E+03	3.665E−02	2.6316E−03
CO	1.03115	0.040	1.124E+04	3.665E−02	2.6316E−03
CO_2	2.7374	4.9909E−02	4.35200E−02	1.3681E−01	7.2105E−03
CO_2	2.51604	4.48842E−02	1.474405E+05	1.3688E−01	4.12381E−03
CO_2[a]	2.7634	4.5628E−02	1.1333E+05	5.1689E−02	3.0819E−03
SO_2	7.08538	0.10896	4.43966E+05	6.87046E−02	1.93727E−03
SO_2	2.12042	2.61817E−02	7.93840E+05	0.844680	1.46531E−02
N_2O[a]	3.0868	5.1953E−02	1.2725E+05	0.10946	3.7755E−03
H_2S	3.10377	3.48471E−02	1.9721E+05	0.144984	4.42477E−03
NH_3	3.78928	5.16461E−02	1.78567E+05	0.10354	7.19561E−03
Methane	1.8550	4.2600E−02	2.257E+04	0.0494	3.38004E−03
Methane	1.79894	4.54625E−02	3.18382E+04	0.04352	2.52033E−03
Ethane	4.15556	6.27724E−02	1.79592E+05	0.34516	1.1122E−02
Ethylene	3.33958	5.56833E−02	1.31140E+05	0.259	0.00860
Acetylene	1.5307	5.5851E−03	2.1586E+05	−0.10001	−3.7810E−05
Propane	6.87225	9.7313E−02	5.08256E+05	0.9477	0.0255
Propyne[a]	5.10806	6.9779E−02	6.40624E+05	0.69714	1.4832E−02
Isobutane	10.23264	1.37544E−01	8.49943E+05	1.93763	4.24352E−02

Butane	10.0847	1.24361E−01	9.9283E+05	1.88231	3.99983E−02
Isobutylene	8.95325	1.16025E−01	9.2728E+05	1.6927	3.48156E−02
Pentane	12.1794	0.156751	2.12121E+06	4.0748	6.6812E−02
Isopentane	12.7959	0.160053	1.74632E+06	3.7562	6.6812E−02
Neopentane[a]	14.9413	0.19534	1.07186E+06	2.72334	5.71607E−02
Neopentane[a]	7.06955	5.17798E−02	1.62085E+06	2.06202	4.62003E−02
Hexane	14.4373	1.77813E−01	3.31935E+06	7.11671	1.09131E−01
Heptane	17.5206	1.99005E−01	4.75574E+06	10.36475	1.51954E−01
Nonane	−41.456199	−9.64946E−01	2.75136E+06	37.17914	6.04989E−01
Decane	−19.38795	−9.46923E−01	3.43152E+06	59.87797	9.86288E−01

				Range of Validity		
Substance	c	α	γ	**Temperature (°C)**	**To d_T**	**P_{max} (atm)**
Hydrogen[a]	−2.4613E+02	−3.4215E−06	1.89E−03	(0)−(150)	2.5	2500
Nitrogen	5.48064E+02	2.91545E−04	7.5E−03	(−163)−(200)	1.25	600
Nitrogen	5.47364E+02	7.093E−05	4.5E−03	(−170)−(100)	2.0	
Oxygen	−2.0376E+02	8.641E−06	3.59E−03	(−110)−(125)	0.8	
CO	1.04E+03	1.350E−04	0.006	(−140)−(−25)		
CO	1.04E+03	1.350E−04	0.006	(−25)−(200)		
CO_2	1.49183E+04	8.4658E−05	5.393E−03	(10)−(150)		1000
CO_2	1.49183E+04	8.4685E−05	5.253E−03	(150)−(250)		1000
CO_2[a]	7.0762E+03	1.1271E−04	4.94E−03	(0)−(275)	2.1	700
SO_2	5.85038E+04	5.86479E−04	8.687E−03	(10)−(250)	2.0	700
SO_2	1.13356E+05	7.1951E−05	5.923E−03	(10)−(250)	2.0	200
N_2O[a]	1.3794E+04	9.377E−05	5.301E−03	(−30)−(150)	2.0	200
H_2S	1.87032E+04	7.0316E−05	4.555E−03	(5)−(170)	2.2	200
NH_3	1.57536E+02	4.651890E−06	1.980E−02	(0)−(300)	1.5	700

(Continued)

Table 2.5 Values of Constants of BWR EOS, Taken From Novak et al. (1972)—cont'd

Substance	c	α	γ	Range of Validity		
				Temperature (°C)	To d_T	P_{max} (atm)
Methane	2.454E+03	1.24359E−04	0.006	(−70)−(200)	1.8	400
Methane	3.5878E+03	3.30E−04	1.05E−02	(0)−(350)	1.8	400
Ethane	3.2726E+04	2.43389E−04	1.18E−02	(0)−(275)	1.6	300
Ethylene	2.112E+04	1.78E−04	9.23E−03	(0)−(200)	1.6	300
Acetylene	6.0162E+03	−5.549E−05	7.14E−03	(20)−(250)	1.6	150
Propane	1.29E+05	6.07175E−04	0.022	(100)−(275)	1.75	
Propyne[a]	1.09855E+05	2.73630E−04	1.245E−02	(50)−(200)		
Isobutane	2.8601E+05	1.07408E−03	0.034	(100)−(240)	1.8	
Butane	3.1640E+05	1.10132E−03	3.4E−02	(150)−(300)	1.8	
Isobutylene	2.7492E+05	9.10889E−04	2.96E−02	(150)−(275)	1.8	
Pentane	8.2417E+05	1.810E−03	4.75E−02	(140)−(280)	1.5	200
Isopentane	6.95E+05	1.70E−03	4.63E−02	(130)−(280)	1.5	200
Neopentane[a]	4.73969E+05	2.24898E−03	5.352E−02	(160)−(275)	2.1	250
Neopentane[a]	4.31017E+05	2.51254E−03	5.342E−02	(30)−(200)	1.7	70
Hexane	1.51276E+06	2.81086E−03	6.668E−02	(275)−(350)	1.8	
Heptane	2.47E+06	4.35611E−03	9E−02	(275)−(350)	1.8	
Nonane	2.516085E+00	3.230516E−03	1.223E−01	(40)−(250)		700
Decane	7.8223E+06	4.35394E−03	1.53E−01	(40)−(250)		700

P (atm), V (liters/mol), T (K), $R = 0.08206$.
[a]Constants calculated with the aid of the standard program at the Department of Physical Chemistry, Institute of Chemical Technology, Prague, using literature data.

$$\frac{\rho_{\mathrm{Mc}}^{2} c}{R T_{\mathrm{c}}^{3}}=A_{20}+A_{21} \omega \tag{2.65}$$

$$\rho_{\mathrm{Mc}}^{2} \gamma=A_{22}+A_{23} \omega \tag{2.66}$$

The generalized coefficients of Eqs. (2.56) through (2.66) are given in Table 2.6.

The BWRS EOS is suitable for light hydrocarbons and reservoir fluids (Riazi, 2005). The accuracy of predicted volumetric data from BWRS EOS are better than cubic EOS; however, the BWRS EOS demands high computational time and is not suitable when successive equilibrium calculations are required.

Table 2.6 The Coefficients of Eqs. (2.56) Through (2.66)

i	A_i
1	0.443690
2	0.115449
3	1.28438
4	−0.920731
5	0.356306
6	1.70871
7	0.0307452
8	0.179433
9	0.006450
10	−0.022143
11	−3.8
12	0.528629
13	0.349261
14	0.484011
15	0.754130
16	0.07322828
17	0.463492
18	0.0705233
19	−0.044448
20	0.504087
21	1.32245
22	0.544979
23	−0.270896

Example 2.14

Dieterici proposed a noncubic EOS in 1899. The Dieterici EOS is expressed as follows:

$$P = \frac{RT}{V-b}\exp\left(\frac{-a}{RTV}\right)$$

The Dieterici EOS gives a more realistic critical compressibility factor ($Z_c = 0.2707$) in comparison with other equations such as Van der Waals, RK, SRK, and PR (Speakman and Partington, 1950; Glasstone, 1951; Hirschfelder et al., 1954; Atkinz and Paula, 2006) and has been revisited by a number of authors (Sadus, 2001, 2002, 2003; Roman et al., 2004).

1. Determine a and b in terms of critical temperature and critical volume by applying the condition at critical point.
2. Determine the critical compressibility factor.

Solution

1. The derivative of pressure with respect to molar volume at the critical point is 0. Taking derivative from P with respect to V at constant temperature gives:

$$\left(\frac{\partial P}{\partial V}\right)_{T_c} = P\left[\frac{aV - ab - RTV^2}{RTV^2(V-b)}\right] \tag{a}$$

$$\left(\frac{\partial^2 P}{\partial V^2}\right)_{T_c} = \left(\frac{\partial P}{\partial V}\right)_{T_c}\left[\frac{aV - ab - RTV^2}{RTV^2(V-b)}\right] + P\left[\frac{-2aV^2 + 4Vab + RTV^3 - 2ab^2}{RTV^3(V-b)^2}\right] \tag{b}$$

Eqs. (a) and (b) are equal to 0 at critical point, which gives Eq. (c) and (d).

$$aV_c - ab - RT_cV_c^2 = 0 \tag{c}$$

$$-2aV_c^2 + 4V_cab + RT_cV_c^3 - 2ab^2 = 0 \tag{d}$$

From Eq. (c) we have:

$$RT_cV_c^3 = aV_c^2 - abV_c \tag{e}$$

Substituting Eq. (e) into Eq. (d) results in

$$V_c^2 - 3bV_c + 2b^2 = 0 \tag{f}$$

Solving Eq. (f) gives two real roots for V_c, $V_c = b$, and $V_c = 2b$. $V_c = b$ is rejected because it is a singularity point for Dieterici EOS, hence:

$$b = \frac{V_c}{2} \tag{g}$$

Substituting Eq. (g) into Eq. (f) results in

$$a = 2RT_cV_c \tag{h}$$

2. Setting the Dieterici EOS at critical point.

$$P_c = \frac{RT_c}{V_c - b}\exp\left(\frac{-a}{RT_cV_c}\right) \tag{i}$$

Substituting Eqs. (g) and (h) in Eq. (i) gives:

$$P_c = \frac{2RT_c}{V_c}\exp(-2) \tag{j}$$

$$Z_c = \frac{P_cV_c}{RT_c} = 2\exp(-2) = 0.2707 \tag{k}$$

2.4 CORRESPONDING STATE CORRELATIONS

According to the law of corresponding states (or principle of corresponding states), all fluids with the same reduced temperature and reduced pressure have almost the same deviation from ideal gas. In other words, according to the law of corresponding states, all fluids with the same reduced temperature and reduced pressure have almost the same compressibility factor. This principle was originally stated by Van der Waals in 1873. The principle of corresponding states can be expressed in a mathematical form by Eq. (2.67)

$$Z = f(T_r, P_r) \tag{2.67}$$

in which Z is the compressibility factor, T_r is the reduced temperature, and P_r is the reduced pressure. The correlations in the form of Eq. (2.67) usually called *two-parameter corresponding states*. Consider the RK EOS; at the critical point ($T_r = P_r = 1$) the critical compressibility factor for all components is 0.333, however, the critical compressibility factor only for normal fluids such as N_2, CH_4, O_2, and Ar is relatively constant. For this reason, the RK EOS is relatively accurate only for normal fluids. Standing and Katz (1942) presented a graphical chart for estimation of the compressibility factor for sweet natural gas. They developed the chart based on experimental data for methane binary mixtures with ethane, propane, butane, and other natural gases and suitable for sweet natural gas with a molecular weight less than 40 (Danesh, 1998). A number of investigators have attempted to fit an equation that gives the original data of the Standing–Katz chart (Hall and

Yarborough, 1973; Dranchuk and Kassem, 1975; Brill and Beggs, 1984). Takacs (1976) reviewed and compared eight equations which represented the Standing–Katz chart. Hall–Yarborough (Hall and Yarborough, 1973) and Dranchuk–Kassem (Dranchuk and Kassem, 1975) equations give results that are more accurate. These equations give reasonable results over a wide range of reduced temperature and reduced pressure (reduced temperature between 1 and 3, reduced pressure between 0.2 and 25) (Whitson and Brulé, 2000). The equation that proposed by Hall–Yarborough is in the following form:

$$Z = 0.06125 P_r T_r^{-1} \xi^{-1} \exp\left[-1.2\left(1 - T_r^{-1}\right)^2\right] \tag{2.68}$$

in which T_r and P_r are the reduced temperature and reduced pressure, respectively. The ξ is a dimensionless parameter that is obtained by solving Eq. (2.69).

$$\begin{aligned} F(\xi) = & -0.06125 P_r T_r^{-1} \exp\left[-1.2\left(1 - T_r^{-1}\right)^2\right] \\ & + \frac{\xi + \xi^2 + \xi^3 - \xi^4}{(1-\xi)^3} - \left(14.76 T_r^{-1} - 9.76 T_r^{-2} + 4.58 T_r^{-3}\right)\xi^2 \\ & + \left(90.7 T_r^{-1} - 242.2 T_r^{-2} + 42.4 T_r^{-3}\right)\xi^{\left(2.18 + 2.82 T_r^{-1}\right)} = 0 \end{aligned} \tag{2.69}$$

The Hall–Yarborough equation (or, in other words, the Standing–Katz chart) is suitable for natural gases and light hydrocarbons (Whitson and Brulé, 2000; Riazi, 2005). For heavier fluids at the identical reduced temperature and reduced pressure, the deviations are not the same and the two-parameter corresponding states are no longer valid. Pitzer et al. (1955) defined a new parameter, *acentric factor*, to extend the principle of corresponding states to other components that are not normal. The acentric factor is defined in terms of reduced vapor pressure at $T_r = 0.7$ as follows:

$$\omega = -\log_{10} P_r^{sat}\Big|_{T_r = 0.7} - 1.0 \tag{2.70}$$

in which ω is the acentric factor, and P_r^{sat} is the reduced vapor pressure and equal to $P_r^{sat} = \frac{P^{sat}|_{T_r=0.7}}{P_c}$. The values of acentric factor for normal fluids (spherical molecules) are zero or near zero (the acentric factor for N_2, CH_4, and O_2 are 0.0403, 0.0115, and 0.0218, respectively). Authors represent the properties in the following general form:

$$M = M^{(0)} + \omega M^{(1)} + \omega^2 M^{(2)} + \ldots \tag{2.71}$$

in which M presented any property such as compressibility factor, enthalpy, entropy, and fugacity coefficient. $M^{(0)}$, $M^{(1)}$, $M^{(2)}$,… in Eq. (2.71) are functions of reduced temperature and reduced pressure. This new theorem known as *three parameter corresponding states* can be expressed as "all fluids with the same reduced temperature, reduced pressure and acentric factor have almost the same deviation from ideal gas" and can be expressed by Eq. (2.72).

$$Z = f(T_r, P_r, \omega) \tag{2.72}$$

Eq. (2.71) is usually truncated after second term. For instance, the general form equation for compressibility factor after truncated is in the following form:

$$Z = Z^{(0)} + \omega Z^{(1)} \tag{2.73}$$

For normal fluid with an acentric factor near zero, the compressibility factor is approximately equal to $Z^{(0)}$. Hence, $Z^{(0)}$ by itself represents a two-parameter corresponding state. A number of Pitzer type correlations are available. The most accurate Pitzer-type correlation was presented by Lee and Kesler in 1975. Lee—Kesler is a modified form of BWR EOS which takes a table form. The original table that was presented by Lee and Kesler covered a wide range of reduced pressure between 0.01 and 10. The $Z^{(0)}$ and $Z^{(1)}$ for reduced pressure ranging up to 14 are available in the API technical data book (Daubert and Danner, 1997).

Example 2.15

Estimate the acentric factor for water. The reported acentric factor for water is 0.3449. The Antoine equation for water is as follows (Abbott et al., 2001):

$$\ln P^{sat} = 16.2620 - \frac{3799.89}{T - 46.80}$$

in which P^{sat} is the saturate vapor pressure in kPa and T is the temperature in K.

Solution

The critical temperature and critical pressure for water are 647.13K and 22,055 kPa, respectively. The temperature at which the reduced temperature is $T = 0.7T_C = 0.7(647.13) = 452.99\text{K}$. The vapor pressure at the temperature 452.99K is determined as follows:

$$\ln P^{sat} = 16.2620 - \frac{3799.89}{452.99 - 46.80} = 6.9071$$

$$P^{sat} = \exp(6.9071) = 999.31 \text{ kPa}$$

The reduced vapor pressure at $T_r = 0.7$ is 999.31/22,055 = 0.0453. Hence, the acentric factor is calculated by Eq. (2.70) as follows:

$$\omega = -\log_{10}(0.0453) - 1.0 = 0.3438$$

Example 2.16

Pitzer et al. proposed the following equation for a second virial coefficient of normal fluids (Abbott et al., 2001).

$$\frac{BP_c}{RT_c} = B^{(0)} + \omega B^{(1)}$$

$B^{(0)}$ and $B^{(1)}$ are defined as follows:

$$B^{(0)} = 0.083 - \frac{0.422}{T_r^{1.6}}$$

$$B^{(1)} = 0.139 - \frac{0.172}{T_r^{4.2}}$$

in which T_r is the reduced temperature.

1. Estimate the reduced Boyle temperature (i.e., T_{Boyle}/T_c) for normal fluids.
2. Estimate the Boyle temperature for methane.

Solution

1. As mentioned before, at the Boyle temperature the second virial coefficient is equal to zero; hence, this can be written:

$$B^{(0)} + \omega B^{(1)} = 0$$

For normal fluids, the acentric factor is near zero, hence the previous equation is reduced to the following equation:

$$B^{(0)} = 0$$

Substitute the equation that has been given for $B^{(0)}$.

$$0.083 - \frac{0.422}{T_r^{1.6}} = 0$$

Solving the previous equation for reduced temperature gives $T_{Br} = 2.763$ (in which T_{Br} is the reduced Boyle temperature). The reduced Boyle temperature from Van der Waals EOS is equal to 3.375.

2. The critical temperature for methane is 190.56K (Danesh, 1998); hence, the Boyle temperature for methane is $T_B = 2.763 \times 190.56 = 526.53$, which is close enough to the expected value (i.e., 510K). When $B^{(1)}$ is considered, the reduced Boyle temperature for methane is 2.729.

No.	Chemical Formula	Component	*A*	*B*	*C*	*D*	*E*	T_{min} (K)	T_{max} (K)
1	$C_4H_{10}O$	1-Butanol (*n*-Butanol)	39.6673	−4.0017E+03	−1.0295E+01	−3.2572E−10	8.6672E−07	183.9	562.9
2	C_4H_8	1-Butene	27.3116	−1.9235E+03	−7.2064E+00	7.4852E−12	3.6481E−06	87.8	419.6
3	C_3H_8O	1-Propanol (*n*-Propanol)	31.5155	−3.4570E+03	−7.5235E+00	−4.2870E−11	1.3029E−07	147.0	536.7
4	C_3H_6O	Acetone	28.5884	−2.4690E+03	−7.3510E+00	2.8025E−10	2.7361E−06	178.5	508.2
5	NH_3	Ammonia	37.1575	−2.0277E+03	−1.1601E+01	7.4625E−03	−9.5811E−12	195.4	405.7
6	C_6H_6	Benzene	31.7718	−2.7254E+03	−8.4443E+00	−5.3534E−09	2.7187E−06	276.7	562.2
7	CS_2	Carbon disulfide	25.1475	−2.0439E+03	−6.7794E+00	3.4828E−03	−1.0105E−14	161.6	552.0
8	CO_2	Carbon dioxide	35.0169	−1.5119 E+03	−1.1334 E+01	9.3368E−03	1.7136E−09	216.6	304.2
9	CO	Carbon monoxide	51.8145	−7.8824E+02	−2.2734E+01	5.1225E−02	6.1896E−11	68.2	132.9
10	C_6H_{12}	Cyclohexane	48.5529	−3.0874E+03	−1.5521E+01	7.3830E−03	6.3563E−12	279.7	553.5
11	C_8H_{10}	Ethylbenzene	36.1998	−3.3402E+03	−9.7970E+00	−1.1467E−11	2.5758E−06	178.2	617.2
12	C_2H_4	Ethylene	18.7964	−9.9962E+02	−4.5788E+00	9.9746E−11	6.7880E−06	104.0	282.4
13	H_2	Hydrogen	3.4132	−4.1316E+01	1.0947E+00	−6.6896E−10	1.4589E−04	14.0	33.2
14	CH_4	Methane	14.6667	−5.7097E+02	−3.3373E+00	2.1999E−09	1.3096E−05	90.7	190.6
15	C_8H_{10}	*m*-Xylene	34.6803	−3.2981E+03	−9.2570E+00	−4.3563E−10	2.4103E−06	225.3	617.1
16	$C_{10}H_8$	Naphthalene	34.9161	−3.9357E+03	−9.0648E+00	−2.0672E−09	1.5550E−06	353.4	748.4
17	C_4H_{10}	*n*-Butane	27.0441	−1.9049E+03	−7.1805E+00	−6.6845E−11	4.2190E−06	134.9	425.2
18	$C_{10}H_{22}$	*n*-Decane	26.5125	−3.3584E+03	−6.1174E+00	−3.3225E−10	4.8554E−07	243.5	618.5
19	C_7H_{16}	*n*-Heptane	65.0257	−3.8188E+03	−2.1684E+01	1.0387E−02	1.0206E−14	182.6	540.3
20	C_6H_{14}	*n*-Hexane	69.7378	−3.6278E+03	−2.3927E+01	1.2810E−02	−1.6844E−13	177.8	507.4

(Continued)

21	N_2	Nitrogen	23.8572	−4.7668E+02	−8.6689E+00	2.0128E−02	−2.4139E−11	63.2	126.1
22	C_9H_{20}	*n*-Nonane	8.8817	−2.8042E+03	1.5262E+00	−1.0464E−02	5.7972E−06	219.6	595.7
23	C_8H_{18}	*n*-Octane	29.0948	−3.0114E+03	−7.2653E+00	−2.2696E−11	1.4680E−06	216.4	568.8
24	C_5H_{12}	*n*-Pentane	33.3239	−2.4227E+03	−9.2354E+00	9.0199E−11	4.1050E−06	143.4	469.7
25	O_2	Oxygen	20.6695	−5.2697E+02	−6.7062E+00	1.2926E−02	−9.8832E−13	54.4	154.6
26	C_8H_{10}	*o*-Xylene	37.2413	−3.4573E+03	−1.0126E+01	9.0676E−11	2.6123E−06	248.0	630.4
27	C_8H_{10}	*p*-Xylene	60.0531	−4.0159E+03	−1.9441E+01	8.2881E−03	−2.3647E−12	286.4	616.3
28	SO_2	Sulfur dioxide	19.7418	−1.8132E+03	−4.1458E+00	−4.4284E−09	8.4918E−07	197.7	430.8
29	C_7H_8	Toluene	34.0775	−3.0379E+03	−9.1635E+00	1.0289E−11	2.7035E−06	178.2	591.8
30	H_2O	Water	29.8605	−3.1522E+03	−7.3037E+00	2.4247E−09	1.8090E−06	273.2	647.1

Example 2.17

The vapor pressure of pure components may be expressed by the following equation:

$$\log_{10}P^{sat} = A + \frac{B}{T} + C\log_{10}T + DT + ET^2$$

in which P^{sat} is the saturation pressure in mmHg; T is the temperature in K; and A, B, C, D, and E are constants. The constants and range of validity for several pure components are reported in the following table (Coker, 2007).

Develop a linear relationship between critical compressibility factor and acentric factor. Lee and Kesler (1975) proposed the following relationship between critical compressibility factor and acentric factor:

$$Z_c = 0.2901 - 0.0879\omega$$

The critical properties for components are given in the following table (Danesh, 1998; Abbott et al., 2001).

No.	Component	P_c (MPa)	T_c (K)	Z_c
1	1-Butanol (*n*-Butanol)	4.423	563.1	0.260
2	1-Butene	4.02	419.59	0.2765
3	1-Propanol (*n*-Propanol)	5.175	536.8	0.254
4	Acetone	4.701	508.2	0.233
5	Ammonia	11.28	405.7	0.242
6	Benzene	4.898	562.16	0.2714
7	Carbon disulfide	7.9	552	0.275
8	Carbon dioxide	7.382	304.19	0.2744
9	Carbon monoxide	3.499	132.92	0.2948
10	Cyclohexane	4.075	553.54	0.2726
11	Ethylbenzene	3.609	617.17	0.2629
12	Ethylene	5.032	282.36	0.2767
13	Hydrogen	1.313	33.18	0.3053
14	Methane	4.599	190.56	0.2862
15	*m*-Xylene	3.541	617.57	0.2594
16	Naphthalene	4.051	748.4	0.269
17	*n*-Butane	3.796	425.12	0.2739
18	*n*-Decane	2.11	617.7	0.2465
19	*n*-Heptane	2.74	540.2	0.2611
20	*n*-Hexane	3.025	507.6	0.2659
21	Nitrogen	3.394	126.1	0.2917
22	*n*-Nonane	2.29	594.6	0.2520
23	*n*-Octane	2.49	568.7	0.2559
24	*n*-Pentane	3.37	469.7	0.2701
25	Oxygen	5.043	154.58	0.2880
26	*o*-Xylene	3.734	630.37	0.2630
27	*p*-Xylene	3.511	616.26	0.2598
28	Sulfur dioxide	7.884	430.75	0.2686
29	Toluene	4.109	591.79	0.2637
30	Water	22.055	647.13	0.2294

(Continued)

Solution

The acentric factor is calculated similar to the previous example. Note that to convert mmHg to MPa, multiply the number of mmHg by 0.101325/760. The results are given in the following table.

No.	Component	T(@ $T_r = 0.7$) = $0.7T_c$	P^{sat} (mmHg)	P^{sat} (MPa)	Acentric Factor Eq. (2.70)
1	1-Butanol (*n*-Butanol)	394.17	846	0.113	0.594
2	1-Butene	293.71	1961	0.261	0.187
3	1-Propanol (*n*-Propanol)	375.76	915	0.122	0.627
4	Acetone	355.74	1741	0.232	0.306
5	Ammonia	283.99	4742	0.632	0.251
6	Benzene	393.51	2261	0.301	0.211
7	Carbon disulfide	386.40	4625	0.617	0.108
8	Carbon dioxide	212.93	3284	0.438	0.227
9	Carbon monoxide	93.04	2253	0.300	0.066
10	Cyclohexane	387.48	1877	0.250	0.212
11	Ethylbenzene	432.02	1346	0.179	0.304
12	Ethylene	197.65	3102	0.414	0.085
13	Hydrogen	23.23	1616	0.215	−0.215
14	Methane	133.39	3366	0.449	0.011
15	*m*-Xylene	432.30	1265	0.169	0.322
16	Naphthalene	523.88	1517	0.202	0.302
17	*n*-Butane	297.58	1798	0.240	0.200
18	*n*-Decane	432.39	515	0.069	0.488
19	*n*-Heptane	378.14	914	0.122	0.352
20	*n*-Hexane	355.32	1124	0.150	0.305
21	Nitrogen	88.27	2321	0.309	0.040
22	*n*-Nonane	416.22	618	0.082	0.444
23	*n*-Octane	398.09	747	0.100	0.398
24	*n*-Pentane	328.79	1427	0.190	0.248
25	Oxygen	108.21	3598	0.480	0.022
26	*o*-Xylene	441.26	1364	0.182	0.313
27	*p*-Xylene	431.38	1244	0.166	0.326
28	Sulfur dioxide	301.53	3364	0.448	0.245
29	Toluene	414.25	1678	0.224	0.264
30	Water	452.99	7478	0.997	0.345

Linear regression using least squares method yields the following relation between critical compressibility factor and acentric factor:

$$Z_c = 0.2857 - 0.0749\omega$$

The correlation coefficient and percent average absolute are 0.766% and 2.61%, respectively. Actually, this equation (and the relation that was proposed by Lee and Kelser) is not in good agreement with the experimental data. These equations are suitable for hydrocarbons. If the nonhydrocarbon components are not considered, linear regression using least square method yields the following relation between critical compressibility factor and acentric factor:

$$Z_c = 0.2881 - 0.0803\omega$$

The correlation coefficient and percent average absolute deviation are 0.983% and 0.70%, respectively, which shows this correlation more accurately for hydrocarbon. The predicted critical compressibility factor and relative deviation for hydrocarbon components are reported in the following table.

Component	Z_c Experimental	Z_c Predicted	R.D. %
1-Butene	0.2765	0.2731	−1.23
Benzene	0.2714	0.2712	−0.08
Cyclohexane	0.2726	0.2711	−0.55
Ethylbenzene	0.2629	0.2637	0.31
Ethylene	0.2767	0.2813	1.65
Hydrogen	0.3053	0.3054	0.02
Methane	0.2862	0.2872	0.37
m-Xylene	0.2594	0.2622	1.09
Naphthalene	0.2690	0.2639	−1.91
n-Butane	0.2739	0.2721	−0.67
n-Decane	0.2465	0.2489	0.98
n-Heptane	0.2611	0.2598	−0.49
n-Hexane	0.2659	0.2636	−0.86
n-Nonane	0.2520	0.2525	0.18
n-Octane	0.2559	0.2561	0.09
n-Pentane	0.2701	0.2682	−0.72
o-Xylene	0.2630	0.2630	0.00
p-Xylene	0.2598	0.2619	0.82
Toluene	0.2637	0.2669	1.21

2.5 MIXING RULES

Most of the EOSs were originally developed for pure components. Each EOS has a number of parameters which are usually based on the properties of pure components such as critical properties and acentric factor. Extending the equations that had been developed for pure components for mixtures is important because most practical problems are encountered with multicomponent mixtures. There are three approaches to extending equations to mixtures (Riazi, 2005). The first approach is determination of an input parameter such as critical temperature, critical pressure, and acentric factor for mixtures (usually called pseudocritical properties). Then, the parameters of an EOS are calculated by the properties of the mixture, and these parameters are substituted into the EOS that had been developed for pure components. The second approach is determination of required properties (such as molar volume) for all pure components that were presented in this approach gives good results but it demands high computational time. Hence, the second approach is not suitable for the mixture composed from many components, particularly when successive equilibrium calculations are required. The third approach is calculated with the parameters of EOS for mixtures using those values for pure components and the mole fraction or weight fraction of components. The third approach is the most widely used.

Several mixing rules developed for mixture (Hirschfelder et al., 1954; Huron and Vidal, 1979; Kwak and Mansoori, 1986; Stryjek and Vera, 1986a,b,c; Economou and Tsonopoulos, 1997; Prausnitz et al., 1998). The cubic EOS is usually extended to mixtures by the *quadratic mixing rule*. Peng and Robinson (1976), Redlich and Kwong (1949) and Soave (1972) used the quadratic mixing rule in their papers. This type of mixing rule is suitable for mixtures of nonpolar components (Soave, 1972). There are a number of mixing rules which have been discussed by authors (Huron and Vidal, 1979; Mathias et al., 1991; Schwartzentruber and Renon, 1991).

For a given mixture (vapor or liquid) with mole fraction z_i, the parameters a (let $a = a_c\alpha$) and b for mixtures are calculated by the quadratic mixing rule as follows:

$$a = \sum_{i=1}^{N} \sum_{j=1}^{N} z_i z_j a_{ij} \tag{2.74}$$

$$b = \sum_{i=1}^{N} z_i b_i \tag{2.75}$$

in which N is the total number of components in a mixture and a_{ij} is defined by Eq. (2.76).

$$a_{ij} = (a_i a_j)^{0.5}(1 - k_{ij}) \tag{2.76}$$

in which k_{ij} is the binary interaction parameter, $k_{ii} = 0$, and $k_{ij} = k_{ji}$. The binary interaction parameters are found from experiment by minimization between predicted and experimental data. The binary interaction parameters that are used with a given EOS are different from the suitable binary interaction parameters for other EOSs. In other words, the binary interaction parameter is developed for particular EOSs and only should be used for those EOSs. Some typical values and correlation for binary interaction parameters are presented in Chapter 7.

The second and third virial coefficients for a given mixture with molar composition z_i are given as follows (Prausnitz et al., 1998):

$$B = \sum_{i=1}^{N} \sum_{j=1}^{N} z_i z_j B_{ij} \tag{2.77}$$

$$C = \sum_{i=1}^{N} \sum_{j=1}^{N} \sum_{k=1}^{N} z_i z_j z_k C_{ijk} \tag{2.78}$$

B_{ii} and B_{jj} are the second virial coefficients for components i and j, respectively. B_{ij} ($B_{ij} = B_{ji}$) can be determined using the following equation (Reid et al., 1987; Abbott et al., 2001):

$$B_{ij} = \frac{RT_{cij}}{P_{cij}} \left(B^{(0)} + \omega_{ij} B^{(1)} \right) \tag{2.79}$$

$B^{(0)}$ and $B^{(1)}$ are evaluated from Eqs. (2.49) and (2.50) or the equations that were presented in Example 2.16 through $T_{rij} = T/T_{cij}$. The parameters T_{cij}, P_{cij}, V_{cij}, Z_{cij}, and ω_{cij} are calculated by Eqs. (2.80) through (2.84).

$$T_{cij} = (T_{ci} T_{cj})^{0.5} (1 - k_{ij}) \tag{2.80}$$

$$P_{cij} = \frac{Z_{cij} R T_{cij}}{V_{cij}} \tag{2.81}$$

$$Z_{cij} = \frac{Z_{ci} + Z_{cj}}{2} \tag{2.82}$$

$$V_{cij} = \left(\frac{V_{ci}^{1/3} + V_{cj}^{1/3}}{2} \right)^3 \tag{2.83}$$

$$\omega_{ij} = \frac{\omega_i + \omega_j}{2} \tag{2.84}$$

Orbey and Vera (1983) proposed the following relation for determination of C_{ijk}:

$$C_{ijk} = \left(C_{ij}C_{ik}C_{jk}\right)^{1/3} \tag{2.85}$$

in which C_{ij} is calculated by Eq. (2.51) using T_{cij}, P_{cij}, and ω_{ij}. T_{cij}, P_{cij}, and ω_{ij} are calculated as for B_{ij}.

To use the Lee–Kesler correlation for mixtures, the required properties of mixtures (i.e., critical temperature, critical pressure, and acentric factor) may be calculated by molar averaging, but in most cases using molar averaging leads to considerable errors. Lee and Kesler (1975) proposed the following set of equations to evaluate the required properties for mixtures.

$$V_{ci} = \frac{Z_{ci}RT_{ci}}{P_{ci}} \tag{2.86}$$

$$Z_{ci} = 0.2905 - 0.085\omega_i \tag{2.87}$$

$$V_c = \frac{1}{8}\sum_{i=1}^{N}\sum_{j=1}^{N} z_i z_j \left(V_{ci}^{1/3} + V_{cj}^{1/3}\right)^3 \tag{2.88}$$

$$T_c = \frac{1}{8V_c}\sum_{i=1}^{N}\sum_{j=1}^{N} z_i z_j \left(V_{ci}^{1/3} + V_{cj}^{1/3}\right)^3 \left(T_{ci}T_{cj}\right)^{0.5} \tag{2.89}$$

$$\omega = \sum_{i=1}^{N} z_i\omega_i \tag{2.90}$$

$$P_c = \frac{Z_c RT_c}{V_c} = (0.2905 - 0.085\omega)\frac{RT_c}{V_c} \tag{2.91}$$

Example 2.18

Estimate the molar volume of an equimolar mixture of methane and ethane at temperature 1000K and pressure 50 MPa by virial EOS. Neglect the third- and higher-order coefficients and set all k_{ij} to zero. The critical properties reported in the following table (Danesh, 1998):

Component	T_c (K)	P_c (Pa)	V_c (m^3/mol)	Z_c	w
Methane	190.56	4559000	0.0000986	0.2862	0.0115
Ethane	184.55	4872000	0.0001445	0.2862	0.0995

Solution

The second virial coefficient is calculated by equations that have been presented in Example 2.16. To calculate the second virial coefficient, the value of B_{12} should be known. To evaluate the parameter B_{12}, the values of T_{c12}, P_{c12}, V_{c12}, Z_{c12}, and ω_{12} are required which are calculated using Eqs. (2.80) through (2.84).

$$T_{c12} = (190.56 \times 184.55)^{0.5}(1 - 0) = 187.53\text{K}$$

$$Z_{c12} = \frac{0.2862 + 0.2793}{2} = 0.2828$$

$$\omega_{12} = \frac{0.0115 + 0.0995}{2} = 0.0555$$

$$V_{cij} = \left(\frac{0.0000986^{1/3} + 0.0001445^{1/3}}{2}\right)^3 = 0.0001201\ \text{m}^3/\text{mol}$$

$$P_{c12} = \frac{0.2828 \times 8.314 \times 187.53}{0.0001201} = 3670891\ \text{Pa}$$

The $T_{r12} = 1000/187.53 = 5.33$; hence, the B_{12} determined as below:

$$B_{12}^{(0)} = 0.083 - \frac{0.422}{5.33^{1.6}} = 0.0540$$

$$B_{12}^{(1)} = 0.139 - \frac{0.172}{5.33^{4.2}} = 0.1388$$

$$B_{12} = \frac{8.314 \times 187.53}{3670891}(0.0540 + 0.0555 \times 0.1388) = 2.3270 \times 10^{-5}\ \text{m}^3/\text{mol}$$

The other calculation tabulated in below table.

(Continued)

ij	z_iz_j	Z_c	V_c (m^3/mol)	T_{cij} (K)	T_r	P_{cij} (Pa)	$B^{(0)}$	$B^{(1)}$	ω	B_{ij} (m^3/mol)	$Z_iZ_jB_{ij}$ (m^3/mol)
11	0.25	0.2862	0.0000986	190.56	5.25	4559000	0.0533	0.1388	0.0115	1.9063E−05	4.7657E−06
22	0.25	0.2793	0.0001445	184.55	5.42	4872000	0.0547	0.1389	0.0995	2.1592E−05	5.3980E−06
12	0.25	0.2828	0.0001201	187.53	5.33	3670891	0.0540	0.1388	0.0555	2.6213E−05	6.5533E−06
											$z_1{}^2B_{11}+z_2{}^2B_{22}+2z_1z_2B_{12}=2.3270E{-}05$ [m^3/mol]

Solving Eq. (2.41) gives the molar volume of the mixture.

$$\frac{50 \times 10^6 \times V}{8.314 \times 1000} - 1 - \frac{2.3270 \times 10^{-5}}{V} = 0$$

$$V = 1.8697 \times 10^{-4}\ \mathrm{m^3/mol}$$

Problems

2.1 Prove that the critical compressibility factor as predicted by PR EOS is 0.307 for all components.

2.2 Using the condition at critical point, prove Eqs. (2.15) and (2.16).

2.3 Derive an expression for virial expansion of SRK EOS.

2.4 Determine the Boyle temperature of methane using PR EOS.

2.5 The molar volume of a given gas is calculated by Eqs. (2.41) and (2.42). Is the calculated molar volume exactly the same value that you get from the equation? Why?

2.6 Show that there are the following relations between the coefficients of two forms of virial EOS.

$$B' = \frac{B}{RT}$$

$$C' = \frac{C - B^2}{(RT)^2}$$

2.7 Repeat Example 2.6 using PR EOS.

2.8 Repeat Example 2.18 using Lee–Kesler and PR equations. Set the binary interaction parameters to zero.

2.9 Show that the reduced form of PR EOS is as follows

$$\frac{BP_c}{RT_c} = B^{(0)} + \omega B^{(1)}$$

REFERENCES

Abbott, M., Smith, J., Van Ness, H., 2001. Introduction to Chemical Engineering Thermodynamics. McGraw-Hill.

Abramowitz, M., Stegun, I.A., 1966. Handbook of mathematical functions. Applied Mathematics Series 55, 62.

Androulakis, I., Kalospiros, N., Tassios, D., 1989. Thermophysical properties of pure polar and nonpolar compounds with a modified VdW-711 equation of state. Fluid Phase Equilibria 45 (2), 135–163.

Atkins, P., De Paula, J., 2006. Atkins' Physical Chemistry, eighth ed. WH Freeman and Company, New York.

Atkinz, P., Paula, J., 2006. Physical Chemistry. Freeman, WH Company, New York, p. 212.
Bendict, M., Webb, G., Rubin, L., 1940. An empirical equation for hydrocarbon thermodynamic properties of light hydrocarbons and their mixtures: methane, ethane, propane, and n-butane. Journal of Chemical Physics 8, 334–345.
Boston, J., Mathias, P., 1980. Phase equilibria in a third-generation process simulator. In: Proceedings of the 2nd International Conference on Phase Equilibria and Fluid Properties in the Chemical Process Industries, West Berlin.
Brill, J.P., Beggs, H.D., 1984. Two-phase Flow in Pipes (publisher not identified).
Carrier, B., Rogalski, M., Péneloux, A., 1988. Correlation and prediction of physical properties of hydrocarbons with the modified Peng-Robinson equation of state. 1. Low and medium vapor pressures. Industrial & Engineering Chemistry Research 27 (9), 1714–1721.
Coker, A.K., 2007. Ludwig's Applied Process Design for Chemical and Petrochemical Plants. Elsevier Gulf Professional Pub.
Danesh, A., 1998. PVT and Phase Behaviour of Petroleum Reservoir Fluids. Elsevier.
Daubert, T.E., Danner, R.P., 1997. API Technical Data Book-petroleum Refining. American Petroleum Institute (API), Washington, DC.
Dieterici, C., 1899. Ueber den kritischen Zustand. Annalen der Physik 305 (11), 685–705.
Dranchuk, P., Kassem, H., 1975. Calculation of Z Factors for Natural Gases Using Equations of State.
Dymond, J.H., Smith, E.B., 1980. Virial coefficients of pure gases and mixtures. A Critical Compilation.
Economou, I.G., Tsonopoulos, C., 1997. Associating models and mixing rules in equations of state for water/hydrocarbon mixtures. Chemical Engineering Science 52 (4), 511–525.
Farrokh-Niae, A., Moddarress, H., Mohsen-Nia, M., 2008. A three-parameter cubic equation of state for prediction of thermodynamic properties of fluids. The Journal of Chemical Thermodynamics 40 (1), 84–95.
Forero, G.L.A., Velásquez, J.A., 2013. A modified Patel–Teja cubic equation of state: part I–generalized model for gases and hydrocarbons. Fluid Phase Equilibria 342, 8–22.
Gasem, K., Gao, W., Pan, Z., Robinson, R., 2001. A modified temperature dependence for the Peng–Robinson equation of state. Fluid Phase Equilibria 181 (1), 113–125.
Glasstone, S., 1951. Textbook of Physical Chemistry.
Haghtalab, A., Mahmoodi, P., Mazloumi, S.H., 2011. A modified Peng–Robinson equation of state for phase equilibrium calculation of liquefied, synthetic natural gas, and gas condensate mixtures. The Canadian Journal of Chemical Engineering 89 (6), 1376–1387.
Hall, K., Yarborough, L., 1973. A new EOS for z-factor calculations. Oil & Gas Journal 82.
Harmens, A., Knapp, H., 1980. Three-parameter cubic equation of state for normal substances. Industrial & Engineering Chemistry Fundamentals 19 (3), 291–294.
Hirschfelder, J.O., Curtiss, C.F., Bird, R.B., Mayer, M.G., 1954. Molecular Theory of Gases and Liquids. Wiley, New York.
Huron, M.-J., Vidal, J., 1979. New mixing rules in simple equations of state for representing vapour-liquid equilibria of strongly non-ideal mixtures. Fluid Phase Equilibria 3 (4), 255–271.
Jeffrey, A., 2005. Complex Analysis and Applications. CRC Press.
Joffe, J., Schroeder, G.M., Zudkevitch, David, 1970. Vapor–liquid equilibria with the redlich-kwong equation of state. AIChE Journal 16.3, 496–498.
Kwak, T., Mansoori, G., 1986. Van der Waals mixing rules for cubic equations of state. Applications for supercritical fluid extraction modelling. Chemical Engineering Science 41 (5), 1303–1309.
Lee, B.I., Kesler, M.G., 1975. A generalized thermodynamic correlation based on three-parameter corresponding states. AIChE Journal 21 (3), 510–527.
Leland, T.W., Chappelear, P.S., 1968. The corresponding states principle—a review of current theory and practice. Industrial & Engineering Chemistry 60 (7), 15–43.

Mason, E.A., Spurling, T.H., 1969. The Virial Equation of State. Pergamon.
Mathias, P.M., 1983. A versatile phase equilibrium equation of state. Industrial & Engineering Chemistry Process Design and Development 22 (3), 385–391.
Mathias, P.M., Copeman, T.W., 1983. Extension of the Peng-Robinson equation of state to complex mixtures: evaluation of the various forms of the local composition concept. Fluid Phase Equilibria 13, 91–108.
Mathias, P.M., Klotz, H.C., Prausnitz, J.M., 1991. Equation-of-state mixing rules for multicomponent mixtures: the problem of invariance. Fluid Phase Equilibria 67, 31–44.
McGlashan, M., Potter, D., 1962. An apparatus for the measurement of the second virial coefficients of vapours; the second viral coefficients of some n-alkanes and of some mixtures of n-alkanes. In: Proceedings of the Royal Society of London A: Mathematical, Physical and Engineering Sciences, The Royal Society.
McGlashan, M., Wormald, C., 1964. Second virial coefficients of some alk-1-enes, and of a mixture of propene+ hept-1-ene. Transactions of the Faraday Society 60, 646–652.
McGuire, G., O'Farrell, A.G., 2002. Maynooth Mathematical Olympiad Manual. Logic Press.
Nishiumi, H., 1980. An improved generalized BWR equation of state with three polar parameters applicable to polar substances. Journal of Chemical Engineering of Japan 13 (3), 178–183.
Nishiumi, H., Kura, S., Yokoyama, T., 1991. Extended BWR equation of state for fluorocarbons, chloroform and carbon tetrachloride. Fluid Phase Equilibria 69, 141–153.
Nishiumi, H., Saito, S., 1975. An improved generalized BWR equation of state applicable to low reduced temperatures. Journal of Chemical Engineering of Japan 8 (5), 356–360.
Novak, J., Malijevsky, A., Sobr, J., Matous, J., 1972. Plyny a plynne smesi. In: Stavove chovani a termodynamicke vlastnosti (Gases and Gas Mixtures, Behaviour of State and Thermodynamic Properties) Academia, Prague.
Onnes, H.K., 1901. Expression of the equation of state of gases and liquids by means of series. In: KNAW, Proceedings.
Orbey, H., Vera, J., 1983. Correlation for the third virial coefficient using T_c, P_c and ω as parameters. AIChE Journal 29 (1), 107–113.
Patel, N.C., Teja, A.S., 1982. A new cubic equation of state for fluids and fluid mixtures. Chemical Engineering Science 37 (3), 463–473.
Peng, D.-Y., Robinson, D.B., 1976. A new two-constant equation of state. Industrial & Engineering Chemistry Fundamentals 15 (1), 59–64.
Perry, R.H., Green, D., Maloney, J., 1997. In: Perry's Handbook of Chemical Engineering.
Pitzer, K.S., Lippmann, D.Z., Curl Jr., R., Huggins, C.M., Petersen, D.E., 1955. The volumetric and thermodynamic properties of fluids. II. Compressibility factor, vapor pressure and entropy of vaporization 1. Journal of the American Chemical Society 77 (13), 3433–3440.
Prausnitz, J.M., Lichtenthaler, R.N., de Azevedo, E.G., 1998. Molecular Thermodynamics of Fluid-phase Equilibria. Pearson Education.
Redlich, O., Kwong, J.N., 1949. On the thermodynamics of solutions. V. An equation of state. Fugacities of gaseous solutions. Chemical Reviews 44 (1), 233–244.
Reid, R.C., Prausnitz, J.M., Poling, B.E., 1987. The Properties of Gases and Liquids.
Riazi, M., 2005. Characterization and Properties of Petroleum Fractions. ASTM International.
Roman, F., Mulero, A., Cuadros, F., 2004. Simple modifications of the van der Waals and Dieterici equations of state: vapour–liquid equilibrium properties. Physical Chemistry Chemical Physics 6 (23), 5402–5409.
Sadus, R.J., 2001. Equations of state for fluids: the Dieterici approach revisited. The Journal of Chemical Physics 115 (3), 1460–1462.
Sadus, R.J., 2002. The Dieterici alternative to the van der Waals approach for equations of state: second virial coefficients. Physical Chemistry Chemical Physics 4 (6), 919–921.

Sadus, R.J., 2003. New Dieterici-type equations of state for fluid phase equilibria. Fluid Phase Equilibria 212 (1), 31–39.
Schmidt, G., Wenzel, H., 1980. A modified van der Waals type equation of state. Chemical Engineering Science 35 (7), 1503–1512.
Schreiber, D.R., Pitzer, K.S., 1989. Equation of state in the acentric factor system. Fluid Phase Equilibria 46 (2), 113–130.
Schwartzentruber, J., Renon, H., 1991. Equations of state: how to reconcile flexible mixing rules, the virial coefficient constraint and the "Michelsen-Kistenmacher syndrome" for multicomponent systems. Fluid Phase Equilibria 67, 99–110.
Soave, G., 1972. Equilibrium constants from a modified Redlich-Kwong equation of state. Chemical Engineering Science 27 (6), 1197–1203.
Soave, G., 1993. 20 years of Redlich-Kwong equation of state. Fluid Phase Equilibria 82, 345–359.
Soave, G., Barolo, M., Bertucco, A., 1993. Estimation of high-pressure fugacity coefficients of pure gaseous fluids by a modified SRK equation of state. Fluid Phase Equilibria 91 (1), 87–100.
Soave, G.S., 1995. A noncubic equation of state for the treatment of hydrocarbon fluids at reservoir conditions. Industrial & Engineering Chemistry Research 34 (11), 3981–3994.
Speakman, J., Partington, J., 1950. An advanced treatise on physical chemistry. In: Fundamental Principles, the Properties of Gases, JSTOR, vol. I.
Standing, M.B., Katz, D.L., 1942. Density of natural gases. Transactions of the AIME 146 (01), 140–149.
Starling, K., 1966. A new approach for determining equation-of-state parameters using phase equilibria data. Society of Petroleum Engineers Journal 6 (04), 363–371.
Starling, K.E., 1973. Fluid Thermodynamic Properties for Light Petroleum Systems. Gulf Pub. Co.
Stryjek, R., Vera, J., 1986a. PRSV2: a cubic equation of state for accurate vapor—liquid equilibria calculations. The Canadian Journal of Chemical Engineering 64 (5), 820–826.
Stryjek, R., Vera, J., 1986b. PRSV—an improved Peng-Robinson equation of state with new mixing rules for strongly nonideal mixtures. The Canadian Journal of Chemical Engineering 64 (2), 334–340.
Stryjek, R., Vera, J., 1986c. PRSV: an improved Peng-Robinson equation of state for pure compounds and mixtures. The Canadian Journal of Chemical Engineering 64 (2), 323–333.
Takacs, G., 1976. Comparisons made for computer Z-factor calculations. Oil and Gas Journal 64–66.
Tsonopoulos, C., 1974. An empirical correlation of second virial coefficients. AIChE Journal 20 (2), 263–272.
Twu, C.H., Bluck, D., Cunningham, J.R., Coon, J.E., 1991. A cubic equation of state with a new alpha function and a new mixing rule. Fluid Phase Equilibria 69, 33–50.
Twu, C.H., Coon, J.E., Cunningham, J.R., 1995a. A new generalized alpha function for a cubic equation of state. Part 1. Peng-Robinson equation. Fluid Phase Equilibria 105 (1), 49–59.
Twu, C.H., Coon, J.E., Cunningham, J.R., 1995b. A new generalized alpha function for a cubic equation of state. Part 2. Redlich-Kwong equation. Fluid Phase Equilibria 105 (1), 61–69.
Wang, S., Xiang, H., Han, B., 2001. The modification and generalization of BWR equation. Fluid Phase Equilibria 181 (1), 71–82.
Whitson, C.H., Brulé, M.R., 2000. Phase Behavior, Henry L. Doherty Memorial Fund of AIME, Society of Petroleum Engineers.
Yu, J.-M., Lu, B.C.-Y., 1987. A three-parameter cubic equation of state for asymmetric mixture density calculations. Fluid Phase Equilibria 34 (1), 1–19.
Zudkevitch, D., Joffe, J., 1970. Correlation and prediction of vapor-liquid equilibria with the Redlich-Kwong equation of state. AIChE Journal 16 (1), 112–119.

CHAPTER THREE

Plus Fraction Characterization

M. Mesbah[1], A. Bahadori[2,3]
[1]Sharif University of Technology, Tehran, Iran
[2]Southern Cross University, Lismore, NSW, Australia
[3]Australian Oil and Gas Services Pty Ltd, Lismore, NSW, Australia

3.1 INTRODUCTION

The prediction of phase behavior of petroleum fluids like bubble/dew point pressure calculation, isothermal flash, and gas to oil ratio calculation is important in some applications such as petroleum production, processing, and transportation (Riazi, 2003). The cubic equation of state (EOS) is extensively used for predicting the phase behavior. These equations were developed using experimental data for pure components but can also be used for multicomponent systems by applying mixing rules. The reservoir fluid, comprised of thousands components and experimental analysis, cannot identify these components. The accurate prediction of phase behavior also needs an accurate representation of the critical properties and acentric factor, while the direct measurement of the critical properties for heavy fraction is not practical. Therefore a full description of reservoir fluid may not possible. On the other hand, performing phase behavior calculation by a large number of components requires a high computational time. The mixture composition usually presented in terms of the mole percent of pure hydrocarbons up to C_6 and for heavier grouped into single carbon number (SCN) group. A conventional laboratory report contains mole fractions of well-defined components (H_2S, N_2, CO_2, C_1, C_2, C_3, i-C_4, n-C_4, i-C_5, n-C_5, and n-C_6), the plus fraction and molecular weight, and the specific gravity of the plus fraction (Luo et al., 2010). In practice, heavy fraction of a reservoir fluid is approximated by experimental and mathematical methods. The characterization of heavy fraction can be divided into three major steps (Yarborough, 1979; Pedersen et al., 1983; Whitson, 1983): (1) splitting the plus fraction into a number of fractions with known molar compositions; (2) assigning the molecular weight and boiling point to each splitting fraction; and (3) predicting the critical properties, acentric factor, binary interaction

Fluid Phase Behavior for Conventional and Unconventional Oil and Gas Reservoirs
http://dx.doi.org/10.1016/B978-0-12-803437-8.00003-8

coefficient, and other parameters used in EOS for each fraction. This chapter outlines some methods that cover these three steps.

3.2 EXPERIMENTAL METHODS

Experimental methods are the most reliable way of characterizing the plus fraction (Whitson and Brulé, 2000). True boiling point (TBP) distillation and chromatography are the most commonly used procedures. Experimental data obtained from TBP distillation are the most accurate way to characterize the plus fraction, particularly when the specific gravity of each cut is calculated (Riazi, 2005). TBP analysis gives the most important data, such as the boiling point, specific gravity, and molecular weight for each cut. TBP distillation is costly and takes about 48 h.

Simulated distillation by gas chromatography (GC), presented in the American Society for Testing and Materials (ASTM) 2887, is a simple method of characterization of the plus fraction. Simulated distillation means producing a distillation curve by GC. Simulated distillation by GC requires less time and amount of sample than TBP distillation (Austad et al., 1983; Chorn, 1984; MacAllister and DeRuiter, 1985). In fact, the amount of mass of each carbon number fraction measured by GC and these data converted to boiling point. TBP curves represent the actual boiling point, and simulated distillation curves represent the boiling point at atmospheric pressure; however, these two curves are very close to each other (Danesh, 1998; Riazi, 2005).

In most cases, simulated distillation gives the required information for plus fraction characterization with much less time and cost compared with a complete TBP distillation. However, it is recommended that at least one complete TBP analysis is performed for the cases of a gas condensate reservoir and a reservoir encounter with gas injection (Whitson and Brulé, 2000).

3.2.1 True Boiling Point Distillation Method

The TBP distillation method is a reliable method for separating stock-tank liquid (oil or condensate) into fractions (cuts) by boiling point distribution. The distillation is conducted using a tray column with 15–100 theoretical stages with a relatively high reflux ratio (i.e., 5 or greater). High theoretical stages and reflux ratio lead to high separation degrees for TBP distillation. TBP cuts can be assumed to be a pure component with unique properties. This assumption is more valid for a cut with a narrow boiling point range.

The standard procedure for TBP distillation, including operating conditions such as the amount of sample, reflux ratio, and equipment specification, is fully described in ASTM 2892.

Distillation begins at atmospheric pressure. As the light components vaporize, the concentration of heavier fractions increases. To avoid the high temperature that can cause thermal cracking of components, the pressure changes to subatmospheric pressure to vaporize the heavier fraction. Usually the sample is distilled at an atmospheric pressure of up to n-C_9 (or the boiling point temperature equal to 151.3°C) and the pressure is lowered to 26.6 mbar from C_{10} to C_{19} and 2.66 mbar from C_{20} to C_{29}. Table 3.1

Table 3.1 Atmospheric Equivalent Boiling Point and Reduced Pressure Boiling Point of the Hydrocarbon Groups (Roenningsen et al., 1989)

Hydrocarbon Group	Normal Alkane AEBP (°C)	Normal Alkane RPBP (°C)	
		26.6 mbar	2.66 mbar
C_6	69.2		
C_7	98.9		
C_8	126.1		
C_9	151.3	51.3	
C_{10}	174.6	70.7	
C_{11}	196.4	88.6	46.5
C_{12}	216.8	105.3	61.8
C_{13}	235.9	121.5	76.5
C_{14}	253.9	136.7	90.1
C_{15}	271.1	151.1	103.3
C_{16}	287	164	115
C_{17}	303	178	128
C_{18}	317	190	139
C_{19}	331	202	151
C_{20}	344	214	161
C_{21}	357	225	172
C_{22}	369	235	181
C_{23}	381	245	191
C_{24}	392	255	199
C_{25}	402	264	208
C_{26}	413	274	217
C_{27}	423	283	225
C_{28}	432	291	233
C_{29}	441	299	241
C_{30}	450	307	248

AEBP, atmospheric equivalent boiling point; *RPBP*, reduced pressure boiling point.

gives the atmospheric equivalent boiling point and the reduced pressure boiling point for the hydrocarbon group (Roenningsen et al., 1989). The boiling point temperature at the subatmospheric pressure is converted to the normal boiling point by correlations. The pressure correction in the correlation usually appears as the logarithm of the subatmospheric pressure to the atmospheric pressure. The most widely used correlation for converting the boiling point temperature at the subatmospheric or superatmospheric pressure to the normal boiling point is the Maxwell and Bonnell (1955) correlation (Wauquier, 1995; Daubert and Danner, 1997; Riazi, 2005). This correlation can be applied for a wide range of pressures (less than 2 mmHg to greater than 760 mmHg).

The sample used in this method usually contains very heavy hydrocarbons such as asphaltenes. These heavy components do not boil off and will be left as the residue. Pedersen et al. (1984) proposed a correlation to extrapolate the TBP curve to 100% distillate. However, the most reliable method for characterizing residue is dividing it into a number of fractions.

The boiling point range for fraction is not specified in the ASTM method. In practice, the boiling point ranges for fractions are listed in Table 3.2. The boiling point range for each fraction in this table for the C_n group is from the normal boiling point of normal alkane with an $n-1$ carbon number (plus 0.5°C) to the normal boiling point of normal alkane with an n carbon number (plus 0.5°C).

Table 3.3 gives TBP results for a North Sea condensate. Table 3.3 also gives the molecular weight and specific gravity for each fraction. The average boiling point for each cut is usually read at the midvolume percent from the TBP curve. For example, the second cut boils from 208.4 to 258.8°F. The initial volume percent is 15.95 and the final volume percent is 27.35; the midvolume is therefore $(15.95 + 27.35)/2 = 21.65$. At this volume percent, the boiling point from the TBP curve is about 235°F.

There are available correlations for estimation of the critical properties, acentric factor, and other properties in terms of the boiling point, molecular weight, and specific gravity. If the boiling point, molecular weight, and specific gravity for each cut is specified then the other properties can be estimated. As mentioned before, the boiling point for each cut is taken from the TBP curve at the midvolume percent. In Table 3.3 the residue is reported as C_{21+}, because the last drop of distillate is collected at the normal boiling point of nC_{20} (plus 0.5°C). Sometimes in the laboratory report of TBP distillation the boiling point ranges are not reported. In these cases the normal paraffin boiling point range is used.

Table 3.2 Boiling Point Range of Petroleum Fractions (Katz and Firoozabadi, 1978)

Hydrocarbon Group	Boiling Range (°C)	Boiling Range (°F)
C_6	36.5 to 69.2	97.9 to 156.7
C_7	69.2 to 98.9	156.7 to 210.1
C_8	98.9 to 126.1	210.1 to 259.1
C_9	126.1 to 151.3	259.1 to 304.4
C_{10}	151.3 to 174.6	304.4 to 346.4
C_{11}	174.6 to 196.4	346.4 to 385.5
C_{12}	196.4 to 216.8	385.5 to 422.2
C_{13}	216.8 to 235.9	422.2 to 456.7
C_{14}	235.9 to 253.9	456.7 to 489.2
C_{15}	253.9 to 271.1	489.2 to 520
C_{16}	271.1 to 287.3	520 to 547
C_{17}	287 to 303	547 to 577
C_{18}	303 to 317	577 to 603
C_{19}	317 to 331	603 to 628
C_{20}	331 to 344	628 to 652
C_{21}	344 to 357	652 to 675
C_{22}	357 to 369	675 to 696
C_{23}	369 to 381	696 to 717
C_{24}	381 to 392	717 to 737
C_{25}	392 to 402	737 to 756
C_{26}	402 to 413	756 to 775
C_{27}	413 to 423	775 to 793
C_{28}	423 to 432	793 to 810
C_{29}	432 to 441	810 to 828
C_{30}	441 to 450	828 to 842
C_{31}	450 to 459	842 to 857
C_{32}	459 to 468	857 to 874
C_{33}	488 to 476	874 to 888
C_{34}	476 to 483	888 to 901
C_{35}	483 to 491	901 to 915

The molecular weight of each cut is measured by a cryoscopic method based on freezing point depression. The concentration of oil in the solvent is about 0.15 kg per kg of solvent (Danesh, 1998). The specific gravity can be calculated by weighting a known volume amount of fraction, pycnometer, or electronic densitometer.

Osjord et al. (1985) plotted the density for 11 North Sea crude oil and condensate versus the carbon number. Their results show that it is not necessary for all properties of the SCN groups to follow the same trend.

Table 3.3 Experimental True Boiling Point Results for a North Sea Condensate (Whitson and Brulé, 2000)

Fraction	Upper T_b (°F)	Ave. T_b[a] (°F)	m_i (g)	SG_i[b]	MW_i	V_i (cm^3)	n_i (mol)	w_i (%)	x_{Vi} (%)	x_i (%)	Sw_i (%)	S x_{Vi} (%)	K_W
C_7	208.4	194.0	90.2	0.7283	96	123.9	0.94	4.35	4.8	7.8	4.35	4.8	11.92
C_8	258.8	235.4	214.6	0.7459	110	287.7	1.951	10.35	11.15	16.19	14.70	15.95	11.88
C_9	303.8	282.2	225.3	0.7658	122	294.2	1.847	10.87	11.4	15.33	25.57	27.35	11.82
C_{10}	347.0	325.4	199.3	0.7711	137	258.5	1.455	9.61	10.02	12.07	35.18	37.37	11.96
C_{11}	381.2	363.2	128.8	0.783	151	164.5	0.853	6.21	6.37	7.08	41.4	43.74	11.97
C_{12}	420.8	401.1	136.8	0.7909	161	173	0.85	6.6	6.7	7.05	48	50.44	12.03
C_{13}	455.0	438.8	123.8	0.8047	181	153.8	0.684	5.97	5.96	5.68	53.97	56.41	11.99
C_{14}	492.8	474.8	120.5	0.8221	193	146.6	0.624	5.81	5.68	5.18	59.78	62.09	11.89
C_{15}	523.4	509.0	101.6	0.8236	212	123.4	0.479	4.9	4.78	3.98	64.68	66.87	12.01
C_{16}	550.4	537.8	74.1	0.8278	230	89.5	0.322	3.57	3.47	2.67	68.26	70.33	12.07
C_{17}	579.2	564.8	76.8	0.829	245	92.6	0.313	3.7	3.59	2.6	71.96	73.92	12.16
C_{18}	604.4	591.8	58.2	0.8378	259	69.5	0.225	2.81	2.69	1.87	74.77	76.62	12.14
C_{19}	629.6	617.0	50.2	0.8466	266	59.3	0.189	2.42	2.3	1.57	77.19	78.91	12.11
C_{20}	653.0	642.2	45.3	0.8536	280	53.1	0.162	2.19	2.06	1.34	79.37	80.97	12.10
C_{21+}			427.6	0.8708	370	491.1	1.156	20.63	19.03	9.59	100	100	
Sum			2073.1			2580.5	12.05	100	100	100			
Ave.			0.8034		172								11.98

Reflux ratio = 1:5; reflux cycle = 18 s; distillation pressure = 201.2–347°F; distillation at 100 mm Hg = 347–471.2°F; and distillation at 10 mm Hg = 471.2–653°F.

$V_i = m_i/SG_i/0.9991$; $n_i = 100 \times m_i/2073.1$; $x_{Vi} = 100 \times V_i/2580.5$; $x_i = 100 \times n_i/12.05$; and $K_W = (T_{bi} + 460)^{1/3}/SG_i$.

[a]Average taken at the midvolume point.

[b]Water = 1.

Although the molecular weight increases with an increasing carbon number, the density of an SCN group can be lower than the previous group.

Some references give procedures for the determination properties from TBP analyses (Organick and Golding, 1952; Katz, 1959; Robinson and Peng, 1978; Austad et al., 1983; Campbell, 1984; Riazi and Daubert, 1986, 1987).

3.2.2 Chromatography

GC is a laboratory technique that is used to identify and measure concentration. The mixture is separated into its components based on the relative attraction to the two phases, one stationary phase (the coating) and one moving phase (the carrier gas). The carrier gas should be chemically inert. Nitrogen, helium, argon, and carbon dioxide are the most commonly used carrier gases. The carrier gas system can contain a molecular sieve to remove water or other possible impurities. This technique is also applied to identify and measure the concentration of the liquid phase. In this case, the moving phase is a liquid phase called liquid chromatography.

The sample is carried by the mobile phase into a column (the stationary phase) and then enters to a detector where the concentration of each component can be determined. A flame ionization detector and a thermal conductivity detector are the two most commonly used detectors. The flame ionization detector is highly sensitive to any compound that creates ions in flame (i.e., all organic compounds), but it cannot detect inorganic compounds and gases such as N_2, H_2O, CO_2, and O_2. The response of this type of detector is proportional to the mass concentration of the ionized compound. The thermal conductivity detector is sensitive to almost all compounds and is used when the gas mixture contains nonhydrocarbon components. The flame ionization detectors are more sensitive than the thermal conductivity detectors.

Many columns are available to separate the mixture. The most commonly used columns are the capillary column and the packed column. The packed column contains finely divided inert materials to maximize its area. The efficiency of packed columns ranges from tens to hundreds of equilibrium stages (Danesh, 1998). The operating condition is adjusted so that the equilibrium stages in the column equal the stages in TBP distillation. The results from this method are known as simulated distillation. The capillary columns treated as many thousands theoretical equilibrium stage and used in preferences to the packed columns. Table 3.4 shows a comparison between results calculated by TBP distillation and simulated distillation

Table 3.4 Comparison of Hydrocarbon Group Property Results Obtained by True Boiling Point Distillation and Capillary Gas Chromatography Analysis (Osjord et al., 1985)

Hydrocarbon Group	TBP Distillation Wt%	MW	Density (g/cm^3)	Simulated Distillation Wt%	MW	Density (g/cm^3)
C_5	0.886	65	0.621	0.792	63.1	0.597
C_6	0.737	82	0.695	0.699	84.8	0.669
C_7	2.371	91	0.751	2	89.4	0.754
C_8	2.825	103	0.778	3.237	102	0.779
C_9	2.539	116	0.793	2.429	116.3	0.799
C_{10+}	90.642	306	0.869	90.846	300.3	0.868
C_{10}	2.479	132	0.798	2.437	133.6	0.801
C_{11}	1.916	147	0.803	2.191	148	0.803
C_{12}	2.352	163	0.81 7	2.523	161.5	0.81 2
C_{13}	2.091	175	0.836	3.106	175.3	0.827
C_{14}	3.667	190	0.843	3.124	189.8	0.84
C_{15}	3.722	205	0.849	3.984	204.8	0.845
C_{16}	2.034	215	0.853	3.383	217.9	0.851
C_{17}	4.135	237	0.844	4.244	235.1	0.842
C_{18}	3.772	251	0.846	3.201	149.8	0.845
C_{19}	3.407	262	0.857	3.523	261.2	0.854
C_{20+}	61.057	426	0.885	59.13	421.6	0.888

MW, molecular weight; *TBP*, true boiling point.

(Osjord et al., 1985). As mentioned before, the results from simulated distillation are in good agreement with the results of TBP distillation.

GC analysis reports the weight percent of the detected component (the GC technique provides no information about the molecular weight or density). The weight fraction is converted to the mole fraction using molecular weight. Note that the GC technique identifies the components (up to the C_9 fraction), and the molecular weights can be extracted from references. Table 3.5 shows a detailed gas phase analysis (Osjord and Malthe-Sørenssen, 1983). As in Table 3.5, standard analysis will identify iso and normal butane and iso and normal pentane, while heavier hydrocarbons are grouped based on their carbon number. Table 3.6 gives a chromatography analysis for a liquid sample (Osjord et al., 1985). The analytical conditions for liquid chromatography are given in Table 3.7. As can be seen from Tables 3.5 and 3.6, the components are identified up to the C_9 fraction. The liquid composition analysis for hydrocarbon up to C_9 is performed using capillary column chromatography. For the hydrocarbon group

Table 3.5 Chromatography Analysis of a Gas Sample (Osjord and Malthe-Sørenssen, 1983)

Component	Formula	Wt%	MW	T_b (°C)	Fraction
Nitrogen	N_2	1.6542	28.013	195.8	N_2
Carbon dioxide	CO_2	2.3040	44.010	−78.5	CO_2
Methane	CH_4	60.5818	16.043	161.5	C_1
Ethane	C_2H_6	15.5326	30.070	−88.5	C_2
Propane	C_3H_8	12.3819	44.097	−42.1	C_3
Isobutane	C_4H_{10}	2.0616	58.124	−11.9	i-C_4
n-Butane	C_4H_{10}	3.2129	58.124	−0.5	n-C_4
2,2-Dimethylpropane	C_5H_{12}	0.0074	72.151	9.5	i-C_5
2-Methylbutane	C_5H_{12}	0.7677	72.151	27.9	i-C_5
n-Pentane	C_5H_{12}	0.6601	72.151	36.1	n-C_5
Cyclopentane	C_5H_{12}	0.0395	70.135	49.3	C_6
2,2-Dimethylbutane	C_6H_{14}	0.0059	86.178	49.8	C_6
2,3-Dimethylbutane	C_6H_{14}	0.0212	86.178	58.1	C_6
2-Methylpentane	C_6H_{14}	0.1404	86.178	60.3	C_6
3-Methylpentane	C_6H_{14}	0.0603	86.178	63.3	C_6
n-Heaxane	C_6H_{14}	0.1302	86.178	68.8	C_6
Methylcyclopentane	C_6H_{14}	0.0684	84.162	71.9	C_7
2,2-Dimethyl pentane	C_7H_{16}	0.0001	100.205	79.3	C_7
Benzene	C_6H_6	0.0648	78.114	80.2	C_7
3,3-Dimethylpentane	C_7H_{16}	0.0005	100.205	80.6	C_7
Cyclohexane	C_6H_{12}	0.0624	82.146	83.0	C_7
3,3-Dimethylpentane	C_7H_{16}	0.0005	100.205	86.1	C_7
1,1-Dimethylcyclo-pentane	C_7H_{14}	0.0025	98.189	87.9	C_7
2,3-Dimethylpentane	C_7H_{16}	0.0045	100.205	89.8	C_7
2-Methylhexane	C_7H_{16}	0.0145	100.205	90.1	C_7
3-Methylhexane	C_7H_{16}	0.0125	100.205	91.9	C_7
1,*cis*-3-Dimethylcyclo-pentane	C_7H_{14}	0.0060	98.189	–	C_7
1,*trans*-3-Dimethylcyclo-pentane	C_7H_{14}	0.0060	98.189	–	C_7
1,*trans*-2-Dimethylcyclo-pentane	C_7H_{14}	0.0094	98.189	91.9	C_7
n-Heptane	C_7H_{16}	0.0290	100.205	98.5	C_7
Methylcyclohexane	C_7H_{14}	0.0565	98.189	101	C_8
Ethylcyclopentane	C_7H_{14}	0.0035	98.189	103.5	C_8
1,*trans*-2,*cis*-4-Tri-methylcyclo-pentane	C_8H_{16}	0.0004	112.216	–	C_8
1,*trans*-2,*cis*-3-Tri-methylcyclo-pentane	C_8H_{16}	0.0002	112.216	–	C_8
Toluene	C_7H_8	0.0436	92.141	110.7	C_8

(Continued)

Table 3.5 Chromatography Analysis of a Gas Sample (Osjord and Malthe-Sørenssen, 1983)—cont'd

Component	Formula	Wt%	MW	T_b (°C)	Fraction
2-Methylheptane	C_8H_{18}	0.0039	114.232	117.7	C_8
3-Methylheptane	C_8H_{18}	0.0025	114.232	119	C_8
1,*trans*-4-Dimethylcyclohexane	C_8H_{16}	0.0022	112.216	119.4	C_8
1,*cis*-3- Dimethylcyclohexane	C_8H_{16}	0.0044	112.216	123.5	C_8
n-Octane	C_8H_{18}	0.0099	114.232	125.7	C_8
m+*p*-Xylene	C_8H_{10}	0.0029	106.168	138.8	C_9
o-Xylene	C_8H_{10}	0.0029	106.168	144.5	C_9
n-Nonane	C_9H_{20}	0.0137	128.259	150.9	C_9
Unidentified decanes	$(C_{10}H_{22})$	0.0081	(142.286)	(174.2)	(C_{10})

MW, molecular weight.

Table 3.6 Capillary Chromatography Analysis of a Liquid Sample (Osjord et al., 1985)

Component	Wt%	Mol %	Volume %	MW	Density (g/cm^3)	Fraction
C_2	0.007	0.058	0.017	30.070	0.3580	C_2
C_3	0.072	0.412	0.122	44.097	0.5076	C_3
i-C_4	0.051	0.222	0.078	58.124	0.5633	*i*-C_4
n-C_4	0.189	0.816	0.276	58.124	0.5847	*n*-C_4
i-C_5	0.188	0.653	0.257	72.151	0.6246	*i*-C_5
n-C_5	0.285	0.991	0.386	72.151	0.6309	*n*-C_5
2,2-DM-C_4	0.012	0.034	0.015	86.178	0.6539	C_6
Cy-C_5	0.052	0.185	0.059	70.135	0.7502	C_6
2,2-DM-C_4	0.028	0.081	0.036	86.178	0.6662	C_6
2-M-C_5	0.165	0.480	0.214	86.178	0.6577	C_6
3-M-C_5	0.102	0.298	0.131	86.178	0.6688	C_6
n-C_6	0.341	0.993	0.440	86.178	0.6638	C_6
M-Cy-C_5	0.231	0.689	0.262	84.162	0.7534	C_7
2,4-DM-C_5	0.015	0.038	0.019	100.205	0.6771	C_7
Benzene	0.355	1.140	0.343	78.114	0.8840	C_7
Cy-C_6	0.483	1.440	0.528	84.162	0.7831	C_7
1,1-DM-Cy-C_5	0.116	0.298	0.131	98.189	0.7590	C_7
3-M-C_6	0.122	0.307	0.152	100.205	0.6915	C_7
1,*trans*-3-DM-Cy-C_5	0.052	0.133	0.059	98.189	0.7532	C_7
1,*trans*-2-DM-Cy-C_5	0.048	0.122	0.054	98.189	0.7559	C_7
n-C_7	0.405	1.014	0.504	100.205	0.6880	C_7
Unspecified C_7	0.171	0.427	0.215	100.205	0.6800	C_7
M-Cy-C_6	0.918	2.348	1.016	98.189	0.7737	C_8
1,1,3-TM-Cy-C_5	0.027	0.061	0.031	112.216	0.7526	C_8
2,2,3-TM-Cy-C_5	0.042	0.093	0.050	114.232	0.7200	C_8
2,5-DM-C_6	0.018	0.039	0.022	114.232	0.6977	C_8

Table 3.6 Capillary Chromatography Analysis of a Liquid Sample (Osjord et al., 1985)—cont'd

Component	Wt%	Mol %	Volume %	MW	Density (g/cm^3)	Fraction
3,3-DM-C_6	0.026	0.057	0.031	114.232	0.7141	C_8
1,*trans*-2,*cis*-3-TM-Cy-C_5	0.025	0.056	0.028	112.216	0.7579	C_8
Toluene	0.958	2.610	0.941	92.143	0.8714	C_8
2,3-DM-C_6	0.033	0.073	0.040	114.232	0.7163	C_8
2-M-C_7	0.137	0.300	0.167	114.232	0.7019	C_8
2-M-C_7	0.094	0.206	0.113	114.232	0.7099	C_8
1,*cis*-3-DM-Cy-C_6	0.190	0.425	0.211	112.216	0.7701	C_8
1,*trans*-4-DM-CY-C_6	0.072	0.162	0.081	112.216	0.7668	C_8
Unspecified naphthene	0.028	0.062	0.031	112.216	0.7700	C_8
Unspecified naphthene	0.013	0.028	0.014	112.216	0.7700	C_8
Unspecified naphthene	0.011	0.025	0.012	112.216	0.7700	C_8
DM-Cy-C_6	0.031	0.069	0.034	112.216	0.7700	C_8
1,*trans*-2-DM-Cy-C_6	0.089	0.199	0.098	112.216	0.7799	C_8
n-C_8	0.434	0.954	0.526	114.232	0.7065	C_8
Unspecified C_8	0.086	0.190	0.105	114.232	0.7000	C_8
Unspecified naphthene	0.047	0.094	0.051	126.243	0.7900	C_9
2,2-DM-C_7	0.009	0.018	0.011	128.259	0.7144	C_9
2,4-DM-C_7	0.017	0.033	0.020	128.259	0.7192	C_9
1,*cis*-2-DM-Cy-C_6	0.024	0.054	0.026	112.216	0.8003	C_9
E-Cy- C_6-1,1,3-TM-Cy-C_6	0.281	0.599	0.305	118.000	0.7900	C_9
Unspecified naphthene	0.047	0.093	0.051	126.243	0.7900	C_9
3,5-DM-C_7	0.017	0.034	0.020	128.259	0.7262	C_9
2,5-DM-C_7	0.003	0.006	0.004	128.259	0.7208	C_9
Ethylbenzene	0.114	0.270	0.112	106.168	0.8714	C_9
Unspecified naphthene	0.027	0.054	0.029	126.243	0.7900	C_9
m+*p*-Xylene	0.697	1.649	0.687	106.168	0.8683	C_9
4-M-C_8	0.020	0.039	0.024	128.259	0.7242	C_9
2-M-C_8	0.054	0.106	0.064	128.259	0.7173	C_9
Unspecified naphthene	0.009	0.018	0.010	126.243	0.7900	C_9
Unspecified naphthene	0.082	0.163	0.089	126.243	0.7900	C_9
Unspecified naphthene	0.007	0.014	0.008	126.243	0.7900	C_9
Ortho-xylene	0.230	0.545	0.223	106.168	0.8844	C_9
3-M-C_8	0.023	0.045	0.027	128.259	0.7242	C_9
1-M,3-E-Cy-C_6	0.078	0.155	0.083	126.243	0.8000	C_9
1-M,4-E-Cy-C_6	0.034	0.068	0.037	126.243	0.7900	C_9
Unspecified naphthene	0.006	0.013	0.007	126.243	0.7900	C_9
Unspecified naphthene	0.004	0.007	0.004	126.243	0.7900	C_9
n-C_9	0.471	0.923	0.559	128.259	0.7214	C_9
Unspecified C_9	0.124	0.243	0.148	128.259	0.7200	C_9

MW, molecular weight.

Table 3.7 Analytical Condition for Liquid Chromatography (Pedersen et al., 1985)

Column:	Type: Fused silica capillary
	Liquid phase: Chrompack CP Sil5
	Carrier gas: He, $u = 22$ cm/s
	Length: 50 m
	Inside diameter: 0.23 mm
	Film thickness: 0.4 μm
Detector:	Type: Flame ionization
	Fuel gas: Hydrogen, 30 ml/min
	Makeup gas: Nitrogen, 30 ml/min
	Temperature: 350°C
Sampling:	Syringe: Hamilton 7001 N
	Sample size: 0.1−0.5 μL
Injector:	Type: Split injector
	Ratio: 1:100
	Liner: Packed Jennings tube
	Temperature: 300°C
Oven:	Type: Temperature programmed
	Initial value: +10°C, 2 min
	Rate 1: 3 degrees/min → 115°C
	Rate 2: 10 degrees/min → 300°C
	Final value: 300°C, 60 min
	Total time: ~2 h

in the range of C_{10} to C_{20+} it is performed using a mini distillation apparatus at subatmospheric pressure to avoid thermal cracking of the sample (Pedersen et al., 1989). If an extended composition up to C_{30+} is required, fractionation may be obtained by distillation at a high temperature (up to 550°C) or at a low pressure (about 2 mmHg); however, this process is time-consuming and difficult (Pedersen et al., 1989).

Table 3.8 shows a typical GC analysis from a commercial pressure−volume−temperature laboratory (Pedersen et al., 2014). In Table 3.8 for hydrocarbon groups in the range of C_{11} to C_{35}, all components are grouped in the carbon number fraction in terms of the boiling point ranges from Table 3.2.

3.3 SPLITTING METHODS

A lot of hydrocarbon and nonhydrocarbon components such as N_2, CO_2, and H_2S comprise reservoir fluid. Laboratory fluid properties reports usually describe the components as heavier than hexane as a pseudo component by molecular weight and specific gravity. This pseudo

Table 3.8 Typical Gas Chromatographic Analysis to C_{36+} (Pedersen et al., 2014)

Formula	Component	Wt%	Mol%
N_2	Nitrogen	0.080	0.338
CO_2	Carbon dioxide	0.210	0.565
C_1	Methane	4.715	34.788
C_2	Ethane	2.042	8.039
C_3	Propane	2.453	6.583
i-C_4	*i*-Butane	0.601	1.223
n-C_4	*n*-Butane	1.852	3.771
i-C_5	Neopentane	0.001	0.002
i-C_5	*i*-Pentane	0.991	1.626
n-C_5	*n*-Pentane	1.341	2.200
C_6	Hexanes	2.433	3.341
	M-C-Pentane	0.310	0.436
	Benzene	0.070	0.106
	Cyclohexane	0.240	0.338
C_7	Heptane	2.142	2.641
	M-C-Heane	0.380	0.459
	Toluene	0.280	0.360
C_8	Octanes	2.433	2.691
	E-Benzene	0.190	0.212
	m/*p*-Xylene	0.390	0.435
	o-Xylene	0.190	0.212
C_9	Nonane	2.353	2.301
	1,2,4-TMB	0.230	0.227
C_{10}	Decane	2.843	2.511
C_{11}	Undecane	2.793	2.249
C_{12}	Dodecane	2.573	1.891
C_{13}	Tridecane	2.513	1.700
C_{14}	Tertadecane	2.312	1.441
C_{15}	Pentadecane	2.322	1.334
C_{16}	Hexanadecane	2.212	1.180
C_{17}	Heptadecane	2.032	1.015
C_{18}	Octadecane	1.962	0.925
C_{19}	Nanadecane	1.992	0.897
C_{20}	Eicosane	1.812	0.780
C_{21}	Heneicosane	1.722	0.700
C_{22}	Docosane	1.662	0.645
C_{23}	Tricosane	1.552	0.578
C_{24}	Tetracosane	1.472	0.526
C_{25}	Pentacosane	1.401	0.481
C_{26}	Hexacosane	1.321	0.436
C_{27}	Heptacosane	1.281	0.406
C_{28}	Octacosane	1.261	0.385

(*Continued*)

Table 3.8 Typical Gas Chromatographic Analysis to C_{36+} (Pedersen et al., 2014)—cont'd

Formula	Component	Wt%	Mol%
C_{29}	Nonacosane	1.241	0.365
C_{30}	Triacontane	1.221	0.348
C_{31}	Hentriacontane	1.221	0.336
C_{32}	Dotriacontane	1.131	0.302
C_{33}	Tritriacontane	1.091	0.282
C_{34}	Tetratriacontane	1.031	0.259
C_{35}	Pentatriaontane	1.021	0.249
C_{36+}	Hexatriacontane plus	29.074	4.885

‡ The C_{7+} molecular weight and density are 276 and 0.8651 g/cm^3, respectively.

component is called the heptane plus fraction. In general, there are two techniques to the characterization of the plus fraction: the pseudocomponent approach and the continuous approach. The pseudocomponent approach refers to an approach in which the plus fraction splits into a number of pseudocomponents with a known mole fraction, molecular weight, specific gravity, and boiling point (Ahmed, 1989; Benmekki and Mansoori, 1989; Riazi, 1989, 1997).

The continuous approach uses a distribution function to describe the mole fraction of components. Characterizing the heavy fraction without splitting this fraction into a number of SCN groups is difficult (Ahmed et al., 1985). Proposed splitting methods are based on observations, for example, that the molar distribution of the condensate system is exponential while the molar distribution of the block oil or crude oil shows left-skewed behavior. Table 3.10 gives extended composition data for a North Sea gas condensate, a North Sea black oil, and a North Sea volatile oil (Pedersen et al., 1989). Table 3.11 presents extended composition and paraffin, naphthene, and aromatic distribution for two North Sea gas condensates.

When no sufficient data are available for a sample, generalized SCN data can be used. Katz and Firoozabadi (1978) reported the molecular weight, specific gravity, and boiling point for SCN groups. Whitson (1983) showed that the molecular weight reported by Katz and Firoozabadi for carbon numbers greater than 22 was inconsistent (Riazi and Al-Sahhaf, 1996). Whitson (1983) revised the generalized data of Katz and Firoozabadi (1978) to improve the consistency of the reported molecular weight. The generalized data for SCN groups are reported in Table 3.9.

Table 3.9 Generalized Single Carbon Group Properties (Danesh, 1998)

SCN	Molecular Weight (Kg/kg mol)	Specific Gravity Relative Density at 288K	Boiling Point (K)	Watson Characterization Factor	T_c (K)	P_c (MPa)	V_c (m^3/kg mol)	Z_c	Acentric Factor
C_6	84	0.690	337	12.27	510	3.271	0.348	0.268	0.251
C_7	96	0.727	366	11.97	547	3.071	0.392	0.265	0.280
C_8	107	0.749	390	11.87	574	2.877	0.433	0.261	0.312
C_9	121	0.768	416	11.82	603	2.665	0.484	0.257	0.352
C_{10}	134	0.782	439	11.82	627	2.481	0.532	0.253	0.389
C_{11}	147	0.793	461	11.85	649	2.310	0.584	0.250	0.429
C_{12}	161	0.804	482	11.86	670	2.165	0.635	0.247	0.467
C_{13}	175	0.815	501	11.85	689	2.054	0.681	0.244	0.501
C_{14}	190	0.826	520	11.84	708	1.953	0.727	0.241	0.536
C_{15}	206	0.836	539	11.84	727	1.853	0.777	0.238	0.571
C_{16}	222	0.843	557	11.87	743	1.752	0.830	0.235	0.610
C_{17}	237	0.851	573	11.87	758	1.679	0.874	0.233	0.643
C_{18}	251	0.856	586	11.89	770	1.614	0.914	0.231	0.672
C_{19}	263	0.861	598	11.90	781	1.559	0.951	0.229	0.698
C_{20}	275	0.866	612	11.93	793	1.495	0.997	0.226	0.732
C_{21}	291	0.871	624	11.93	804	1.446	1.034	0.224	0.759
C_{22}	300	0.876	637	11.95	815	1.393	1.077	0.221	0.789
C_{23}	312	0.881	648	11.95	825	1.356	1.110	0.220	0.815
C_{24}	324	0.885	659	11.96	834	1.314	1.147	0.217	0.841
C_{25}	337	0.888	671	11.99	844	1.263	1.193	0.215	0.874
C_{26}	349	0.892	681	12.00	853	1.230	1.226	0.213	0.897
C_{27}	360	0.896	691	12.00	862	1.200	1.259	0.211	0.944
C_{28}	372	0.889	701	12.02	870	1.164	1.296	0.209	0.968
C_{29}	382	0.902	709	12.03	877	1.140	1.323	0.207	0.985
C_{30}	394	0.905	719	12.04	885	1.107	1.361	0.205	1.008
C_{31}	404	0.909	728	12.04	893	1.085	1.389	0.203	1.026
C_{32}	415	0.912	737	12.05	901	1.060	1.421	0.201	1.046
C_{33}	426	0.915	745	12.05	907	1.039	1.448	0.199	1.063
C_{34}	437	0.917	753	12.07	914	1.013	1.480	0.197	1.082
C_{35}	445	0.920	760	12.07	920	0.998	1.502	0.196	1.095
C_{36}	456	0.922	768	12.08	926	0.974	1.534	0.194	1.114
C_{37}	464	0.925	774	12.07	932	0.964	1.550	0.193	1.124
C_{38}	475	0.927	782	12.09	938	0.941	1.583	0.191	1.142
C_{39}	484	0.929	788	12.09	943	0.927	1.604	0.190	1.154
C_{40}	495	0.931	796	12.11	950	0.905	1.636	0.188	1.172
C_{41}	502	0.933	801	12.11	954	0.896	1.652	0.187	1.181
C_{42}	512	0.934	807	12.13	959	0.877	1.680	0.185	1.195
C_{43}	521	0.936	813	12.13	964	0.864	1.701	0.184	1.207
C_{44}	531	0.938	821	12.14	970	0.844	1.733	0.181	1.224
C_{45}	539	0.940	826	12.14	974	0.835	1.749	0.180	1.232

T_c, P_c, and V_c: Calculated from Twu (1984) correlations.
Z_c: Calculated from $P_{cvc} = Z_c R T_c$.
Acentric factor: Calculated from Lee and Kesler (1975) and Kesler and Lee (1976) correlations.
SCN, single carbon number.

Table 3.10 Extended Composition Data for a North Sea Gas Condensate, a North Sea Black Oil, and a North Sea Volatile Oil

	North Sea Gas Condensate				North Sea Black Oil				North Sea Volatile Oil			
Component	Wt%	Mol %	MW	Density (g/cm^3) at 15°C	Wt%	Mol %	MW	Density (g/cm^3) at 15°C	Wt%	Mol %	MW	Density (g/cm^3) at 15°C
N_2	0.571	0.60			0.145	0.56			0.258	0.58		
CO_2	5.031	3.34			1.450	3.55			2.297	3.27		
C_1	40.667	74.16			6.757	45.34			13.780	53.89		
C_2	8.126	7.90			1.531	5.48			4.108	8.57		
C_3	6.254	4.15			1.516	3.70			4.254	6.05		
i-C_4	1.401	0.71			0.378	0.70			0.969	1.05		
n-C_4	2.855	1.44			0.891	1.65			2.263	2.44		
i-C_5	1.306	0.53			0.489	0.73			1.013	0.88		
n-C_5	1.637	0.66			0.580	0.87			1.348	1.17		
C_6	2.355	0.81			1.043	1.33			1.970	1.45		
C_7	3.749	1.20	91.2	0.746	2.276	2.73	89.9	0.757	3.489	2.38	91.9	0.742
C_8	4.100	1.15	104.0	0.770	3.125	3.26	103.2	0.777	4.331	2.59	104.7	0.765
C_9	2.577	0.63	119.0	0.788	2.342	2.14	117.7	0.796	3.329	1.75	119.2	0.788
C_{10}	2.329	0.50	133.0	0.795	2.379	1.94	133.0	0.796	3.173	1.50	131.0	0.791
C_{11}	1.466	0.29	144.0	0.790	2.205	1.62	147.0	0.800	3.666	1.55	147.0	0.796
C_{12}	1.458	0.27	155.0	0.802	2.179	1.47	160.0	0.815	2.408	0.93	161.0	0.811
C_{13}	1.624	0.28	168.0	0.814	2.693	1.69	172.0	0.833	3.125	1.13	171.0	0.826
C_{14}	1.413	0.22	181.0	0.824	2.789	1.62	186.0	0.843	2.952	1.01	182.0	0.836
C_{15}	1.165	0.17	195.0	0.833	2.937	1.59	200.0	0.849	2.521	0.80	195.0	0.843
C_{16}	1.057	0.15	204.0	0.836	2.553	1.30	213.0	0.858	2.878	0.86	208.0	0.848
C_{17}	1.096	0.14	224.0	0.837	2.388	1.11	233.0	0.851	2.211	0.60	228.0	0.844
C_{18}	0.729	0.09	234.0	0.839	2.885	1.26	247.0	0.856	2.701	0.68	247.0	0.848
C_{19}	1.137	0.13	248.0	0.844	2.571	1.07	258.0	0.868	2.184	0.54	252.0	0.859
C_{20+}	5.896	0.47	362.0	0.877	51.898	13.32	421.0	0.914	28.773	4.34	411.0	0.903

MW, molecular weight.

Table 3.11 Extended Composition Data for Two North Sea Gas Condensates With Paraffin, Naphthene, and Aromatic Distribution

	North Sea Gas Condensate (Pedersen et al., 1985)						North Sea Gas Condensate (Pedersen et al., 1989)					
			Density (g/cm^3)	PNA Distribution Mol%					Density (g/cm^3)	PNA Distribution Mol%		
Component	Mol%	MW	at 15°C	*P*	*N*	*A*	Mol%	MW	at 15°C	*P*	*N*	*A*
N_2	0.120						0.64					
CO_2	2.490						9.16					
C_1	76.430						68.80					
C_2	7.460						8.43					
C_3	3.120						5.11					
i-C_4	0.590						0.81					
n-C_4	1.210						1.45					
i-C_5	0.500						0.52					
n-C_5	0.590						0.53					
C_6	0.790						0.63					
C_7	0.950	95	0.726	0.564	0.361	0.076	0.83	0.741	96	0.50	0.42	0.08
C_8	1.080	106	0.747	0.113	0.611	0.277	0.95	0.780	107	0.45	0.38	0.17
C_9	0.780	116	0.769	0.483	0.311	0.206	0.52	0.807	121	0.48	0.27	0.25
C_{10}	0.592	133	0.781	0.530	0.275	0.195	0.26	0.819	134	0.47	0.30	0.23
C_{11}	0.467	152	0.778	0.681	0.193	0.126	0.20	0.810	147	0.56	0.27	0.17
C_{12}	0.345	164	0.785	0.757	0.123	0.120	0.17	0.828	161	0.55	0.24	0.21
C_{13}	0.375	179	0.802	0.709	0.183	0.108	0.16	0.849	175	0.54	0.22	0.24
C_{14}	0.304	193	0.815	0.635	0.209	0.156	0.15	0.857	190	0.49	0.27	0.24
C_{15}	0.237	209	0.817	0.729	0.168	0.103	0.11	0.868	206	0.52	0.20	0.28
C_{16}	0.208	218	0.824	0.624	0.232	0.144	0.086	0.872	222	0.55	0.19	0.26
C_{17}	0.220	239	0.825	0.668	0.185	0.147	0.078	0.859	237	0.57	0.20	0.23

(Continued)

Table 3.11 Extended Composition Data for Two North Sea Gas Condensates With Paraffin, Naphthene, and Aromatic Distribution—cont'd

	North Sea Gas Condensate (Pedersen et al., 1985)						North Sea Gas Condensate (Pedersen et al., 1989)					
			Density (g/cm^3)	PNA Distribution Mol%					Density (g/cm^3)	PNA Distribution Mol%		
Component	Mol%	MW	at 15°C	*P*	*N*	*A*	Mol%	MW	at 15°C	*P*	*N*	*A*
C_{18}	0.169	250	0.831	0.675	0.192	0.133	0.068	0.854	251	0.70:C_{18+}	0.11:C_{18+}	0.19:C_{18+}
C_{19}	0.140	264	0.841	0.652	0.190	0.158	0.050	0.866	263			
C_{20}	0.833:C_{20+}	377:C_{20+}	0.87:C_{20+}	0.519:C_{20+}	0.320:C_{20+}	0.161:C_{20+}	0.046	0.873	339:C_{20+}			
C_{21}							0.035	0.876				
C_{22}							0.025	0.876				
C_{23}							0.034	0.875				
C_{24}							0.023	0.877				
C_{25}							0.017	0.876				
C_{26}							0.018	0.878				
C_{27}							0.014	0.882				
C_{28}							0.012	0.886				
C_{29}							0.013	0.889				
C_{30+}							0.047	0.908				

MW, molecular weight; *PNA*, paraffin, naphthene, and aromatic.

A set of requirements must be satisfied for each splitting method listed below (assuming that the plus fraction is grouped as heptane plus):

$$\sum_{n=7}^{N} z_{C_n} = z_{C_{7+}} \tag{3.1}$$

$$\sum_{n=7}^{N} z_{C_n} MW_{C_n} = z_{C_{7+}} MW_{C_{7+}} \tag{3.2}$$

$$\sum_{n=7}^{N} z_{C_n} \frac{MW_{C_n}}{SG_{C_n}} = z_{C_{7+}} \frac{MW_{C_{7+}}}{SG_{C_{7+}}} \tag{3.3}$$

3.3.1 Katz Method

The simplest method for splitting the plus fraction is the Katz (1983) method. The proposed method is in the form of an exponential function and is suitable for a condensate system (Riazi, 2005):

$$z_{C_n} = 1.38205\ z_{C_{7+}} \exp(-0.25903n) \tag{3.4}$$

where z_{C_n} is the mole fraction of the SCN group, C_n; $z_{C_{7+}}$ is the mole fraction of the heptane plus; and n is the carbon number.

Direct use of the Katz splitting method usually does not give appropriate results. In order to improve this method, the equation is modified as follows:

$$z_{C_n} = A z_{C_{7+}} \exp(-Bn) \tag{3.5}$$

where the A and B parameters are obtained in a way that satisfies Eqs. (3.1) and (3.2). Therefore

$$A \sum_{n=7}^{N} \exp(-Bn) - 1 = 0 \tag{3.6}$$

$$\frac{\sum_{n=7}^{N} \frac{MW_{C_n}}{SG_{C_n}} \exp(-Bn)}{\sum_{n=7}^{N} \exp(-Bn)} - \frac{MW_{C_{7+}}}{SG_{C_{7+}}} = 0 \tag{3.7}$$

This method is suitable when little or no compositional analysis of the C_{7+} is available.

Example 3.1

The total concentration of the C_{7+} fraction in a condensate is 5.45% (Al-Meshari, 2005). The specific gravity and molecular weight of the C_{7+} fraction are 0.7964 and 158, respectively. Describe the C_{7+} fraction, by SCN groups, extended to C_{30+} using the improved Katz splitting method.

Solution

The specific gravity and molecular weight of the SCN group are assumed to be those in Table 3.9, and the heaviest fraction is C_{45}. Solving Eq. (3.7) by the Newton–Raphson method for B results in

$$B = 0.1872$$

Substituting the B value in Eq. (3.6), the value of A is determined as

$$A = \frac{1}{\left[\sum_{n=7}^{45} \exp(-0.1872n)\right]} = 0.6335$$

Substituting A and B in Eq. (3.5), the mole fraction of the SCN groups is determined as in the following table.

SCN group	C_7	C_8	C_9	C_{10}	C_{11}	C_{12}	C_{13}	C_{14}	C_{15}	C_{16}
Mol%	0.9311	0.7721	0.6403	0.5310	0.4403	0.3652	0.3028	0.2511	0.2082	0.1727
SCN group	C_{17}	C_{18}	C_{19}	C_{20}	C_{21}	C_{22}	C_{23}	C_{24}	C_{25}	C_{26}
Mol%	0.1432	0.1188	0.0985	0.0817	0.0677	0.0562	0.0466	0.0386	0.0320	0.0266
SCN group	C_{27}	C_{28}	C_{29}	C_{30}	C_{31}	C_{32}	C_{33}	C_{34}	C_{35}	C_{36}
Mol%	0.0220	0.0183	0.0151	0.0126	0.0104	0.0086	0.0072	0.0059	0.0049	0.0041
SCN group	C_{37}	C_{38}	C_{39}	C_{40}	C_{41}	C_{42}	C_{43}	C_{44}	C_{45}	
Mol%	0.0034	0.0028	0.0023	0.0019	0.0016	0.0013	0.0011	0.0009	0.0008	

SCN, single carbon number.

The C_{30+} mole fraction and molecular weight are found as

$$z_{C_{30+}} = \sum_{C_{30}}^{C_{45}} z_{C_n} = 0.0699\%$$

$$z_{C_{30+}} MW_{C_{30+}} = \sum_{C_{30}}^{C_{45}} z_{C_n} MW_{C_n}$$

$$MW_{C_{30+}} = 435$$

If the volume of the C_{30+} fraction is assumed to be equal to the sum of all of the components volume, then the specific gravity of the C_{30+} fraction is found as

$$\frac{z_{C_{30+}} MW_{C_{30+}}}{SG_{C_{30+}}} = \sum_{C_{30}}^{C_{45}} \frac{z_{C_n} MW_{C_n}}{SG_{C_n}}$$

$$SG_{C_{30+}} = 0.917$$

3.3.2 Pedersen Method

Pedersen et al. (1984, 1992) proposed a linear relationship between the SCN and the logarithm of the heavy fraction concentration for a gas condensate system. Their correlation is written as

$$\ln z_{C_n} = A + Bn \tag{3.8}$$

where A and B are the constant parameters that would be determined for each fluid. They evaluated this equation for 17 North Sea oil mixtures with molar compositions to C_{80+}. According to their results, this relationship fits the C_{7+} distribution of all mixtures, and there is no advantage of having measured a compositional analysis beyond C_{20+}. A similar distribution function was suggested by Yarborough (1979).

In phase behavior calculations, the carbon number is not directly used, so it is better to replace the carbon number by a physical property such as molecular weight (Danesh, 1998). The relation between the carbon number and molecular weight can be described by the following equation (Pedersen et al., 1984, 1992):

$$MW_{C_n} = 14n - 4 \tag{3.9}$$

Based on Eq. (3.9), Eq. (3.8) can be written in terms of molecular weight as follows:

$$\ln z_{C_n} = A + BMW_{C_n} \tag{3.10}$$

where A and B can be determined by regression (using a least-square method or other methods) for a partial analysis of C_{7+}.

Example 3.2

The total concentration of the C_{7+} fraction of a condensate is 5.45%. The partial analysis of C_{7+} is available and has been represented in Table 3.12. Describe the C_{7+} fraction by SCN groups, extended to C_{30+} using the Pedersen splitting method.

(Continued)

Table 3.12 Partial Analysis of Heavy End (Example 3.2) (Al-Meshari, 2005)

Component	Mol%	MW	SG
C_7	18.17	96	0.727
C_8	9.36	107	0.749
C_9	20.5	121	0.768
C_{10}	9.36	134	0.782
C_{11}	6.97	147	0.793
C_{12}	5.50	161	0.804
C_{13}	5.14	175	0.815
C_{14}	4.04	190	0.826
C_{15}	3.67	206	0.836
C_{16}	2.75	222	0.843
C_{17}	2.20	237	0.851
C_{18}	2.20	251	0.856
C_{19}	1.83	263	0.861
C_{20+}	8.26	338	0.889

MW, molecular weight; *SG*, specific gravity.

Solution

Fig. 3.1 shows the relation between mole fraction and molecular weight. According to this figure, the assumption of a linear relationship between the logarithm of the mole fraction and the molecular weight is reasonable for this fluid.

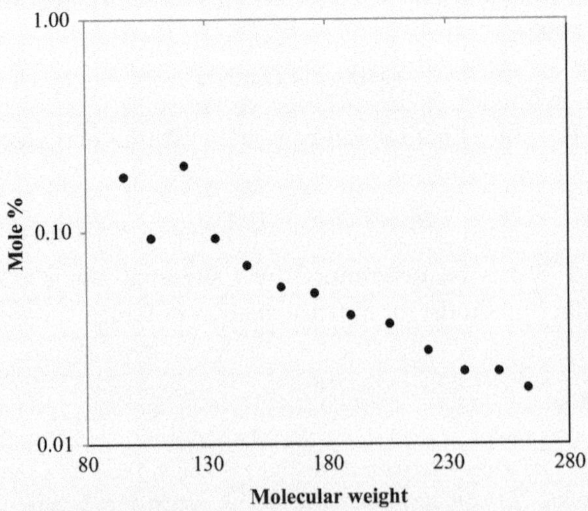

Figure 3.1 Relation between the mol% and molecular weight of single carbon number groups.

The A and B constants are determined by the least-square fitting method (excluding C_{20+}):

$$\ln x_{C_n} = -0.5456 - 0.0135 MW_{C_n}$$

where

$$x_{C_n} = \frac{z_{C_n}}{0.0545}$$

Similar to the previous example, the molecular weight and specific gravity of SCN groups are assumed to be the same as those in Table 3.9. By substituting the molecular weight in the above equation, the mole fraction of each SCN group (x_{C_n}) is obtained. The results are represented in the following table.

SCN	x_{C_n}	MW	SG	$MW_{C_n} x_{C_n}$	$\frac{MW_{C_n} x_{C_n}}{SG_{C_n}}$
C_7	0.1817	96	0.727	17.4385	23.9870
C_8	0.0936	107	0.749	10.0128	13.3683
C_9	0.2055	121	0.768	24.8661	32.3777
C_{10}	0.0936	134	0.782	12.5394	16.0351
C_{11}	0.0697	147	0.793	10.2495	12.9250
C_{12}	0.0550	161	0.804	8.8624	11.0229
C_{13}	0.0514	175	0.815	8.9908	11.0317
C_{14}	0.0404	190	0.826	7.6697	9.2854
C_{15}	0.0367	206	0.836	7.5596	9.0426
C_{16}	0.0275	222	0.843	6.1101	7.2480
C_{17}	0.0220	237	0.851	5.2183	6.1320
C_{18}	0.0220	251	0.856	5.5266	6.4563
C_{19}	0.0183	263	0.861	4.8257	5.6047
C_{20}	0.0141	275	0.866	3.8826	4.4834
C_{21}	0.0114	291	0.871	3.3100	3.8002
C_{22}	0.0101	300	0.876	3.0217	3.4494
C_{23}	0.0086	312	0.881	2.6723	3.0333
C_{24}	0.0073	324	0.885	2.3598	2.6665
C_{25}	0.0061	337	0.888	2.0592	2.3189
C_{26}	0.0052	349	0.892	1.8134	2.0330
C_{27}	0.0045	360	0.896	1.6123	1.7994
C_{28}	0.0038	372	0.889	1.4167	1.5936
C_{29}	0.0033	382	0.902	1.2710	1.4091
Sum	0.9918			153.3	191.10

MW, molecular weight; *SCN*, single carbon number; *SG*, specific gravity.

The mole fraction of C_{30+} is calculated as

$$x_{C_{30+}} = 1 - \sum_{C_7}^{C_{29}} x_{C_n} = 1 - 0.9918 = 0.0082$$

(Continued)

The molecular weight of the C_{7+} fraction should not be changed when the analysis is extended toC_{30+}, hence the molecular weight of the C_{7+} fraction is calculated as

$$MW_{C_{7+}} = \sum_{C_7}^{C_{20+}} x_{C_n} MW_{C_n} = 157.8 = \sum_{C_7}^{C_{29}} x_{C_n} MW_{C_n} + x_{C_{30+}} MW_{C_{30+}}$$

$$= 153.3 + (0.0082 \times MW_{C_{30+}})$$

$$MW_{C_{30+}} = 547$$

Similar to the previous example, the volume of the C_{30+} fraction is the sum of the volumes of all of its components. In this manner, by using a similar approach as for molecular weight, the specific gravity of the C_{30+} fraction is calculated as follows:

$$\frac{MW_{C_{7+}}}{SG_{C_{7+}}} = \sum_{C_7}^{C_{20+}} \frac{x_{C_n} MW_{C_n}}{SG_{C_n}} = 195.91 = \frac{157.8}{SG_{C_{7+}}}$$

$$SG_{C_{7+}} = 0.805$$

$$\frac{MW_{C_{7+}}}{SG_{C_{7+}}} = \sum_{C_7}^{C_{29}} \frac{x_{C_n} MW_{C_n}}{SG_{C_n}} + \frac{x_{C_{30+}} MW_{C_{30+}}}{SG_{C_{30+}}} = 195.91 = 191.10 + \frac{0.0082 \times 547}{SG_{C_{30+}}}$$

$$SG_{C_{30+}} = 0.937$$

3.3.3 Gamma Distribution Method

Instead of characterizing the heavy fraction by SCN groups, it could be characterized by a continuous description. A number of SCN groups characterize this continuous function, but it is valid at a discrete carbon number. The mathematical form of this function is as follows:

$$\int_{I_{n-1}}^{I_n} p(I)dI = x_{C_n} \tag{3.11}$$

where x_{C_n} is the normalized mole fraction of each SCN, and $p(I)$ is called the distribution function. The variable I could be any property that characterizes the compounds of the fluid such as the molecular weight or the boiling point (Whitson et al., 1990). If we take $I \equiv \mathrm{MW}$ then

$$\int_{\mathrm{MW}_{n-1}}^{\mathrm{MW}_n} p(\mathrm{MW})d\mathrm{MW} = x_{\mathrm{C}_n} \tag{3.12}$$

Similarly, the molecular weight of each SCN group is found as

$$\int_{\mathrm{MW}_{n-1}}^{\mathrm{MW}_n} \mathrm{MW}\cdot p(\mathrm{MW})d\mathrm{MW} = \mathrm{MW}_{\mathrm{C}_n} x_{\mathrm{C}_n} \tag{3.13}$$

The more general model used for distribution function is the three-parameter gamma model (Danesh, 1998). The gamma distribution function could be applied to a wide range of fluids including black oils, bitumen, and petroleum residues. The three-parameter gamma model discussed by Whitson (1983) is as follows:

$$p(\mathrm{MW}) = \frac{(\mathrm{MW}-\eta)^{\alpha-1}\exp\left(-\dfrac{\mathrm{MW}-\eta}{\beta}\right)}{\beta^{\alpha}\Gamma(\alpha)} \tag{3.14}$$

where α and β define the form of the distribution function, and η is the minimum molecular weight present in the heavy fraction. The average and variance values of this function are $(\alpha\beta + \eta)$ and $(\alpha\beta^2)$, respectively. If the heavy fraction is present as C_{n+}, then the average value of the distribution function is $\mathrm{MW}_{\mathrm{C}_{n+}}$, so

$$\beta = \frac{\mathrm{MW}_{\mathrm{C}_{n+}} - \eta}{\alpha} \tag{3.15}$$

$\Gamma(\alpha)$ is the gamma function defined as

$$\Gamma(\alpha) = \int_0^{\infty} x^{\alpha-1}e^{-x}dx \tag{3.16}$$

The gamma function, $\Gamma(\alpha)$, for α between 1 and 2 could be estimated by the following expression (Abramowitz and Stegun, 1966):

$$\Gamma(\alpha+1) = 1 + \sum_{i=1}^{8} A_i\alpha^i \quad 0 \le \alpha \le 1 \tag{3.17}$$

where $A_1 = -0.577191625$; $A_2 = 0.988205891$; $A_3 = -0.897056937$; $A_4 = 0.918206857$; $A_5 = -0.756704078$; $A_6 = 0.482199394$; $A_7 = -0.193527818$; and $A_8 = 0.035868343$.

The recurrence formula is

$$\Gamma(\alpha + 1) = \alpha\Gamma(\alpha) \tag{3.18}$$

Eq. (3.18) is used to evaluate $\Gamma(\alpha)$ when α is the outside range of Eq. (3.17).

The value of α usually ranges from 0.5 to 2.5 for reservoir fluids (Whitson et al., 1990; Danesh, 1998). The application of the gamma distribution function to heavy oils, bitumen, and petroleum residues shows that α has a range of 25–30 (Burle et al., 1985). Fig. 3.2 shows a typical distribution function for a heavy fraction with $MW_{C_{7+}} = 200$, $\eta = 90$ at different values of α. As shown in Fig. 3.2 for $\alpha \leq 1$, Eq. (3.14) exhibits exponential behavior with a continuous reduction of concentration that is suitable for gas condensate systems (Danesh, 1998; Riazi, 2005).

Parameter η is physically defined as the lightest component present in the heavy fraction. This parameter should be considered as a mathematical constant rather than a physical property. Several suggestions for η range from 86 to 94 (Al-Meshari, 2005). Different values for η are reported in Table 3.13 (Al-Meshari, 2005). Note that the molecular weight of the SCN groups are assumed to be the same as those in Table 3.9.

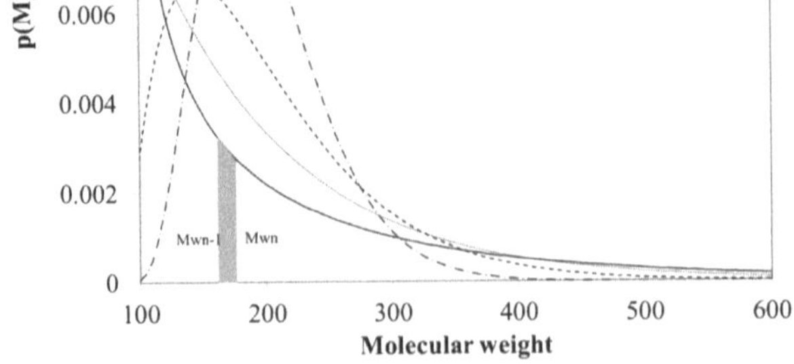

Figure 3.2 Gamma distribution function for different values of α, $MW_{C_{7+}} = 200, \eta = 90$.

Table 3.13 Different Methods to Calculate η (Assuming the Plus Fraction is Heptane Plus)

Method	η
Midpoint between the molecular weight of SCN6 and SCN7	(MW SCN6 + MW SCN7)/2 = 90
Midpoint between the molecular weight of normal hexane and normal heptane	$(MW_{nC_6} + MW_{nC_7})/2 = 93$
Molecular weight of SCN6	84
Molecular weight of normal hexane	86

SCN, single carbon number.

Whitson et al. (1990) proposed a relation that correlates η to α. This correlation was obtained for 44 samples of a stabilized petroleum liquid (stack tank oil and condensate).

$$\eta = 110\left[1 - \frac{1}{1 + \frac{4.043}{\alpha^{0.723}}}\right] \tag{3.19}$$

The mole fraction of the pseudo component that includes all of the components with molecular weights between MW_{n-1} and MW_n is recognized by the shaded area between the p(M) function and the molecular weight axis, like in Fig. 3.2.

Substituting Eq. (3.14) in Eq. (3.12), the normalized mole fraction of each SCN x_{C_n} group calculates as

$$x_{C_n} = \int_{MW_{n-1}}^{MW_n} p(MW)dMW = P(MW_n) - P(MW_{n-1}) \tag{3.20}$$

where MW_n and MW_{n-1} are the upper and lower molecular weight boundaries for the SCN group n. Note that the lower molecular weight boundary for each SCN group is the same as the upper molecular weight boundary for the preceding SCN group. $P(MW)$ could be obtained in terms of an infinite series:

$$P(MW) = \exp(-\zeta)\sum_{i=0}^{\infty}\left[\frac{\zeta^{\alpha+i}}{\Gamma(\alpha+i+1)}\right] \tag{3.21}$$

where

$$\zeta = \frac{\mathrm{MW} - \eta}{\beta} \tag{3.22}$$

The summation of Eq. (3.21) should be performed until the difference between the two successive terms is less than 10^{-8}.

Eq. (3.13) would be rewritten as

$$\mathrm{MW}_{C_n} = \frac{1}{x_{C_n}}\left[\int_{\eta}^{\mathrm{MW}_n} \mathrm{MW}\cdot p(\mathrm{MW})d\mathrm{MW} - \int_{\eta}^{\mathrm{MW}_{n-1}} \mathrm{MW}\cdot p(\mathrm{MW})d\mathrm{MW}\right] \tag{3.23}$$

which can be shown as the average molecular weight of each SCN group equal to

$$\mathrm{MW}_{C_n} = \eta + \alpha\beta\frac{P_1(\mathrm{MW}_n) - P_1(\mathrm{MW}_{n-1})}{P(\mathrm{MW}_n) - P(\mathrm{MW}_{n-1})} \tag{3.24}$$

The function $P_1(\mathrm{MW})$ is also evaluated by Eq. (3.21), but with the summation starting from $i = 1$.

The weight fraction of each SCN group is found as

$$w_{C_n} = \frac{x_{C_n}\mathrm{MW}_{C_n}}{\eta + \alpha\beta} \tag{3.25}$$

In cases with a lack of partial analysis of the heavy fraction, the value of α is assumed to be equal to 1. The values of η and β can be determined from Eqs. (3.19) and (3.16), respectively.

Eq. (3.14) with a value of $\alpha = 1$ reduces to an exponential function:

$$p(\mathrm{MW}) = \frac{\exp\left(-\frac{\mathrm{MW} - \eta}{\beta}\right)}{\beta} \tag{3.26}$$

substituting Eq. (3.26) in Eq. (3.12) results in

$$x_{C_n} = -\exp\left(\frac{\eta}{\beta}\right)\left[\exp\left(-\frac{\mathrm{MW}_n}{\beta}\right) - \exp\left(-\frac{\mathrm{MW}_{n-1}}{\beta}\right)\right] \tag{3.27}$$

and substituting Eq. (3.27) in Eq.(3.13) gives the average molecular weight of each SCN group.

$$\mathrm{MW}_{C_n}=\frac{-\beta\exp\left(\frac{\eta}{\beta}\right)\left[\left(\frac{\mathrm{MW}_n}{\beta}+1\right)\exp\left(-\frac{\mathrm{MW}_n}{\beta}\right)-\left(\frac{\mathrm{MW}_{n-1}}{\beta}+1\right)\exp\left(-\frac{\mathrm{MW}_{n-1}}{\beta}\right)\right]}{x_{C_n}} \tag{3.28}$$

Example 3.3

Prove that if we assume that $MW_n - MW_{n-1} = 14$, Eq. (3.27) leads to Eq. (3.10).

Solution

Substituting $MW_{n-1} = MW_n - 14$ in Eq. (3.27)

$$x_{C_n}=-\exp\left(\frac{\eta}{\beta}\right)\left[\exp\left(-\frac{MW_n}{\beta}\right)-\exp\left(-\frac{MW_n-14}{\beta}\right)\right]$$

or

$$x_{C_n}=\exp\left(\frac{\eta-MW_n}{\beta}\right)\left[\exp\left(\frac{14}{\beta}\right)-1\right]$$

By taking the logarithm of both sides of the above equation, we have

$$\ln x_{C_n}=\left[\frac{\eta}{\beta}+\ln\left(\exp\left(\frac{14}{\beta}\right)-1\right)\right]-\frac{1}{\beta}MW_n$$

which is similar to Eq. (3.10), with *A* and *B* as follows

$$A=\left[\frac{\eta}{\beta}+\ln\left(\exp\left(\frac{14}{\beta}\right)-1\right)\right]$$

$$B=-\frac{1}{\beta}$$

Example 3.4

Describe the C_{7+} fraction of the condensate in Example 3.1 by a continuous function, and estimate the mole fraction and molecular weight of the SCN groups. Extend the heavy fraction to C_{45+}.

(Continued)

Solution

In the absence of a partial analysis of the C_{7+} fraction we assumed that $\alpha = 1$.

In order to continue we need to specify the lower and upper molecular weight boundaries. Usually two methods are used to calculate the lower and upper molecular weight boundaries (Danesh, 1998; Al-Meshari, 2005). We used these two methods to solve the problem.

Normal Cut method: In this method, the molecular weights of the normal paraffins are used to specify the lower and upper molecular weights. Alternatively, we can write

$$MW_n - MW_{n-1} = 14$$

and

$$MW_n = 14n + 2$$

where n is the carbon number.

The value of η is equal to the normal hexane molecular weight. It is recommended that when the normal cut method is used, the value of η is equal to the molecular weight of the normal alkane, which is smaller than the plus fraction (Al-Meshari, 2005).The β parameter as found by Eq. (3.15).

$$\beta = \frac{MW_{C_{7+}} - \eta}{\alpha} = \frac{158 - 86}{1} = 72$$

The distribution function of the C_{7+} fraction in terms of the molecular weight is obtained by substituting the β and η parameters in Eq. (3.26).

$$p(MW) = \frac{\exp\left(-\frac{MW - 86}{72}\right)}{72} = 0.0459xp\left(-\frac{MW}{72}\right)$$

Using the result of previous example and the above equation, the normalized mole fraction of each SCN group can be determined as

$$x_{C_n} = \exp\left[\left(\frac{\eta - 2}{\beta} + \ln\left(\exp\left(\frac{14}{\beta}\right) - 1\right)\right) - \frac{14n}{\beta}\right]$$
$$= \exp(-0.37215 - 0.19444n)$$

The normalized mole fraction of the C_7 group is

$$x_{C_n} = \exp[-0.37215 - (0.19444 \times 7)] = 0.1767$$

To calculate the molecular weight of the C_7 group, the lower and upper molecular weight boundaries are

$$MW_6 = (14 \times 6) + 2 = 86$$

$$MW_7 = (14 \times 7) + 2 = 100$$

Therefore the average molecular weight of the C_7 group from Eq. (3.28) is

$$MW_{C_7} = \frac{-72 \times \exp\left(\frac{86}{72}\right)\left[\left(\frac{100}{72}+1\right)\exp\left(-\frac{100}{72}\right) - \left(\frac{86}{72}+1\right)\exp\left(-\frac{86}{72}\right)\right]}{0.1767} = 93$$

The normalized mole fraction and the average molecular weight for other SCN groups is similarly calculated, as shown in the following table.

SCN	Upper Molecular Weight Boundary MW_n	Lower Molecular Weight Boundary MW_{n-1}	x_{C_n}	MW_{C_n} Eq. (3.28)
C_7	100	$86 = \eta$	0.1767	93
C_8	114	100	0.1455	107
C_9	128	114	0.1198	121
C_{10}	142	128	0.0986	135
C_{11}	156	142	0.0812	149
C_{12}	170	156	0.0668	163
C_{13}	184	170	0.0550	177
C_{14}	198	184	0.0453	191
C_{15}	212	198	0.0373	205
C_{16}	226	212	0.0307	219
C_{17}	240	226	0.0253	233
C_{18}	254	240	0.0208	247
C_{19}	268	254	0.0171	261
C_{20}	282	268	0.0141	275
C_{21}	296	282	0.0116	289
C_{22}	310	296	0.0096	303
C_{23}	324	310	0.0079	317
C_{24}	338	324	0.0065	331
C_{25}	352	338	0.0053	345
C_{26}	366	352	0.0044	359
C_{27}	380	366	0.0036	373
C_{28}	394	380	0.0030	387
C_{29}	408	394	0.0025	401
C_{30}	422	408	0.0020	415

(Continued)

—cont'd

SCN	Upper Molecular Weight Boundary MW_n	Lower Molecular Weight Boundary MW_{n-1}	x_{C_n}	MW_{C_n} Eq. (3.28)
C_{31}	436	422	0.0017	429
C_{32}	450	436	0.0014	443
C_{33}	464	450	0.0011	457
C_{34}	478	464	0.0009	471
C_{35}	492	478	0.0008	485
C_{36}	506	492	0.0006	499
C_{37}	520	506	0.0005	513
C_{38}	534	520	0.0004	527
C_{39}	548	534	0.0004	541
C_{40}	562	548	0.0003	555
C_{41}	576	562	0.0002	569
C_{42}	590	576	0.0002	583
C_{43}	604	590	0.0002	597
C_{44}	618	604	0.0001	611
C_{45}	632	618	0.0001	625

SCN, single carbon group.

The average molecular weight calculated by this approach is considerably different from the molecular weight of the SCN group in a generalized table.

Midpoint method: In this method, the upper molecular weight boundary of the group n is the midpoint between the SCN molecular weights of the n and $n+1$ groups that are given in a generalized table. For example, the upper molecular boundary of group 10 is calculated as

$$MW_{10} = \frac{\text{MWSCN10} + \text{MW SCN11}}{2} = \frac{134 + 147}{2} = 140.5$$

The value of η is equal to 90, and the β parameter as determined by Eq. (3.15) is equal to 68. Similarly, the distribution function is

$$p(MW) = \frac{\exp\left(-\frac{MW - 90}{68}\right)}{68} = 0.0552xp\left(-\frac{MW}{68}\right)$$

The normalized mole fraction and the average molecular weight are determined from Eqs. (3.27) and (3.28), respectively. The results are reported in the following table. The average molecular weights of the SCN groups determined by this approach are close to those values in a generalized table. The normalized mole fraction and the molecular weight of the C_{45+} fraction are calculated similarly to Example 3.1 (a more accurate value for the molecular weight of C_{7+} is 158.4).

SCN	Upper Molecular Weight Boundary MW_n	Lower Molecular Weight Boundary MW_{n-1}	x_{C_n} Eq. (3.27)	MW_{C_n} Eq. (3.28)	MW_{C_n} (Table 3.9)
C_7	101.5	$90 = \eta$	0.1556	96	96
C_8	114	101.5	0.1418	108	107
C_9	127.5	114	0.1265	121	121
C_{10}	140.5	127.5	0.1002	134	134
C_{11}	154	140.5	0.0857	147	147
C_{12}	168	154	0.0726	161	161
C_{13}	182.5	168	0.0610	175	175
C_{14}	198	182.5	0.0523	190	190
C_{15}	214	198	0.0428	206	206
C_{16}	229.5	214	0.0329	221	222
C_{17}	244	229.5	0.0247	236	237
C_{18}	257	244	0.0181	250	251
C_{19}	269	257	0.0139	263	263
C_{20}	283	269	0.0134	276	275
C_{21}	295.5	283	0.0098	289	291
C_{22}	306	295.5	0.0070	301	300
C_{23}	318	306	0.0068	312	312
C_{24}	330.5	318	0.0059	324	324
C_{25}	343	330.5	0.0049	337	337
C_{26}	354.5	343	0.0038	349	349
C_{27}	366	354.5	0.0032	360	360
C_{28}	377	366	0.0026	371	372
C_{29}	388	377	0.0022	382	382
C_{30}	399	388	0.0019	393	394
C_{31}	409.5	399	0.0015	404	404
C_{32}	420.5	409.5	0.0014	415	415
C_{33}	431.5	420.5	0.0012	426	426
C_{34}	441	431.5	0.0009	436	437
C_{35}	450.5	441	0.0007	446	445
C_{36}	460	450.5	0.0006	455	456
C_{37}	469.5	460	0.0006	465	464
C_{38}	479.5	469.5	0.0005	474	475
C_{39}	489.5	479.5	0.0004	484	484
C_{40}	498.5	489.5	0.0003	494	495
C_{41}	507	498.5	0.0003	503	502
C_{42}	516.5	507	0.0003	512	512
C_{43}	526	516.5	0.0002	521	521
C_{44}	535	526	0.0002	530	531
C_{45+}	–	–	0.0013	783	–

SCN, single carbon group.

If there is sufficient data available regarding the partial analysis of the heavy fraction, then the parameters of the distribution function could be optimized. Whitson et al. (1990) suggest a procedure to optimize the parameters of the distribution function. A procedure similar to their procedure is provided here.

1. Determine the experimental fraction weight using the following relation:

$$w_{C_n} = \frac{(x_{C_n} MW_{C_n})_{\text{experimental}}}{\sum_{i=7}^{N} (x_{C_n} MW_{C_n})_{\text{experimental}}} \tag{3.29}$$

2. For the first guess, assume that $\alpha = 1$ and that the values of η and β are estimated from Eqs. (3.19) and (3.15), respectively.
3. Assume an upper molecular weight boundary, MW_n, for a group. Calculate the $P(MW_n)$ and x_{C_n} from Eqs. (3.21) and (3.20). Then calculate the average molecular weight and normalized weight fraction for groups from Eqs. (3.24) and (3.25).
4. If the calculated weight fraction does not match the experimental weight fraction within an acceptable tolerance (e.g., 10^{-7}), modify the upper molecular weight boundary and return to step 3. Use the Newton or Chord method to solve the problem.
5. Repeat steps 3 and 4 for all groups, excluding the last one. Determine the sum of the square errors.

$$\text{error}(\alpha, \beta, \eta) = \frac{1}{N-1} \sum_{i=1}^{N-1} \left[(MW_{C_n})_{\text{exp.}} - (MW_{C_n})_{\text{model}} \right]^2 \tag{3.30}$$

6. If we use Eq. (3.19) to calculate η and Eq. (3.15) to calculate β, the only optimizing parameter is α. Minimize the error by adjusting α.
7. Calculate the average molecular weight for each group by using the new parameters α, β, and η.
8. Compare the model molecular weight and the mole fraction with the experimental values. If the model values do not match with the experimental values within the acceptable tolerance, return to step 3.

Example 3.5

The partial analysis of the fraction of an oil sample is shown in Table 3.14 (Hoffman et al., 1953).

Describe the heavy fraction by a continuous distribution function (optimize the parameters of the distribution function).

Table 3.14 Partial Analysis of Heavy End

Component	$(z_{C_n})_{exp.}$	$(MW_{C_n})_{exp.}$
C_7	0.0263	99
C_8	0.0234	110
C_9	0.0235	121
C_{10}	0.0224	132
C_{11}	0.0241	145
C_{12}	0.0246	158
C_{13}	0.0266	172
C_{14}	0.0326	186
C_{15}	0.0363	203
C_{16}	0.0229	222
C_{17}	0.0171	238
C_{18}	0.0143	252
C_{19}	0.013	266
C_{20}	0.0108	279
C_{21}	0.0087	290
C_{22}	0.0072	301
C_{23}	0.0058	315
C_{24}	0.0048	329
C_{25}	0.0039	343
C_{26}	0.0034	357
C_{27}	0.0028	371
C_{28}	0.0025	385
C_{29}	0.0023	399
C_{30+}	0.0091	444
Total	0.3684	

Solution

According to this procedure, we calculated the weight fraction from Eq. (3.29).

(Continued)

Component	$(z_{C_n})_{exp.}$	$(x_{C_n})_{exp.} = \frac{(z_{C_n})_{exp.}}{0.3684}$	$(MW_{C_n})_{exp.}$	$(x_{C_n} MW_{C_n})_{exp.}$	$w_{C_n} = \frac{(x_{C_n} MW_{C_n})_{exp.}}{198.70}$
C_7	0.0263	0.0714	99	7.07	0.0356
C_8	0.0234	0.0635	110	6.99	0.0352
C_9	0.0235	0.0638	121	7.72	0.0388
C_{10}	0.0224	0.0608	132	8.03	0.0404
C_{11}	0.0241	0.0654	145	9.49	0.0477
C_{12}	0.0246	0.0668	158	10.55	0.0531
C_{13}	0.0266	0.0722	172	12.42	0.0625
C_{14}	0.0326	0.0885	186	16.46	0.0828
C_{15}	0.0363	0.0985	203	20.00	0.1007
C_{16}	0.0229	0.0622	222	13.80	0.0694
C_{17}	0.0171	0.0464	238	11.05	0.0556
C_{18}	0.0143	0.0388	252	9.78	0.0492
C_{19}	0.013	0.0353	266	9.39	0.0472
C_{20}	0.0108	0.0293	279	8.18	0.0412
C_{21}	0.0087	0.0236	290	6.85	0.0345
C_{22}	0.0072	0.0195	301	5.88	0.0296
C_{23}	0.0058	0.0157	315	4.96	0.0250
C_{24}	0.0048	0.0130	329	4.29	0.0216
C_{25}	0.0039	0.0106	343	3.63	0.0183
C_{26}	0.0034	0.0092	357	3.29	0.0166
C_{27}	0.0028	0.0076	371	2.82	0.0142
C_{28}	0.0025	0.0068	385	2.61	0.0131
C_{29}	0.0023	0.0062	399	2.49	0.0125
C_{30+}	0.0091	0.0247	444	10.97	0.0552
Total	0.3684	1		$198.70 = MW_{C7+}$	1

Assume that $\alpha = 1$. The value of η calculated from Eq. (3.19) is equal to 88.2. β as determined using Eq. (3.15) is equal to 110.5. The results of steps 3 to 5 are reported in the following table.

Component	MW_n Upper Molecular Weight Boundary Eq. (3.24)	$P(MW_n)$ Eq. (3.21)	$P_1(MW_n)$	$(x_{C_n})_{model}$ Eq. (3.20)	$(MW_{C_n})_{model}$ Eq. (3.24)	$(w_{C_n})_{model}$ Eq. (3.25)	$(w_{C_n})_{exp.}$
C_6	88.2	0	0				
C_7	97.0	0.0764	0.0030	0.0764	92.5	0.0356	0.0356
C_8	105.6	0.1454	0.0111	0.0690	101.2	0.0352	0.0352
C_9	115.0	0.2155	0.0251	0.0700	110.2	0.0388	0.0388
C_{10}	124.9	0.2824	0.0443	0.0670	119.9	0.0404	0.0404
C_{11}	136.7	0.3550	0.0722	0.0726	130.7	0.0477	0.0477
C_{12}	150.1	0.4287	0.1089	0.0737	143.2	0.0531	0.0531
C_{13}	166.4	0.5073	0.1585	0.0786	158.0	0.0625	0.0625
C_{14}	189.4	0.6000	0.2335	0.0927	177.5	0.0828	0.0828
C_{15}	220.5	0.6979	0.3363	0.0979	204.2	0.1007	0.1007
C_{16}	244.7	0.7574	0.4137	0.0594	232.1	0.0694	0.0694
C_{17}	266.4	0.8006	0.4792	0.0433	255.2	0.0556	0.0556
C_{18}	288.0	0.8360	0.5395	0.0353	276.8	0.0492	0.0492
C_{19}	311.4	0.8673	0.5994	0.0314	299.3	0.0472	0.0472
C_{20}	334.9	0.8927	0.6532	0.0253	322.7	0.0412	0.0412
C_{21}	357.4	0.9125	0.6993	0.0198	345.7	0.0345	0.0345
C_{22}	379.7	0.9285	0.7398	0.0160	368.2	0.0296	0.0296
C_{23}	401.3	0.9412	0.7745	0.0127	390.1	0.0250	0.0250
C_{24}	422.8	0.9516	0.8050	0.0104	411.7	0.0216	0.0216
C_{25}	443.9	0.9600	0.8312	0.0084	433.0	0.0183	0.0183
C_{26}	465.9	0.9672	0.8552	0.0072	454.5	0.0166	0.0166
C_{27}	488.0	0.9731	0.8760	0.0059	476.6	0.0142	0.0142
C_{28}	511.9	0.9784	0.8955	0.0052	499.5	0.0131	0.0131
C_{29}	539.3	0.9831	0.9142	4.74E−03	525.0	0.0125	0.0125

(Continued)

Use Eq. (3.30) to determine the error:

$$\text{error}(\alpha_1 = 1) = \frac{1}{29}\sum_{i=1}^{29}\left[(MW_{C_n})_{\text{exp.}} - (MW_{C_n})_{\text{model}}\right]^2 = 3607.53$$

To optimize the α value we can use a simple interpolation between the α value and error. Guess another value for α (e.g., $\alpha = 1.1$) and repeat steps 3 to 5.

$$\text{error}(\alpha_2 = 1.1) = \frac{1}{29}\sum_{i=1}^{29}\left[(MW_{C_n})_{\text{exp.}} - (MW_{C_n})_{\text{model}}\right]^2 = 2611.35$$

The new value for α is found as follows:

$$\alpha_{n+2} = \alpha_{n+1} - \frac{(\alpha_{n+1} - \alpha_n)\text{error}(\alpha_{n+1})}{\text{error}(\alpha_{n+1}) - \text{error}(\alpha_n)}$$

$$\alpha_3 = \alpha_2 - \frac{(\alpha_2 - \alpha_1)\text{error}(\alpha_2)}{\text{error}(\alpha_2) - \text{error}(\alpha_1)} = 1.1 - \frac{(1.1 - 1) \times 2611.35}{22611.35 - 3607.53} = 1.362$$

Continue trial and error until the error given is an acceptable value (e.g., error < 10). For $\alpha = 2.227$ we have

$$\text{error}(\alpha = 2.277) = \frac{1}{29}\sum_{i=1}^{29}\left[(MW_{C_n})_{\text{exp.}} - (MW_{C_n})_{\text{model}}\right]^2 = 6.97$$

The absolute average residual (AAR) is

$$\begin{aligned}\text{AAR} &= \frac{1}{N-1}\sum_{i=1}^{N-1}\left|(MW_{C_n})_{\text{exp.}} - (MW_{C_n})_{\text{model}}\right| \\ &= \frac{1}{29}\sum_{i=1}^{29}\left|(MW_{C_n})_{\text{exp.}} - (MW_{C_n})_{\text{model}}\right| = 2.23\end{aligned}$$

which shows that the obtained value for α is desirable. The optimized value of α is from Whitson et al. (1990) research result is 2.259.

The final results are reported in the following table. The η and β values are 75.9 and 53.9, respectively.

Component	MW_n Upper Molecular Weight Boundary Eq. (3.24)	$P(MW_n)$ Eq. (3.21)	$P_1(MW_n)$	$(x_{C_n})_{model}$ Eq. (3.20)	$(MW_{C_n})_{model}$ Eq. (3.24)	$(MW_{C_n})_{exp.}$	$(w_{C_n})_{model}$ Eq. (3.25)	$(w_{C_n})_{exp.}$
C_6	$75.9 = \eta$	0	0	—	—	—	—	—
C_7	106.9	0.0732	0.0123	0.0732	96.6	99	0.0356	0.0356
C_8	119.3	0.1349	0.0311	0.0617	113.3	110	0.0352	0.0352
C_9	130.1	0.1967	0.0557	0.0619	124.7	121	0.0388	0.0388
C_{10}	139.9	0.2562	0.0843	0.0594	135.0	132	0.0404	0.0404
C_{11}	150.5	0.3215	0.1211	0.0653	145.2	145	0.0477	0.0477
C_{12}	161.7	0.3891	0.1653	0.0676	156.1	158	0.0531	0.0531
C_{13}	174.5	0.4630	0.2207	0.0739	168.1	172	0.0625	0.0625
C_{14}	191.5	0.5530	0.2991	0.0900	182.8	186	0.0828	0.0828
C_{15}	213.1	0.6520	0.4008	0.0990	201.9	203	0.1007	0.1007
C_{16}	229.2	0.7145	0.4745	0.0625	220.9	222	0.0694	0.0694
C_{17}	243.3	0.7613	0.5356	0.0468	236.1	238	0.0556	0.0556
C_{18}	257.1	0.8004	0.5911	0.0391	250.0	252	0.0492	0.0492
C_{19}	271.8	0.8359	0.6455	0.0355	264.2	266	0.0472	0.0472
C_{20}	286.2	0.8653	0.6940	0.0293	278.8	279	0.0412	0.0412
C_{21}	299.9	0.8887	0.7353	0.0234	292.9	290	0.0345	0.0345
C_{22}	313.4	0.9079	0.7714	0.0192	306.5	301	0.0296	0.0296
C_{23}	326.3	0.9234	0.8022	0.0155	319.6	315	0.0250	0.0250
C_{24}	339.0	0.9363	0.8291	0.0129	332.4	329	0.0216	0.0216
C_{25}	351.3	0.9468	0.8522	0.0105	345.0	343	0.0183	0.0183
C_{26}	364.2	0.9560	0.8733	0.0092	357.6	357	0.0166	0.0166
C_{27}	377.0	0.9636	0.8916	0.0076	370.4	371	0.0142	0.0142
C_{28}	390.7	0.9704	0.9087	0.0068	383.6	385	0.0131	0.0131
C_{29}	406.4	0.9767	0.9251	6.25E−03	398.3	399	0.0125	0.0125

The mole fraction and molecular weight of C_{30+} are calculated similarly to Example 3.2.

$$x_{C_{30+}} = 1 - \sum_{C_7}^{C_{29}} x_{C_n} = 0.0233$$

$$MW_{C_{7+}} = \sum_{C_7}^{C_{20+}} x_{C_n} MW_{C_n} = 198.70 = \sum_{C_7}^{C_{29}} x_{C_n} MW_{C_n} + x_{C_{30+}} MW_{C_{30+}}$$

$$= 187.73 + (0.0233 \times MW_{C_{30+}})$$

$$MW_{C_{30+}} = 470.3$$

3.4 PROPERTIES ESTIMATION

In the absence of experimental data, the properties of the heavy fraction must be estimated. These properties are specific gravity, boiling point, molecular weight, critical properties, or even characterization factors such as the Watson factor. Several methods are available for calculating these properties. Some properties, such as the specific gravity, boiling point, and molecular weight, can be estimated from a generalized table. In addition, several correlations have been suggested by different authors to calculate these properties. In this section the most widely used correlations are reviewed. The units of temperature, pressure, and volume are K, MPa, and m^3/kg mol, respectively. The specific gravity is calculated at 15.5°C.

3.4.1 Watson Characterization Factor Estimation

To classify the petroleum fraction, the petroleum industries usually use a characterization parameter. The Watson characterization factor is the most widely used characterization parameter, as presented by Watson et al. (1935). This parameter is defined as

$$K_W = \frac{(1.8T_b)^{\frac{1}{3}}}{SG} \tag{3.31}$$

where T_b is the normal boiling point in K, and SG is the specific gravity. Based on this definition, for pure hydrocarbon we have (Danesh, 1998)

$12.5 < K_W \leq 13.5$	Paraffins
$11 < K_W \leq 12.5$	Naphtenes
$8.5 < K_W \leq 11$	Aromatics

Watson characterization factor can be determined from the molecular weight and specific gravity using the Riazi–Daubert (Austad et al., 1983) correlation.

$$K_w = 4.5579\text{MW}^{0.15178}\text{SG}^{-0.84573} \tag{3.32}$$

This relation is suitable for the last fraction when the normal boiling point is not available. Experience shows that the above equation is more reliable for fractions lighter than C_{20}. In most cases, for a given field the variation of the Watson characterization factor is relatively small, particularly for a heavy fraction (Austad et al., 1983).

3.4.2 Boiling Point Estimation

Riazi–Daubert correlation (Riazi and Daubert, 1987)

$$T_b = 3.76587\left[\exp\left(3.7741 \times 10^{-3}\text{MW} + 2.98404\text{SG} - 4.25288 \times 10^{-3}\text{MWSG}\right)\right]\text{MW}^{0.40167}\text{SG}^{-1.58262} \quad \text{for } 70 < \text{MW} < 300 \tag{3.33}$$

Riazi–Daubert correlation (Riazi, 2005)

$$T_b = 9.3369\left[\exp\left(1.6514 \times 10^{-4}\text{MW} + 1.4103\text{SG} - 7.5152 \times 10^{-4}\text{MWSG}\right)\right]\text{MW}^{0.5369}\text{SG}^{-0.7276} \quad \text{for } 300 < \text{MW} < 700 \tag{3.34}$$

Eq. (3.34) can be used for molecular weights between 70 and 300 with less accuracy (Riazi, 2005).

Soreide correlation (Soreide, 1989)

$$T_b = 1071.28 - 9.417 \times 10^4 \left[\exp\left(-4.922 \times 10^{-3}\text{MW} - 4.7685\text{SG} - 3.462 \times 10^{-3}\text{MWSG}\right)\right] \text{MW}^{-0.03522}\text{SG}^{3.266} \quad \text{for } 90^\circ C < T_b < 560^\circ\text{C} \tag{3.35}$$

In this correlation, the boiling point of very large molecules approaches 1071.28K. Soreide compared a given correlation for the boiling point for a petroleum fraction with molecular weights in the range of 70–450 and found that Eq. (3.33) overestimates the boiling point. Eqs. (3.34) and (3.35) have the same error, with an average absolute deviation percent (AAD%) of about 1% (Soreide, 1989).

Example 3.6

Estimate the normal boiling point of a heavy fraction with the specifications in Example 3.1.

1. Using the Watson characterization factor
2. Using the Riazi–Daubert correlation

Solution

1. Use Eq. (3.32) to determine the Watson number:

$$K_W = 4.5579(158)^{0.15178}(0.7964)^{-0.84573} = 11.91$$

Rewrite Eq. (3.31) in the following form to calculate the normal boiling point:

$$T_b = \frac{(K_W \times SG)^3}{1.8}$$

$$T_b = \frac{(11.91 \times 0.7964)^3}{1.8} = 474.68\text{K}$$

2. For molecular weights in the range of 70–300, the normal boiling point is estimated by Eq. (3.33), which results in

$$T_b = 472.20\text{K}$$

3.4.3 Critical Properties and Acentric Factor Estimation

The critical properties that comprise a group of reservoir fluids, especially the critical temperature, critical pressure, and acentric factor, are required for tuning EOS. These properties are usually related to the specific gravity and boiling point. The critical temperature correlations are more reliable than other correlations. The most widely used correlations are given below.

Lee–Kesler correlations (Lee and Kesler, 1975; Kesler and Lee, 1976)

$$T_c = 189.8 + 450.6SG + (0.4244 + 0.1174SG)T_b + (0.1441 - 1.0069SG) \times 10^5 T_b^{-1} \quad (3.36)$$

$$\ln P_c = 3.3864 - \frac{0.0566}{SG} - \left[\left(0.43639 + \frac{4.1216}{SG} + \frac{0.21343}{SG^2}\right) \times 10^{-3} T_b\right] + \left[\left(0.47579 + \frac{1.182}{SG} + \frac{0.15302}{SG^2}\right) \times 10^{-6} T_b^2\right] - \left[\left(2.4505 + \frac{9.9099}{SG^2}\right) \times 10^{-10} T_b^3\right] \quad (3.37)$$

$$\omega = \frac{\left[\ln P_{br} - 5.92714 + 6.09648 T_{br}^{-1} + 1.28862 \ln T_{br} - 0.169347 T_{br}^{6}\right]}{\left[15.2518 - 15.6875 T_{br}^{-1} - 13.4721 \ln T_{br} + 0.43577 T_{br}^{6}\right]}$$
$$\text{for } T_{br} \leq 0.8 \tag{3.38}$$

$$\omega = -7.904 + 0.1352 K_w - 0.007465 K_w^2 + 8.359 T_{br} + (1.408 - 0.01063 K_w) T_{br}^{-1} \quad \text{for } T_{br} > 0.8 \tag{3.39}$$

$$Z_c = 0.2905 - 0.085\omega \tag{3.40}$$

where $T_{br} = T_b/T_c$, $P_{br} = P_b/P_c$, and P_b is the pressure at which T_b is measured. For example, for a normal boiling point, $P_b = 0.101325$ MPa. These correlations are recommended for molecular weight ranges of 70–700 by the authors (Riazi, 2005).

Cavett correlations (Cavett, 1962)

$$\begin{aligned} T_c = 426.7062278 &+ \left[(9.5187183 \times 10^{-1})(1.8T_b - 459.67)\right] \\ &- \left[(6.01889 \times 10^{-4})(1.8T_b - 459.67)^2\right] \\ &- \left[(4.95625 \times 10^{-3})(\text{API})(1.8T_b - 459.67)\right] \\ &+ \left[(2.160588 \times 10^{-7})(1.8T_b - 459.67)^3\right] \\ &+ \left[(2.949718 \times 10^{-6})(\text{API})(1.8T_b - 459.67)^2\right] \\ &+ \left[(1.817311 \times 10^{-8})(\text{API}^2)(1.8T_b - 459.67)^2\right] \end{aligned} \tag{3.41}$$

$$\begin{aligned} \log(10P_c) = 1.6675956 &+ \left[(9.412011 \times 10^{-4})(1.8T_b - 459.67)\right] \\ &- \left[(3.047475 \times 10^{-6})(1.8T_b - 459.67)^2\right] \\ &- \left[(2.087611 \times 10^{-5})(\text{API})(1.8T_b - 459.67)\right] \\ &+ \left[(1.5184103 \times 10^{-9})(1.8T_b - 459.67)^3\right] \\ &- \left[(1.1047899 \times 10^{-8})(\text{API})(1.8T_b - 459.67)^2\right] \\ &- \left[(4.8271599 \times 10^{-8})(\text{API}^2)(1.8T_b - 459.67)\right] \\ &+ \left[(1.3949619 \times 10^{-10})(\text{API}^2)(1.8T_b - 459.67)^2\right] \end{aligned} \tag{3.42}$$

where API gravity is a measure comes from American Petroleum Institute of how heavy or light a petroleum liquid is compared to water. It is defined as

$$\mathrm{API} = \frac{141.5}{\mathrm{SG}} - 131.5 \tag{3.43}$$

Riazi–Daubert correlations (Riazi and Daubert, 1980, 1987; Aladwani and Riazi, 2005; Riazi, 2005)

$$T_c = 19.06232 T_b^{0.58848} \mathrm{SG}^{0.3596} \quad \text{for } C_5 \text{ to } C_{20} \text{ or } 70 < \mathrm{MW} < 300 \tag{3.44}$$

$$T_c = 9.5233\left[\exp\left(-9.314 \times 10^{-4} T_b - 0.544442\mathrm{SG} + 6.4791 \times 10^{-4} T_b \mathrm{SG}\right)\right] T_b^{0.81067} \mathrm{SG}^{0.53691} \tag{3.45}$$

for C_5 toC_{20} or $70 < \mathrm{MW} < 300$

$$T_c = 35.9413\left[\exp\left(-6.9 \times 10^{-4} T_b - 1.4442\mathrm{SG} + 4.91 \times 10^{-4} T_b \mathrm{SG}\right)\right] T_b^{0.7293} \mathrm{SG}^{1.2771} \tag{3.46}$$

for C_{20} to C_{50} or $300 < \mathrm{MW} < 700$

$$P_c = 5.53027 \times 10^6 T_b^{-2.3125} \mathrm{SG}^{2.3201} \quad \text{for } C_5 \text{ to } C_{20} \text{ or } 70 < \mathrm{MW} < 300 \tag{3.47}$$

$$P_c = 3.1958 \times 10^4\left[\exp\left(-8.505 \times 10^{-3} T_b - 4.8014\mathrm{SG} + 5.749 \times 10^{-3} T_b \mathrm{SG}\right)\right] T_b^{-0.4844} \mathrm{SG}^{4.0846} \tag{3.48}$$

for C_5 to C_{20} or $70 < \mathrm{MW} < 300$

$$P_c = 0.69575\left[\exp\left(-1.35 \times 10^{-2} T_b - 0.3129\mathrm{SG} + 9.174 \times 10^{-3} T_b \mathrm{SG}\right)\right] T_b^{0.6791} \mathrm{SG}^{-0.6807} \tag{3.49}$$

for C_{20} to C_{50} or $300 < \mathrm{MW} < 700$

$$V_C = 1.7842 \times 10^{-7} T_b^{2.3829} \mathrm{SG}^{-1.683} \quad \text{for } C_5 \text{ to } C_{20} \text{ or } 70 < \mathrm{MW} < 300 \tag{3.50}$$

$$V_c = 6.1677 \times 10^7\left[\exp\left(-7.583 \times 10^{-3} T_b - 28.5524\mathrm{SG} + 1.172 \times 10^{-2} T_b \mathrm{SG}\right)\right] T_b^{1.20493} \mathrm{SG}^{17.2074} \tag{3.51}$$

for C_5 to C_{20} or $70 < \mathrm{MW} < 300$

Note that the accuracies of Eqs. (3.46) and (3.49) are greater than the accuracies of Eqs. (3.45) and (3.48), respectively. Eqs. (3.46) and (3.49) can be used for hydrocarbon in the range of C_5-C_{20} with acceptable accuracy (Riazi, 2005).

Twu correlations (based on the Perturbation Expansion) (Twu, 1984)

For normal alkanes

$$T_c^{n-\text{alkane}} = T_b\left[0.533272 + 0.343831 \times 10^{-3} T_b + 2.526167 \times 10^{-7} T_b^2 - 1.65848 \times 10^{-10} T_b^3 + 4.60774 \times 10^{24} T_b^{-13}\right]^{-1} \tag{3.52}$$

$$P_c^{n-\text{alkane}} = \left[0.318317 + 0.099334\varphi^{0.5} + 2.89698\varphi + 3.00546\varphi^2 + 8.65163\varphi^4\right]^2 \tag{3.53}$$

$$V_c^{n-\text{alkane}} = \left[0.82055 + 0.715468\varphi + 2.21266\varphi^3 + 13411.1\varphi^{14}\right]^{-8} \tag{3.54}$$

$$\text{SG}^{n-\text{alkane}} = 0.843593 - 0.128624\varphi - 3.36159\varphi^3 - 13749.5\varphi^{12} \tag{3.55}$$

where

$$\varphi \equiv 1 - \frac{T_b}{T_c^{n-\text{alkane}}} \tag{3.56}$$

$$T_b = \exp\left[5.12640 + 2.71579\psi - 0.286590\psi^2 - 39.8544\psi^{-1} - 0.122488\psi^{-2}\right] - 13.7512\psi + 19.6197\psi^2 \tag{3.57}$$

and

$$\psi = \ln \text{MW}^{n-\text{alkane}} \tag{3.58}$$

Eq. (3.57) could be solved for molecular weight by following an initial guess.

$$\text{MW}^{n-\text{alkane}} = \frac{T_b}{5.800 - 0.0052 T_b} \tag{3.59}$$

For petroleum fractions

$$T_c = T_c^{n-\text{alkane}} \left[\frac{1 + 2f_T}{1 - 2f_T} \right]^2 \tag{3.60}$$

$$f_T = \Delta S_T \left[-0.270159 T_b^{-0.5} + \left(0.0398285 - 0.706691 T_b^{-0.5} \right) \Delta S_T \right] \tag{3.61}$$

$$\Delta S_T = \exp\left[5 \left(\text{SG}^{n-\text{alkane}} - \text{SG} \right) \right] - 1 \tag{3.62}$$

$$V_c = V_c^{n-\text{alkane}} \left[\frac{1 + 2f_V}{1 - 2f_V} \right]^2 \tag{3.63}$$

$$f_V = \Delta S_V \left[0.347776 T_b^{-0.5} + \left(-0.182421 + 2.24890 T_b^{-0.5} \right) \Delta S_V \right] \tag{3.64}$$

$$\Delta S_V = \exp\left[4 \left(\left(\text{SG}^{n-\text{alkane}} \right)^2 - \text{SG}^2 \right) \right] - 1 \tag{3.65}$$

$$P_c = P_c^{n-\text{alkane}} \left(\frac{T_c}{P_c^{n-\text{alkane}}} \right) \left(\frac{V_c^{n-\text{alkane}}}{V_c} \right) \left[\frac{1 + 2f_P}{1 - 2f_P} \right]^2 \tag{3.66}$$

$$\begin{aligned} f_P = \Delta S_P \big[&\left(2.53262 - 34.4321 T_b^{-0.5} - 0.00230193 T_b \right) \\ &+ \left(-11.4277 + 187.934 T_b^{-0.5} + 0.00414963 T_b \right) \Delta S_P \big] \end{aligned} \tag{3.67}$$

$$\Delta S_P = \exp\left[0.5 \left(\text{SG}^{n-\text{alkane}} - \text{SG} \right) \right] - 1 \tag{3.68}$$

Winn–Mobil (Sim–Daubert) correlations (Riazi, 1979; Sim and Daubert, 1980)

$$\ln T_c = -0.58779 + 4.2009 T_b^{0.08615} \text{SG}^{0.04614} \tag{3.69}$$

$$P_c = 6.148341 \times 10^6 T_b^{-2.3177} \text{SG}^{2.4853} \tag{3.70}$$

Hall–Yarborough correlation (Hall and Yarborough, 1971)

$$V_C = 1.56 \times 10^{-3} \text{MW}^{1.15} \text{SG}^{-0.7935} \tag{3.71}$$

Edmister correlation (Edmister, 1958)

$$\omega = \frac{3}{7} \left(\frac{T_{br}}{1 - T_{br}} \right) \left[\log_{10} \left(P_{br}^{-1} \right) \right] - 1 \tag{3.72}$$

Korsten correlation (Korsten, 2000)

$$\omega = 0.5899\left(\frac{T_{br}^{1.3}}{1 - T_{br}^{1.3}}\right)\left[\log_{10}\left(P_{br}^{-1}\right)\right] - 1 \tag{3.73}$$

where T_{br} and P_{br} in Eqs. (3.71) and (3.72) are defined the same as in Eq. (3.38).

Example 3.7

Estimate the critical volume of n-Tetradecylbenzene using the Lee–Kesler method. The normal boiling point, specific gravity, and critical volume of n-Tetradecylbenzene are 627.15K, 0.8587, and 1.030 m^3/kg mol, respectively (data are taken from the API technical data book Daubert and Danner, 1997).

Solution

The determined values for the critical temperature, critical pressure, acentric factor, and critical compressibility factor are given in the following table.

Variable	T_c(K)	P_c(MPa)	T_{br}	P_{br}	ω	Z_c
Equation	(3.36)	(3.37)	—	—	(3.38)	(3.40)
Value	791.23	1.3143	0.793	0.077	0.8762	0.2160

The Lee–Kesler correlation does not provide a correlation for estimating the critical volume. The critical volume can be calculated from

$$V_c = \frac{Z_c R T_c}{P_c} = \frac{0.2159 \times 0.008314 \times 791.23}{1.3143} = 1.081\,\frac{\text{m}^3}{\text{kg mol}}$$

Example 3.8

Predict the acentric factor for n-Pentylcyclopentane by following the set of T_c–P_c–ω correlations. Then obtain the AAD% for each set of equations.

Set 1: T_c–P_c, API technical data book; ω-Edmister method

Set 2: T_c–P_c, ω; Lee–Kesler method

Set1: T_c–P_c, Winn–Mobil (Sim–Daubert) method; ω-Edmister method

The normal boiling point and critical properties from the API technical data book are as follows (Daubert and Danner, 1997): $T_b = 453.65$K (normal boiling point), SG = 0.7954, $T_c = 643.80$K, $P_c = 2.45$MPa, and $\omega = 0.4184$.

(Continued)

Solution

Set 1: Use the critical temperature and critical pressure from the API technical data book to calculate T_{br} and P_{br}. Note that P_b is the atmospheric pressure, equal to 0.101325 MPa.

$$T_{br} = 453.65/643.80 = 0.705$$

$$P_{br} = 0.101325/2.45 = 0.041$$

The acentric factor is determined by Eq. (3.72) as follows:

$$\omega = \frac{3}{7}\left(\frac{0.705}{1 - 0.705}\right)\left[\log_{10}\left(0.041^{-1}\right)\right] - 1 = 0.4145$$

Set 2: Estimate critical properties with the Lee–Kesler method as given below.

Variable	T_c (K)	P_c (MPa)	T_{br}	P_{br}
Equation	(3.36)	(3.37)	–	–
Value	638.32	2.4545	0.711	0.041

The reduced boiling point temperature is less than 0.8, so the acentric factor is determined by Eq. (3.38), which results in

$$\omega = 0.4611$$

Set 3: The results of estimating the critical temperature and critical pressure using the Winn–Mobill correlations are reported in the following table.

Variable	T_c (K)	P_c (MPa)	T_{br}	P_{br}
Equation	(3.69)	(3.70)	–	–
Value	634.71	2.4222	0.715	0.042

Acentric factor is obtained by using the Edmister correlation, which results in

$$\omega = 0.4802$$

The AAD% for each set is presented below.

	Set 1	Set 2	Set 3
AAD%	0.93	10.21	14.77

AAD%, average absolute deviation percent.

3.4.4 Molecular Weight Estimation

The molecular weight of a petroleum fraction can be estimated from a generalized table or correlations. In most cases, the molecular weight correlates in terms of the boiling point and the specific gravity. The following equations are presented to estimate the molecular weight.

Riazi–Daubert correlation (Riazi and Daubert, 1980)

$$\mathrm{MW} = 1.6607 \times 10^{-4} T_b^{2.1962} \mathrm{SG}^{-1.0164} \quad \text{for } T_b < 670\mathrm{K} \tag{3.74}$$

Riazi–Daubert correlation (Riazi, 2005)

$$\mathrm{MW} = 42.965\left[\exp\left(2.097 \times 10^{-4} T_b - 7.78712\mathrm{SG} + 2.08476 \times 10^{-3} T_b \mathrm{SG}\right)\right] T_b^{1.26007} \mathrm{SG}^{4.98308} \quad \text{for } 300K < T_b < 850\mathrm{K} \tag{3.75}$$

Lee–Kesler correlation (Kesler and Lee, 1976)

$$\begin{aligned}\mathrm{MW} = & -12272.6 + 9486.4\mathrm{SG} + (8.3741 - 5.9917\mathrm{SG})T_b \\ & + \left[\left(1 - 0.77084\mathrm{SG} - 0.02058\mathrm{SG}^2\right) \times \left(0.7465 - 222.466 T_b^{-1}\right) 10^7 T_b^{-1}\right] \\ & + \left[\left(1 - 0.80882\mathrm{SG} + 0.02226\mathrm{SG}^2\right) \times \left(0.3228 - 17.335 T_b^{-1}\right) 10^{12} T_b^{-3}\right] \quad \text{for } T_b < 750\mathrm{K}\end{aligned} \tag{3.76}$$

Winn–Mobil (Sim–Daubert) correlation (Riazi, 1979; Sim and Daubert, 1980)

$$\mathrm{MW} = 2.70579 \times 10^{-5} T_b^{2.4966} \mathrm{SG}^{-1.174} \tag{3.77}$$

Twu correlation (based on the Perturbation Expansion) (Twu, 1984)

$$\ln\mathrm{MW} = \ln\mathrm{MW}^{n-\mathrm{alkane}} \left[\frac{1 + 2f_M}{1 - 2f_M}\right]^2 \tag{3.78}$$

$$f_M = \Delta S_M\left[\Psi + \left(-0.0175691 + 0.143979 T_b^{-0.5}\right)\Delta S_M\right] \tag{3.79}$$

$$\Psi = \left|0.0123420 - 0.244541 T_b^{-0.5}\right| \tag{3.80}$$

$$\Delta S_M = \exp\left[5\left(\mathrm{SG}^{n-\mathrm{alkane}} - \mathrm{SG}\right)\right] - 1 \tag{3.81}$$

Example 3.9

Estimate the molecular weight of n-Tetradecylbenzene using Eq. (3.57). Then correct the calculating value using Eq. (3.78). The normal boiling point, molecular weight, and specific gravity of n-Tetradecylbenzene are 627.15K, 274.5, and 0.8587, respectively (data taken from the API technical data book Daubert and Danner, 1997).

Solution

To estimate the molecular weight, Eq. (3.57) should be solved for the ψ parameter. Using Eq. (3.59) for finding an initial guess

$$MW^{n-\text{alkane}} = \frac{627.15}{5.800 - 0.0052(627.15)} = 247.02$$

$$\psi^{\text{initialguess}} = \ln(247.02) = 5.51$$

With this initial guess and the Newton–Raphson method, solve the following equation:

$$\exp\left[5.12640 + 2.71579\psi - 0.286590\psi^2 - 39.8544\psi^{-1} - 0.122488\psi^{-2}\right]$$
$$-13.7512\psi + 19.6197\psi^2 = T_b = 627.15$$

$$\psi = 5.68$$

$$MW^{n-\text{alkane}} = \exp(5.68) = 292.54$$

Now we correct this value. First the properties of the normal alkane with the same boiling point are calculated.

The critical temperature of the normal alkane is calculated by Eq. (3.52):

$$T_c^{n-\text{alkane}} = 776.80\text{K}$$

The φ value and the specific gravity are determined by Eqs. (3.56) and (3.55), respectively.

$$\varphi = 1 - \frac{627.15}{776.80} = 0.193$$

$$SG^{n-\text{alkane}} = 0.843593 - 0.128624(0.193) - 3.36159(0.193)^3$$
$$- 13749.5(0.193)^{12} = 0.795$$

Using Eqs. (3.78)–(3.81) the corrected value is determined as follows:

$$\Delta S_M = \exp[5(0.795 - 0.8587)] - 1 = -0.274$$

$$\Psi = \left|0.0123420 - 0.244541(627.15)^{-0.5}\right| = 0.002577$$

$$f_M = -0.274\left[0.002577 + \left(-0.0175691 + 0.143979(627.15)^{-0.5}\right) - 0.274\right] = -0.00159$$

$$MW = \exp\left\{\ln(292.54)\left[\frac{1 + 2(-0.00159)}{1 - 2(-0.00159)}\right]^2\right\} = 272.27$$

The average absolute deviation percent is −0.8%, which shows good accuracy.

3.4.5 Specific Gravity Estimation

Most correlations that are used to predict critical properties are functions of the normal boiling point and specific gravity. In the characterization procedure the specific gravity of each SCN group must be calculated. The specific gravity of each SCN group can be calculated by the assumption of a constant Watson characterization factor. Solve Eq. (3.32) for a specific gravity by assuming a constant Watson characterization factor (Aguilar Zurita and McCain Jr., 2002).

$$SG = \left[\frac{K_W}{4.5579 MW^{0.15178}}\right]^{-\frac{1}{0.84573}} \tag{3.82}$$

Eq. (3.3) can be rewritten in the following form:

$$SG_{C_{7+}} = \frac{z_{C_{7+}} MW_{C_{7+}}}{\sum_{C_7}^{C_N} z_{C_n} \frac{MW_{C_n}}{SG_{C_n}}} \tag{3.83}$$

Substituting Eq. (3.82) in Eq. (3.83) gives

$$SG_{C_{7+}} = \frac{z_{C_{7+}} MW_{C_{7+}}}{\sum_{C_7}^{C_N}\left(z_{C_n} MW_{C_n}\left[\frac{4.5579 MW_{C_n}^{0.15178}}{K_W}\right]^{-\frac{1}{0.84573}}\right)} \tag{3.84}$$

Solving Eq. (3.84) for the Watson characterization factor gives

$$K_w = \left[\frac{\zeta\,SG_{C_{7+}}}{z_{C_{7+}} MW_{C_{7+}}}\right]^{-0.84573} \tag{3.85}$$

where ζ is constant and calculated as follows

$$\zeta = \sum_{C_7}^{C_N} z_{C_n} MW_{C_n} \left(4.5579 MW_{C_n}^{0.15178}\right)^{-\frac{1}{0.84573}} \tag{3.86}$$

Use the Watson characterization factor calculated from Eq. (3.85) to calculate the specific gravity of each SCN group. The plus fraction is usually extended into 45 SCN groups (Aguilar Zurita and McCain Jr., 2002; Al-Meshari, 2005).

Example 3.10

The extended molar composition of an oil is given in Table 3.15. Use the experimental mole fraction and the molecular weight in the generalized table to estimate the specific gravity of each SCN group from Eq. (3.82) and compare with the experimental values.

Solution

To calculate the specific gravity using Eq. (3.82), the mole fraction, molecular weight, and specific gravity of C_{7+} should be calculated. Assume that the molecular weight of each SCN group is the same as those in Table 3.9. Similar to Example 3.1 the mole fraction and molecular weight of C_{7+} fraction calculate as follows.

$$z_{C_{7+}} = \sum_{C_7}^{C_{30+}} z_{C_n} = 0.9426$$

$$z_{C_{7+}} MW_{C_{7+}} = \sum_{C_7}^{C_{30+}} z_{C_n} MW_{C_n}$$

$$MW_{C_{30+}} = 258.7$$

Table 3.15 Composition of an Oil Sample (Example 3.10) (Pedersen et al., 1992)

Component	z_{C_n}	MW	SG
C_1	0.0013		
C_2	0.0050		
C_3	0.0047		
i-C_4	0.0055		
n-C_4	0.0062		
i-C_5	0.0108		
n-C_5	0.0050		
C_6	0.0189		
C_7	0.0534		0.749
C_8	0.0854		0.768
C_9	0.0704		0.793
C_{10}	0.0680		0.808
C_{11}	0.0551		0.815
C_{12}	0.0500		0.836
C_{13}	0.0558		0.850
C_{14}	0.0508		0.861
C_{15}	0.0466		0.873
C_{16}	0.0380		0.882
C_{17}	0.0267		0.873
C_{18}	0.0249		0.875
C_{19}	0.0214		0.885
C_{20}	0.0223		0.903
C_{21}	0.0171		0.898
C_{22}	0.0142		0.898
C_{23}	0.0163		0.899
C_{24}	0.0150		0.900
C_{25}	0.0125		0.905
C_{26}	0.0145		0.907
C_{27}	0.0133		0.911
C_{28}	0.0123		0.915
C_{29}	0.0115		0.920
C_{30+}	0.1471	624	0.953

MW, molecular weight; *SG*, specific gravity.

As in Example 3.1, the volume of the C_{7+} fraction is assumed to be equal to the sum of all of the component volumes and the specific gravity of the C_{7+} fraction can be found as

$$\frac{z_{C_{7+}} MW_{C_{7+}}}{SG_{C_{7+}}} = \sum_{C_7}^{C_{30+}} \frac{z_{C_n} MW_{C_n}}{SG_{C_n}}$$

$$SG_{C_{7+}} = 0.890$$

(Continued)

The value of ζ is determined as below.

SCN	z_{C_n}	MW (Table 3.9)	$z_{C_n} MW_{C_n} \left(4.5579 MW_{C_n}^{0.15178}\right)^{-\frac{1}{0.84573}}$
C_7	0.0534	96	0.376
C_8	0.0854	107	0.657
C_9	0.0704	121	0.599
C_{10}	0.0680	134	0.629
C_{11}	0.0551	147	0.550
C_{12}	0.0500	161	0.538
C_{13}	0.0558	175	0.643
C_{14}	0.0508	190	0.626
C_{15}	0.0466	206	0.614
C_{16}	0.0380	222	0.532
C_{17}	0.0267	237	0.395
C_{18}	0.0249	251	0.386
C_{19}	0.0214	263	0.344
C_{20}	0.0223	275	0.372
C_{21}	0.0171	291	0.299
C_{22}	0.0142	300	0.255
C_{23}	0.0163	312	0.302
C_{24}	0.0150	324	0.287
C_{25}	0.0125	337	0.247
C_{26}	0.0145	349	0.294
C_{27}	0.0133	360	0.277
C_{28}	0.0123	372	0.263
C_{29}	0.0115	382	0.251
C_{30+}	0.1471	624	4.811
Sum			$\zeta = 14.548$

MW, molecular weight; *SCN*, single carbon number.

The Watson characterization factor is calculate by Eq. (3.85).

$$K_W = \left[\frac{14.548 \times 0.890}{0.9426 \times 258.7}\right]^{-0.84573} = 11.98$$

Now the specific gravity of each SCN group can be calculated by Eq. (3.82). The results are given in the following table.

SCN	z_{C_n}	MW	$SG^{exp.}$	$SG^{cal.}$ Eq. (3.82)	$\frac{\lvert SG^{cal.} - SG^{exp.}\rvert}{SG^{exp.}} \times 100$
C_7	0.0534	96	0.749	0.724	3.36
C_8	0.0854	107	0.768	0.738	3.90
C_9	0.0704	121	0.793	0.755	4.85
C_{10}	0.068	134	0.808	0.768	4.89
C_{11}	0.0551	147	0.815	0.781	4.13

—cont'd

SCN	z_{C_n}	MW	$SG^{exp.}$	$SG^{cal.}$ Eq. (3.82)	$\frac{\lvert SG^{cal.} - SG^{exp.}\rvert}{SG^{exp.}} \times 100$
C_{12}	0.05	161	0.836	0.794	5.00
C_{13}	0.0558	175	0.85	0.806	5.15
C_{14}	0.0508	190	0.861	0.818	4.97
C_{15}	0.0466	206	0.873	0.830	4.91
C_{16}	0.038	222	0.882	0.841	4.61
C_{17}	0.0267	237	0.873	0.851	2.49
C_{18}	0.0249	251	0.875	0.860	1.70
C_{19}	0.0214	263	0.885	0.867	1.99
C_{20}	0.0223	275	0.903	0.874	3.17
C_{21}	0.0171	291	0.898	0.883	1.64
C_{22}	0.0142	300	0.898	0.888	1.10
C_{23}	0.0163	312	0.899	0.894	0.52
C_{24}	0.015	324	0.9	0.900	0.05
C_{25}	0.0125	337	0.905	0.907	0.20
C_{26}	0.0145	349	0.907	0.913	0.61
C_{27}	0.0133	360	0.911	0.918	0.73
C_{28}	0.0123	372	0.915	0.923	0.88
C_{29}	0.0115	382	0.92	0.927	0.81
C_{30+}	0.1471	624	0.953	1.013	6.28

MW, molecular weight; *SCN*, single carbon number; *SG*, specific gravity.

Example 3.11

The following equation is suitable for characterizing the plus fraction (Riazi, 2005):

$$F^* = \left[\frac{A}{B}\ln\left(\frac{1}{x^*}\right)\right]^{\frac{1}{B}}$$

where

$$F^* = \frac{F - F_0}{F_0}, x^* = 1 - x_{cumulative}$$

F is a property such as the molecular weight, specific gravity, or normal boiling point. $x_{cumulative}$ is the cumulative mole, weight, or volume fraction. Usually the cumulative mole fraction is used to express the molecular weight distribution, the cumulative weight fraction is used to express the boiling point distribution, and the cumulative volume fraction is used to express the specific gravity distribution. F_0 is the value of F at $x_{cumulative} = 0$ or $x^* = 1$. F_0 is physically represented as the value of the property P for the lightest component in the mixture; however, it is determined as a mathematical constant. The initial guess for F_0 is a value, which should be lower than the first value of F in the data set.

(Continued)

Then the F_0 is adjusted to minimize the root mean square error (RMS). The RMS is defined as

$$\text{error}(F_0) = \left[\frac{1}{N}\sum_{i=1}^{N}\left(F_i^{\text{cal.}} - F_i^{\text{exp.}}\right)^2\right]^{0.5}$$

A and B can be determined by a linear regression when sufficient data is available for the property F. The cumulative fraction is calculated by the following equation:

$$x_{c_i} = x_{c_{i-1}} + \frac{x_{i-1} + x_i}{2}$$

where x_{c_i} is the cumulative fraction for group i, and $x_{c_0} = 0$. In this example x_{mn}, x_{wn}, and x_{vn} are the normalized mole fraction, normalized weight fraction, and normalized volume fraction, respectively, and x_{cmn}, x_{cwn}, and x_{cvn} are the cumulative mole fraction, cumulative weight fraction, and cumulative volume fraction, respectively.

The first equation in this example can be rewritten in the following linear form:

$$Y = C_1 + C_2 X$$

where $Y = \ln P^*$; $X = \ln[\ln(1/x^*)]$; $B = 1/C_2$; and $A = B\exp(C_1 B)$. C_1 and C_2 are determined by the following equation derived from the least squares linear regression method:

$$C_2 = \frac{\sum X_i \sum Y_i - N\sum (X_i Y_i)}{\left(\sum X_i\right)^2 - N\sum (X_i^2)}$$

$$C_1 = \frac{\sum Y_i - C_2\sum X_i}{N}$$

where N is the number of data. Note that all cumulative fractions are determined in terms of the normalized fraction. Obtain (1) the molecular weight distribution function, (2) the specific gravity distribution function, and (3) the boiling point distribution function for the North Sea gas condensate in Table 3.10.

Solution

The gravity of water at 15°C is around 1; therefore the value of the specific gravity and density are the same. First, the values of the volume fraction, normalized fraction, and cumulative fraction are calculated. The results are given in the following table.

SCN	MW	SG	w_{C_n}	z_{C_n}	$w_{C_n}SG_n$	Normalized Fraction $x_{wn}=\frac{w_{C_n}}{0.298}$	$x_{mn}=\frac{z_{C_n}}{0.057}$	$x_{vn}=\frac{w_{C_n}SG_n}{0.368}$	Cumulative Fraction x_{cwn}	x_{cmn}	x_{cvn}
C_7	91.2	0.746	0.037	0.012	0.050	0.126	0.211	0.136	0.063	0.105	0.068
C_8	104.0	0.770	0.041	0.012	0.053	0.138	0.202	0.145	0.195	0.312	0.209
C_9	119.0	0.788	0.026	0.006	0.033	0.086	0.111	0.089	0.307	0.468	0.325
C_{10}	133.0	0.795	0.023	0.005	0.029	0.078	0.088	0.080	0.389	0.568	0.409
C_{11}	144.0	0.790	0.015	0.003	0.019	0.049	0.051	0.050	0.453	0.637	0.474
C_{12}	155.0	0.802	0.015	0.003	0.018	0.049	0.047	0.049	0.502	0.686	0.524
C_{13}	168.0	0.814	0.016	0.003	0.020	0.055	0.049	0.054	0.553	0.735	0.576
C_{14}	181.0	0.824	0.014	0.002	0.017	0.047	0.039	0.047	0.604	0.779	0.626
C_{15}	195.0	0.833	0.012	0.002	0.014	0.039	0.030	0.038	0.648	0.813	0.669
C_{16}	204.0	0.836	0.011	0.002	0.013	0.035	0.026	0.034	0.685	0.841	0.705
C_{17}	224.0	0.837	0.011	0.001	0.013	0.037	0.025	0.036	0.721	0.866	0.740
C_{18}	234.0	0.839	0.007	0.001	0.009	0.024	0.016	0.024	0.752	0.887	0.769
C_{19}	248.0	0.844	0.011	0.001	0.013	0.038	0.023	0.037	0.783	0.906	0.799
C_{20+}	362.0	0.877	0.059	0.005	0.067	0.198	0.083	0.182	0.901	0.959	0.909
Sum	—	—	0.298	0.057	0.368	1	1	1	—	—	—

MW, molecular weight; *SCN*, single carbon number; *SG*, specific gravity.

1. Assume that the molecular weight distribution function is as follows:

$$MW^* = \frac{MW - MW_0}{MW_0} = \left[\frac{A_M}{B_M}\ln\left(\frac{1}{x^*}\right)\right]^{\frac{1}{B_M}}$$

To obtain the molecular weight distribution function, the values of MW_0, A_M, and B_M should be calculated. Based on the values of the molecular weight and the cumulative mole fraction, first the values of MW* and x^* are calculated. Then X and Y are determined using the given equations in the problem. In calculating MW*, a value for MW_0 is needed. The first initial guess should be less than the molecular weight of the lightest fraction in the mixture (i.e., 91.9). Assume that the first initial guess is 88. C_1 and C_2 calculate from the linear regression. Then A and B calculate based on C_1 and C_2.

$$\left.\begin{aligned} C_2 &= \frac{(1.663 \times -4.351) - 14(12.295)}{1.663^2 - 14(10.111)} = 1.292 \\ C_1 &= \frac{-4.351 - 1.292(1.663)}{14} = -0.464 \end{aligned}\right\} \rightarrow \begin{cases} B_M = \dfrac{1}{1.292} = 0.540 \\ A_M = 0.540\exp(-0.464 \times 0.540) = 0.774 \end{cases}$$

Using trial and error (similar to Example 3.5), the adjusted value for MW_0 is 89. The results are presented in the following table.

			$MW_0 = 88, C_1 = -0.464, C_2 = 1.292$, $A_M = 0.540, B_M = 0.774$, RMS = 7.73					$MW_0 = 89, C_1 = -0.529, C_2 = 1.387$, $A_M = 0.492, B_M = 0.721, RMS = 5.94$				
MW	$x^* = 1 - x_{cmn}$	X	MW*	Y	X_iY_i	X_i^2	$MW_i^{cal.}$	MW*	Y	X_iY_i	X_i^2	$MW_i^{cal.}$
91.2	0.895	−2.194	0.036	−3.314	7.272	4.815	91.2	0.025	−3.700	8.119	4.815	91.5
104	0.688	−0.984	0.182	−1.705	1.677	0.968	103.5	0.169	−1.781	1.752	0.968	102.4
119	0.532	−0.459	0.352	−1.043	0.479	0.211	118.6	0.337	−1.087	0.499	0.211	116.7
133	0.432	−0.176	0.511	−0.671	0.118	0.031	132.1	0.494	−0.704	0.124	0.031	130.1
144	0.363	0.013	0.636	−0.452	−0.006	0.000	144.3	0.618	−0.481	−0.006	0.000	142.4
155	0.314	0.148	0.761	−0.273	−0.040	0.022	155.0	0.742	−0.299	−0.044	0.022	153.4
168	0.265	0.283	0.909	−0.095	−0.027	0.080	167.7	0.888	−0.119	−0.034	0.080	166.6
181	0.221	0.411	1.057	0.055	0.023	0.169	182.0	1.034	0.033	0.014	0.169	181.6
195	0.187	0.516	1.216	0.195	0.101	0.267	195.8	1.191	0.175	0.090	0.267	196.3
204	0.159	0.609	1.318	0.276	0.168	0.371	209.5	1.292	0.256	0.156	0.371	211.0
224	0.134	0.700	1.545	0.435	0.305	0.490	224.6	1.517	0.417	0.292	0.490	227.3
234	0.113	0.778	1.659	0.506	0.394	0.605	239.2	1.629	0.488	0.380	0.605	243.2
248	0.094	0.860	1.818	0.598	0.514	0.740	256.2	1.787	0.580	0.499	0.740	261.9
362	0.041	1.159	3.114	1.136	1.316	1.343	335.4	3.067	1.121	1.299	1.343	350.5
Sum	–	1.663	–	−4.351	12.295	10.111	–	–	−5.102	13.140	10.111	–

MW, molecular weight; *RMS*, root mean square error.

Based on the results the molecular weight distribution function is

$$MW = 89\left[1 + \left(0.683\ \ln\left(\frac{1}{1 - x_{cmn}}\right)\right)^{1.387}\right]$$

2. Assume the specific gravity distribution function given by the following relation:

$$SG^* = \frac{SG - SG_0}{SG_0} = \left[\frac{A_{SG}}{B_{SG}} \ln\left(\frac{1}{x^*}\right)\right]^{\frac{1}{B_{SG}}}$$

Similar to the previous example, SG_0, A_{SG}, and B_{SG} should be calculated. The initial guess for SG_0 should be less than the specific gravity of the C_7 fraction. Take 0.700 for the initial guess of SG_0. The adjusted value for SG_0 is equal to 0.719 and the specific gravity distribution function is

$$SG = 0.719\left[1 + \left(0.020\ \ln\left(\frac{1}{1 - x_{cvn}}\right)\right)^{0.498}\right]$$

The results are given in the following table.

3. Estimate the boiling point using Eq. (3.33). Use the calculated boiling point to choose a suitable initial guess for T_{b0}. The initial guess must be less than 359; choose 350 as an initial guess for T_{b0}. The adjusted value for T_{b0} is 340. The results are shown in the following table.

SG	$x^* = 1 - x_{cvn}$	X	$SG_0 = 0.700, C_1 = -1.750, C_2 = 0.380$ $A_{SG} = 0.026, B_{SG} = 2.630, RMS = 0.005$					$SG_0 = 0.719, C_1 = -1.958, C_2 = 0.498$ $A_{SG} = 0.039, B_{SG} = 2.008, RMS = 0.004$				
			SG*	Y	X_iY_i	X_i^2	$SG_i^{cal.}$	SG*	Y	X_iY_i	X_i^2	$SG_i^{cal.}$
0.746	0.932	−2.650	0.066	−2.722	7.215	7.02	0.744	0.037	−3.297	8.739	7.02	0.747
0.77	0.791	−1.452	0.100	−2.303	3.344	2.109	0.770	0.070	−2.654	3.855	2.109	0.769
0.788	0.675	−0.933	0.126	−2.074	1.934	0.870	0.785	0.095	−2.350	2.192	0.870	0.783
0.795	0.591	−0.641	0.136	−1.997	1.281	0.411	0.795	0.105	−2.253	1.445	0.411	0.793
0.79	0.526	−0.441	0.129	−2.051	0.905	0.195	0.803	0.098	−2.321	1.025	0.195	0.801
0.802	0.476	−0.297	0.146	−1.926	0.573	0.088	0.809	0.115	−2.164	0.644	0.088	0.807
0.814	0.424	−0.153	0.163	−1.815	0.278	0.023	0.815	0.132	−2.029	0.311	0.023	0.813
0.824	0.374	−0.016	0.177	−1.731	0.027	0.000	0.821	0.145	−1.928	0.030	0.000	0.820
0.833	0.331	0.099	0.190	−1.661	−0.165	0.010	0.826	0.158	−1.846	−0.183	0.010	0.826
0.836	0.295	0.199	0.194	−1.638	−0.325	0.039	0.831	0.162	−1.820	−0.361	0.039	0.831
0.837	0.260	0.297	0.196	−1.631	−0.484	0.088	0.836	0.163	−1.811	−0.538	0.088	0.837
0.839	0.231	0.383	0.199	−1.617	−0.619	0.146	0.841	0.166	−1.794	−0.687	0.146	0.842
0.844	0.201	0.474	0.206	−1.581	−0.749	0.224	0.846	0.173	−1.753	−0.830	0.224	0.848
0.877	0.091	0.873	0.253	−1.375	−1.200	0.762	0.870	0.219	−1.518	−1.326	0.762	0.876
Sum		−4.260		−26.122	12.015	11.992			−29.540	14.316	11.992	

RMS, root mean square error; *SG*, specific gravity.

			$T_{b0} = 350, C_1 = -0.761, C_2 = 1.012$ $A_{T_b} = 0.466, B_{T_b} = 0.988$, RMS = 13.29					$T_{b0} = 340, C_1 = -0.668, C_2 = 0.830$ $A_{T_b} = 0.539, B_{T_b} = 1.204$, *RMS* = 2.82				
T_b Eq. (3.33)	$x^* = 1 - x_{cwn}$	X	T_b^*	Y	X_iY_i	X_i^2	$T_{bi}^{cal.}$	T_b^*	Y	X_iY_i	X_i^2	$T_{bi}^{cal.}$
359.0	0.937	−2.734	0.026	−3.656	9.995	7.47	360.3	0.056	−2.882	7.879	7.47	358.0
385.6	0.805	−1.530	0.102	−2.286	3.498	2.342	384.7	0.134	−2.009	3.075	2.342	388.9
413.4	0.693	−1.004	0.181	−1.708	1.716	1.009	409.2	0.216	−1.533	1.540	1.009	415.7
436.1	0.611	−0.708	0.246	−1.402	0.993	0.501	429.9	0.283	−1.263	0.894	0.501	436.8
451.4	0.547	−0.506	0.290	−1.239	0.627	0.256	447.9	0.328	−1.116	0.565	0.256	454.5
468.9	0.498	−0.361	0.340	−1.079	0.390	0.131	463.4	0.379	−0.970	0.350	0.131	469.1
488.5	0.447	−0.215	0.396	−0.927	0.200	0.046	481.5	0.437	−0.829	0.178	0.046	485.8
506.8	0.396	−0.075	0.448	−0.803	0.061	0.006	501.5	0.491	−0.712	0.054	0.006	503.7
525.3	0.352	0.042	0.501	−0.691	−0.029	0.002	520.7	0.545	−0.607	−0.026	0.002	520.5
536.4	0.315	0.144	0.533	−0.630	−0.091	0.021	539.2	0.578	−0.549	−0.079	0.021	536.5
559.4	0.279	0.244	0.598	−0.513	−0.126	0.060	559.4	0.645	−0.438	−0.107	0.060	553.5
570.7	0.248	0.332	0.631	−0.461	−0.153	0.110	578.7	0.678	−0.388	−0.129	0.110	569.6
586.0	0.217	0.424	0.674	−0.394	−0.167	0.180	601.1	0.723	−0.324	−0.137	0.180	587.9
687.6	0.099	0.839	0.965	−0.036	−0.030	0.703	732.0	1.022	0.022	0.019	0.703	689.8
		−5.110		−15.827	16.883	12.840			−13.597	14.077	12.840	

RMS, root mean square error.

Table 3.16 Comparing Different Methods for Predicting the Molecular Weight of Petroleum Fractions

Method	Equation	AAD%
Riazi–Daubert	(3.75)	3.9
Lee–Kesler	(3.76)	8.2
Winn–Mobil (Sim–Daubert)	(3.77)	5.4
Twu	(3.78)	5.0

AAD%, average absolute deviation percent.

Table 3.17 Comparing Different Methods for Estimating the Critical Temperature and Critical Pressure

Method	Equation	AAD%	
		T_c	P_c
Lee–Kesler	(3.36) and (3.37)	0.7	4
Cavett	(3.41) and (3.42)	3.0	5.5
Riazi–Daubert	(3.44) and (3.47)	1.1	3.1
Twu	(3.60) and (3.66)	0.6	3.9
Winn–Mobil (Sim–Daubert)	(3.69) and (3.70)	1.0	4.5

AAD%, average absolute deviation percent.

The boiling point distribution function is

$$T_b = 340\left[1 + \left(0.447 \ln\left(\frac{1}{1 - x_{cwn}}\right)\right)^{0.830}\right]$$

Riazi and Daubert (1987) evaluates different methods for estimating the molecular weight of the petroleum fraction for 625 fractions from the Penn State database on petroleum fractions. The AAD% for different methods is given in Table 3.16.

In addition, they evaluated different correlations for estimating the critical temperature and critical pressure for 138 hydrocarbons from different families. The AAD% of different methods are reported in Table 3.17.

3.5 RECOMMENDED PLUS FRACTION CHARACTERIZATION PROCEDURE

1. Use $\alpha = 1$ and calculate the mole fraction of each SCN group using Eq. (3.27). The plus fraction is extended into 45 SCN groups. Specify the upper molecular weight boundary using the midpoint method.

2. Assign the molecular weight for each SCN group from Table 3.9. In some cases it is better to calculate the molecular weight of each SCN group using Eq. (3.28).
3. Calculate the mole fraction and molecular weight of the C_{45+} fraction using the following equations:

$$z_{C_{45+}} = z_{C_{7+}} - \sum_{n=7}^{44} z_{C_n} \tag{3.87}$$

$$MW_{C_{45+}} = \frac{z_{C_{7+}} MW_{C_{7+}} - \sum_{C_7}^{C_{44}} z_{C_n} MW_{C_n}}{z_{C_{45+}}} \tag{3.88}$$

4. Assume a constant Watson characterization factor. Determine the Watson characterization factor from Eq. (3.85) and use Eq. (3.82) to calculate the specific gravity of each SCN group.
5. Assign the boiling point for each SCN group from Table 3.9, and estimate the boiling point of C_{45+} using Eq. (3.34).
6. Predict the critical temperature and critical pressure of each SCN group using Eqs. (3.46) and (3.49), respectively. Then estimate the acentric factor from Eqs. (3.38) and (3.39).

Example 3.12

The molar composition of an oil sample is shown in Table 3.18 (Pedersen et al., 1992).

The molecular weight and specific gravity of the C_{7+} fraction are 211.5 and 0.846, respectively. Characterize the heavy fraction using the recommended procedure.

Table 3.18 Molar Composition of an Oil Sample (Example 3.12)

Component	Mol%
N_2	0.69
CO_2	0.12
C_1	47.09
C_2	5.69
C_3	4.39
i-C_4	0.95
n-C_4	2.42
i-C_5	1.11
n-C_5	1.46
C_6	2.26
C_{7+}	33.82

Solution

Set $\alpha = 1$ and estimate η using Eq. (3.19).

$$\eta = 110\left[1 - \frac{1}{1 + \frac{4.043}{1^{0.723}}}\right] = 88.2$$

The β parameter is determined using Eq. (3.15) equal to 123.3. The normalized mole fraction is determined using Eq. (3.27) as follows:

$$x_{C_n} = -\exp\left(\frac{88.2}{123.3}\right)\left[\exp\left(-\frac{MW_n}{123.3}\right) - \exp\left(-\frac{MW_{n-1}}{123.3}\right)\right]$$

The results of steps 1 to 3 are reported in the following table. The value of $z_{C_n} MW_{C_n}\left(4.5579 MW_{C_n}^{0.15178}\right)^{\frac{1}{0.84573}}$ is reported in the last column.

SCN	Upper Molecular Weight Boundary MW_n	Lower Molecular Weight Boundary MW_{n-1}	MW_{C_n} (Table 3.9)	x_{C_n} Eq. (3.27)	$z_{C_n} = x_{C_n} \times 0.3382$	Parameter of Eq. (3.86)
C_7	101.5	$88.2 = \eta$	96	0.1023	0.0346	0.2436
C_8	114	101.5	108	0.0865	0.0293	0.2269
C_9	127.5	114	121	0.0841	0.0284	0.2421
C_{10}	140.5	127.5	134	0.0727	0.0246	0.2277
C_{11}	154	140.5	147	0.0678	0.0229	0.2291
C_{12}	168	154	161	0.0629	0.0213	0.2290
C_{13}	182.5	168	175	0.0581	0.0196	0.2263
C_{14}	198	182.5	190	0.0550	0.0186	0.2292
C_{15}	214	198	206	0.0499	0.0169	0.2225
C_{16}	229.5	214	221	0.0426	0.0144	0.2010
C_{17}	244	229.5	236	0.0353	0.0119	0.1757
C_{18}	257	244	250	0.0283	0.0096	0.1477
C_{19}	269	257	263	0.0236	0.0080	0.1284
C_{20}	283	269	276	0.0248	0.0084	0.1403
C_{21}	295.5	283	289	0.0199	0.0067	0.1168
C_{22}	306	295.5	301	0.0152	0.0051	0.0924
C_{23}	318	306	312	0.0159	0.0054	0.0993
C_{24}	330.5	318	324	0.0150	0.0051	0.0966
C_{25}	343	330.5	337	0.0135	0.0046	0.0902
C_{26}	354.5	343	349	0.0113	0.0038	0.0774
C_{27}	366	354.5	360	0.0103	0.0035	0.0724
C_{28}	377	366	371	0.0090	0.0030	0.0648
C_{29}	388	377	382	0.0082	0.0028	0.0607
C_{30}	399	388	393	0.0075	0.0025	0.0568

(Continued)

C_{31}	409.5	399	404	0.0066	0.0022	0.0508
C_{32}	420.5	409.5	415	0.0063	0.0021	0.0499
C_{33}	431.5	420.5	426	0.0058	0.0020	0.0466
C_{34}	441	431.5	436	0.0046	0.0015	0.0378
C_{35}	450.5	441	446	0.0042	0.0014	0.0356
C_{36}	460	450.5	455	0.0039	0.0013	0.0335
C_{37}	469.5	460	465	0.0036	0.0012	0.0316
C_{38}	479.5	469.5	474	0.0035	0.0012	0.0312
C_{39}	489.5	479.5	484	0.0033	0.0011	0.0293
C_{40}	498.5	489.5	494	0.0027	0.0009	0.0248
C_{41}	507	498.5	503	0.0024	0.0008	0.0222
C_{42}	516.5	507	512	0.0025	0.0008	0.0234
C_{43}	526	516.5	521	0.0023	0.0008	0.0219
C_{44}	535	526	530	0.0020	0.0007	0.0196
C_{45+}	–	–	653	0.0267	0.0090	0.3066
			Eq. (3.88)	Eq. (3.87)		
Sum						$\zeta = 4.462$

SCN, single carbon number.

The Watson characterization factor is found by Eq. (3.85).

$$K_w = \left[\frac{4.462 \times 0.846}{0.3382 \times 211.5}\right]^{-0.84573} = 12.04$$

Use the calculated Watson characterization factor and Eq. (3.82) to determine the specific gravity of each SCN group. Assign the boiling point for each SCN group from Table 3.9, and predict the boiling point of the C_{45+} fraction using Eq. (3.34). Then estimate the critical temperature of each SCN group using Eqs. (3.46) and (3.49), respectively, and estimate the acentric factor from Eqs. (3.38) and (3.39). The results of steps 4 to 6 are reported in the following table.

SCN	**T_b (K) (Table 3.1)**	**Specific Gravity Eq. (3.82)**	**T_c (K) Eq. (3.46)**	**P_c (MPa) Eq. (3.49)**	**$T_{br} = \frac{T_b}{T_c}$**	**ω Eqs. (3.38) and (3.39)**
C_7	366	0.719	546.68	3.06	0.669	0.2786
C_8	390	0.735	572.41	2.81	0.681	0.3175
C_9	416	0.750	599.22	2.56	0.694	0.3631
C_{10}	439	0.764	622.56	2.37	0.705	0.4050
C_{11}	461	0.777	644.33	2.21	0.715	0.4476
C_{12}	482	0.789	664.89	2.08	0.725	0.4902
C_{13}	501	0.801	683.29	1.97	0.733	0.5309
C_{14}	520	0.813	701.43	1.88	0.741	0.5744
C_{15}	539	0.825	719.30	1.79	0.749	0.6209

—cont'd

SCN	T_b (K) (Table 3.1)	Specific Gravity Eq. (3.82)	T_c (K) Eq. (3.46)	P_c (MPa) Eq. (3.49)	$T_{br} = \frac{T_b}{T_c}$	ω Eqs. (3.38) and (3.39)
C_{16}	557	0.836	735.80	1.72	0.757	0.6684
C_{17}	573	0.845	750.45	1.66	0.764	0.7133
C_{18}	586	0.854	762.42	1.63	0.769	0.7519
C_{19}	598	0.862	773.31	1.59	0.773	0.7896
C_{20}	612	0.870	785.49	1.55	0.779	0.8358
C_{21}	624	0.877	796.04	1.52	0.784	0.8777
C_{22}	637	0.883	807.04	1.48	0.789	0.9249
C_{23}	648	0.889	816.34	1.45	0.794	0.9669
C_{24}	659	0.895	825.68	1.43	0.798	1.0111
C_{25}	671	0.901	835.73	1.40	0.803	0.9599
C_{26}	681	0.907	844.16	1.38	0.807	0.9852
C_{27}	691	0.912	852.38	1.36	0.811	1.0108
C_{28}	701	0.917	860.54	1.34	0.815	1.0365
C_{29}	709	0.922	867.25	1.33	0.818	1.0562
C_{30}	719	0.926	875.28	1.31	0.821	1.0818
C_{31}	728	0.931	882.56	1.30	0.825	1.1045
C_{32}	737	0.936	889.78	1.28	0.828	1.1271
C_{33}	745	0.940	896.28	1.28	0.831	1.1467
C_{34}	753	0.944	902.63	1.26	0.834	1.1667
C_{35}	760	0.948	908.28	1.26	0.837	1.1837
C_{36}	768	0.951	914.46	1.24	0.840	1.2040
C_{37}	774	0.955	919.38	1.24	0.842	1.2181
C_{38}	782	0.958	925.50	1.23	0.845	1.2382
C_{39}	788	0.962	930.36	1.23	0.847	1.2523
C_{40}	796	0.965	936.51	1.22	0.850	1.2721
C_{41}	801	0.968	940.58	1.22	0.852	1.2836
C_{42}	807	0.972	945.28	1.22	0.854	1.2980
C_{43}	813	0.975	949.96	1.21	0.856	1.3123
C_{44}	821	0.978	955.92	1.20	0.859	1.3322
C_{45+}	849 Eq. (3.34)	1.015	983.25	1.39	0.864	1.3786

SCN, single carbon number.

Problems

3.1 Consider the gas condensates in Table 3.11. Calculate the weight fraction for all of the components present in the mixture.

3.2 The following table shows an example of true boiling point results (Pedersen et al., 2014). Use the Maxwell and Bonnell (1955) correlation to convert the boiling point at the subatmospheric pressure to the normal boiling point.

Fraction	Actual Temperature (°C)	Density (g/cm^3)	MW	Weight %	Cumulative Weight %
P = 1.01 bar					
Gas	–	–	33.5	0.064	0.064
$<C_6$	36.5	0.598	62.5	3.956	4.020
C_6	69.2	0.685	82.0	2.016	6.036
C_7	98.9	0.737	98.9	6.125	12.161
C_8	126.1	0.754	126.1	4.606	16.767
C_9	151.3	0.774	151.3	5.046	21.813
P = 26.6 mbar					
C_{10}	70.9	0.789	134.7	4.020	25.833
C_{11}	88.7	0.794	150.3	3.953	29.786
C_{12}	105.7	0.806	166.4	4.061	33.847
C_{13}	121.8	0.819	181.4	3.800	37.647
C_{14}	136.9	0.832	194.0	4.421	42.068
C_{15}	151.2	0.834	209.4	3.765	45.833
C_{16}	164.3	0.844	222.4	2.969	48.802
C_{17}	178	0.841	240.9	3.800	52.602
C_{18}	191	0.847	256.0	2.813	55.415
C_{19}	203	0.86	268.2	3.364	58.779
P = 2.66 mbar					
C_{20}	161	0.874	269.4	1.115	59.894
C_{21}	172	0.87	282.5	2.953	62.847
C_{22}	181	0.872	297.7	2.061	64.908
C_{23}	191	0.875	310.1	1.797	66.705
C_{24}	199	0.877	321.8	1.421	68.126
C_{25}	208	0.881	332.4	2.083	70.209
C_{26}	217	0.886	351.1	1.781	71.990
C_{27}	226	0.888	370.8	1.494	73.484
C_{28}	234	0.895	381.6	1.625	75.109
C_{29}	241	0.898	393.7	1.233	76.342
C_{30+}	-	0.935	612.0	23.658	100.000

MW, molecular weight.

The Maxwell and Bonnell correlation is as follows (Maxwell and Bonnell, 1955; Riazi, 2005):

$$T_b' = \frac{748.1QT}{1 + T(0.3861Q - 0.00051606)}$$

$$Q = \frac{6.761560 - 0.987672 \log_{10}P}{3000.538 - 43.00 \log_{10}P} \quad (P < 2 \ \text{mmHg})$$

$$Q = \frac{5.994296 - 0.972546 \log_{10}P}{2663.129 - 95.76 \log_{10}P} \quad (2 \leq P \leq 760 \quad \text{mmHg})$$

$$T_b = T_b' + 1.3889F(K_W - 12)\log_{10}\frac{P}{760}$$

$$F = -3.2985 + 0.009T_b$$

where P is the pressure at which the distillation data is available in mmHg; T is the boiling point originally available at pressure P in Kelvin; T_b' is the normal boiling point corrected to $K_W = 12$ in Kelvin; T_b is the normal boiling point in Kelvin; and F is the correction factor for the fraction with a K_W different from 12.

Hint: The Watson characterization factor can be estimated from an estimated value of T_b'.

3.3 Consider the extended composition data for the gas condensate in Table 3.10. Extend the analysis to C_{30+} using the Pedersen splitting method.

3.4 Consider the extended composition data for the North Sea black oil in Table 3.10. Describe the plus fraction by a gamma distribution function (optimized by the parameters of the distribution function).

3.5 Predict the molecular weight of n-Heneicosane using the Twu and Riazi–Daubert correlations. The normal boiling point, molecular weight, and specific gravity of n-Heneicosane are 356.5°C, 296.6, and 0.7954, respectively (data taken from the API technical data book Daubert and Danner, 1997).

3.6 Predict the critical properties and acentric factor for 2-Methylhexane by following a set of T_c–P_c–ω correlations, and then obtain the AAD% for each set of equations.

Set 1: T_c–P_c; Twu method; ω-Edmister method

Set 2: T_c–P_c; Riazi–Daubert method; ω-Korsten method

The normal boiling point and critical properties from the API technical data book are as follows (Daubert and Danner, 1997): $T_b = 363.25\text{K}$ (normal boiling point), $SG = 0.7954$, $T_c = 530.37\text{K}$, $P_c = 2.73$ MPa, and $\omega = 0.3277$.

3.7 Obtain the molecular weight distribution function in the form that is presented in Example 3.11 for the North Sea gas condensates in Table 3.11.

3.8 The molar composition of a gas condensate from Iran is given in the following table (Firoozabadi et al., 1978).

Component	Mol%
N_2	0.08
CO_2	2.44
C_1	82.10
C_2	5.78
C_3	2.87
i-C_4	0.56
n-C_4	1.23
i-C_5	0.52
n-C_5	0.60
C_6	0.72
C_{7+}	3.10

The molecular weight and specific gravity of the C_{7+} fraction are 132 and 0.774, respectively. Characterize the C_{7+} fraction using the recommended procedure.

REFERENCES

Abramowitz, M., Stegun, I.A., 1966. Handbook of mathematical functions. Applied Mathematics Series 55, 62.

Aguilar Zurita, R.A., McCain Jr., W.D., 2002. An efficient tuning strategy to calibrate cubic EOS for compositional simulation. In: SPE Annual Technical Conference and Exhibition, Society of Petroleum Engineers.

Ahmed, T., Cady, G., Story, A., 1985. A generalized correlation for characterizing the hydrocarbon heavy fractions. In: SPE Annual Technical Conference and Exhibition, Society of Petroleum Engineers.

Ahmed, T.H., 1989. Hydrocarbon Phase Behavior. Gulf Publishing Company.

Al-Meshari, A.A., 2005. New Strategic Method to Tune Equation-of-state to Match Experimental Data for Compositional Simulation. Texas A&M University.

Aladwani, H., Riazi, M., 2005. Some guidelines for choosing a characterization method for petroleum fractions in process simulators. Chemical Engineering Research and Design 83 (2), 160–166.

Austad, T., Hvidsten, J., Norvik, H., Whitson, C., 1983. Practical aspects of characterizing petroleum fluids. In: Conference on North Sea Condensate Reservoirs and Their Development, London.

Benmekki, E., Mansoori, G., 1989. Pseudoization technique and heavy fraction characterization with equation of state models. Advanced Thermodynamics 1, 57–78.

Burle, M., Kumar, K., Watanasiri, S., 1985. Characterization methods improve phase-behavior predictions. Oil & Gas Journal (United States) 83 (6).

Campbell, J.M., 1984. Gas conditioning and processing: advanced techniques and applications. In: Campbell Petroleum Series.
Cavett, R.H., 1962. Physical data for distillation calculations. In: Vapor-liquid Equilibria Proceedings.
Chorn, L., 1984. Simulated distillation of petroleum crude oil by gas chromatography. Characterizing the heptanes-plus fraction. Journal of Chromatographic Science 22 (1), 17–21.
Danesh, A., 1998. PVT and Phase Behaviour of Petroleum Reservoir Fluids. Elsevier.
Daubert, T.E., Danner, R.P., 1997. API Technical Data Book-petroleum Refining. American Petroleum Institute (API), Washington, DC.
Edmister, W., 1958. Applied hydrocarbon thermodynamics, part 4: compressibility factors and equation of state. Petroleum Refiner 37 (8), 173–179.
Firoozabadi, A., Hekim, Y., Katz, D.L., 1978. Reservoir depletion calculations for gas condensates using extended analyses in the Peng-Robinson equation of state. The Canadian Journal of Chemical Engineering 56 (5), 610–615.
Hall, K., Yarborough, L., 1971. New, simple correlation for predicting critical volume. Chemical Engineering 78 (25), 76.
Hoffman, A., Crump, J., Hocott, C., 1953. Equilibrium constants for a gas-condensate system. Journal of Petroleum Technology 5 (01), 1–10.
Katz, D., Firoozabadi, A., 1978. Predicting phase behavior of condensate/crude-oil systems using methane interaction coefficients. Journal of Petroleum Technology 30 (11), 1649–1655.
Katz, D.L., 1983. Overview of phase behavior in oil and gas production. Journal of Petroleum Technology 35 (06), 1205–1214.
Katz, D.L.V., 1959. Handbook of Natural Gas Engineering. McGraw-Hill.
Kesler, M.G., Lee, B.I., 1976. Improve prediction of enthalpy of fractions. Hydrocarbon Processing 55 (3), 153–158.
Korsten, H., 2000. Internally consistent prediction of vapor pressure and related properties. Industrial & Engineering Chemistry Research 39 (3), 813–820.
Lee, B.I., Kesler, M.G., 1975. A generalized thermodynamic correlation based on three-parameter corresponding states. AIChE Journal 21 (3), 510–527.
Luo, P., Wang, X., Gu, Y., 2010. Characterization of asphaltenes precipitated with three light alkanes under different experimental conditions. Fluid Phase Equilibria 291 (2), 103–110.
MacAllister, D., DeRuiter, R., 1985. Further development and application of simulated distillation for enhanced oil recovery. In: SPE Annual Technical Conference and Exhibition, Society of Petroleum Engineers.
Maxwell, J., Bonnell, L., 1955. Vapor Pressure Charts for Petroleum Hydrocarbons. Esso Research and Engineering Company.
Organick, E., Golding, B., 1952. Prediction of saturation pressures for condensate-gas and volatile-oil mixtures. Journal of Petroleum Technology 4 (05), 135–148.
Osjord, E., Rønningsen, H., Tau, L., 1985. Distribution of weight, density, and molecular weight in crude oil derived from computerized capillary GC analysis. Journal of High Resolution Chromatography 8 (10), 683–690.
Osjord, E.H., Malthe-Sørenssen, D., 1983. Quantitative analysis of natural gas in a single run by the use of packed and capillary columns. Journal of Chromatography A 279, 219–224.
Pedersen, K., Thomassen, P., Fredenslund, A., 1983. SRK-EOS calculation for crude oils. Fluid Phase Equilibria 14, 209–218.
Pedersen, K.S., Blilie, A.L., Meisingset, K.K., 1992. PVT calculations on petroleum reservoir fluids using measured and estimated compositional data for the plus fraction. Industrial & Engineering Chemistry Research 31 (5), 1378–1384.
Pedersen, K.S., Christensen, P.L., Shaikh, J.A., 2014. Phase Behavior of Petroleum Reservoir Fluids. CRC Press.

Pedersen, K.S., Fredenslund, A., Thomassen, P., 1989. Properties of Oils and Natural Gases. Gulf Pub. Co., Book Division.

Pedersen, K.S., Thomassen, P., Fredenslund, A., 1984. Thermodynamics of petroleum mixtures containing heavy hydrocarbons. 1. Phase envelope calculations by use of the Soave-Redlich-Kwong equation of state. Industrial & Engineering Chemistry Process Design and Development 23 (1), 163–170.

Pedersen, K.S., Thomassen, P., Fredenslund, A., 1985. Thermodynamics of petroleum mixtures containing heavy hydrocarbons. 3. Efficient flash calculation procedures using the SRK equation of state. Industrial & Engineering Chemistry Process Design and Development 24 (4), 948–954.

Riazi, M.-R., 1979. Prediction of Thermophysical Properties of Petroleum Fractions (Ph.D. thesis).

Riazi, M., 2003. Improved phase behavior calculations of reservoir fluids using a new distribution model. Scientia Iranica 10 (3), 341–345.

Riazi, M., 2005. Characterization and Properties of Petroleum Fractions. ASTM International.

Riazi, M., Daubert, T., 1980. Simplify property predictions. Hydrocarbon Processing 60 (3), 115–116.

Riazi, M.R., 1989. Distribution model for properties of hydrocarbon-plus fractions. Industrial & Engineering Chemistry Research 28 (11), 1731–1735.

Riazi, M.R., 1997. A continuous model for C_{7+} fraction characterization of petroleum fluids. Industrial & Engineering Chemistry Research 36 (10), 4299–4307.

Riazi, M.R., Al-Sahhaf, T.A., 1996. Physical properties of heavy petroleum fractions and crude oils. Fluid Phase Equilibria 117 (1), 217–224.

Riazi, M.R., Daubert, T.E., 1986. Analytical correlations interconvert distillation-curve types. Oil & Gas Journal 84 (34), 51–57.

Riazi, M.R., Daubert, T.E., 1987. Characterization parameters for petroleum fractions. Industrial & Engineering Chemistry Research 26 (4), 755–759.

Robinson, D.B., Peng, D.-Y., 1978. The Characterization of the Heptanes and Heavier Fractions for the GPA Peng-Robinson Programs. Gas Processors Association.

Roenningsen, H.P., Skjevrak, I., Osjord, E., 1989. Characterization of North Sea petroleum fractions: hydrocarbon group types, density and molecular weight. Energy & Fuels 3 (6), 744–755.

Sim, W.J., Daubert, T.E., 1980. Prediction of vapor-liquid equilibria of undefined mixtures. Industrial & Engineering Chemistry Process Design and Development 19 (3), 386–393.

Soreide, I., 1989. Improved Phase Behavior Predictions of Petroleum Reservoir Fluids from Cubic Equation of State. (Dr. Eng. dissertation). Norwegian Institute of Technology, Trondheim, Norway.

Twu, C.H., 1984. An internally consistent correlation for predicting the critical properties and molecular weights of petroleum and coal-tar liquids. Fluid Phase Equilibria 16 (2), 137–150.

Watson, K., Nelson, E., Murphy, G.B., 1935. Characterization of petroleum fractions. Industrial & Engineering Chemistry 27 (12), 1460–1464.

Wauquier, J.-P., 1995. Petroleum refining. In: Crude Oil, Petroleum Products, Process Flowsheets, vol. 1. Editions Technip, Paris.

Whitson, C.H., 1983. Characterizing hydrocarbon plus fractions. Society of Petroleum Engineers Journal 23 (04), 683–694.

Whitson, C.H., Anderson, T.F., Søreide, I., 1990. Application of the gamma distribution model to molecular weight and boiling point data for petroleum fractions. Chemical Engineering Communications 96 (1), 259–278.

Whitson, C.H., Brulé, M.R., 2000. Phase behavior. In: Henry L. Doherty Memorial Fund of AIME, Society of Petroleum Engineers.

Yarborough, L., 1979. Application of a Generalized Equation of State to Petroleum Reservoir Fluids.

CHAPTER FOUR

Tuning Equations of State

M. Mesbah[1], A. Bahadori[2,3]
[1]Sharif University of Technology, Tehran, Iran
[2]Southern Cross University, Lismore, NSW, Australia
[3]Australian Oil and Gas Services Pty Ltd, Lismore, NSW, Australia

Equations of state (EOSs) are most widely used in phase behavior predictions. EOSs could be used in phase equilibrium calculations, especially for mixtures containing heavy hydrocarbons. Unfortunately, the prediction using EOSs may not be accurate. For example, the errors in saturation are commonly about ±10% or ±5% in density (Whitson and Brulé, 2000). EOSs also predict a bubble point pressure instead of a dew point pressure at reservoir conditions, or vice versa. The poor capability in the prediction of properties by EOSs may be raised from errors in determining heavy fraction properties, insufficient data for heavy fraction, usage of unsuitable value for the binary interaction parameter, or inaccurate determination of overall composition.

Some authors (Whitson, 1984; Coats, 1985; Coats and Smart, 1986; Agarwal et al., 1987; Pedersen et al. 1988a,b; Soreide, 1989; Aguilar Zurita and McCain Jr, 2002) give procedures for improving the EOS characterization. At first the experimental data and fluid composition should be checked. If the experimental data and fluid composition appear reliable, adjusting the parameters of EOSs is necessary. Some references presented the methods for modifying the cubic EOS. These methods usually modify the properties of plus fraction including critical temperature, acentric factor, binary interaction parameters between methane and plus fraction, or constant parameters of EOSs. The binary interaction parameters between plus fraction and nonhydrocarbon components may be chosen as a tuning parameter when the injection gas contains considerable amounts of nonhydrocarbons.

Based on the proposed procedure by Aguilar and McCain, tuning of EOS could be divided into the following steps (Aguilar Zurita and McCain Jr, 2002; Al-Meshari, 2005):

- Splitting plus fraction up to single carbon number (SCN) 45;
- Assigning critical temperature, critical pressure, and acentric factor for extended groups;
- Matching the saturation pressure using extended groups;

Fluid Phase Behavior for Conventional and Unconventional Oil and Gas Reservoirs
http://dx.doi.org/10.1016/B978-0-12-803437-8.00004-X

- Grouping SCN to multiple carbon number (MCN) group;
- Assigning critical temperature, critical pressure, and acentric factor for each MCN group;
- Matching the saturation pressure using grouped composition;
- Matching volumetric data.

The first two steps have been discussed in the previous chapter. Remaining steps are discussed in this chapter.

4.1 MATCHING THE SATURATION PRESSURE USING THE EXTENDED GROUPS

The prediction of gas–liquid phase behavior strongly depends on the accuracy of inputs that are used for pressure–volume–temperature simulation. For example, if Peng–Robinson EOS (PR EOS) is used for the prediction of saturation pressure (bubble point pressure or dew point pressure), critical properties, molecular weights, binary interaction coefficients, and mole fractions are provided as inputs. Properties of well-defined components (H_2S, N_2, CO_2, C_1, C_2, C_3, i-C_4, n-C_4, i-C_5, n-C_5, and n-C_6) are available, and it is not reasonable that these properties are selected as tuning parameters. However, the properties of plus fraction are estimated from empirical correlation, and it is possible that these correlations have poor accuracy. In addition, the calculation of the molecular weight of plus fraction approximately contains 5–20% of experimental errors (Pedersen et al., 1989; Aguilar Zurita and McCain Jr, 2002; Al-Meshari, 2005). It is recommended to choose the properties of plus fraction as adjustable parameters rather than adjusting the parameter of EOS. Several authors have proposed the selection of molecular weight of plus fraction as an adjustable parameter (Thomassen et al., 1987; Guo and Du, 1989; Pedersen et al., 1989; Wang, 1989; Aguilar Zurita and McCain Jr, 2002; Al-Meshari, 2005).

Laboratory analysis measures the weight fraction. EOS needs mole fractions as inputs. If the mole fraction of the plus fraction is changed, the molecular weight of the mixture also changes and the mole fraction of all components must be recalculated. Molecular weight of well-defined components is not changed during tuning process and just the molecular weight of plus fraction is changed. Therefore modifying the molecular weight of mixture has the same effect as modifying plus fraction. Aguilar Zurita and

McCain Jr (2002) proposed a strategy for matching saturation pressure. The proposed approach for calculating saturation pressure is based on a slightly different approach proposed by Aguilar and McCain.

1. Calculate the molecular weight of the mixture using Eq. (4.1).

$$\mathrm{MW}_{\mathrm{mix.}} = \sum_i z_i \mathrm{MW}_i \tag{4.1}$$

2. Determine the weight fraction for each component using the reported mole fractions. Calculated weight fraction does not change by modifying the molecular weight of the mixture.

$$w_i = \frac{z_i \mathrm{MW}_i}{\mathrm{MW}_{\mathrm{mix.}}} \tag{4.2}$$

3. Characterize heavy fraction using procedure that has been described in Chapter 3.
4. Calculate the saturation pressure by PR EOS using the extended groups.
5. If the saturation pressure does not match, modify the molecular weight of the mixture and recalculate the mole fraction of all components except plus fraction using the following equation.

$$z_i = \frac{w_i \mathrm{MW}_{\mathrm{mix.}}}{\mathrm{MW}_i} \tag{4.3}$$

6. Then the mole fraction and molecular weight of plus fraction are determined as follows (assuming plus fraction is grouped as heptane plus).

$$z_{C_{7+}} = 1 - \sum_{i=1}^{N-1} z_i \tag{4.4}$$

$$\mathrm{MW}_{C_{7+}} = \frac{\mathrm{MW}_{\mathrm{mix.}} - \sum_{i=1}^{N-1} z_i \mathrm{MW}_i}{z_{C_{7+}}} \tag{4.5}$$

Summation of Eqs. (4.4) and (4.5) included all components except the plus fraction (i.e., H_2S, N_2, CO_2, C_1, C_2, C_3, i-C_4, n-C_4, i-C_5, n-C_5, and n-C_6).

7. Repeat steps 3–6 until the saturation pressure is matched.

It is obvious that this procedure can be applied for other EOSs. The continuation of this section, a brief discussion on the calculation of the saturation pressure by PR EOS, has been represented.

Peng and Robinson (1976) proposed their EOS in the following form:

$$P = \frac{RT}{V-b} - \frac{a}{V(V+b)+b(V-b)} \tag{4.6}$$

where

$$a = \alpha a_c \tag{4.7}$$

$$a_c = 0.45724\frac{R^2T_c^2}{P_c} \tag{4.8}$$

$$b = 0.07780\frac{RT_c}{P_c} \tag{4.9}$$

α is defined by Eqs. (4.10) and (4.11).

$$\alpha = \left[1+\kappa\left(1-\left(\frac{T}{T_c}\right)^{0.5}\right)\right]^2 \tag{4.10}$$

$$\kappa = 0.37464 + 1.54226\omega - 0.26992\omega^2 \tag{4.11}$$

Robinson and Peng (1978) proposed a modified expression for heavier components ($\omega > 0.49$).

$$\kappa = 0.3796 + 1.485\omega - 0.1644\omega^2 + 0.016667\omega^3 \tag{4.12}$$

PR EOS can be represented in terms of compressibility factor by the following equation.

$$Z^3 - (1-B)Z^2 + \left(A-3B^2-2B\right)Z - \left(AB-B^2-B^3\right) = 0 \tag{4.13}$$

where A and B are defined as

$$A = \frac{aP}{R^2T^2} \tag{4.14}$$

$$B = \frac{bP}{RT} \tag{4.15}$$

The parameters a and b in the PR EOS are determined for a given mixture as

$$a = \sum_i \sum_j z_i z_j (a_i a_j)^{0.5} (1 - k_{ij}) \tag{4.16}$$

$$b = \sum_i z_i b_i \tag{4.17}$$

where k_{ij} is the binary interaction coefficient where $k_{ii} = 0$ and $k_{ij} = k_{ji}$. Usually $k_{ij} = 0$ for most hydrocarbon/hydrocarbon pairs except for methane/plus fraction pair (Whitson and Brulé, 2000; Riazi, 2005). The binary interaction coefficient for methane and plus fraction components could be estimated from the Whitson (1983) correlation. This correlation is based on data presented by Katz and Firoozabadi (1978).

$$k_{C_1-C_n} = 0.14SG_{C_n} - 0.0668 \quad \text{for } n \geq 6 \tag{4.18}$$

Eq. (4.18) should be only used with PR EOS (Whitson, 1983).

The binary interaction coefficient for nonhydrocarbon/hydrocarbon pairs is usually different from zero (Whitson, 1983; Pedersen et al., 1989; Aguilar Zurita and McCain Jr, 2002; Al-Meshari, 2005; Riazi, 2005). Table 4.1 represents the binary interaction coefficient for nonhydrocarbon/hydrocarbon pairs (Whitson and Brulé, 2000).

Table 4.1 Binary Interaction Coefficients for the Peng–Robinson Equation of State (Whitson and Brulé, 2000)

Component	N_2	CO_2	H_2S
N_2	—	0.000	0.130
CO_2	0.000	—	0.135
H_2S	0.130	0.135	—
C_1	0.025	0.105	0.070
C_2	0.010	0.130	0.085
C_3	0.090	0.125	0.080
i-C_4	0.095	0.120	0.075
C_4	0.095	0.115	0.075
i-C_5	0.100	0.115	0.070
C_5	0.110	0.115	0.070
C_6	0.110	0.115	0.055
C_{7+}	0.110	0.115	0.050[a]

[a]Should decrease gradually with the increasing carbon number.

Fugacity of the mixture and fugacity of each component in the mixture are given by Eqs. (4.19) and (4.20), respectively.

$$\ln\frac{f}{P} = \ln\phi = Z - 1 - \ln(Z - B) + \frac{A}{2\sqrt{2}B}\ln\left[\frac{Z + (1-\sqrt{2})B}{Z + (1+\sqrt{2})B}\right] \tag{4.19}$$

$$\begin{aligned}\ln\frac{f_i}{z_iP} = \ln\phi_i &= \frac{b_i}{b}(Z-1) - \ln(Z-B)\\ &+ \frac{A}{2\sqrt{2}B}\left(\frac{2}{a}\sum_{j=1}^{N} z_j a_{ij} - \frac{b_i}{b}\right)\ln\left[\frac{Z + (1-\sqrt{2})B}{Z + (1+\sqrt{2})B}\right]\end{aligned} \tag{4.20}$$

where $a_{ij} = (a_i a_j)^{0.5}(1 - k_{ij})$, Z is the compressibility factor of mixture, z_i is the mole fraction of component i in the mixture, f is the fugacity of the mixture, f_i is the fugacity of component i in the mixture, ϕ is the fugacity coefficient of the mixture, and ϕ_i is the fugacity coefficient of component i in the mixture. Eqs. (4.6) to (4.20) are used for both liquid and vapor phases.

Equilibrium ratio is defined by Eq. (4.21).

$$K_i = \frac{y_i}{x_i} = \frac{\phi_i^{\mathrm{V}}}{\phi_i^{\mathrm{L}}} \tag{4.21}$$

where y_i and x_i are the mole fraction of the component i in the vapor and liquid phases, respectively. Superscripts V and L stand for vapor and liquid phases, respectively. Initial guess for equilibrium ratio is usually estimated from Wilson correlation (Danesh, 1998; Whitson and Brulé, 2000).

$$K_i = \frac{P_{ci}}{P}\exp\left[5.37(1+\omega_i)\left(1 - \frac{T_{ci}}{T}\right)\right] \tag{4.22}$$

The following approach has been proposed to calculate the saturation pressure. The required inputs are mole fractions (z_i), properties of components (T_{ci},P_{ci},ω_i), and the binary interaction coefficient that is computed using Eq. (4.18) and Table 4.1.

1. Calculate the parameters of PR EOS using Eqs. (4.7) to (4.12).

2. Predict a value for pressure, and put $x_i = z_i$ (for bubble point calculation) or put $y_i = z_i$ (for dew point calculation).
3. Estimate the equilibrium ratio from Eq. (4.22) for all components.
4. Calculate the mole fractions in the vapor phase using $y_i = K_i x_i$ (for bubble point calculation) or calculate mole fractions in the liquid phase using $x_i = y_i / K_i$ (for dew point calculation).
5. Normalize the mole fractions.
6. Determine A and B in Eq. (4.13) using Eqs. (4.14) and (4.7) to (4.17).
7. Solve Eq. (4.13) for both vapor and liquid phases. If three roots are obtained by solving Eq. (4.13), select the biggest root for the vapor phase and the smallest root for the liquid phase.
8. Calculate the fugacity of all components for vapor and liquid phases using Eq. (4.20).
9. Calculate the error using Eq. (4.23)

$$\text{error} = \sum_{i=1}^{N} \left(1 - \frac{f_i^{\mathrm{L}}}{f_i^{\mathrm{V}}}\right)^2 \tag{4.23}$$

10. If the error is greater than 10^{-12}, adjust the equilibrium ratio using Eq. (4.21), correct the pressure, then return to step 3.

Example 4.1

Estimate the bubble point pressure of the mixture in Example 3.12 using PR EOS (using the proposed approach) at the temperature 345.8K. The experimental value at the given temperature is 23.74 MPa.

Solution

The properties of well-defined components (i.e., H_2S, N_2, CO_2, C_1, C_2, C_3, i-C_4, n-C_4, i-C_5, n-C_5, and n-C_6) are extracted from Danesh (1998). The properties of extended SCN groups are calculated in Example 3.12. A bubble point pressure of 15 MPa is assumed as the initial guess. The results of steps 1–5 are reported in the following table.

(Continued)

Component	x_i	T_c (K)	P_c (MPa)	ω	K_i Eq. (4.22)	$y_i = K_i x_i$	Normalized y_i	a_{ci} Eq. (4.8) (Pa m^6/mol^2)	b_i Eq. (4.9) (m^3/mol)	α Eq. (4.10)	$a_i = a_{ci}\alpha$ (Pa m^6/mol^2)
N_2	0.0069	126.10	3.39	0.0403	7.862	0.0542	0.0309	0.15	2.41E−05	0.509	0.076
CO_2	0.0012	304.19	7.38	0.2276	1.088	0.0013	0.0007	0.40	2.67E−05	0.908	0.360
C_1	0.4709	190.56	4.60	0.0115	3.513	1.6543	0.9410	0.25	2.68E−05	0.746	0.186
C_2	0.0569	305.32	4.87	0.0995	0.648	0.0369	0.0210	0.61	4.06E−05	0.934	0.565
C_3	0.0439	369.83	4.25	0.1523	0.184	0.0081	0.0046	1.02	5.63E−05	1.040	1.058
i-C_4	0.0095	408.14	3.65	0.1770	0.078	0.0007	0.0004	1.44	7.23E−05	1.104	1.593
n-C_4	0.0242	425.12	3.80	0.2002	0.058	0.0014	0.0008	1.50	7.24E−05	1.136	1.708
i-C_5	0.0111	460.43	3.38	0.2275	0.025	2.81E−04	1.60E−04	1.98	8.81E−05	1.199	2.377
n-C_5	0.0146	469.70	3.37	0.2515	0.020	2.95E−04	1.68E−04	2.07	9.02E−05	1.223	2.531
C_6	0.0226	510.00	3.27	0.3013	0.008	1.78E−04	1.02E−04	2.51	1.01E−04	1.308	3.290
C_7	0.0346	546.68	3.06	0.2786	3.78E−03	1.31E−04	7.44E−05	3.09	1.16E−04	1.346	4.156
C_8	0.0293	572.41	2.81	0.3175	1.82E−03	5.32E−05	3.03E−05	3.69	1.32E−04	1.408	5.188
C_9	0.0284	599.22	2.56	0.3631	7.99E−04	2.27E−05	1.29E−05	4.43	1.51E−04	1.479	6.557
C_{10}	0.0246	622.56	2.37	0.4050	3.77E−04	9.27E−06	5.27E−06	5.17	1.70E−04	1.546	7.990
C_{11}	0.0229	644.33	2.21	0.4476	1.79E−04	4.11E−06	2.34E−06	5.94	1.89E−04	1.614	9.582
C_{12}	0.0213	664.89	2.08	0.4902	8.61E−05	1.83E−06	1.04E−06	6.72	2.07E−04	1.686	11.325
C_{13}	0.0196	683.29	1.97	0.5309	4.30E−05	8.44E−07	4.80E−07	7.49	2.24E−04	1.754	13.141
C_{14}	0.0186	701.43	1.88	0.5744	2.10E−05	3.90E−07	2.22E−07	8.27	2.41E−04	1.828	15.119
C_{15}	0.0169	719.30	1.79	0.6209	9.86E−06	1.67E−07	9.48E−08	9.14	2.60E−04	1.907	17.423
C_{16}	0.0144	735.80	1.72	0.6684	4.69E−06	6.75E−08	3.84E−08	9.95	2.77E−04	1.988	19.780
C_{17}	0.0119	750.45	1.66	0.7133	2.34E−06	2.78E−08	1.58E−08	10.72	2.92E−04	2.066	22.151
C_{18}	0.0096	762.42	1.63	0.7519	1.30E−06	1.25E−08	7.09E−09	11.27	3.03E−04	2.133	24.045
C_{19}	0.0080	773.31	1.59	0.7896	7.34E−07	5.87E−09	3.34E−09	11.89	3.15E−04	2.199	26.146
C_{20}	0.0084	785.49	1.55	0.8358	3.72E−07	3.12E−09	1.78E−09	12.58	3.28E−04	2.280	28.690
C_{21}	0.0067	796.04	1.52	0.8777	2.01E−07	1.35E−09	7.68E−10	13.18	3.39E−04	2.354	31.026
C_{22}	0.0051	807.04	1.48	0.9249	1.01E−07	5.18E−10	2.94E−10	13.91	3.53E−04	2.438	33.919
C_{23}	0.0054	816.34	1.45	0.9669	5.54E−08	2.99E−10	1.70E−10	14.53	3.64E−04	2.514	36.518
C_{24}	0.0051	825.68	1.43	1.0111	2.95E−08	1.51E−10	8.57E−11	15.07	3.74E−04	2.594	39.087

C_{25}	0.0046	835.73	1.40	0.9599	3.12E−08	1.44E−10	8.16E−11	15.77	3.86E−04	2.545	40.127
C_{26}	0.0038	844.16	1.38	0.9852	1.96E−08	7.43E−11	4.23E−11	16.32	3.96E−04	2.597	42.393
C_{27}	0.0035	852.38	1.36	1.0108	1.22E−08	4.28E−11	2.44E−11	16.89	4.05E−04	2.651	44.759
C_{28}	0.0030	860.54	1.34	1.0365	7.61E−09	2.28E−11	1.30E−11	17.47	4.15E−04	2.705	47.247
C_{29}	0.0028	867.25	1.33	1.0562	5.21E−09	1.46E−11	8.29E−12	17.88	4.22E−04	2.748	49.115
C_{30}	0.0025	875.28	1.31	1.0818	3.22E−09	8.04E−12	4.57E−12	18.49	4.32E−04	2.803	51.809
C_{31}	0.0022	882.56	1.30	1.1045	2.09E−09	4.59E−12	2.61E−12	18.94	4.39E−04	2.852	54.021
C_{32}	0.0021	889.78	1.28	1.1271	1.34E−09	2.82E−12	1.60E−12	19.55	4.50E−04	2.902	56.743
C_{33}	0.0020	896.28	1.28	1.1467	9.15E−10	1.83E−12	1.04E−12	19.84	4.53E−04	2.947	58.455
C_{34}	0.0015	902.63	1.26	1.1667	6.13E−10	9.19E−13	5.23E−13	20.44	4.63E−04	2.992	61.148
C_{35}	0.0014	908.28	1.26	1.1837	4.37E−10	6.12E−13	3.48E−13	20.70	4.66E−04	3.031	62.727
C_{36}	0.0013	914.46	1.24	1.2040	2.91E−10	3.79E−13	2.16E−13	21.32	4.77E−04	3.077	65.588
C_{37}	0.0012	919.38	1.24	1.2181	2.17E−10	2.61E−13	1.48E−13	21.55	4.80E−04	3.110	67.019
C_{38}	0.0012	925.50	1.23	1.2382	1.46E−10	1.75E−13	9.94E−14	22.01	4.87E−04	3.157	69.484
C_{39}	0.0011	930.36	1.23	1.2523	1.08E−10	1.19E−13	6.77E−14	22.24	4.89E−04	3.191	70.971
C_{40}	0.0009	936.51	1.22	1.2721	7.22E−11	6.50E−14	3.70E−14	22.72	4.97E−04	3.237	73.559
C_{41}	0.0008	940.58	1.22	1.2836	5.62E−11	4.50E−14	2.56E−14	22.92	4.99E−04	3.265	74.850
C_{42}	0.0008	945.28	1.22	1.2980	4.16E−11	3.33E−14	1.89E−14	23.15	5.01E−04	3.300	76.406
C_{43}	0.0008	949.96	1.21	1.3123	3.05E−11	2.44E−14	1.39E−14	23.57	5.08E−04	3.335	78.623
C_{44}	0.0007	955.92	1.20	1.3322	2.03E−11	1.42E−14	8.07E−15	24.07	5.15E−04	3.383	81.418
C_{45+}	0.0090	983.25	1.39	1.3786	5.51E−12	4.96E−14	2.82E−14	21.99	4.58E−04	3.528	77.560

The parameters A and B in Eq. (4.13) are determined using Eqs. (4.14) and (4.7) to (4.17) for both vapor and liquid phases. The binary interaction coefficient is determined by Eq. (4.18) and Table 4.1.

For liquid phase:

$$a^L = \sum_i \sum_j x_i x_j (a_i a_j)^{0.5}(1 - k_{ij}) = 3.2055 \text{ Pa m}^6/\text{mol}^2$$

$$b^L = \sum_i x_i b_i = 1.0735 \times 10^{-4} \text{ m}^3/\text{mol}$$

$$A^L = 5.8165, B^L = 0.5601$$

For vapor phase:

$$a^V = \sum_i \sum_j y_i y_j (a_i a_j)^{0.5}(1 - k_{ij}) = 0.1916 \text{ Pa m}^6/\text{mol}^2$$

$$b^V = \sum_i y_i b_i = 2.7233 \times 10^{-5} \text{ m}^3/\text{mol}$$

$$A^V = 0.3477, B^V = 0.1421$$

Solve Eq. (4.13) for compressibility factor using the calculated parameter for each phase:

$$Z^L = 0.7026, Z^V = 0.8879$$

The fugacity of each component for both phases should be calculated by Eq. (4.20) and the error should be checked. The results of steps 8–10 are presented in the following table.

Component	f_i^L (MPa) Eq. (4.20)	f_i^V (MPa) Eq. (4.20)	$\phi_i^L = \frac{f_i^L}{z_i P}$	$\phi_i^V = \frac{f_i^V}{z_i P}$	K_i Eq. (4.21)	$K_i x_i$
N_2	0.506	0.500	4.888	1.081	4.524	0.031
CO_2	0.018	0.008	0.991	0.725	1.368	0.002
C_1	14.356	12.092	2.032	0.857	2.373	1.117
C_2	0.430	0.184	0.504	0.583	0.864	0.049
C_3	0.131	0.030	0.198	0.433	0.458	0.020
i-C_4	1.47E−02	2.17E−03	1.03E−01	3.45E−01	3.00E−01	2.85E−03
n-C_4	2.86E−02	3.83E−03	7.87E−02	3.21E−01	2.45E−01	5.93E−03
i-C_5	6.68E−03	6.11E−04	4.01E−02	2.55E−01	1.58E−01	1.75E−03
n-C_5	7.19E−03	6.05E−04	3.28E−02	2.40E−01	1.37E−01	2.00E−03
C_6	4.94E−03	3.03E−04	1.46E−02	1.99E−01	7.34E−02	1.66E−03
C_7	3.94E−03	1.82E−04	7.60E−03	1.63E−01	4.65E−02	1.61E−03
C_8	1.71E−03	5.97E−05	3.89E−03	1.32E−01	2.95E−02	8.64E−04
C_9	7.66E−04	1.98E−05	1.80E−03	1.02E−01	1.76E−02	5.00E−04
C_{10}	3.22E−04	6.38E−06	8.72E−04	8.07E−02	1.08E−02	2.66E−04
C_{11}	1.44E−04	2.23E−06	4.19E−04	6.36E−02	6.60E−03	1.51E−04

—cont'd

Component	f_i^L (MPa) Eq. (4.20)	f_i^V(MPa) Eq. (4.20)	$\phi_i^L = \frac{f_i^L}{z_i P}$	$\phi_i^V = \frac{f_i^V}{z_i P}$	K_i Eq. (4.21)	$K_i x_i$
C_{12}	6.27E−05	7.81E−07	1.96E−04	4.99E−02	3.93E−03	8.36E−05
C_{13}	2.79E−05	2.85E−07	9.48E−05	3.96E−02	2.39E−03	4.70E−05
C_{14}	1.22E−05	1.04E−07	4.36E−05	3.11E−02	1.40E−03	2.60E−05
C_{15}	4.79E−06	3.42E−08	1.89E−05	2.40E−02	7.86E−04	1.33E−05
C_{16}	1.75E−06	1.07E−08	8.10E−06	1.86E−02	4.35E−04	6.27E−06
C_{17}	6.42E−07	3.46E−09	3.60E−06	1.46E−02	2.47E−04	2.94E−06
C_{18}	2.57E−07	1.27E−09	1.78E−06	1.20E−02	1.49E−04	1.42E−06
C_{19}	1.07E−07	4.90E−10	8.95E−07	9.78E−03	9.15E−05	7.30E−07
C_{20}	4.90E−08	2.05E−10	3.89E−07	7.69E−03	5.06E−05	4.24E−07
C_{21}	1.82E−08	7.13E−11	1.81E−07	6.20E−03	2.92E−05	1.96E−07
C_{22}	5.84E−09	2.13E−11	7.64E−08	4.83E−03	1.58E−05	8.13E−08
C_{23}	2.86E−09	9.88E−12	3.53E−08	3.87E−03	9.10E−06	4.88E−08
C_{24}	1.20E−09	3.99E−12	1.56E−08	3.10E−03	5.04E−06	2.55E−08
C_{25}	1.20E−09	3.77E−12	1.74E−08	3.08E−03	5.65E−06	2.58E−08
C_{26}	5.42E−10	1.65E−12	9.50E−09	2.59E−03	3.66E−06	1.40E−08
C_{27}	2.70E−10	7.95E−13	5.14E−09	2.18E−03	2.36E−06	8.21E−09
C_{28}	1.24E−10	3.55E−13	2.75E−09	1.82E−03	1.51E−06	4.58E−09
C_{29}	6.97E−11	1.98E−13	1.66E−09	1.59E−03	1.05E−06	2.90E−09
C_{30}	3.27E−11	9.06E−14	8.71E−10	1.32E−03	6.59E−07	1.67E−09
C_{31}	1.61E−11	4.42E−14	4.86E−10	1.13E−03	4.31E−07	9.58E−10
C_{32}	8.42E−12	2.28E−14	2.67E−10	9.49E−04	2.82E−07	6.00E−10
C_{33}	4.77E−12	1.30E−14	1.59E−10	8.33E−04	1.91E−07	3.72E−10
C_{34}	2.06E−12	5.56E−15	9.18E−11	7.09E−04	1.29E−07	2.01E−10
C_{35}	1.21E−12	3.29E−15	5.77E−11	6.31E−04	9.15E−08	1.31E−10
C_{36}	6.42E−13	1.73E−15	3.29E−11	5.35E−04	6.15E−08	8.17E−11
C_{37}	3.95E−13	1.08E−15	2.20E−11	4.84E−04	4.54E−08	5.58E−11
C_{38}	2.26E−13	6.19E−16	1.26E−11	4.15E−04	3.03E−08	3.62E−11
C_{39}	1.38E−13	3.81E−16	8.34E−12	3.75E−04	2.23E−08	2.46E−11
C_{40}	6.38E−14	1.78E−16	4.73E−12	3.21E−04	1.47E−08	1.35E−11
C_{41}	4.01E−14	1.13E−16	3.35E−12	2.95E−04	1.13E−08	9.18E−12
C_{42}	2.64E−14	7.55E−17	2.20E−12	2.66E−04	8.30E−09	6.97E−12
C_{43}	1.70E−14	4.90E−17	1.42E−12	2.35E−04	6.04E−09	4.70E−12
C_{44}	8.31E−15	2.42E−17	7.92E−13	2.00E−04	3.95E−09	2.70E−12
C_{45+}	2.60E−14	7.63E−17	1.92E−13	1.80E−04	1.07E−09	9.64E−12
	$\text{Error} = \sum_{i=1}^{N}\left(1 - \frac{f_i^L}{f_i^V}\right)^2 = 3.21 \times 10^6$					Sum = 1.237

The pressure is modified for the next iteration as follows:

$$P_{new} = P_{old} \sum_i K_i x_i$$

$$P_{new} = 15 \times 1.237 = 18.557 \text{ MPa}$$

(Continued)

Now, with the new pressure and adjusted equilibrium ratio repeat steps 4–10. The error, $\sum K_i x_i$, and modified pressure for a few iterations are given in the following table.

Iteration Number	1	2	3	4	27	49
$\sum_i K_i x_i$	1.23714	1.12041	1.06968	1.04331	1.00001	1.00000
Modified pressure (MPa)	18.5771	20.7917	22.2405	23.2036	25.3170	25.3179
Error	3.21E+06	3.43E+03	2.18E+02	5.19E+01	3.59E−06	5.99E−13

The relative deviation percent is

$$\frac{23.74 - 25.32}{23.74} \times 100 = -6.65\%$$

If all binary interaction coefficients are set to zero, the relative deviation is 13.45%, which shows the significant impact of the binary interaction coefficient on phase behavior calculation.

Example 4.2

Match the saturation pressure in Example 4.1 (using the results of Example 4.1)

Solution

At first the weight fraction of each component should be calculated [molecular weight of well-defined components have been extracted from Danesh (1998)].

Component	z_i	MW (g/mol)	$z_i MW_i$ (g/mol)	w_i Eq. (4.2)
N_2	0.0069	28.0	0.193	0.00218
CO_2	0.0012	44.0	0.053	0.00060
C_1	0.4709	16.0	7.555	0.08513
C_2	0.0569	30.1	1.711	0.01928
C_3	0.0439	44.1	1.936	0.02181
i-C_4	0.0095	58.1	0.552	0.00622
n-C_4	0.0242	58.1	1.407	0.01585
i-C_5	0.0111	72.2	0.801	0.00902
n-C_5	0.0146	72.2	1.053	0.01187
C_6	0.0226	86.2	1.948	0.02195
C_{7+}	0.3382	211.5	71.537	0.80609
Sum	1	–	$MW_{mix.} = 88.74$ g/mol	1

The weight fraction is not changed when matching saturation pressure. From Example 4.1 we know that the calculated bubble point pressure is 25.32 MPa if molecular weight of mixture is 88.74 g/mol. Guess another value for molecular weight of mixture, 88.00 g/mol. Recalculate the mole fraction of all components except plus fraction using Eq. (4.3).

$$z_{N_2} = \frac{0.00218 \times 88.00}{28.0} = 0.0068, z_{CO_2} = 0.0012, z_{C_1} = 0.4669, z_{C_2} = 0.0564, z_{C_3}$$
$$= 0.0435, z_{iC_4} = 0.0094, z_{C_4} = 0.0240, z_{iC_5} = 0.0110, z_{C_5} = 0.0145, z_{C_6}$$
$$= 0.0224$$

Now determine the mole fraction and molecular weight of plus fraction using Eqs. (4.4) and (4.5).

$$z_{C_{7+}} = 1 - \sum_{i=1}^{N-1} z_i = 1 - 0.6562 = 0.3438$$

$$MW_{C_{7+}} = \frac{MW_{mix.} - \sum_{i=1}^{N-1} z_i MW_i}{z_{C_{7+}}} = \frac{88.00 - 17.06}{0.3438} = 206.36$$

The new value for η is 118.2. Using the recommended characterization procedure in Chapter 3 recalculate the mole fraction and other properties (critical temperature, critical pressure, and acentric factor) of the extended group. Note that when the mole fraction of SCN group is changed, according to Eq. 3.84, the specific gravity of each SCN group changes, which causes critical temperature, critical pressure, and acentric factor to change. The binary interaction parameters between methane and plus fraction are also changed. The calculated bubble point pressure (calculated similar to Example 4.1) is 25.09 MPa. Adjust the molecular weight of mixture by a simple linear interpolation, that is, 83.66 g/mol. Bubble point pressure in this case is 23.98 MPa, which is close enough to experimental value, and the mole fraction and molecular weight of C_{7+} are 0.3761 and 179.3 g/mol, respectively. The mole fraction and properties of extended groups are given in the following table.

Component	z_i	Specific Gravity	T_c (K)	P_c (MPa)	ω	$k_{C_1-C_N}$
N_2	0.0065	–	126.10	3.39	0.0403	–
CO_2	0.0011	–	304.19	7.38	0.2276	–
C_1	0.4439	–	190.56	4.60	0.0115	–
C_2	0.0536	–	305.32	4.87	0.0995	–

(Continued)

—cont'd

Component	z_i	Specific Gravity	T_c (K)	P_c (MPa)	ω	$k_{C_1-C_N}$
C_3	0.0414	–	369.83	4.25	0.1523	–
i-C_4	0.0090	–	408.14	3.65	0.1770	–
n-C_4	0.0228	–	425.12	3.80	0.2002	–
i-C_5	0.0105	–	460.43	3.38	0.2275	–
n-C_5	0.0138	–	469.70	3.37	0.2515	–
C_6	0.0213	–	510.00	3.27	0.3013	0.0261
C_7	0.0511	0.745	553.37	3.23	0.2513	0.0375
C_8	0.0417	0.761	579.19	2.98	0.2920	0.0397
C_9	0.0390	0.776	606.12	2.74	0.3397	0.0419
C_{10}	0.0325	0.791	629.53	2.56	0.3839	0.0439
C_{11}	0.0292	0.804	651.38	2.40	0.4291	0.0457
C_{12}	0.0260	0.817	671.99	2.27	0.4748	0.0476
C_{13}	0.0230	0.829	690.42	2.17	0.5187	0.0493
C_{14}	0.0209	0.842	708.58	2.08	0.5657	0.0511
C_{15}	0.0181	0.854	726.47	2.00	0.6162	0.0528
C_{16}	0.0148	0.865	743.01	1.93	0.6674	0.0543
C_{17}	0.0117	0.875	757.66	1.88	0.7162	0.0557
C_{18}	0.0090	0.884	769.63	1.85	0.7585	0.0570
C_{19}	0.0073	0.892	780.50	1.82	0.7996	0.0581
C_{20}	0.0074	0.900	792.72	1.78	0.8494	0.0592
C_{21}	0.0057	0.908	803.29	1.75	0.8948	0.0603
C_{22}	0.0042	0.914	814.33	1.72	0.9455	0.0612
C_{23}	0.0042	0.920	823.68	1.69	0.9907	0.0620
C_{24}	0.0039	0.926	833.04	1.67	1.0384	0.0629
C_{25}	0.0034	0.933	843.13	1.64	1.0926	0.0638
C_{26}	0.0027	0.939	851.57	1.63	1.1405	0.0646
C_{27}	0.0024	0.944	859.84	1.61	0.9864	0.0654
C_{28}	0.0020	0.949	868.03	1.59	1.0118	0.0661
C_{29}	0.0018	0.954	874.75	1.59	1.0314	0.0668
C_{30}	0.0016	0.959	882.83	1.57	1.0567	0.0675
C_{31}	0.0014	0.964	890.14	1.56	1.0791	0.0681
C_{32}	0.0013	0.968	897.40	1.55	1.1015	0.0688
C_{33}	0.0011	0.973	903.92	1.55	1.1210	0.0694
C_{34}	0.0009	0.977	910.32	1.54	1.1407	0.0700
C_{35}	0.0008	0.981	915.98	1.53	1.1576	0.0706
C_{36}	0.0007	0.985	922.22	1.52	1.1775	0.0710
C_{37}	0.0006	0.988	927.14	1.53	1.1916	0.0716
C_{38}	0.0006	0.992	933.33	1.51	1.2114	0.0721

—cont'd

Component	z_i	Specific Gravity	T_c (K)	P_c (MPa)	ω	$k_{C_1-C_N}$
C_{39}	0.0005	0.996	938.20	1.52	1.2254	0.0726
C_{40}	0.0004	0.999	944.40	1.51	1.2449	0.0731
C_{41}	0.0004	1.002	948.47	1.52	1.2563	0.0735
C_{42}	0.0004	1.006	953.19	1.51	1.2705	0.0740
C_{43}	0.0003	1.009	957.90	1.51	1.2846	0.0744
C_{44}	0.0003	1.012	963.93	1.50	1.3042	0.0749
C_{45+}	0.0028	1.032	980.35	1.62	1.3388	0.0777

The relative deviation percent for plus fraction molecular weight during molecular weight adjustment is −15.24%, which is acceptable.

Example 4.3

Composition of a gas condensate from Iran reported in Table 4.2 (Firoozabadi et al., 1978).

The molecular weight and specific gravity of C_{7+} fraction are 132 g/mol and 0.774, respectively. The experimental dew point pressure at temperature 355.65K is 28.1 MPa. Match the saturation pressure using the recommended procedure.

Solution

At first, the plus fraction characterized by the recommended procedure is similar to Example 3.12. Note that the molecular weight of SCN groups is determined by Eq. 3.28 (why?). The results of characterization procedure are given in the following table.

Table 4.2 Composition of a Gas Condensate From Iran (Firoozabadi et al., 1978) (Example 4.3)

Component	Mol%
N_2	0.08
CO_2	2.44
C_1	82.10
C_2	5.78
C_3	2.87
i-C_4	0.56
n-C_4	1.23
i-C_5	0.52
n-C_5	0.60
C_6	0.72
C_{7+}	3.10

(Continued)

SCN	z_i	T_b (K) Table 3.1	Molecular Weight (g/mol) Eq. (3.28)	Specific Gravity Eq. (3.82)	T_c (K) Eq. (3.46)	P_c (MPa) Eq. (3.49)	$T_{br}=\frac{T_b}{T_c}$	ω Eq. (3.38) and Eq (3.39)
C_7	0.0008	366	94.5	0.724	547.94	3.09	0.668	0.2731
C_8	0.0244	390	107.5	0.741	574.07	2.85	0.679	0.3108
C_9	0.8210	416	120.4	0.756	600.92	2.60	0.692	0.3568
C_{10}	0.0578	439	133.7	0.770	624.37	2.42	0.703	0.3989
C_{11}	0.0287	461	146.9	0.784	646.22	2.26	0.713	0.4420
C_{12}	0.0056	482	160.6	0.796	666.74	2.12	0.723	0.4854
C_{13}	0.0123	501	174.9	0.808	685.21	2.02	0.731	0.5268
C_{14}	0.0052	520	189.8	0.820	703.35	1.93	0.739	0.5712
C_{15}	0.0060	539	205.5	0.832	721.18	1.84	0.747	0.6188
C_{16}	0.0072	557	221.3	0.843	737.83	1.77	0.755	0.6671
C_{17}	0.0081	573	236.4	0.853	752.49	1.72	0.761	0.7130
C_{18}	0.0057	586	250.2	0.862	764.43	1.68	0.767	0.7526
C_{19}	0.0046	598	262.7	0.870	775.25	1.65	0.771	0.7911
C_{20}	0.0032	612	275.6	0.877	787.43	1.60	0.777	0.8381
C_{21}	0.0025	624	289.0	0.885	798.03	1.58	0.782	0.8810
C_{22}	0.0019	637	300.5	0.891	808.99	1.54	0.787	0.9289
C_{23}	0.0014	648	311.7	0.897	818.33	1.51	0.792	0.9718
C_{24}	0.0011	659	324.0	0.903	827.70	1.48	0.796	1.0169
C_{25}	7.74E−04	671	336.5	0.909	837.71	1.46	0.801	0.9534
C_{26}	5.23E−04	681	348.5	0.915	846.15	1.44	0.805	0.9786
C_{27}	3.47E−04	691	360.0	0.920	854.44	1.42	0.809	1.0040
C_{28}	2.27E−04	701	371.3	0.925	862.64	1.40	0.813	1.0295
C_{29}	1.58E−04	709	382.3	0.930	869.35	1.39	0.816	1.0492
C_{30}	1.37E−04	719	393.3	0.935	877.39	1.38	0.819	1.0747
C_{31}	9.02E−05	728	404.0	0.940	884.66	1.36	0.823	1.0974
C_{32}	5.82E−05	737	414.8	0.944	891.87	1.35	0.826	1.1200
C_{33}	5.15E−05	745	425.8	0.948	898.37	1.34	0.829	1.1396
C_{34}	4.06E−05	753	436.1	0.953	904.77	1.33	0.832	1.1594
C_{35}	3.05E−05	760	445.6	0.956	910.37	1.32	0.835	1.1765
C_{36}	2.13E−05	768	455.1	0.960	916.62	1.31	0.838	1.1965
C_{37}	1.64E−05	774	464.6	0.963	921.49	1.31	0.840	1.2108
C_{38}	1.21E−05	782	474.3	0.967	927.70	1.30	0.843	1.2306
C_{39}	9.44E−06	788	484.3	0.971	932.56	1.30	0.845	1.2447
C_{40}	7.34E−06	796	493.8	0.974	938.68	1.29	0.848	1.2645
C_{41}	5.48E−06	801	502.6	0.977	942.73	1.29	0.850	1.2761
C_{42}	4.50E−06	807	511.6	0.980	947.44	1.29	0.852	1.2904
C_{43}	3.50E−06	813	521.1	0.984	952.17	1.29	0.854	1.3045
C_{44}	2.39E−06	821	530.3	0.987	958.16	1.27	0.857	1.3243
C_{45+}	1.92E−06	829 Eq. 3.34	578.8	1.002	967.33	1.37	0.857	1.3345

Using somehow similar procedure that is described for calculating bubble point pressure in Example 4.1, dew point pressure is calculated; however, the pressure is corrected by a different method. Choose 20 MPa as the initial guess for pressure; at this pressure $\sum y_i/K_i = 5.156$. Guess another value for dew point pressure, 22 MPa; at this pressure, $\sum y_i/K_i = 2.945$. The errors for pressures 20 and 22 MPa are 38.982 and 24.817, respectively. In this example the pressure is corrected by using linear interpolation as follows:

$$P_{\text{new}} = \frac{(20-22)\times 10^6}{5.156-2.945}(1-5.156) + \left(20\times 10^6\right) = 23.76\times 10^6 \text{ Pa}$$

The value of $\sum y_i/K_i$ at pressure 23.76 MPa is 2.044 and the error is 14.599. Repeat this procedure. At pressure 32.98 MPa, the value of $\sum y_i/K_i$ is equal to 1 and the error is equal to 5.57E−013. Hence the dew point pressure at the temperature 355.65K is 32.98 MPa.

Now the weight fraction of each component should be calculated.

Component	z_i	MW (g/mol)	z_iW_i (g/mol)	w_i Eq. (4.2)
N_2	0.08	28.0	0.022	0.00094
CO_2	2.44	44.0	1.074	0.04506
C_1	82.10	16.0	13.171	0.55267
C_2	5.78	30.1	1.738	0.07293
C_3	2.87	44.1	1.266	0.05310
i-C_4	0.56	58.1	0.325	0.01366
n-C_4	1.23	58.1	0.715	0.03000
i-C_5	0.52	72.2	0.375	0.01574
n-C_5	0.60	72.2	0.433	0.01816
C_6	0.72	86.2	0.620	0.02604
C_{7+}	3.10	132.0	4.092	0.17170
Sum	1	–	$MW_{\text{mix.}} = 23.83$ g/mol	1

As mentioned before the weight fraction is not changed when matching saturation pressure. The calculated dew point pressure is 32.76 MPa if molecular weight of mixture is 23.83 g/mol. Similar to the previous example guess another value for molecular weight of mixture, 23.80 g/mol. Recalculate the mole fraction of all components except plus fraction using Eq. (4.3).

$$z_{N_2} = \frac{0.00094\times 23.80}{28.0} = 0.0008, z_{CO_2} = 0.0244, z_{C_1} = 0.8199, z_{C_2} = 0.0577, z_{C_3} = 0.0287, z_{iC4} = 0.0056, z_{C_4} = 0.0123, z_{iC_5} = 0.0052, z_{C5} = 0.0060, z_{C_6} = 0.0072$$

Now determine the mole fraction and molecular weight of plus fraction using Eqs. (4.4) and (4.5).

(Continued)

$$z_{C_{7+}} = 1 - \sum_{i=1}^{N-1} z_i = 1 - 0.9677 = 0.0323$$

$$MW_{C_{7+}} = \frac{MW_{mix.} - \sum_{i=1}^{N-1} z_i MW_i}{z_{C_{7+}}} = \frac{23.80 - 19.71}{0.0323} = 126.5$$

The new value for η is 38.3. Using recommended characterization procedure in Chapter 6, recalculate the mole fraction and other properties. The calculated dew point pressure is 30.22 MPa. Adjust the molecular weight of the mixture by a simple linear interpolation, that is, 23.78 g/mol. Dew point pressure in this case is 28.53 MPa, which is close enough to experimental value, and the molecular weight of C_{7+} is 123.28 g/mol. The mole fraction and properties of extended groups after matching the saturation pressure are given in the following table.

Component	z_i	Molecular Weight (g/mol)	Specific Gravity	T_c (K)	P_c (MPa)	ω	$k_{C_1-C_N}$
N_2	0.0008	28.014	–	126.10	3.39	0.0403	–
CO_2	0.0243	44.01	–	304.19	7.38	0.2276	–
C_1	0.8192	16.043	–	190.56	4.60	0.0115	–
C_2	0.0577	30.07	–	305.32	4.87	0.0995	–
C_3	0.0286	44.096	–	369.83	4.25	0.1523	–
n-C_4	0.0056	58.123	–	408.14	3.65	0.1770	–
i-C_4	0.0123	58.123	–	425.12	3.80	0.2002	–
n-C_5	0.0052	72.15	–	460.43	3.38	0.2275	–
i-C_5	0.0060	72.15	–	469.70	3.37	0.2515	–
C_6	0.0072	86.177	–	510.00	3.27	0.3013	0.0261
C_7	0.0105	94.4	0.734	550.65	3.16	0.2618	0.0360
C_8	0.0068	107.4	0.751	576.81	2.92	0.3003	0.0384
C_9	0.0051	120.3	0.767	603.71	2.67	0.3472	0.0405
C_{10}	0.0033	133.6	0.781	627.19	2.49	0.3903	0.0426
C_{11}	0.0024	146.8	0.795	649.08	2.33	0.4344	0.0444
C_{12}	0.0017	160.5	0.807	669.62	2.20	0.4791	0.0462
C_{13}	0.0012	174.8	0.820	688.10	2.10	0.5218	0.0480
C_{14}	8.05E−04	189.7	0.832	706.24	2.01	0.5676	0.0497
C_{15}	5.31E−04	205.4	0.844	724.08	1.93	0.6167	0.0514
C_{16}	3.28E−04	221.2	0.855	740.74	1.86	0.6666	0.0529
C_{17}	2.00E−04	236.3	0.865	755.41	1.81	0.7140	0.0544
C_{18}	1.21E−04	250.1	0.874	767.35	1.77	0.7551	0.0556
C_{19}	7.82E−05	262.7	0.882	778.17	1.74	0.7950	0.0567
C_{20}	6.31E−05	275.5	0.890	790.37	1.70	0.8434	0.0577
C_{21}	3.85E−05	288.9	0.897	800.98	1.67	0.8877	0.0588
C_{22}	2.33E−05	300.5	0.904	811.96	1.63	0.9370	0.0597

—cont'd

Component	z_i	Molecular Weight (g/mol)	Specific Gravity	T_c (K)	P_c (MPa)	ω	$k_{C_1-C_N}$
C_{23}	1.93E−05	311.7	0.910	821.32	1.60	0.9812	0.0605
C_{24}	1.42E−05	323.9	0.916	830.69	1.58	1.0278	0.0614
C_{25}	9.96E−06	336.4	0.922	840.72	1.56	1.0804	0.0623
C_{26}	6.50E−06	348.4	0.928	849.17	1.54	0.9687	0.0631
C_{27}	4.69E−06	359.9	0.933	857.48	1.52	0.9941	0.0639
C_{28}	3.25E−06	371.2	0.939	865.69	1.50	1.0195	0.0646
C_{29}	2.38E−06	382.2	0.943	872.41	1.50	1.0391	0.0653
C_{30}	1.74E−06	393.2	0.948	880.46	1.48	1.0645	0.0660
C_{31}	1.22E−06	404.0	0.953	887.75	1.47	1.0871	0.0666
C_{32}	9.42E−07	414.7	0.957	894.97	1.45	1.1096	0.0672
C_{33}	6.88E−07	425.7	0.962	901.49	1.45	1.1291	0.0679
C_{34}	4.43E−07	436.0	0.966	907.90	1.44	1.1488	0.0684
C_{35}	3.38E−07	445.5	0.970	913.51	1.43	1.1659	0.0690
C_{36}	2.58E−07	455.0	0.973	919.78	1.42	1.1858	0.0695
C_{37}	1.97E−07	464.5	0.977	924.66	1.42	1.2000	0.0700
C_{38}	1.57E−07	474.3	0.981	930.88	1.41	1.2197	0.0705
C_{39}	1.18E−07	484.3	0.984	935.75	1.42	1.2337	0.0710
C_{40}	8.10E−08	493.8	0.988	941.90	1.41	1.2534	0.0715
C_{41}	5.96E−08	502.6	0.991	945.95	1.41	1.2649	0.0719
C_{42}	5.16E−08	511.5	0.994	950.66	1.41	1.2792	0.0724
C_{43}	3.94E−08	521.0	0.997	955.40	1.41	1.2932	0.0728
C_{44}	2.86E−08	530.3	1.001	961.43	1.40	1.3129	0.0733
C_{45+}	9.80E−08	570.1	1.014	969.84	1.48	1.3255	0.0751

The relative deviation percent for plus fraction molecular weight during molecular weight adjustment is −6.20%, which is well acceptable.

4.2 GROUPING METHODS

Computational time required for phase behavior simulation increases considerably by increasing the number of components that are used for describing reservoir fluid. The number of sufficient pseudocomponents to describe reservoir fluid depends mainly on the process being simulated (Whitson and Brulé, 2000). For example, the number of sufficient components for the prediction of phase behavior of reservoir fluid under pressure depletion is two (Danesh, 1998) although simulation of miscibility in a slim-tube needs 12−15 components (Whitson and Brulé,

2000). Many authors (Jacoby et al., 1959; Lee et al., 1981; Hong, 1982; Whitson, 1983; Montel and Gouel, 1984; Gonzales et al., 1986; Schlijper, 1986; Chorn and Mansoori, 1989; Pedersen et al., 1989; Wu and Fish, 1989; Danesh et al., 1992; Neau et al., 1993; Manafi et al., 1999; Shariati et al., 1999) have given recommendations on grouping SCN fraction into MCN groups. In this section the most widely used methods are presented.

4.2.1 Whitson Method

Whitson (1983) proposed that heavy fraction (C_{7+} fraction) can be grouped into N_p pseudocomponents, where N_p is given by Eq. (4.24).

$$N_p = \text{Integer}[1 + 3.3 \log(N - 7)] \tag{4.24}$$

where N is the last carbon group number in the original fluid. The groups are separated based on molecular weight. The molecular weight boundary is given by:

$$\text{MW}_k = \text{MW}_{C_7} \left\{ \exp\left[\left(\frac{1}{N_p} \right) \ln \left(\frac{\text{MW}_{C_N}}{\text{MW}_{C_7}} \right) \right] \right\}^k \tag{4.25}$$

where k is the group number and $k = 1, 2, 3, \ldots N_p$. The components of the original fluid with the molecular weight between MW_{k-1} and MW_k fall within group k. This method should be used when N is greater than 20 (Whitson and Brulé, 2000). MW_{C_N} is the molecular weight of last carbon group number (which may actually be a plus fraction).

Example 4.4

Describe the fluid in Example 3.12 by a number of pseudocomponents using Whitson method.

Solution

The last carbon group number describing the heavy fraction is 45. So the number of pseudocomponents is calculated as follows by Eq. (4.24).

$$N_p = \text{Integer}[1 + 3.3 \log(45 - 7)] = 6$$

As mentioned before, the carbon numbers that are grouped into a pseudocomponent are specified by molecular weight boundaries, which are determined by Eq. (4.25). Using Eq. (4.25), the upper molecular weight boundary for the first pseudocomponent is determined.

$$MW_1 = 96\left\{\exp\left[\left(\frac{1}{6}\right)\ln\left(\frac{653}{96}\right)\right]\right\}^1 = 132$$

where 653 g/mol is the molecular weight of C_{45+} fraction (which is a plus fraction). The upper molecular weight boundary for first pseudocomponent is 132 g/mol, hence C_7, C_8, and C_9 are grouped into the first pseudocomponent (the molecular weight of C_{10} is 134 g/mol, which is greater than 132 g/mol, so the first pseudocomponent is not included in C_{10}). Similarly, the upper molecular weight for other pseudocomponent is calculated and given in the following table.

Pseudocomponent Group, *k*	Upper Molecular Weight Boundary for *k*th Pseudocomponent, MW_k (g/mol)	Components in Group
1	132	C_7–C_9
2	182	C_{10}–C_{13}
3	250	C_{14}–C_{18}
4	345	C_{19}–C_{25}
5	474	C_{26}–C_{38}
6	653	C_{39}–C_{45+}

4.2.2 Pedersen et al. Method (Equal Weight Method)

Pedersen et al. (1984) proposed that components of the original fluid are grouped based on mass where pseudocomponents are of approximately the same weight. In this method the critical temperature, critical pressure, and acentric factor of each pseudocomponent are found by weight mean average.

Example 4.5

Extended composition and critical properties of a North Sea gas condensate are given in Table 4.3 (Pedersen et al., 2014).

Describe the C_{7+} fraction by three pseudocomponent using equal weight method.

Solution

The weight of each component is equal to mole fraction of component times its molecular weight ($z_i MW_i$). The weight of 100 kg moles of this sample is 3120.81 and the weight of C_{7+} fraction in 100 kg moles of this sample is 1167.65. Hence the objective weight of each group is 1167.65/3 = 389.22. Adding components from C_7 downward until the weight reaches a value around 389.22. Group-I

(Continued)

Table 4.3 Extended Composition and Critical Properties of a North Sea Gas Condensate (Example 4.5)

Component	Mol% z_i	MW (g/mol)	Specific Gravity	T_c (°C)	P_c (bar)	Omega
N_2	0.12	28.014	—	−146.95	33.94	0.04
CO_2	2.49	44.01	—	31.05	73.76	0.225
C_1	76.43	16.043	—	−82.55	46	0.008
C_2	7.46	30.07	—	32.25	48.84	0.098
C_3	3.12	44.097	—	96.65	42.46	0.152
n-C_4	1.21	58.124	—	152.05	38	0.193
i-C_4	0.59	58.124	—	134.95	36.48	0.176
n-C_5	0.59	72.151	—	196.45	33.74	0.251
i-C_5	0.5	72.151	—	187.25	33.84	0.227
C_6	0.79	87.178	0.664	234.25	29.69	0.296
C_7	0.95	95	0.726	258.7	31.44	0.465
C_8	1.08	106	0.747	278.4	28.78	0.497
C_9	0.78	116	0.769	295.6	27.22	0.526
C_{10}	0.592	133	0.781	318.8	23.93	0.574
C_{11}	0.467	152	0.778	339.8	20.58	0.626
C_{12}	0.345	164	0.785	353.6	19.41	0.658
C_{13}	0.375	179	0.802	371.4	18.65	0.698
C_{14}	0.304	193	0.815	386.8	18.01	0.735
C_{15}	0.237	209	0.817	401.7	16.93	0.775
C_{16}	0.208	218	0.824	410.8	16.66	0.798
C_{17}	0.22	239	0.825	428.7	15.57	0.849
C_{18}	0.169	250	0.831	438.7	15.31	0.874
C_{19}	0.14	264	0.841	451.5	15.11	0.907
C_{20}	0.101	275	0.845	460.8	14.87	0.932
C_{21}	0.0888	291	0.849	473.6	14.48	0.966
C_{22}	0.078	305	0.853	484.7	14.21	0.969
C_{23}	0.0686	318	0.857	494.8	13.99	1.023
C_{24}	0.0603	331	0.860	504.7	13.8	1.049
C_{25}	0.053	345	0.864	515.1	13.61	1.075
C_{26}	0.0465	359	0.867	525.4	13.43	1.101
C_{27}	0.0409	374	0.870	536.1	13.26	1.128
C_{28}	0.0359	388	0.873	546.0	13.12	1.151
C_{29}	0.0316	402	0.876	555.8	12.99	1.174
C_{30}	0.0277	416	0.879	565.5	12.88	1.195
C_{31}	0.0244	430	0.881	575.0	12.77	1.216
C_{32}	0.0214	444	0.884	584.4	12.68	1.235
C_{33}	0.0188	458	0.887	593.7	12.59	1.253
C_{34}	0.0165	472	0.889	602.9	12.52	1.270
C_{35}	0.0145	486	0.891	612.0	12.44	1.285
C_{36}	0.0128	500	0.894	621.0	12.38	1.300
C_{37}	0.0112	514	0.898	630.0	12.32	1.313
C_{38}	0.00986	528	0.891	638.8	12.26	1.325

Table 4.3 Extended Composition and Critical Properties of a North Sea Gas Condensate (Example 4.5)—cont'd

Component	Mol% z_i	MW (g/mol)	Specific Gravity	T_c (°C)	P_c (bar)	Omega
C_{39}	0.00866	542	0.900	647.6	12.21	1.335
C_{40}	0.00761	556	0.902	656.3	12.17	1.344
C_{41}	0.00609	570	0.904	664.9	12.12	1.352
C_{42}	0.00588	584	0.906	673.5	12.09	1.359
C_{43}	0.00517	598	0.908	682.0	12.05	1.364
C_{44}	0.00454	612	0.910	690.5	12.02	1.368
C_{45}	0.00399	626	0.912	698.9	11.99	1.371
C_{46}	0.00351	640	0.914	707.3	11.96	1.372
C_{47}	0.00308	654	0.916	715.6	11.93	1.372
C_{48}	0.00271	668	0.917	723.8	11.91	1.371
C_{49}	0.00238	682	0.919	732.0	11.89	1.369
C_{50}	0.00209	696	0.921	740.2	11.87	1.365
C_{51}	0.00183	710	0.922	748.3	11.85	1.359
C_{52}	0.00161	724	0.924	756.4	11.84	1.353
C_{53}	0.00142	738	0.926	764.4	11.82	1.345
C_{54}	0.00128	752	0.927	772.4	11.81	1.335
C_{55}	0.00109	766	0.929	780.4	11.8	1.325
C_{56}	0.000962	780	0.930	788.3	11.78	1.313
C_{57}	0.000845	794	0.932	796.2	11.77	1.300
C_{58}	0.000743	808	0.933	804.1	11.77	1.286
C_{59}	0.000653	822	0.934	811.9	11.76	1.270
C_{60}	0.000574	836	0.936	819.7	11.75	1.253
C_{61}	0.000504	850	0.937	827.5	11.75	1.236
C_{62}	0.000443	864	0.939	835.2	11.74	1.216
C_{63}	0.000389	878	0.940	843.0	11.74	1.196
C_{64}	0.000342	892	0.941	850.6	11.73	1.175
C_{65}	0.0003	906	0.942	858.3	11.73	1.152
C_{66}	0.000264	920	0.944	866.0	11.73	1.129
C_{67}	0.000232	934	0.945	873.6	11.72	1.104
C_{68}	0.000204	948	0.946	881.2	11.72	1.078
C_{69}	0.000179	962	0.947	888.7	11.72	1.052
C_{70}	0.000157	976	0.949	896.3	11.72	1.024
C_{71}	0.000138	990	0.950	903.8	11.72	0.995
C_{72}	0.000122	1004	0.951	911.3	11.72	0.965
C_{73}	0.000107	1018	0.952	918.8	11.72	0.935
C_{74}	0.0000939	1032	0.953	926.3	11.73	0.903
C_{75}	0.0000825	1046	0.954	933.7	11.73	0.871
C_{76}	0.0000725	1060	0.955	941.2	11.73	0.838
C_{77}	0.0000637	1074	0.956	948.6	11.73	0.804
C_{78}	0.000056	1088	0.957	956.0	11.74	0.769
C_{79}	0.0000492	1102	0.959	963.4	11.74	0.734
C_{80}	0.0000432	1116	0.956	970.7	11.74	0.697

(Continued)

consists of C_7–C_{10} with a weight of 373.95. Similarly, Group-II consists of C_{11}–C_{17} with a weight of 400.82 and Group-III consists of C_{18}–C_{80} with a weight of 392.89. Details of calculations are reported in the following table.

Component	Mol% z_i	MW_i (g/mol)	Z_iMW_i (g/mol)	Group-I	Group-II	Group-III
N_2	0.12	28.014	–			
CO_2	2.49	44.01	–			
C_1	76.43	16.043	–			
C_2	7.46	30.07	–			
C_3	3.12	44.097	–			
n-C_4	1.21	58.124	–			
i-C_4	0.59	58.124	–			
n-C_5	0.59	72.151	–			
i-C_5	0.5	72.151	–			
C_6	0.79	87.178	–			
C_7	0.95	95	90.25	90.25		
C_8	1.08	106	114.48	114.48		
C_9	0.78	116	90.48	90.48		
C_{10}	0.592	133	78.736	78.736		
C_{11}	0.467	152	70.984		70.984	
C_{12}	0.345	164	56.58		56.58	
C_{13}	0.375	179	67.125		67.125	
C_{14}	0.304	193	58.672		58.672	
C_{15}	0.237	209	49.533		49.533	
C_{16}	0.208	218	45.344		45.344	
C_{17}	0.22	239	52.58		52.58	
C_{18}	0.169	250	42.25			42.25
C_{19}	0.14	264	36.96			36.96
C_{20}	0.101	275	27.775			27.775
C_{21}	0.0888	291	25.8408			25.8408
C_{22}	0.078	305	23.79			23.79
C_{23}	0.0686	318	21.8148			21.8148
C_{24}	0.0603	331	19.9593			19.9593
C_{25}	0.053	345	18.285			18.285
C_{26}	0.0465	359	16.6935			16.6935
C_{27}	0.0409	374	15.2966			15.2966
C_{28}	0.0359	388	13.9292			13.9292
C_{29}	0.0316	402	12.7032			12.7032
C_{30}	0.0277	416	11.5232			11.5232
C_{31}	0.0244	430	10.492			10.492
C_{32}	0.0214	444	9.5016			9.5016
C_{33}	0.0188	458	8.6104			8.6104
C_{34}	0.0165	472	7.788			7.788

—cont'd

Component	Mol% z_i	MW_i (g/mol)	Z_iMW_i (g/mol)	Group-I	Group-II	Group-III
C_{35}	0.0145	486	7.047			7.047
C_{36}	0.0128	500	6.4			6.4
C_{37}	0.0112	514	5.7568			5.7568
C_{38}	0.00986	528	5.20608			5.20608
C_{39}	0.00866	542	4.69372			4.69372
C_{40}	0.00761	556	4.23116			4.23116
C_{41}	0.00609	570	3.4713			3.4713
C_{42}	0.00588	584	3.43392			3.43392
C_{43}	0.00517	598	3.09166			3.09166
C_{44}	0.00454	612	2.77848			2.77848
C_{45}	0.00399	626	2.49774			2.49774
C_{46}	0.00351	640	2.2464			2.2464
C_{47}	0.00308	654	2.01432			2.01432
C_{48}	0.00271	668	1.81028			1.81028
C_{49}	0.00238	682	1.62316			1.62316
C_{50}	0.00209	696	1.45464			1.45464
C_{51}	0.00183	710	1.2993			1.2993
C_{52}	0.00161	724	1.16564			1.16564
C_{53}	0.00142	738	1.04796			1.04796
C_{54}	0.00128	752	0.96256			0.96256
C_{55}	0.00109	766	0.83494			0.83494
C_{56}	0.000962	780	0.75036			0.75036
C_{57}	0.000845	794	0.67093			0.67093
C_{58}	0.000743	808	0.600344			0.600344
C_{59}	0.000653	822	0.536766			0.536766
C_{60}	0.000574	836	0.479864			0.479864
C_{61}	0.000504	850	0.4284			0.4284
C_{62}	0.000443	864	0.382752			0.382752
C_{63}	0.000389	878	0.341542			0.341542
C_{64}	0.000342	892	0.305064			0.305064
C_{65}	0.0003	906	0.2718			0.2718
C_{66}	0.000264	920	0.24288			0.24288
C_{67}	0.000232	934	0.216688			0.216688
C_{68}	0.000204	948	0.193392			0.193392
C_{69}	0.000179	962	0.172198			0.172198
C_{70}	0.000157	976	0.153232			0.153232
C_{71}	0.000138	990	0.13662			0.13662
C_{72}	0.000122	1004	0.122488			0.122488
C_{73}	0.000107	1018	0.108926			0.108926
C_{74}	0.0000939	1032	0.096905			0.096905

(Continued)

—cont'd

Component	Mol% z_i	MW_i (g/mol)	Z_iMW_i (g/mol)	Group-I	Group-II	Group-III
C_{75}	0.0000825	1046	0.086295			0.086295
C_{76}	0.0000725	1060	0.07685			0.07685
C_{77}	0.0000637	1074	0.068414			0.068414
C_{78}	0.000056	1088	0.060928			0.060928
C_{79}	0.0000492	1102	0.054218			0.054218
C_{80}	0.0000432	1116	0.048211			0.048211
Sum	–	–	1167.65	373.95	400.82	392.89
Weight%				11.98	12.84	12.59

4.2.3 The Cotterman and Prausnitz Method (Equal Mole Method)

Cotterman and Prausnitz (1985) proposed that components of the original fluid are grouped based on mole percent, where pseudocomponents contain approximately the same mole percent.

Example 4.6

Repeat Example 4.5 by using the Cotterman and Prausnitz method.

Solution

The total mole percent of three groups is 6.7%, with an objective value 6.7/3 = 2.23. Add components from C_7 downward until the mole percent reaches a value of around 2.23. Group-I consists of C_7–C_8 with a mole percent of 2.03. Similarly, Group-II consists of C_9–C_{13} with a mole percent of 2.56 and Group-III consists of C_{14}–C_{80} with a mole percent of 2.11. Details of calculations are reported in the following table.

Component	Mol% z_i	Group-I	Group-II	Group-III
N_2	0.12			
CO_2	2.49			
C_1	76.43			
C_2	7.46			
C_3	3.12			
n-C_4	1.21			
i-C_4	0.59			
n-C_5	0.59			
i-C_5	0.5			
C_6	0.79			
C_7	0.95	0.95		

—cont'd

Component	Mol% z_i	Group-I	Group-II	Group-III
C_8	1.08	1.08		
C_9	0.78		0.78	
C_{10}	0.592		0.592	
C_{11}	0.467		0.467	
C_{12}	0.345		0.345	
C_{13}	0.375		0.375	
C_{14}	0.304			0.304
C_{15}	0.237			0.237
C_{16}	0.208			0.208
C_{17}	0.22			0.22
C_{18}	0.169			0.169
C_{19}	0.14			0.14
C_{20}	0.101			0.101
C_{21}	0.0888			0.0888
C_{22}	0.078			0.078
C_{23}	0.0686			0.0686
C_{24}	0.0603			0.0603
C_{25}	0.053			0.053
C_{26}	0.0465			0.0465
C_{27}	0.0409			0.0409
C_{28}	0.0359			0.0359
C_{29}	0.0316			0.0316
C_{30}	0.0277			0.0277
C_{31}	0.0244			0.0244
C_{32}	0.0214			0.0214
C_{33}	0.0188			0.0188
C_{34}	0.0165			0.0165
C_{35}	0.0145			0.0145
C_{36}	0.0128			0.0128
C_{37}	0.0112			0.0112
C_{38}	0.00986			0.00986
C_{39}	0.00866			0.00866
C_{40}	0.00761			0.00761
C_{41}	0.00609			0.00609
C_{42}	0.00588			0.00588
C_{43}	0.00517			0.00517
C_{44}	0.00454			0.00454
C_{45}	0.00399			0.00399
C_{46}	0.00351			0.00351
C_{47}	0.00308			0.00308
C_{48}	0.00271			0.00271

(Continued)

—cont'd

Component	Mol% z_i	Group-I	Group-II	Group-III
C_{49}	0.00238			0.00238
C_{50}	0.00209			0.00209
C_{51}	0.00183			0.00183
C_{52}	0.00161			0.00161
C_{53}	0.00142			0.00142
C_{54}	0.00128			0.00128
C_{55}	0.00109			0.00109
C_{56}	0.000962			0.000962
C_{57}	0.000845			0.000845
C_{58}	0.000743			0.000743
C_{59}	0.000653			0.000653
C_{60}	0.000574			0.000574
C_{61}	0.000504			0.000504
C_{62}	0.000443			0.000443
C_{63}	0.000389			0.000389
C_{64}	0.000342			0.000342
C_{65}	0.0003			0.0003
C_{66}	0.000264			0.000264
C_{67}	0.000232			0.000232
C_{68}	0.000204			0.000204
C_{69}	0.000179			0.000179
C_{70}	0.000157			0.000157
C_{71}	0.000138			0.000138
C_{72}	0.000122			0.000122
C_{73}	0.000107			0.000107
C_{74}	0.0000939			0.0000939
C_{75}	0.0000825			0.0000825
C_{76}	0.0000725			0.0000725
C_{77}	0.0000637			0.0000637
C_{78}	0.000056			0.000056
C_{79}	0.0000492			0.0000492
C_{80}	0.0000432			0.0000432
Sum	–	2.03	2.56	2.11

4.2.4 Danesh et al. Method

A grouping method based on concentration and molecular weight of components in a mixture was proposed by Danesh et al. (1992). This method proposed that the components of the original fluid are arranged in the order of their normal boiling point temperatures and are grouped together in an ascending order to form N_P groups so that the sum of the mole fractions times the logarithm of the molecular weight becomes approximately equal for each groups.

Example 4.7

Repeat Example 4.5 by using Danesh et al. method.

Solution

The value of $z_i \ln(MW_i)$ is calculated for each component. The total value for C_{7+} fraction is 33.83. So the objective value for each group is 33.83/3 = 11.28. Add components from C_7 downward until the value of $\sum z_i \ln(MW_i)$ reaches a value around 11.23. Group-I consists of C_7–C_8 with $\sum z_i \ln(MW_i) = 9.36$. Similarly, Group-II consists of C_9–C_{13} with $\sum z_i \ln(MW_i) = 12.65$ and Group-III consists of C_{14}–C_{80} with $\sum z_i \ln(MW_i) = 11.81$. Details of calculations are given in the following table.

Component	Mol% z_i	MW_i (g/mol)	$z_i \ln(MW_i)$ (g/mol)	Group-I	Group-II	Group-III
N_2	0.12	28.014	–			
CO_2	2.49	44.01	–			
C_1	76.43	16.043	–			
C_2	7.46	30.07	–			
C_3	3.12	44.097	–			
n-C_4	1.21	58.124	–			
i-C_4	0.59	58.124	–			
n-C_5	0.59	72.151	–			
i-C_5	0.5	72.151	–			
C_6	0.79	87.178	–			
C_7	0.95	95	4.326183	4.326183		
C_8	1.08	106	5.036514	5.036514		
C_9	0.78	116	3.7078		3.7078	
C_{10}	0.592	133	2.895087		2.895087	
C_{11}	0.467	152	2.346152		2.346152	
C_{12}	0.345	164	1.759454		1.759454	
C_{13}	0.375	179	1.94527		1.94527	
C_{14}	0.304	193	1.599858			1.599858
C_{15}	0.237	209	1.266133			1.266133
C_{16}	0.208	218	1.119975			1.119975
C_{17}	0.22	239	1.204822			1.204822
C_{18}	0.169	250	0.933127			0.933127
C_{19}	0.14	264	0.780633			0.780633
C_{20}	0.101	275	0.567294			0.567294
C_{21}	0.0888	291	0.503791			0.503791

(Continued)

—cont'd

Component	Mol% z_i	MW_i (g/mol)	$z_i \ln(MW_i)$ (g/mol)	Group-I	Group-II	Group-III
C_{22}	0.078	305	0.446184			0.446184
C_{23}	0.0686	318	0.395277			0.395277
C_{24}	0.0603	331	0.349868			0.349868
C_{25}	0.053	345	0.309708			0.309708
C_{26}	0.0465	359	0.273574			0.273574
C_{27}	0.0409	374	0.242302			0.242302
C_{28}	0.0359	388	0.214			0.214
C_{29}	0.0316	402	0.189488			0.189488
C_{30}	0.0277	416	0.16705			0.16705
C_{31}	0.0244	430	0.147956			0.147956
C_{32}	0.0214	444	0.130451			0.130451
C_{33}	0.0188	458	0.115185			0.115185
C_{34}	0.0165	472	0.10159			0.10159
C_{35}	0.0145	486	0.0897			0.0897
C_{36}	0.0128	500	0.079547			0.079547
C_{37}	0.0112	514	0.069913			0.069913
C_{38}	0.00986	528	0.061813			0.061813
C_{39}	0.00866	542	0.054517			0.054517
C_{40}	0.00761	556	0.048101			0.048101
C_{41}	0.00609	570	0.038645			0.038645
C_{42}	0.00588	584	0.037455			0.037455
C_{43}	0.00517	598	0.033055			0.033055
C_{44}	0.00454	612	0.029132			0.029132
C_{45}	0.00399	626	0.025693			0.025693
C_{46}	0.00351	640	0.02268			0.02268
C_{47}	0.00308	654	0.019968			0.019968
C_{48}	0.00271	668	0.017627			0.017627
C_{49}	0.00238	682	0.01553			0.01553
C_{50}	0.00209	696	0.01368			0.01368
C_{51}	0.00183	710	0.012014			0.012014
C_{52}	0.00161	724	0.010602			0.010602
C_{53}	0.00142	738	0.009378			0.009378
C_{54}	0.00128	752	0.008477			0.008477
C_{55}	0.00109	766	0.007239			0.007239
C_{56}	0.000962	780	0.006406			0.006406
C_{57}	0.000845	794	0.005642			0.005642
C_{58}	0.000743	808	0.004974			0.004974
C_{59}	0.000653	822	0.004383			0.004383
C_{60}	0.000574	836	0.003862			0.003862
C_{61}	0.000504	850	0.0034			0.0034
C_{62}	0.000443	864	0.002995			0.002995

—cont'd

Component	Mol% z_i	MW_i (g/mol)	z_i ln(MW_i) (g/mol)	Group-I	Group-II	Group-III
C_{63}	0.000389	878	0.002637			0.002637
C_{64}	0.000342	892	0.002323			0.002323
C_{65}	0.0003	906	0.002043			0.002043
C_{66}	0.000264	920	0.001802			0.001802
C_{67}	0.000232	934	0.001587			0.001587
C_{68}	0.000204	948	0.001398			0.001398
C_{69}	0.000179	962	0.00123			0.00123
C_{70}	0.000157	976	0.001081			0.001081
C_{71}	0.000138	990	0.000952			0.000952
C_{72}	0.000122	1004	0.000843			0.000843
C_{73}	0.000107	1018	0.000741			0.000741
C_{74}	0.0000939	1032	0.000652			0.000652
C_{75}	0.0000825	1046	0.000574			0.000574
C_{76}	0.0000725	1060	0.000505			0.000505
C_{77}	0.0000637	1074	0.000445			0.000445
C_{78}	0.000056	1088	0.000392			0.000392
C_{79}	0.0000492	1102	0.000345			0.000345
C_{80}	0.0000432	1116	0.000303			0.000303
Sum	—	—	33.83	9.36	12.65	11.81

4.2.5 The Aguilar and McCain Method

Aguilar Zurita and McCain Jr (2002) proposed that group ethane with propane, iso-butane with normal butane and iso-pentane and normal pentane and normal hexane. In this method the methane, hydrogen sulfide, carbon dioxide, and nitrogen are not grouped with other components and considered as pseudocomponents separately. They proposed that the C_{7+} fraction be grouped into two MCN groups (MCN1 and MCN2). The mole fractions of MCN1 and MCN2 are calculated by the following relations.

$$z_{MCN2} = \frac{0.028686608}{1 + 335.91986 \exp(-56.3452274 z_{C_{7+}})} \tag{4.26}$$

$$z_{MCN1} = z_{C_{7+}} - z_{MCN2} \tag{4.27}$$

If there is no satisfactory agreement between the calculated and experimental data, the MCN1 may split into two MCN groups: MCN1a and MCN1b (Aguilar Zurita and McCain Jr, 2002). For volatile oils MCN1 is split into MCN1a-MCN1b as 40−60 mol% and for gas condensates MCN1 is split into MCN1a-MCN1b as 95-5 mol% (Aguilar Zurita and McCain Jr, 2002).

4.3 COMPOSITION RETRIEVAL

Compositions of the fluid vary significantly in some processes such as gas injection, gas–oil displacement, and gas cycling. For example, the first column of Table 4.4 gives the composition of a black oil. The composition of equilibrated phase in the first contact of a test where the 120 cm^3 of a rich gas is added to 60 cm^3 of the original fluid at the temperature 373K and pressure 20.79 MPa is given in the second and third column of Table 4.4. It can be seen that the composition of the original fluid varied. In these cases, the results of phase behavior prediction obtained from the group properties that were generated from the original fluid may be inaccurate. The accuracy could be improved by retrieving the fluid composition of each phase after calculating

Table 4.4 Equilibrated Phase Composition for Black Oil at 20.79 MPa and 373K (Danesh, 1998)

	Original Fluid Composition	Composition of Equilibrated Phase	
Component	Mol%	Oil	Gas
C_1	46.80	47.198	70.287
C_2	8.77	11.618	11.767
C_3	7.44	11.473	9.041
n-C_4	4.01	7.059	4.341
n-C_5	2.56	1.295	0.634
n-C_6	1.77	0.982	0.389
methylcyclopentane	2.25	1.297	0.461
cyclohexene	2.20	1.301	0.422
n-C_7	0.46	0.279	0.09
methylcyclohexane	2.36	1.463	0.423
Toluene	0.72	0.448	0.125
n-C_8	1.02	0.648	0.174
o-Xylene	1.79	1.199	0.264
n-C_9	1.66	1.112	0.247
n-C_{10}	2.73	1.923	0.353
n-C_{11}	2.37	1.733	0.261
n-C_{12}	2.04	1.545	0.192
n-C_{13}	1.77	1.382	0.144
n-C_{14}	1.53	1.219	0.119
n-C_{15}	1.34	1.089	0.082
n-C_{16}	1.15	0.956	0.061
n-C_{17}	0.99	0.833	0.045
n-C_{18}	0.87	0.735	0.034
n-C_{19}	0.75	0.646	0.025
n-C_{20}	0.65	0.567	0.019

the equilibrium conditions and forming new groups based on retrieving composition. Danesh et al. (1992) proposed a modified form of the Wilson equation. This equation is suitable to describe the variations of equilibrium ratios. The changes in the equilibrium ratio for each components result from the changes in the mixture composition. Danesh et al. suggested that the logarithm of equilibrium ratio is expressed by a linear function as follows:

$$\ln(K_i) = A + B(1+\omega_i)\left[1 - \frac{1}{T_{ri}}\right] \tag{4.28}$$

where K is the equilibrium ratio, ω is the acentric factor, T_r is the reduced temperature, and A and B are constants. The equilibrium information of a few components is sufficient to determine these constants. A and B could be determined by the least squares method; then the equilibrium data for other components are obtained.

Example 4.8

The composition and properties of a volatile oil are reported in Table 4.5. This oil was flashed at the temperature 373K and the pressure 20 MPa. The fluid is described by three component groups using Danesh et al. method and molar-averaged properties. The predicted results using a phase behavior model are given in Table 4.6. Calculate the composition of equilibrated phase in terms of the original components.

Solution

The equilibrium ratio for each component is equal to mole fraction in gas phase over mole fraction in liquid phase. The constants in Eq. (4.28) are determined by using the least squares method.

Component	T_c (K)	Acentric Factor	T_r	$K = y/x$	$(1+\omega_i)\left[1-\frac{1}{T_{ri}}\right]$	$\ln(K_i)$
Group-I (methane)	190.56	0.012	1.96	1.700	0.4947	0.5304
Group-II	378.64	0.160	0.99	0.641	−0.0175	−0.4446
Group-III	643.95	0.511	0.58	0.054	−1.0975	−2.9150

$$\ln(K_i) = 2.183 - 0.4917(1+\omega_i)\left[1 - \frac{1}{T_{ri}}\right]$$

(Continued)

Table 4.5 Composition and Properties of a Volatile Oil (Danesh, 1998) (Example 4.8)

Component	z_i Feed Composition	MW (g/mol)	T_c (K)	P_c (Pa)	Acentric Factor
C_1	74.18	16.04	190.56	4.60E+06	0.0115
C_2	5.32	30.07	305.32	4.87E+06	0.0995
C_3	4.67	44.10	369.83	4.25E+06	0.1523
n-C_4	2.58	58.12	425.12	3.80E+06	0.2002
n-C_5	0.97	72.15	469.7	3.37E+06	0.2515
n-C_6	0.69	86.18	507.6	3.03E+06	0.3013
methylcyclopentane	0.88	84.16	532.79	3.78E+06	0.2302
cyclohexene	0.86	84.16	553.54	4.08E+06	0.2118
nC_7	0.18	100.20	540.2	2.74E+06	0.3495
methylcyclohexane	0.94	98.19	572.19	3.47E+06	0.235
Toluene	0.28	92.14	591.79	4.11E+06	0.2641
n-C_8	0.41	114.23	568.7	2.49E+06	0.3996
o-Xylene	0.72	106.17	630.37	3.73E+06	0.3127
n-C_9	0.66	128.26	594.6	2.29E+06	0.4435
n-C_{10}	1.11	142.29	617.7	2.11E+06	0.4923
n-C_{11}	0.96	156.31	639	1.95E+06	0.5303
n-C_{12}	0.83	170.34	658	1.82E+06	0.5764
n-C_{13}	0.73	184.37	675	1.68E+06	0.6174
n-C_{14}	0.63	198.39	693	1.57E+06	0.643
n-C_{15}	0.56	212.42	708	1.48E+06	0.6863
n-C_{16}	0.48	226.45	723	1.40E+06	0.7174
n-C_{17}	0.42	240.47	736	1.34E+06	0.7697
n-C_{18}	0.36	254.50	747	1.27E+06	0.8114
n-C_{19}	0.32	268.53	758	1.21E+06	0.8522
n-C_{20}	0.27	282.55	768	1.16E+06	0.9069

Table 4.6 Flash Calculation Results (Example 4.8)

Component	z_i Feed Composition	T_c (K)	Acentric Factor	Gas (y)	Oil (x)
Group-I (methane)	0.7418	190.56	0.012	0.857	0.504
Group-II	0.1510	378.64	0.160	0.126	0.197
Group-III	0.1072	643.95	0.511	0.016	0.298

Liquid mole fraction: 0.3326.

The equilibrium ratios of the original components could be calculated by substituting the acentric factor and reduced temperature of each component in the previous equation. Then the mole fraction of each component in liquid and vapor phases is determined using the following equation.

$$x_i = \frac{z_i}{n^L + (1 - n^L)K_i}, \quad n^L = 0.3326 \text{ (Component balance for each component)}$$

$$y_i = K_i x_i \quad \text{(Equilibrium ratio definition)}$$

The composition of liquid and vapor phases (equilibrated phases) are determined as reported in the following table.

Component	z_i	T_c (K)	Acentric Factor	T_r	$(1+\omega_i)\left[1-\frac{1}{T_{ri}}\right]$	K_i	x_i	Normalized x_i	y_i	Normalized y_i
C_1	0.7418	190.56	0.0115	1.957	0.4947	1.8010	0.48338	0.49417	0.87055	0.86121
C_3	0.0532	305.32	0.0995	1.222	0.1995	0.9454	0.05524	0.05648	0.05223	0.05167
C_3	0.0467	369.83	0.1523	1.009	0.0098	0.6248	0.06225	0.06364	0.03889	0.03847
n-C_4	0.0258	425.12	0.2002	0.877	−0.1677	0.4241	0.04184	0.04278	0.01775	0.01755
n-C_5	0.0097	469.7	0.2515	0.794	−0.3245	0.3012	0.01820	0.01860	0.00548	0.00542
n-C_6	0.0069	507.6	0.3013	0.735	−0.4696	0.2194	0.01430	0.01462	0.00314	0.00310
methylcyclopentane	0.0088	532.79	0.2302	0.700	−0.5270	0.1936	0.01895	0.01937	0.00367	0.00363
cyclohexene	0.0086	553.54	0.2118	0.674	−0.5865	0.1700	0.01924	0.01967	0.00327	0.00323
n-C_7	0.0018	540.2	0.3495	0.690	−0.6049	0.1633	0.00414	0.00424	0.00068	0.00067
methylcyclohexane	0.0094	572.19	0.235	0.652	−0.6595	0.1449	0.02178	0.02226	0.00316	0.00312
Toluene	0.0028	591.79	0.2641	0.630	−0.7415	0.1212	0.00682	0.00697	0.00083	0.00082
n-C_8	0.0041	568.7	0.3996	0.656	−0.7343	0.1231	0.00979	0.01001	0.00121	0.00119
o-Xylene	0.0072	630.37	0.3127	0.592	−0.9058	0.0847	0.01840	0.01881	0.00156	0.00154
n-C_9	0.0066	594.6	0.4435	0.627	−0.8576	0.0941	0.01679	0.01717	0.00158	0.00156
n-C_{10}	0.0111	617.7	0.4923	0.604	−0.9790	0.0722	0.02913	0.02978	0.00210	0.00208
n-C_{11}	0.0096	639	0.5303	0.584	−1.0913	0.0565	0.02598	0.02656	0.00147	0.00145
n-C_{12}	0.0083	658	0.5764	0.567	−1.2045	0.0441	0.02301	0.02352	0.00101	0.00100
n-C_{13}	0.0073	675	0.6174	0.553	−1.3095	0.0351	0.02048	0.02093	0.00072	0.00071
n-C_{14}	0.0063	693	0.643	0.538	−1.4095	0.0282	0.01801	0.01841	0.00051	0.00050
n-C_{15}	0.0056	708	0.6863	0.527	−1.5145	0.0224	0.01614	0.01650	0.00036	0.00036
n-C_{16}	0.0048	723	0.7174	0.516	−1.6115	0.0181	0.01404	0.01435	0.00025	0.00025
n-C_{17}	0.0042	736	0.7697	0.507	−1.7223	0.0142	0.01216	0.01243	0.00017	0.00017
n-C_{18}	0.0036	747	0.8114	0.499	−1.8163	0.0116	0.01067	0.01090	0.00012	0.00012
n-C_{19}	0.0032	758	0.8522	0.492	−1.9118	0.0094	0.00932	0.00953	0.00009	0.00009
n-C_{20}	0.0027	768	0.9069	0.486	−2.0194	0.0074	0.00812	0.00830	0.00006	0.00006
Sum	1	—	—	—	—	—	0.97818	1	1.01085	1

4.4 ASSIGNING PROPERTIES TO MULTIPLE CARBON NUMBER

Several methods have been suggested to calculate the properties of pseudocomponents (Chueh and Prausnitz, 1968; Lee and Kesler, 1975; Hong, 1982; Pedersen et al., 1984; Wu and Batycky, 1988; Leibovici, 1993).

Molar averaging is the simplest and most common mixing rule.

$$\theta_k = \frac{\sum_{i(k)} z_i \theta_i}{z_k} \tag{4.29}$$

where z_i is the original mole fraction of components and θ represents the property of components such as critical temperature, critical pressure, critical volume, molecular weight, or acentric factor. z_k is the mole fraction of group k in the mixture and is defined as

$$z_k = \sum_{i(k)} z_i \tag{4.30}$$

Pedersen et al. (1984) suggested the using of mass fraction instead of mole fraction in Eq. (4.29). Wu and Batycky (1988) suggested to calculate the properties of the group by a combination of molar averaging and weight averaging.

Lee and Kesler (1975) proposed the mixing rules in Eqs. (4.31) to (4.34) on the basis of the Chueh and Prausnitz's (1968) arguments.

$$v_{ck} = \frac{1}{8} \sum_{i(k)} \sum_{j(k)} z_i z_j \left(v_{ci}^{\frac{1}{3}} + v_{cj}^{\frac{1}{3}} \right)^3 \tag{4.31}$$

$$T_{ck} = \frac{1}{8 v_{ck}} \sum_{i(k)} \sum_{j(k)} z_i z_j (T_{ci} T_{cj})^{\frac{1}{2}} \left(v_{ci}^{\frac{1}{3}} + v_{cj}^{\frac{1}{3}} \right)^3 \tag{4.32}$$

$$Z_{ck} = 0.2905 - 0.085\omega_k \tag{4.33}$$

$$P_{ck} = \frac{Z_{ck} R T_{ck}}{v_{ck}} \tag{4.34}$$

The acentric factor and molecular weight are determined by molar averaging in the Lee–Kesler method.

The binary interaction parameters between the pseudocomponents k and q could be determined by Eq. (4.35).

$$k_{kq} = \frac{\sum_{i(k)} \sum_{j(q)} z_i z_j k_{ij}}{z_k z_q} \quad k \neq q \tag{4.35}$$

A comparison between these mixing rules did not clear priority for any of them (Danesh et al., 1992).

The selected EOS and the number of group used to describing the original fluid affected the results. In some cases, an improvement in accuracy relative to that using the full composition was also observed probably due to the cancellation of errors.

Example 4.9

Using the results of Example 4.2, described the oil by the Aguilar and McCain method. Then estimate the critical temperature, critical pressure, and acentric factor for each group by molar averaging.

Solution

The mole fraction of C_{7+} after matching saturation pressure using extended group is 0.3761; hence the mole fraction of MCN1 (Group-VI) and MCN2 (Group-VII) is determined by Eqs. (4.26) and (4.27).

$$z_{MCN2} = \frac{0.028686608}{1 + 335.91986 \exp(-56.3452274 \times 0.3761)} = 0.0287$$

$$z_{MCN1} = 0.3761 - 0.0287 = 0.3475$$

Add components from C_7 downward until the mole fraction reaches a value around 0.3475. MCN1 consists of C_7–C_{24} with a weight of 0.3498. Hence the MCN2 consists of C_{25}–C_{45} with a mole fraction 0.0263. Based on the Aguilar and McCain method the other group formed by combining ethane with propane and combine iso-butane with normal butane and iso-pentane and normal pentane and normal hexane. The methane, hydrogen sulfide, carbon dioxide, and nitrogen are considered as pseudocomponents separately.

The properties of pseudocomponents are determined by Eq. (4.29). The results are reported in the following table.

Component	z_i	T_c (K)	P_c (MPa)	w_i	z_iT_c (K)	z_iP_c (MPa)	z_iw_i
N_2	0.0065	126.10	3.39E+06	0.0403	0.820	2.21E+04	2.62E−04
Group-I	**0.0065**	**126.10**	**3.39E+06**	**0.0403**	**0.820**	**2.21E+04**	**2.62E−04**
CO_2	0.0011	304.19	7.38E+06	0.2276	0.344	8.35E+03	2.57E−04
Group-II	**0.0011**	**304.19**	**7.38E+06**	**0.2276**	**0.344**	**8.35E+03**	**2.57E−04**
C_1	0.4439	190.56	4.60E+06	0.0115	84.589	2.04E+06	5.10E−03
Group-III	**0.4439**	**190.56**	**4.60E+06**	**0.0115**	**84.589**	**2.04E+06**	**5.10E−03**
C_2	0.0536	305.32	4.87E+06	0.0995	16.377	2.61E+05	5.34E−03
C_3	0.0414	369.83	4.25E+06	0.1523	15.305	1.76E+05	6.30E−03
Group-IV	**0.0950**	**32.681/0.0950 =333.415**	**4.37E+05/0.0950 =4.60E+06**	**1.16E−02/0.0950 =0.1225**	**31.681**	**4.37E+05**	**1.16E−02**
i-C_4	0.0090	408.14	3.65E+06	3.655	3.27E+04	1.59E−03	3.655
n-C_4	0.0228	425.12	3.80E+06	9.698	8.66E+04	4.57E−03	9.698
i-C_5	0.0105	460.43	3.38E+06	4.818	3.54E+04	2.38E−03	4.818
n-C_5	0.0138	469.70	3.37E+06	6.464	4.64E+04	3.46E−03	6.464
C_6	0.0213	510.00	3.27E+06	10.865	6.97E+04	6.42E−03	10.865
Group-V	**0.0773**	**35.500/0.0773 =459.264**	**2.71E+05/0.0773 =3.50E+06**	**1.84E−02/0.0773 =0.2382**	**35.500**	**2.71E+05**	**1.84E−02**
C_7	0.0511	553.37	3.23E+06	0.2513	28.301	1.65E+05	1.29E−02
C_8	0.0417	579.19	2.98E+06	0.2920	24.137	1.24E+05	1.22E−02
C_9	0.0390	606.12	2.74E+06	0.3397	23.655	1.07E+05	1.33E−02
C_{10}	0.0325	629.53	2.56E+06	0.3839	20.455	8.31E+04	1.25E−02
C_{11}	0.0292	651.38	2.40E+06	0.4291	19.005	7.00E+04	1.25E−02
C_{12}	0.0260	671.99	2.27E+06	0.4748	17.485	5.91E+04	1.24E−02
C_{13}	0.0230	690.42	2.17E+06	0.5187	15.913	5.01E+04	1.20E−02
C_{14}	0.0209	708.58	2.08E+06	0.5657	14.809	4.35E+04	1.18E−02
C_{15}	0.0181	726.47	2.00E+06	0.6162	13.185	3.64E+04	1.12E−02
C_{16}	0.0148	743.01	1.93E+06	0.6674	10.989	2.86E+04	9.87E−03
C_{17}	0.0117	757.66	1.88E+06	0.7162	8.890	2.21E+04	8.40E−03
C_{18}	0.0090	769.63	1.85E+06	0.7585	6.960	1.67E+04	6.86E−03
C_{19}	0.0073	780.50	1.82E+06	0.7996	5.679	1.33E+04	5.82E−03
C_{20}	0.0074	792.72	1.78E+06	0.8494	5.836	1.31E+04	6.25E−03

C_{21}	0.0057	803.29	1.75E+06	0.8948	4.565	9.96E+03	5.08E−03
C_{22}	0.0042	814.33	1.72E+06	0.9455	3.425	7.22E+03	3.98E−03
C_{23}	0.0042	823.68	1.69E+06	0.9907	3.500	7.18E+03	4.21E−03
C_{24}	0.0039	833.04	1.67E+06	1.0384	3.223	6.45E+03	4.02E−03
Group-VI	**0.3498**	**223.012/0.3498 =657.475**	**8.64E+05/0.3498 =2.47E+06**	**1.65E−01/0.3498 =0.4719**	**230.012**	**8.64E+05**	**1.65E−01**
C_{25}	0.0034	843.13	1.64E+06	1.0926	2.844	5.55E+03	3.69E−03
C_{26}	0.0027	851.57	1.63E+06	1.1405	2.316	4.43E+03	3.10E−03
C_{27}	0.0024	859.84	1.61E+06	0.9864	2.061	3.86E+03	2.36E−03
C_{28}	0.0020	868.03	1.59E+06	1.0118	1.759	3.23E+03	2.05E−03
C_{29}	0.0018	874.75	1.59E+06	1.0314	1.571	2.85E+03	1.85E−03
C_{30}	0.0016	882.83	1.57E+06	1.0567	1.405	2.50E+03	1.68E−03
C_{31}	0.0014	890.14	1.56E+06	1.0791	1.202	2.11E+03	1.46E−03
C_{32}	0.0013	897.40	1.55E+06	1.1015	1.128	1.95E+03	1.38E−03
C_{33}	0.0011	903.92	1.55E+06	1.1210	1.007	1.72E+03	1.25E−03
C_{34}	0.0009	910.32	1.54E+06	1.1407	0.783	1.32E+03	9.81E−04
C_{35}	0.0008	915.98	1.53E+06	1.1576	0.709	1.19E+03	8.97E−04
C_{36}	0.0007	922.22	1.52E+06	1.1775	0.644	1.06E+03	8.22E−04
C_{37}	0.0006	927.14	1.53E+06	1.1916	0.583	9.59E+02	7.49E−04
C_{38}	0.0006	933.33	1.51E+06	1.2114	0.555	9.01E+02	7.20E−04
C_{39}	0.0005	938.20	1.52E+06	1.2254	0.500	8.09E+02	6.53E−04
C_{40}	0.0004	944.40	1.51E+06	1.2449	0.408	6.52E+02	5.38E−04
C_{41}	0.0004	948.47	1.52E+06	1.2563	0.352	5.62E+02	4.66E−04
C_{42}	0.0004	953.19	1.51E+06	1.2705	0.358	5.68E+02	4.77E−04
C_{43}	0.0003	957.90	1.51E+06	1.2846	0.324	5.12E+02	4.34E−04
C_{44}	0.0003	963.93	1.50E+06	1.3042	0.279	4.35E+02	3.77E−04
C_{45+}	0.0028	980.35	1.62E+06	1.3388	2.732	4.51E+03	3.73E−03
Group-VII	**0.0263**	**23.520/0.0263 =894.052**	**4.17E+04/0.0263 =1.58E+06**	**2.97E−02/0.0263 =1.1279**	**23.520**	**4.17E+04**	**2.97E−02**

Example 4.10

For the fluid that is described in Example 4.2, determine the binary interaction parameter between (1) Group-I and Group-V, (2) Group-III and Group-VI, and (3) Group-III and Group-VII by Eq. (4.35).

Solution

(1) The binary interaction parameter for the original component is extracted from Table 4.1. According to Eq. (4.35), the binary interaction is determined as follows.

$$k_{I-V} = \frac{\sum_{i(I)} \sum_{j(V)} z_i z_j k_{ij}}{z_I z_V}$$

$$= \frac{1}{0.0065 \times 0.0773} \left[z_{N_2} z_{iC_4} k_{N_2-iC_4} + z_{N_2} z_{nC_4} k_{N_2-nC_4} + z_{N_2} z_{iC_5} k_{N_2-iC_5} + z_{N_2} z_{nC_5} k_{N_2-nC_5} + z_{N_2} z_{C_6} k_{N_2-C_6} \right]$$

	k_{ij}	z_j	$z_i z_j$ ($z_i = 0.0065$)	$z_i z_j k_{ij}$
N_2-i-C_4	0.095	0.0090	6.18E−04	5.53E−06
N_2-n-C_4	0.095	0.0228	6.18E−04	1.41E−05
N_2-i-C_5	0.100	0.0105	6.50E−04	6.80E−06
N_2-n-C_5	0.110	0.0138	7.15E−04	9.84E−06
N_2-C_6	0.110	0.0213	7.15E−04	1.52E−05
–	–	0.0773	–	5.15E−05

$k_{I-V} = 5.15E-05/(0.0065 \times 0.0773) = 0.1025$

(2) and (3)Using the calculated binary interaction parameters in Example 4.2 and Eq. (4.35), the binary interaction parameter between Group-III and Group-VI and Group-III and Group-VII is determined to be similar to the previous part. The results are given in the following table.

	k_{ij}	z_j	$z_i z_j$ ($z_i = 0.4439$)	$z_i z_j k_{ij}$
C_1-C_7	0.0375	0.0511	2.27E−02	8.50E−04
C_1-C_8	0.0397	0.0417	1.85E−02	7.34E−04
C_1-C_9	0.0419	0.0390	1.73E−02	7.26E−04
C_1-C_{10}	0.0439	0.0325	1.44E−02	6.33E−04
C_1-C_{11}	0.0457	0.0292	1.30E−02	5.92E−04
C_1-C_{12}	0.0476	0.0260	1.15E−02	5.50E−04

—cont'd	k_{ij}	z_j	z_iz_j (z_i = 0.4439)	$z_iz_jk_{ij}$
C_1–C_{13}	0.0493	0.0230	1.02E−02	5.05E−04
C_1–C_{14}	0.0511	0.0209	9.28E−03	4.74E−04
C_1–C_{15}	0.0528	0.0181	8.06E−03	4.25E−04
C_1–C_{16}	0.0543	0.0148	6.57E−03	3.56E−04
C_1–C_{17}	0.0557	0.0117	5.21E−03	2.90E−04
C_1–C_{18}	0.0570	0.0090	4.01E−03	2.29E−04
C_1–C_{19}	0.0581	0.0073	3.23E−03	1.88E−04
C_1–C_{20}	0.0592	0.0074	3.27E−03	1.94E−04
C_1–C_{21}	0.0603	0.0057	2.52E−03	1.52E−04
C_1–C_{22}	0.0612	0.0042	1.87E−03	1.14E−04
C_1–C_{23}	0.0620	0.0042	1.89E−03	1.17E−04
C_1–C_{24}	0.0629	0.0039	1.72E−03	1.08E−04
–	–	0.3498	–	7.24E−03
k_{III-VI} = 7.24E−03/(0.4439 × 0.3498) = 0.0466				
C_1–C_{25}	0.0638	0.0034	1.50E−03	9.56E−05
C_1–C_{26}	0.0646	0.0027	1.21E−03	7.80E−05
C_1–C_{27}	0.0654	0.0024	1.06E−03	6.96E−05
C_1–C_{28}	0.0661	0.0020	9.00E−04	5.95E−05
C_1–C_{29}	0.0668	0.0018	7.97E−04	5.32E−05
C_1–C_{30}	0.0675	0.0016	7.07E−04	4.77E−05
C_1–C_{31}	0.0681	0.0014	5.99E−04	4.08E−05
C_1–C_{32}	0.0688	0.0013	5.58E−04	3.84E−05
C_1–C_{33}	0.0694	0.0011	4.95E−04	3.43E−05
C_1–C_{34}	0.0700	0.0009	3.82E−04	2.67E−05
C_1–C_{35}	0.0706	0.0008	3.44E−04	2.43E−05
C_1–C_{36}	0.0710	0.0007	3.10E−04	2.20E−05
C_1–C_{37}	0.0716	0.0006	2.79E−04	2.00E−05
C_1–C_{38}	0.0721	0.0006	2.64E−04	1.90E−05
C_1–C_{39}	0.0726	0.0005	2.37E−04	1.72E−05
C_1–C_{40}	0.0731	0.0004	1.92E−04	1.40E−05
C_1–C_{41}	0.0735	0.0004	1.65E−04	1.21E−05
C_1–C_{42}	0.0740	0.0004	1.67E−04	1.23E−05
C_1–C_{43}	0.0744	0.0003	1.50E−04	1.12E−05
C_1–C_{44}	0.0749	0.0003	1.28E−04	9.62E−06
C_1–C_{45+}	0.0777	0.0028	1.24E−03	9.61E−05
–	–	0.0263	–	8.02E−04
$k_{III-VII}$ = 8.02E−04/(0.4439 × 0.0263) = 0.0686				

Example 4.11

The most widely used correlation for the estimation of the binary interaction parameter for hydrocarbon pairs is that of Chueh and Prausnitz (1968). The Chueh and Prausnitz (1968) equation for prediction of binary interaction parameter is as follows.

$$k_{ij} = A\left[1 - \left(\frac{2V_{ci}^{1/6}V_{cj}^{1/6}}{V_{ci}^{1/3} + V_{cj}^{1/3}}\right)^{B}\right]$$

where V_{ci} and V_{cj} are the critical volume of components of i and j. Originally $A = 1$ and $B = 6$; however, in practical cases the value of B is set to 6. The value of A is usually adjusted to match saturation pressure or other vapor–liquid equilibrium (VLE) data (Danesh, 1998; Li and Englezos, 2003). For this example, the values of A and B for methane/C_{7+} pairs are taken as 0.18 and 6, respectively. For fluid that is described in Example 4.2, determine the binary interaction parameter for methane/C_{7+} pairs (using Twu correlation to determine the critical volume). Then, determine the binary interaction parameter between Group-III and Group-VI and Group-III and Group-VII by Eq. (4.35).

Solution

First, we should determine the critical volume by Twu correlation (described in Section 3.4.3), then the binary interaction parameter for methane/C_{7+} pairs is determined by the Chueh–Prausnitz equation. The critical volume for methane is 0.0986 m^3/kg moles (Danesh, 1998). The results are given in the following table.

Component	T_b	SG	T_{cp}	phi	V_{cp}	Sp	ΔS_V	f_V	V_c	k_{ij}
C_7	366	0.745	534.45	0.315	0.414	0.685	−0.291	−0.011	0.380	0.0251
C_8	390	0.761	560.38	0.304	0.463	0.701	−0.293	−0.011	0.424	0.0290
C_9	416	0.776	587.50	0.292	0.521	0.717	−0.298	−0.011	0.475	0.0332
C_{10}	439	0.791	610.68	0.281	0.577	0.729	−0.311	−0.012	0.523	0.0368
C_{11}	461	0.804	632.19	0.271	0.635	0.740	−0.327	−0.014	0.570	0.0402
C_{12}	482	0.817	652.13	0.261	0.695	0.749	−0.347	−0.015	0.615	0.0433
C_{13}	501	0.829	669.71	0.252	0.752	0.757	−0.370	−0.017	0.657	0.0459
C_{14}	520	0.842	686.88	0.243	0.814	0.764	−0.395	−0.019	0.698	0.0485
C_{15}	539	0.854	703.64	0.234	0.879	0.770	−0.421	−0.021	0.740	0.0509
C_{16}	557	0.865	719.17	0.226	0.944	0.776	−0.443	−0.024	0.781	0.0532
C_{17}	573	0.875	732.71	0.218	1.005	0.781	−0.466	−0.026	0.816	0.0550
C_{18}	586	0.884	743.53	0.212	1.056	0.784	−0.487	−0.028	0.842	0.0564
C_{19}	598	0.892	753.38	0.206	1.106	0.787	−0.506	−0.030	0.867	0.0576
C_{20}	612	0.900	764.72	0.200	1.165	0.791	−0.522	−0.032	0.900	0.0592
C_{21}	624	0.908	774.30	0.194	1.218	0.794	−0.538	−0.034	0.926	0.0604
C_{22}	637	0.914	784.55	0.188	1.278	0.797	−0.552	−0.036	0.957	0.0618
C_{23}	648	0.920	793.12	0.183	1.330	0.799	−0.564	−0.038	0.983	0.0630
C_{24}	659	0.926	801.61	0.178	1.383	0.802	−0.577	−0.039	1.008	0.0640

—cont'd

Component	T_b	SG	T_{cp}	phi	V_{cp}	Sp	ΔS_V	f_V	V_c	k_{ij}
C_{25}	671	0.933	810.76	0.172	1.443	0.804	−0.591	−0.041	1.035	0.0652
C_{26}	681	0.939	818.32	0.168	1.494	0.806	−0.604	−0.043	1.057	0.0660
C_{27}	691	0.944	825.81	0.163	1.546	0.808	−0.615	−0.045	1.080	0.0670
C_{28}	701	0.949	833.23	0.159	1.600	0.810	−0.625	−0.046	1.104	0.0679
C_{29}	709	0.954	839.13	0.155	1.644	0.811	−0.636	−0.048	1.119	0.0685
C_{30}	719	0.959	846.45	0.151	1.700	0.813	−0.645	−0.049	1.143	0.0694
C_{31}	728	0.964	852.98	0.147	1.751	0.814	−0.655	−0.051	1.163	0.0702
C_{32}	737	0.968	859.48	0.143	1.804	0.816	−0.664	−0.052	1.184	0.0709
C_{33}	745	0.973	865.22	0.139	1.851	0.817	−0.673	−0.054	1.200	0.0715
C_{34}	753	0.977	870.92	0.135	1.900	0.818	−0.681	−0.055	1.219	0.0721
C_{35}	760	0.981	875.89	0.132	1.943	0.819	−0.689	−0.057	1.233	0.0726
C_{36}	768	0.985	881.54	0.129	1.993	0.820	−0.696	−0.058	1.253	0.0733
C_{37}	774	0.988	885.75	0.126	2.031	0.821	−0.703	−0.059	1.264	0.0737
C_{38}	782	0.992	891.35	0.123	2.082	0.822	−0.709	−0.060	1.284	0.0744
C_{39}	788	0.996	895.54	0.120	2.121	0.822	−0.716	−0.061	1.295	0.0747
C_{40}	796	0.999	901.09	0.117	2.174	0.823	−0.723	−0.063	1.314	0.0754
C_{41}	801	1.002	904.55	0.114	2.207	0.824	−0.729	−0.064	1.323	0.0756
C_{42}	807	1.006	908.69	0.112	2.248	0.824	−0.735	−0.065	1.335	0.0760
C_{43}	813	1.009	912.81	0.109	2.289	0.825	−0.740	−0.066	1.348	0.0765
C_{44}	821	1.012	918.29	0.106	2.344	0.826	−0.745	−0.067	1.370	0.0771
C_{45+}	844	1.032	933.99	0.096	2.509	0.828	−0.781	−0.073	1.389	0.0777

The k_{III-VI} and $k_{III-VII}$ are 0.0406 and 0.0703, respectively, which are calculated similar to the previous example.

4.5 MATCHING THE SATURATION PRESSURE USING THE GROUPED COMPOSITION

After grouping the extended groups, the match of calculated and experimental saturation pressure may be altered slightly. Aguilar Zurita and McCain Jr (2002) and Al-Meshari (2005) proposed a methodology to match the saturation pressure using only one variable. Aguilar Zurita and McCain Jr (2002) used the ratio of the critical temperature to the critical pressure of the heaviest component, MCN2. Al-Meshari (2005) used the acentric factor of the heaviest component as the adjusting variable.

Example 4.12

Using the results of Example 4.9 and Example 4.10, match the saturation pressure by Al-Meshari (2005) method.

Solution

The mole fractions and critical properties of the groups are given in the following table (from Example 4.9).

Component	z_i	T_c (K)	P_c (Pa)	w_i
Group-I (N_2)	0.0065	126.10	3.39E+06	0.0403
Group-II (CO_2)	0.0011	304.19	7.38E+06	0.2276
Group-III (C_1)	0.4439	190.56	4.60E+06	0.0115
Group-IV (C_2–C_3)	0.0950	333.42	4.60E+06	0.1225
Group-V (C_4–C_6)	0.0773	459.26	3.50E+06	0.2382
Group-VI (C_7–C_{24})	0.3498	657.47	2.47E+06	0.4719
Group-VII (C_{25}–C_{45+})	0.0263	894.05	1.58E+06	1.1279

Binary interaction parameters between grouped compositions are reported in the following table.

	Group-I (N_2)	Group-II (CO_2)	Group-III (C_1)	Group-IV (C_2–C_3)	Group-V (C_4–C_6)	Group-VI (C_7–C_{24})	Group-VII (C_{25}–C_{45+})
Group-I (N_2)	0	0	0.025	0.0448	0.1025	0.11	0.11
Group-II (CO_2)	0	0	0.105	0.1278	0.1156	0.115	0.115
Group-III (C_1)	0.025	0.105	0	0	0	0.0468	0.0694
Group-IV (C_2–C_3)	0.0448	0.1278	0	0	0	0	0
Group-V (C_4–C_6)	0.1025	0.1156	0	0	0	0	0
Group-VI (C_7–C_{24})	0.11	0.115	0.0468	0	0	0	0
Group-VII (C_{25}–C_{45+})	0.11	0.115	0.0694	0	0	0	0

The bubble point pressure at temperature 345.8K is 23.02 MPa (the bubble point pressure is calculated similar to the Example 4.1). Guess another value for the acentric factor of the heaviest component, Group-VII. If we select 1.2500 for the acentric factor of Group-VII, the calculated bubble point pressure is 23.49 MPa. Adjust the acentric factor of the heaviest component by a simple linear interpolation, that is, 1.3149. Bubble point pressure in this case is 23.74 MPa that is matched with the experimental value. The variation of the acentric factor of the heaviest component during acentric factor adjustment is 16.5%.

Example 4.13

The constant composition expansion (CCE) and the constant volume depletion (CVD) are the two most widely used tests at reservoir temperature (Danesh, 1998). The CCE experiment is used for oil samples and gas condensate fluids. Bubble point pressure, iso-thermal oil compressibility, and undersaturated oil density are usually determined by CCE experiment for an oil sample. The total relative volume, which is defined as the volume of gas or gas–oil mixture divided by the volume at dew point pressure and compressibility factor, is determined by CCE experiment for a gas condensate sample. Most CCE experiments are conducted in a visual cell for gas condensates. The system pressure is lowered stepwise, where the equilibrium is obtained at each pressure value. Hence, each step can be modeled by a flash calculation. In gas condensate reservoir, the percentage of the vapor decreases as the pressure declines. This phenomenon is known as the *retrograde condensation*. However, the percentage of vapor can be increased with continued pressure decline.

The composition of a gas condensate sample is reported in Table 4.7.

Table 4.7 Composition of a Gas Condensate (Al-Meshari, 2005) (Example 4.13)

Component	Mol%
N_2	3.12
CO_2	3.23
C_1	69.76
C_2	9.03
C_3	4.02
i-C_4	0.81
n-C_4	1.44
i-C_5	0.6
n-C_5	0.55
C_6	0.96
C_{7+}	6.47

The molecular weight and specific gravity are 163.8 g/mol and 0.804, respectively. The CCE experiment report is given in Table 4.8.

Using the recommended procedure in Chapter 6, characterize the gas condensate sample and match the saturation pressure using extended groups. Then use the Aguilar and McCain method for grouping extended groups and determine the critical temperature, critical pressure, and acentric factor for each group by molar averaging method. Match the saturation pressure using the grouped composition by adjusting the acentric factor of the heaviest

(Continued)

Table 4.8 Total Relative Volume and Compressibility Factor of Vapor Phase From Constant Composition Expansion Experiment at Temperature 424.82K (Al-Meshari, 2005) (Example 4.13)

P (MPa)	V/V_{sat}	Compressibility Factor of Vapor Phase
62.14	0.8390	1.406
60.76	0.8458	1.386
55.25	0.8770	1.307
53.87	0.8859	1.287
52.49	0.8951	1.267
51.11	0.9054	1.248
49.73	0.9161	1.228
48.35	0.9270	1.209
46.97	0.9396	1.190
45.60	0.9526	1.171
44.22	0.9666	1.152
42.84	0.9815	1.133
41.46	0.9977	1.115
41.27[a]	1.0000	1.112
40.08	1.0152	–
38.70	1.0342	–
37.32	1.0549	–
35.95	1.0775	–
34.57	1.1024	–
33.19	1.1298	–
31.81	1.1602	–
30.43	1.1941	–
29.05	1.2332	–
27.67	1.2753	–
26.30	1.3243	–
24.92	1.3808	–
23.54	1.4463	–
22.16	1.5233	–

[a]Dew point pressure.

component. Finally, determine the total relative volume and compressibility factor of vapor phase by tuned EOS.

Solution

The plus fraction is extended up to C_{44} similar to Example 3.12. Then the dew point pressure is matched by molecular weight adjustment. The mole fraction and molecular weight of C_{7+} are 0.0695 and 151.7 g/mol, respectively. The mole fraction and properties of extended groups after matching the saturation by molecular weight adjustment are given in the following table.

$MW_{mix.}$ = 31.38 g/mol, Z_{C7+} = 0.0695, MW_{C7+} = 151.7 g/mol, P_{dew}(calculated) = 41.37 MPa, P_{dew}(experimental) = 41.27 MPa

Component	z_i	MW (g/mol)	Specific Gravity	T_c (K)	P_c (MPa)	ω	$k_{C_1-C_N}$
N_2	0.0310	28.01	–	126.10	3.39	0.0403	–
CO_2	0.0321	44.01	–	304.19	7.38	0.2276	–
C_1	0.6941	16.04	–	190.56	4.60	0.0115	–
C_2	0.0898	30.07	–	305.32	4.87	0.0995	–
C_3	0.0400	44.10	–	369.83	4.25	0.1523	–
i-C_4	0.0081	58.12	–	408.14	3.65	0.1770	–
n-C_4	0.0143	58.12	–	425.12	3.80	0.2002	–
i-C_5	0.0060	72.15	–	460.43	3.38	0.2275	–
n-C_5	0.0055	72.15	–	469.70	3.37	0.2515	–
C_6	0.0096	86.18	–	510.00	3.27	0.3013	0.0261
C_7	0.0131	94.61	0.731	549.86	3.14	0.2650	0.0356
C_8	0.0101	107.55	0.748	576.00	2.90	0.3033	0.0379
C_9	0.0089	120.51	0.764	602.89	2.65	0.3500	0.0401
C_{10}	0.0069	133.78	0.778	626.36	2.47	0.3927	0.0421
C_{11}	0.0058	147.01	0.791	648.23	2.31	0.4365	0.0440
C_{12}	0.0049	160.74	0.804	668.77	2.18	0.4808	0.0458
C_{13}	0.0040	174.97	0.816	687.24	2.08	0.5232	0.0475
C_{14}	0.0034	189.94	0.828	705.39	1.98	0.5685	0.0492
C_{15}	0.0027	205.66	0.840	723.22	1.90	0.6171	0.0509
C_{16}	0.0021	221.44	0.852	739.88	1.83	0.6666	0.0524
C_{17}	0.0015	236.47	0.862	754.53	1.78	0.7135	0.0538
C_{18}	0.0011	250.28	0.871	766.47	1.74	0.7541	0.0551
C_{19}	8.39E−04	262.81	0.878	777.28	1.71	0.7936	0.0562
C_{20}	7.99E−04	275.74	0.886	789.48	1.67	0.8416	0.0572
C_{21}	5.79E−04	289.05	0.893	800.08	1.64	0.8854	0.0583
C_{22}	4.05E−04	300.61	0.900	811.05	1.60	0.9343	0.0592
C_{23}	3.88E−04	311.81	0.906	820.41	1.57	0.9781	0.0600
C_{24}	3.34E−04	324.05	0.912	829.78	1.55	1.0242	0.0609
C_{25}	2.74E−04	336.55	0.918	839.80	1.52	1.0764	0.0617
C_{26}	2.09E−04	348.58	0.924	848.25	1.51	0.9717	0.0625
C_{27}	1.74E−04	360.08	0.929	856.55	1.49	0.9971	0.0633
C_{28}	1.40E−04	371.34	0.934	864.76	1.47	1.0225	0.0640
C_{29}	1.17E−04	382.34	0.939	871.47	1.46	1.0422	0.0647
C_{30}	9.87E−05	393.34	0.944	879.53	1.45	1.0676	0.0654
C_{31}	7.95E−05	404.11	0.949	886.80	1.43	1.0902	0.0660
C_{32}	7.04E−05	414.84	0.953	894.02	1.42	1.1127	0.0666

(Continued)

—cont'd

$MW_{mix.}$ = 31.38 g/mol, Z_{C7+} = 0.0695, MW_{C7+} = 151.7 g/mol, P_{dew}(calculated) = 41.37 MPa, P_{dew}(experimental) = 41.27 MPa

Component	z_i	MW (g/mol)	Specific Gravity	T_c (K)	P_c (MPa)	ω	$k_{C_1-C_N}$
C_{33}	5.92E−05	425.84	0.958	900.53	1.41	1.1323	0.0673
C_{34}	4.35E−05	436.13	0.962	906.94	1.40	1.1520	0.0679
C_{35}	3.75E−05	445.63	0.966	912.55	1.40	1.1691	0.0684
C_{36}	3.23E−05	455.13	0.969	918.81	1.39	1.1891	0.0689
C_{37}	2.78E−05	464.63	0.973	923.69	1.39	1.2033	0.0694
C_{38}	2.51E−05	474.37	0.976	929.90	1.38	1.2230	0.0699
C_{39}	2.14E−05	484.37	0.980	934.77	1.38	1.2371	0.0704
C_{40}	1.66E−05	493.89	0.984	940.91	1.37	1.2568	0.0709
C_{41}	1.37E−05	502.66	0.987	944.96	1.37	1.2684	0.0713
C_{42}	1.33E−05	511.63	0.990	949.67	1.37	1.2826	0.0718
C_{43}	1.14E−05	521.13	0.993	954.41	1.37	1.2967	0.0722
C_{44}	9.35E−06	530.39	0.996	960.43	1.36	1.3164	0.0727
C_{45+}	6.15E−05	598.56	1.018	976.47	1.48	1.3454	0.0757

The mole fractions and critical properties of the groups are given in the following table (after acentric factor adjustment).

Component	z_i	T_c (K)	P_c (MPa)	w_i
Group-I (N_2)	0.0310	126.10	3.39	0.0403
Group-II (CO_2)	0.0321	304.19	7.38	0.2276
Group-III (C_1)	0.6941	190.56	4.60	0.0115
Group-IV (C_2–C_3)	0.1298	325.19	4.68	0.1158
Group-V (C_4–C_6)	0.0434	451.14	3.54	0.2284
Group-VI (C_7–C_{19})	0.0654	628.06	2.53	0.4121
Group-VII (C_{20}–C_{45+})	0.0040	832.74	1.56	1.1900

Binary interaction parameters between grouped compositions are presented in the following table.

	Group-I (N_2)	Group-II (CO_2)	Group-III (C_1)	Group-IV (C_2-C_3)	Group-V (C_4-C_6)	Group-VI (C_7-C_{19})	Group-VII ($C_{20}-C_{45+}$)
Group-I (N_2)	0	0	0.025	0.0346	0.1009	0.11	0.11
Group-II (CO_2)	0	0	0.105	0.1285	0.1159	0.115	0.115
Group-III (C_1)	0.025	0.105	0	0	0	0.0424	0.0612
Group-IV (C_2-C_3)	0.0346	0.1285	0	0	0	0	0
Group-V (C_4-C_6)	0.1009	0.1159	0	0	0	0	0
Group-VI (C_7-C_{19})	0.11	0.115	0.0424	0	0	0	0
Group-VII ($C_{20}-C_{45+}$)	0.11	0.115	0.0612	0	0	0	0

The composition, critical temperature, critical pressure, acentric factor, and binary interaction parameter between grouped compositions are given in the last two tables. Now we can simulate the CCE experiment. As mentioned before, each step can be modeled by flash calculation. The flash calculation by PR EOS is similar to bubble/dew point calculation. We flash the mixture at pressure 31.81 MPa (the flash calculation for other pressure values is similar). Using Eqs. (4.7) to (4.12) the parameters of PR EOS are determined. The equilibrium ratio is estimated by Eq. (4.22). The parameters of PR EOS and equilibrium ratio are reported in the following table.

(Continued)

Component	z_i	T_c (K)	P_c (MPa)	ω	T_r	K_i Eq. (4.22)	a_{ci} Eq. (4.8) (Pa m^6/mol^2)	b_i Eq. (4.9) (m^3/mol)	α Eq. (4.10)	$a_i = a_{ci}\alpha$ (Pa m^6/mol^2)
Group-I (N_2)	0.0310	126.10	3.39	0.0403	3.369	5.422	0.15	2.40E−05	0.404	0.060
Group-II (CO_2)	0.0321	304.19	7.38	0.2276	1.397	1.509	0.40	2.67E−05	0.758	0.300
Group-III (C_1)	0.6941	190.56	4.60	0.0115	2.229	2.890	0.25	2.68E−05	0.651	0.162
Group-IV (C_2–C_3)	0.1298	325.19	4.68	0.1158	1.306	0.600	0.71	4.49E−05	0.849	0.606
Group-V (C_4–C_6)	0.0434	451.14	3.54	0.2284	0.942	0.074	1.82	8.24E−05	1.043	1.894
Group-VI (C_7–C_{19})	0.0654	628.06	2.53	0.4121	0.676	0.002	4.93	1.61E−04	1.372	6.769
Group-VII (C_{20}–C_{45+})	0.0040	832.74	1.56	1.1900	0.510	6.12E−07	14.04	3.45E−04	2.418	33.935

Let 1 mol of the mixture be flashed at pressure 31.81 MPa and temperature 424.82K into n^L moles of liquid and n^V moles of vapor. n^V can be determined by solving the following equation.

$$\sum_{i=1}^{N} \frac{z_i(K_i - 1)}{1 + (K_i - 1)n^V} = 0 \longrightarrow n^V = 0.7958$$

The mole fraction of component i in liquid phase (x_i) and vapor phase (y_i) is determined by the total material balance for system, material balance for component i, and equilibrium definition.

Component	z_i	K_i	$x_i = \frac{z_i}{1+(K_i-1)n^V}$	$y_i = \frac{z_i K_i}{1+(K_i-1)n^V}$
Group-I (N_2)	0.0310	5.422	0.0069	0.0372
Group-II (CO_2)	0.0321	1.509	0.0229	0.0345
Group-III (C_1)	0.6941	2.890	0.2772	0.8011
Group-IV (C_2–C_3)	0.1298	0.600	0.1905	0.1143
Group-V (C_4–C_6)	0.0434	0.074	0.1649	0.0122
Group-VI (C_7–C_{19})	0.0654	0.002	0.3178	0.0007
Group-VII (C_{20}–C_{45+})	0.0040	6.12E–07	0.0198	1.21E–08

The parameters A and B in Eq. (4.13) are determined using Eqs. (4.14) and (4.7) to (4.17) for both vapor and liquid phases.

For liquid phase:

$$a^L = \sum_i \sum_j x_i x_j (a_i a_j)^{0.5}(1 - k_{ij}) = 2.0693 \text{ Pa m}^6/\text{mol}^2$$

$$b^L = \sum_i x_i b_i = 8.8282 \times 10^{-5} \text{ m}^3/\text{mol}$$

$$A^L = 5.2766, B^L = 0.7951$$

For vapor phase:

$$a^V = \sum_i \sum_j y_i y_j (a_i a_j)^{0.5}(1 - k_{ij}) = 0.2080 \text{ Pa m}^6/\text{mol}^2$$

$$b^V = \sum_i y_i b_i = 2.9535 \times 10^{-5} \text{ m}^3/\text{mol}$$

$$A^V = 0.5303, B^V = 0.2660$$

(Continued)

Solve Eq. (4.13) for compressibility factor using the calculated parameter for each phase:

$$Z^L = 0.9996, Z^V = 1.1018$$

The fugacity of each component for both phases should be calculated by Eq. (4.20) and the error is checked.

Component	f_i^L (MPa) Eq. (4.20)	f_i^V (MPa) Eq. (4.20)	$\phi_i^L = \frac{f_i^L}{z_i P}$	$\phi_i^V = \frac{f_i^V}{z_i P}$	K_i Eq. (4.21)
Group-I (N_2)	0.651	1.460	2.977	1.232	2.4165
Group-II (CO_2)	0.737	0.867	1.013	0.790	1.2828
Group-III (C_1)	13.689	23.953	1.553	0.940	1.6517
Group-IV (C_2–C_3)	3.296	2.189	0.544	0.602	0.9031
Group-V (C_4–C_6)	0.806	0.131	0.154	0.337	0.4565
Group-VI (C_7–C_{19})	0.128	0.003	0.013	0.121	0.1048
Group-VII (C_{20}–C_{45+})	4.86E−06	2.15E−09	7.72E−06	5.59E−03	0.0014

$$\text{error} = \sum_{i=1}^{N}\left(1 - \frac{f_i^L}{f_i^V}\right)^2$$

$$= 5.11 \times 10^6$$

Update the equilibrium ratio and repeat the procedure until the error reaches a value less than 10^{-12}. The final results are given in the following table.

$n^L = 0.0309, n^V = 0.9691, Z^L = 1.1506, Z^V = 0.9537$

Component	x_i	y_i	f_i^L (MPa) Eq. (4.20)	f_i^V (MPa) Eq. (4.20)	$\phi_i^L = \frac{f_i^L}{z_i P}$	$\phi_i^V = \frac{f_i^V}{z_i P}$	K_i Eq. (4.21)
Group-I (N_2)	0.0148	0.0316	1.464	1.464	3.1095	1.4580	2.133
Group-II (CO_2)	0.0248	0.0324	0.811	0.811	1.0264	0.7872	1.304
Group-III (C_1)	0.4330	0.7024	22.356	22.356	1.6231	1.0005	1.622
Group-IV (C_2–C_3)	0.1260	0.1300	2.193	2.193	0.5471	0.5304	1.032
Group-V (C_4–C_6)	0.0653	0.0427	0.315	0.315	0.1517	0.2320	0.654
Group-VI (C_7–C_{19})	0.2424	0.0598	0.095	0.095	0.0123	0.0497	0.247
Group-VII (C_{20}–C_{45+})	0.0936	0.0012	2.04E−05	2.04E−05	6.85E−06	5.41E−04	0.013

$$\text{error} = \sum_{i=1}^{N}\left(1 - \frac{f_i^L}{f_i^V}\right)^2$$

$$= 5.74 \times 10^{-13}$$

The total volume (vapor phase and liquid phase) is calculated as follows.

$$V^{\text{total}} = V^{\text{vapor}} + V^{\text{liquid}}$$

$$V^{\text{total}} = \frac{Z^V n^V RT}{P} + \frac{Z^L n^L RT}{P} = \frac{RT}{P}\left[Z^L n^L + Z^V n^V\right]$$

$$V^{\text{total}} = \frac{8.314 \times 424.82}{31810000}[(1.1506 \times 0.0309) + (0.9537 \times 0.9691)]$$
$$= 1.066 \times 10^{-4}\ \text{m}^3$$

The total relative volume and compressibility factor of vapor phase for other pressure values are reported in the following table.

Experimental Results			**Calculated Results**							
P (MPa)	V/V_{sat}	Z^V	n^L	n^V	Z^L	Z^V	V^L (m^3)	V^V (m^3)	V^{total} (m^3)	V/V_{sat}
62.14	0.839	1.406	0.0000	1.0000	–	1.2948	–	7.360E−05	7.360E−05	0.809
60.76	0.8458	1.386	0.0000	1.0000	–	1.2840	–	7.464E−05	7.464E−05	0.821
55.25	0.877	1.307	0.0000	1.0000	–	1.2371	–	7.909E−05	7.909E−05	0.870
53.87	0.8859	1.287	0.0000	1.0000	–	1.2245	–	8.028E−05	8.028E−05	0.883
52.49	0.8951	1.267	0.0000	1.0000	–	1.2115	–	8.152E−05	8.152E−05	0.897
51.11	0.9054	1.248	0.0000	1.0000	–	1.1982	–	8.280E−05	8.280E−05	0.911
49.73	0.9161	1.228	0.0000	1.0000	–	1.1847	–	8.414E−05	8.414E−05	0.925
48.35	0.927	1.209	0.0000	1.0000	–	1.1714	–	8.557E−05	8.557E−05	0.941
46.97	0.9396	1.19	0.0000	1.0000	–	1.1602	–	8.724E−05	8.724E−05	0.959
45.60	0.9526	1.171	0.0000	1.0000	–	1.1194	–	8.670E−05	8.670E−05	0.953
44.22	0.9666	1.152	0.0000	1.0000	–	1.1004	–	8.789E−05	8.789E−05	0.967
42.84	0.9815	1.133	0.0000	1.0000	–	1.0820	–	8.920E−05	8.920E−05	0.981
41.46	0.9977	1.115	0.0000	1.0000	–	1.0640	–	9.059E−05	9.066E−05	0.997
41.27‡	1	1.112	0.0000	1.0000	–	1.0623	–	9.093E−05	9.093E−05 = V_{sat}	1.000
40.08	1.0152	–	0.0058	0.9942	1.3747	1.0466	7.013E−07	9.170E−05	9.240E−05	1.016
38.70	1.0342	–	0.0105	0.9895	1.3422	1.0297	1.284E−06	9.299E−05	9.427E−05	1.037
37.32	1.0549	–	0.0148	0.9852	1.3078	1.0133	1.833E−06	9.448E−05	9.631E−05	1.059
35.95	1.0775	–	0.0189	0.9811	1.2717	0.9976	2.357E−06	9.616E−05	9.852E−05	1.083
34.57	1.1024	–	0.0228	0.9772	1.2333	0.9823	2.876E−06	9.807E−05	1.009E−04	1.110
33.19	1.1298	–	0.0268	0.9732	1.1929	0.9676	3.399E−06	1.002E−04	1.036E−04	1.139
31.81	1.1602	–	0.0309	0.9691	1.1506	0.9537	3.943E−06	1.026E−04	1.066E−04	1.172
30.43	1.1941	–	0.0352	0.9648	1.1067	0.9406	4.528E−06	1.053E−04	1.098E−04	1.208
29.05	1.2332	–	0.0401	0.9599	1.0614	0.9284	5.174E−06	1.083E−04	1.135E−04	1.248
27.67	1.2753	–	0.0456	0.9544	1.0154	0.9173	5.904E−06	1.118E−04	1.177E−04	1.294
26.30	1.3243	–	0.0516	0.9484	0.9698	0.9077	6.721E−06	1.156E−04	1.223E−04	1.345
24.92	1.3808	–	0.0582	0.9418	0.9247	0.8996	7.630E−06	1.201E−04	1.277E−04	1.404
23.54	1.4463	–	0.0650	0.9350	0.8810	0.8931	8.587E−06	1.253E−04	1.339E−04	1.472
22.16	1.5233	–	0.0713	0.9287	0.8390	0.8883	9.536E−06	1.315E−04	1.410E−04	1.551

4.6 VOLUME TRANSLATION

A comparison between the predicted liquid molar volume and the experimental data of pure compounds generally shows a systematic deviation. This deviation is approximately constant over a wide pressure range away from the critical point (Danesh, 1998). This deviation is called *volume translation* or *volume shift parameter*. The volume shift parameter can solve the weakness in molar liquid volumetric predictions by two-constant EOS. Volume translation concept was introduced by Martin in 1979. Péneloux et al. (1982) used the volume shift parameter to improve the volumetric prediction capabilities of the Soave–Redlich–Kwong EOS. They show that the volume shift parameter does not affect equilibrium calculations for pure compounds or mixtures and therefore does not affect the original VLE capabilities of the Soave–Redlich–Kwong EOS, which is considered as the main aspect of their work. Volume shift parameter can be applied equally for other two-constant EOSs. Jhaveri and Youngren (1988) applied the volume shift parameter for PR EOS. Volume shift parameter is applied to the calculated molar volume by EOS in the following form.

$$V = V^{\mathrm{EOS}} - c \tag{4.36}$$

where v is the corrected molar volume, V^{EOS} is the calculated molar volume by EOS, and c is the volume shift parameter. Péneloux et al. (1982) show that the multicomponent VLE is unchanged if the volume shift parameter of the mixture is calculated by molar average mixing rule.

$$V_{\mathrm{Liquid}} = V^{\mathrm{EOS}}_{\mathrm{Liquid}} - \sum_{i=1}^{N} x_i c_i \tag{4.37}$$

$$V_{\mathrm{Vapor}} = V^{\mathrm{EOS}}_{\mathrm{Vapor}} - \sum_{i=1}^{N} y_i c_i \tag{4.38}$$

where $V^{\mathrm{EOS}}_{\mathrm{Liquid}}$ and $V^{\mathrm{EOS}}_{\mathrm{Vapor}}$ are the liquid and vapor molar volumes by EOS, respectively, x_i and y_i are the mole fraction of component i in liquid and vapor phases, respectively, and c_i is the volume shift parameter of component i. The fugacity for liquid and vapor phases by introducing the volume shift parameter is

$$\left(f_i^{\mathrm{Liquid}}\right)_{\mathrm{modified}} = \left(f_i^{\mathrm{Liquid}}\right)_{\mathrm{original}} \exp\left(-c_i \frac{P}{RT}\right) \tag{4.39}$$

$$\left(f_i^{\text{Vapor}}\right)_{\text{modified}} = \left(f_i^{\text{Vapor}}\right)_{\text{original}} \exp\left(-c_i\frac{P}{RT}\right) \tag{4.40}$$

It is obvious that the fugacity ratio remains unchanged.

Péneloux et al. (1982) suggested the volume shift parameter is calculated for each component separately by matching the saturated liquid density at $T_r = 0.7$. They correlate the volume shift parameter as a function of Rackett compressibility factor, critical temperature, and critical pressure.

$$c = 0.40768(0.29441 - Z_{RA})\frac{RT_c}{P_c} \tag{4.41}$$

where Z_{RA} is the Rackett compressibility as developed by Spencer and Danner (1973) in the modified Rackett Eq. (4.42):

$$v^{\text{sat}} = \left(\frac{RT_c}{P_c}\right) Z_{\text{RA}}^{\left[1+(1-T_r)^{\frac{2}{7}}\right]} \tag{4.42}$$

where v_{sat} is the saturated liquid molar volume.

Jhaveri and Youngren (1988) defined a dimensionless shift parameter by dividing the volume shift parameter by the second PR EOS, b.

$$s_i = \frac{c_i}{b_i} \tag{4.43}$$

Jhaveri and Youngren proposed the following equation for heptane plus fraction:

$$s_i = \frac{c_i}{b_i} = 1 - \frac{A_1}{\text{MW}_i^{A_2}} \tag{4.44}$$

The values of A_0 and A_1 are given in Table 4.9.

The dimensionless shift parameter for selected pure components, which is determined by matching saturated liquid density at $T_r = 0.7$, is given in Table 4.10.

Table 4.9 Jhaveri and Youngren (1988) Volume Shift Parameter Correlation for Heptane Plus Fractions With the Peng–Robinson Equation of State

Component Type	A_1	A_2
n-alkanes	2.258	0.1823
n-alkylcyclohexanes	3.004	0.2324
n-alkylbenzenes	2.516	0.2008

Table 4.10 Volume Shift Parameter for Pure Components for the Peng–Robinson Equation of State and the Soave–Redlich–Kwong Equation of State (Whitson and Brulé, 2000)

Component	Peng–Robinson Equation of State	Soave–Redlich–Kwong Equation of State
N_2	−0.1927	−0.0079
CO_2	−0.0817	0.0833
H_2S	−0.1288	0.0466
C_1	−0.1595	0.0234
C_2	−0.1134	0.0605
C_3	−0.0863	0.0825
i-C_4	−0.0844	0.0830
n-C_4	−0.0675	0.0975
i-C_5	−0.0608	0.1022
n-C_5	−0.0390	0.1209
n-C_6	−0.0080	0.1467
n-C_7	0.0033	0.1554
n-C_8	0.0314	0.1794
n-C_9	0.0408	0.1868
n-C_{10}	0.0655	0.2080

In practice, the accuracy of two-constant EOS by using volume shift parameter can improve as three-constant equation of state (Fuller, 1976; Usdin and McAuliffe, 1976; Schmidt and Wenzel, 1980; Patel and Teja, 1982). Hence if there is no good agreement between the calculated and experimental volumetric data, the use of volume shift parameter can significantly improve the calculated volumetric data.

Problems

4.1 Molar composition of a gas condensate is reported in the following table. The molecular weight and specific gravity of C_{7+} fraction are 158 g/mol and 0.8299, respectively. Estimate the dew point pressure of the mixture using PR EOS at the temperature 397.78K. The experimental value at the given temperature is 41.53 MPa (Al-Meshari, 2005). Use the proposed approach in Chapter 6 to characterize the plus fraction.

Component	Mol%
N_2	0.11
CO_2	0.01
C_1	68.93
C_2	8.63
C_3	5.34
i-C_4	1.15
n-C_4	2.33
i-C_5	0.93
n-C_5	0.85
C_6	1.73
C_{7+}	9.99

4.2 Match the saturation pressure in the previous problem by adjusting the molecular weight of the mixture (using the results of Example 4.1).

4.3 Repeat Example 4.5 by the Aguilar and McCain method. Then estimate the critical temperature, critical pressure, and acentric factor for each group by weight average method.

4.4 Using the results of Problem 4.2, describe the sample by the Aguilar and McCain method. Then estimate the critical temperature, critical pressure, and acentric factor for each group by molar averaging.

4.5 Using the results of Problem 4.2, estimate the binary interaction parameter for methane/C_{7+} pairs by Chueh–Prausnitz equation (using Twu correlation for the determination of critical volume).

4.6 Estimate the binary interaction parameter between groups for sample that is described in Problem 4.2 [using Example 4.1 and Eq. (4.18)].

4.7 Using the results of Problems 4.4 and 4.6, match the saturation pressure by the Al-Meshari method.

4.8 Total relative volume and compressibility factor of vapor phase from constant composition expansion experiment at reservoir temperature, 397.78K, for a gas condensate sample in Example 4.1 are given in the following table (Al-Meshari, 2005). Using the results of Problems 4.2, 4.6, and 4.7 determine the total relative volume and compressibility factor of vapor phase by Peng–Robinson equation of state.

Compressibility Factor of Vapor Phase	V/V_{sat}	P (MPa)
1.328	0.9341	7514.7
1.264	0.9523	7014.7
1.199	0.9727	6514.7
1.175	0.9834	6314.7
1.163	0.9891	6214.7
1.15	0.9942	6114.7
1.14	1	6024.7
–	1.0034	5964.7
–	1.0076	5914.7
–	1.0138	5814.7
–	1.0267	5614.7
–	1.0481	5314.7
–	1.0749	5014.7
–	1.1268	4514.7
–	1.2024	4014.7
–	1.3096	3514.7
–	1.4689	3014.7
–	1.7169	2514.7
–	2.091	2114.7
–	2.2747	1874.7
–	2.515	1697.7
–	2.9087	1474.7
–	3.3173	1304.7
–	3.7153	1174.7
–	4.1342	1064.7

REFERENCES

Agarwal, R., Li, Y.-K., Nghiem, L., 1987. A regression technique with dynamic-parameter selection or phase behavior matching. In: SPE California Regional Meeting. Society of Petroleum Engineers.

Aguilar Zurita, R.A., McCain Jr., W.D., 2002. An efficient tuning strategy to calibrate cubic EOS for compositional simulation. In: SPE Annual Technical Conference and Exhibition. Society of Petroleum Engineers.

Al-Meshari, A.A., 2005. New Strategic Method to Tune Equation-of-State to Match Experimental Data for Compositional Simulation. Texas A&M University.

Chorn, L.G., Mansoori, G.A., 1989. C_{7+} Fraction Characterization. Taylor & Francis.

Chueh, P., Prausnitz, J., 1968. Calculation of high-pressure vapor-liquid equilibria. Industrial & Engineering Chemistry 60 (3), 34–52.

Coats, K., Smart, G., 1986. Application of a regression-based EOS PVT program to laboratory data. SPE Reservoir Engineering 1 (03), 277–299.

Coats, K.H., 1985. Simulation of gas condensate reservoir performance. Journal of Petroleum Technology 37 (10), 1,870–871,886.

Cotterman, R.L., Prausnitz, J.M., 1985. Flash calculations for continuous or semicontinuous mixtures by use of an equation of state. Industrial & Engineering Chemistry Process Design and Development 24 (2), 434–443.

Danesh, A., Xu, D.-H., Todd, A.C., 1992. A grouping method to optimize oil description for compositional simulation of gas-injection processes. SPE Reservoir Engineering 7 (03), 343–348.

Danesh, A., 1998. PVT and Phase Behaviour of Petroleum Reservoir Fluids. Elsevier.

Firoozabadi, A., Hekim, Y., Katz, D.L., 1978. Reservoir depletion calculations for gas condensates using extended analyses in the Peng-Robinson equation of state. The Canadian Journal of Chemical Engineering 56 (5), 610–615.

Fuller, G.G., 1976. A modified Redlich-Kwong-Soave equation of state capable of representing the liquid state. Industrial & Engineering Chemistry Fundamentals 15 (4), 254–257.

Gonzales, E., Colonomos, P., Rusinek, I., 1986. A new approach for characterizing oil fractions, and for selecting pseudo-components of hydrocarbons. Journal of Canadian Petroleum Technology 25 (02).

Guo, T.-M., Du, L., 1989. A three-parameter cubic equation of state for reservoir fluids. Fluid Phase Equilibria 52, 47–57.

Hong, K., 1982. Lumped-component characterization of crude oils for compositional simulation. In: SPE Enhanced Oil Recovery Symposium. Society of Petroleum Engineers.

Jacoby, R., Koeller, R., Berry Jr., V., 1959. Effect of composition and temperature on phase behavior and depletion performance of rich gas-condensate systems. Journal of Petroleum Technology 11 (07), 58–63.

Jhaveri, B.S., Youngren, G.K., 1988. Three-parameter modification of the Peng-Robinson equation of state to improve volumetric predictions. SPE Reservoir Engineering 3 (03), 1033–1040.

Katz, D., Firoozabadi, A., 1978. Predicting phase behavior of condensate/crude-oil systems using methane interaction coefficients. Journal of Petroleum Technology 30 (11), 1649–1655.

Lee, B.I., Kesler, M.G., 1975. A generalized thermodynamic correlation based on three-parameter corresponding states. AIChE Journal 21 (3), 510–527.

Lee, S., Jacoby, R., Chen, W., Culham, W., 1981. Experimental and theoretical studies on the fluid properties required for simulation of thermal processes. Society of Petroleum Engineers Journal 21 (05), 535–550.

Leibovici, C.F., 1993. A consistent procedure for the estimation of properties associated to lumped systems. Fluid Phase Equilibria 87 (2), 189–197.

Li, X.-S., Englezos, P., 2003. Vapor-liquid equilibrium of systems containing alcohols using the statistical associating fluid theory equation of state. Industrial & Engineering Chemistry Research 42 (20), 4953–4961.

Manafi, H., Mansoori, G.A., Ghotbi, S., 1999. Phase behavior prediction of petroleum fluids with minimum characterization data. Journal of Petroleum Science and Engineering 22 (1), 67–93.

Martin, J.J., 1979. Cubic equations of state-which? Industrial & Engineering Chemistry Fundamentals 18 (2), 81–97.

Montel, F., Gouel, P., 1984. A new lumping scheme of analytical data for compositional studies. In: SPE Annual Technical Conference and Exhibition. Society of Petroleum Engineers.

Neau, E., Jaubert, J.N., Rogalski, M., 1993. Characterization of heavy oils. Industrial & Engineering Chemistry Research 32 (6), 1196–1203.

Patel, N.C., Teja, A.S., 1982. A new cubic equation of state for fluids and fluid mixtures. Chemical Engineering Science 37 (3), 463–473.

Pedersen, K.S., Thomassen, P., Fredenslund, A., 1984. Thermodynamics of petroleum mixtures containing heavy hydrocarbons. 1. Phase envelope calculations by use of the Soave-Redlich-Kwong equation of state. Industrial & Engineering Chemistry Process Design and Development 23 (1), 163–170.

Pedersen, K.S., Thomassen, P., Fredenslund, A., 1988a. Characterization of Gas Condensate Mixtures. American Institute of Chemical Engineers, New York, NY.
Pedersen, K.S., Thomassen, P., Fredenslund, A., 1988b. On the dangers of "tuning" equation of state parameters. Chemical Engineering Science 43 (2), 269–278.
Pedersen, K.S., Fredenslund, A., Thomassen, P., 1989. Properties of Oils and Natural Gases. Gulf Pub. Co., Book Division.
Pedersen, K.S., Christensen, P.L., Shaikh, J.A., 2014. Phase Behavior of Petroleum Reservoir Fluids. CRC Press.
Péneloux, A., Rauzy, E., Fréze, R., 1982. A consistent correction for Redlich-Kwong-Soave volumes. Fluid Phase Equilibria 8 (1), 7–23.
Peng, D.-Y., Robinson, D.B., 1976. A new two-constant equation of state. Industrial & Engineering Chemistry Fundamentals 15 (1), 59–64.
Riazi, M., 2005. Characterization and Properties of Petroleum Fractions. ASTM International.
Robinson, D.B., Peng, D.-Y., 1978. The Characterization of the Heptanes and Heavier Fractions for the GPA Peng-Robinson Programs. Gas Processors Association.
Schlijper, A., 1986. Simulation of compositional processes: the use of pseudocomponents in equation-of-state calculations. SPE Reservoir Engineering 1 (05), 441–452.
Schmidt, G., Wenzel, H., 1980. A modified van der Waals type equation of state. Chemical Engineering Science 35 (7), 1503–1512.
Shariati, A., Peters, C.J., Moshfeghian, M., 1999. A systematic approach to characterize gas condensates and light petroleum fractions. Fluid Phase Equilibria 154 (2), 165–179.
Soreide, I., 1989. Improved Phase Behavior Predictions of Petroleum Reservoir Fluids from a Cubic Equation of State. Dr. Techn.
Spencer, C.F., Danner, R.P., 1973. Prediction of bubble-point density of mixtures. Journal of Chemical and Engineering Data 18 (2), 230–234.
Thomassen, P., Pedersen, K., Fredenslund, A., 1987. Adjustment of C_{7+} Molecular Weights in the Characterization of Petroleum Mixtures Containing Heavy Hydrocarbons. Paper SPE 16036.
Usdin, E., McAuliffe, J.C., 1976. A one parameter family of equations of state. Chemical Engineering Science 31 (11), 1077–1084.
Wang, P., 1989. Prediction of Phase Behavior for Gas Condensate. Paper SPE 19503.
Whitson, C.H., Brulé, M.R., 2000. Phase Behavior, Henry L. Doherty Memorial Fund of AIME. Society of Petroleum Engineers.
Whitson, C.H., 1983. Characterizing hydrocarbon plus fractions. Society of Petroleum Engineers Journal 23 (04), 683–694.
Whitson, C.H., 1984. Effect of C_{7+} properties on equation-of-state predictions. Society of Petroleum Engineers Journal 24 (06), 685–696.
Wu, R., Batycky, J., 1988. Pseudocomponent characterization for hydrocarbon miscible displacement. SPE Reservoir Engineering 3 (03), 875–883.
Wu, R., Fish, R., 1989. C characterization for fluid properties predictions. Journal of Canadian Petroleum Technology 28 (04).

CHAPTER FIVE

Vapor—Liquid Equilibrium (VLE) Calculations

E. Soroush[1], A. Bahadori[2,3]
[1]Sahand University of Technology, Tabriz, Iran
[2]Southern Cross University, Lismore, NSW, Australia
[3]Australian Oil and Gas Services Pty Ltd, Lismore, NSW, Australia

5.1 AN INTRODUCTION TO EQUILIBRIUM

A system with no tendency to change at the macroscopic scale is said to be in equilibrium. In other words, for a single-phase multicomponent system, there must be no change in temperature, pressure, and compositions of all species to fulfill the conditions of equilibrium. However, what will be the criteria if several phases with different compositions exist together?

When a reversible ideal process takes place in a closed system, at uniform pressure and temperature, the first and second laws of thermodynamics could be combined into Eq. (5.1):

$$\delta S^t \geq \frac{\delta Q^t}{T} \tag{5.1}$$

in which S denotes entropy, Q is a sign for transfer of heat, T indicates system temperature, and superscript t means total property of the system. With the use of thermodynamic relations, one can interpret Eq. (5.1) as other thermodynamic state properties:

$$\delta Q^t - T\delta S^t \leq 0 \tag{5.2}$$

and in reversible process:

$$dQ^t = dU^t + PdV^t \tag{5.3}$$

so in the differential form of Eq. (5.2):

$$dU^t + PdV^t - TdS^t \leq 0 \tag{5.4}$$

$$dU^t + d(PV^t) - d(TS^t) \leq 0 \tag{5.5}$$

Fluid Phase Behavior for Conventional and Unconventional Oil and Gas Reservoirs
http://dx.doi.org/10.1016/B978-0-12-803437-8.00005-1

$$d(U^t + PV^t - TS^t) \leq 0 \tag{5.6}$$

Now recalling two thermodynamics definitions,

$$H \equiv U + PV \tag{5.7}$$

$$G \equiv H - TS \tag{5.8}$$

in which H is system enthalpy and G denotes Gibbs free energy. The combination of Eqs. (5.7) and (5.6) results in

$$d(H^t - TS^t) \leq 0 \tag{5.9}$$

Substituting Eq. (5.8) in Eq. (5.9), we have:

$$dG^t_{P,T} \leq 0 \tag{5.10}$$

in which subscripts P and T denote constant pressure and temperature. Eq. (5.10) states that the Gibbs free energy at constant temperature and pressure tends to remain constant in a reversible process, whereas it tends to decrease in an irreversible process. In a real process, the transition will go on until the system Gibbs energy reaches its global minimum. This is the equilibrium state and any real process will eventually tends to reach this condition. Therefore, the necessary and sufficient condition for equilibrium is minimization of Gibbs free energy.

In fact, the main goal of phase-equilibrium calculations is to determine a state in which $dG(T,P,x_i) = 0$ is satisfied. Therefore, in a multicomponent system with several phases, temperature, pressure, and partial Gibbs energies must be equal to fulfill the conditions of thermodynamic equilibrium.

The minimization of Gibbs free energy for a multiphase closed system could be interpreted in a more appropriate mathematical formulation. Consider a closed container in which phases α, β, … are in equilibrium. This requires that:

$$dG^t_{T,P} = 0 \tag{5.11}$$

which means that the differential of total Gibbs free energy of each phase should be zero:

$$d\left(G^{t\alpha} + G^{t\beta} + \ldots\right) = 0 \tag{5.12}$$

A closed system could be assumed a group of open systems, in which each phase represents an open system and could perform mass transfer to other phases. If n_i be the number of moles of each component, and N

represents the total number of system components, the total Gibbs energy of the open system α would be formulated as:

$$dG^{t\alpha} = -S^{\alpha}dT + V^{\alpha}dP + \sum_{i}^{N}\left(\frac{\partial G^{\alpha}}{\partial n_i}\right)_{T,P,n_{j\neq i}} dn_i \tag{5.13}$$

in which the sigma term is added to describe any mass transfer across the phase boundary. If M is considered as an extensive property, its derivative with respect to number of moles of any component at fixed temperature, pressure, and other mole numbers $\left(\left(\frac{\partial M}{\partial n_i}\right)_{T,P,n\neq j}\right)$ is called partial molar property of that component. The partial molar Gibbs energy of component i, which appears in Eq. (5.13), is also called chemical potential and is shown by μ_i. Therefore, Eq. (5.13) could be rewritten as:

$$dG^{t\alpha} = -S^{\alpha}dT + V^{\alpha}dP + \sum_{i}^{N}\mu_i^{\alpha}dn_i^{\alpha} \tag{5.14}$$

With the use of Eq. (5.14), Eq. (5.12) could be altered to:

$$dG^{t} = \sum_{\varepsilon=1}^{\kappa}\left(-S^{\varepsilon}dT + V^{\varepsilon}dP + \sum_{i}\mu_i^{\varepsilon}dn_i^{\varepsilon}\right) \tag{5.15}$$

in which κ denotes number of phases and ε shows each phase. At constant temperature and pressure, Eq. (5.15) reduced to:

$$(dG^{t})_{P,T} = \sum_{\varepsilon=1}^{\kappa}\sum_{i}\mu_i^{\varepsilon}dn_i^{\varepsilon} \tag{5.16}$$

As it is a closed system, and no chemical reaction is taking place, all species have a constant number of moles inside the system. Eq. (5.17) shows this in mathematical form:

$$\sum_{\varepsilon}^{\kappa}dn_i^{\varepsilon} = 0 \tag{5.17}$$

The combination of Eqs. (5.16) and (5.17) results in:

$$\mu_i^{\alpha} = \mu_i^{\beta} = \ldots = \mu_i^{\kappa} \tag{5.18}$$

Eq. (5.18) states that if the chemical potentials of all components in all phases are equal, the system is in equilibrium. To use this equation for practical purposes, a proper relation should be established between chemical potential and measurable quantities.

As for a pure component, partial molar properties are no different from molar properties; the following analogy could hold for a pure ideal gas:

$$dG = -SdT + VdP \Leftrightarrow dg_i = -s_i dT + v_i dP \tag{5.19}$$

In a uniform temperature condition,

$$\left(\frac{\partial g_i}{\partial P}\right)_T = v_i = \frac{RT}{P} \tag{5.20}$$

Integrating Eq. (5.20) at constant temperature:

$$g_i - g_i^0 = \mu_i - \mu_i^0 = RT \ln\left(P/P^0\right) \tag{5.21}$$

in which subscript 0 shows the reference state. Eq. (5.21) shows the changes in chemical potential of an ideal gas when its pressure varies from P^0 to P in an isothermal manner. Eq. (5.21) could be generalized for real fluids by introducing a new property instead of pressure, called fugacity. This property, which is also called "corrected pressure," has the dimension of pressure and denotes by f:

$$\mu_i - \mu_i^0 = RT \ln\left(f_i / f_i^0\right) \tag{5.22}$$

Considering Eq. (5.22) for component i in all phases of an equilibrated heterogeneous system, at the same temperature and pressure the reference state would be the same for all phases; therefore, the equality of all chemical potentials will result in:

$$f_i^\alpha = f_i^\beta = \ldots = f_i^\kappa \tag{5.23}$$

Eq. (5.23) is another interpretation of equilibrium condition; it has a considerable practical importance because fugacity is a measurable property.

To study the equilibrium systems, introduction of some parameters will come in handy. The ratio of fugacity to pressure is called fugacity coefficient and is denoted by ϕ. In a multicomponent system, the fugacity coefficient of component i will be defined as:

$$\phi_i = \frac{f_i}{P z_i} \tag{5.24}$$

in which z_i represents the mole fraction of component i. Another vital equilibrium concept is activity. The ratio of the fugacity of component i to its reference state shows the fugacity contribution in mixture, is called activity, and is shown by α.

$$\frac{f_i}{f_i^0} = \alpha \tag{5.25}$$

As the fugacity of component i in a mixture depends on its concentration, it can be said that the activity of species i is related to its concentration. The ratio of α_i to its concentration is called activity coefficient and is demonstrated by γ.

$$\frac{\alpha_i}{x_i} = \gamma_i \tag{5.26}$$

which shows that

$$f_i = \gamma_i x_i f_i^0 \tag{5.27}$$

The activity coefficient approaches unity when x_i approaches one. This function is a very important and practical in vapor–liquid equilibrium (VLE) calculations. There are lots of models in literature proposed to find its values for different mixtures.

Imagine a vapor–liquid equilibrium system. The fugacity of component i in both phases will be equal:

$$f_i^{\mathrm{L}} = f_i^{\mathrm{V}} \tag{5.28}$$

With respect to Eq. (5.24), one can write for each phase:

$$f_i^{\mathrm{V}} = y_i P \phi_i^{\mathrm{V}} \tag{5.29}$$

$$f_i^{\mathrm{L}} = x_i P \phi_i^{\mathrm{L}} \tag{5.30}$$

in which y_i is a sign of mole fraction in the gas phase and x_i denotes the mole fraction in the liquid phase. Hence, considering Eqs. (5.25) to (5.27):

$$\frac{y_i}{x_i} = \frac{\phi_i^{\mathrm{L}}}{\phi_i^{\mathrm{V}}} \equiv K_i \tag{5.31}$$

The symbol K_i is known as equilibrium ratio and demonstrates the distribution of component i between phases. This function is a key to all phase-calculation problems and simplifies the equations. The equilibrium ratio, also known as K-value, is a function of temperature, pressure, and composition and could be defined with the help of the activity coefficient:

$$\frac{\gamma_i f_i^0}{P \phi_i^{\mathrm{V}}} = K_i \tag{5.32}$$

Later, we show that each of the definitions in Eqs. (5.31) and (5.32) is important in its own way.

5.2 FLASH CALCULATIONS

Flash calculations are one of the essential parts of all reservoir and process calculations. They determine the amounts and composition of coexisting hydrocarbon gas and liquid phases, at a known temperature and pressure, in a vessel or reservoir. The most common VLE problems are isothermal two-phase flash calculations. Nevertheless, there is a complication that at the given temperature and pressure, whether the mixture forms two phases in equilibrium or remains as a stable single phase. In this section, we will assume that the mixture ether yields equilibrium-phase composition or results in a trivial solution. In principle, the mixture stability should be tested even if the outcomes seem to have physical consistency (Whitson and Brulé, 2000). The discussion on the stability later in this chapter will be presented.

Imagine 1 mol of a hydrocarbon mixture containing N components with the known mole fraction ($z_1, z_2, \ldots, z_N$) is entering a two-phase flash separator at constant temperature and pressure. According to Duhem's law, when the exact amount of N different species mixed to form a closed system of Θ phases, all intensive and extensive properties of each phase will be specified if and only if two independent variables of the system are known (Abbott et al., 2001). As we said, here temperature and pressure are fixed. So the two independent variables required by Duhem's law are given. We should determine the molar amount of each phase and molar composition of each phase after the equilibrium in the separator. Therefore, the unknowns will be $x_1, x_2, \ldots, x_N, y_1, y_2, \ldots, y_N, L, V$ which L and V represent the molar amount of liquid and vapor phase respectively. A simple material balance, suggest that:

$$V + L = 1 \tag{5.33}$$

L and V are not independent. Therefore, we have $2N + 1$ unknowns. To solve this problem $2N + 1$ independent equations are required. From equilibrium relations and material balance we can find the following equations:

$$y_i = K_i x_i \quad i = 1, 2, \ldots, N \quad N \text{ Independent Equations} \tag{5.34}$$

$$z_i = x_i L + y_i(1 - L) \quad i = 1, 2, \ldots, N \quad N \text{ Independent Equations} \tag{5.35}$$

$$\sum x_i = 1 \quad \left(\sum y_i = 1\right) \quad \text{An Independent Equation} \tag{5.36}$$

Eqs. (5.34) to (5.36) give us $2N + 1$ independent variables to solve the problem. Combining Eqs. (5.33–5.35) one can find:

$$x_i = \frac{z_i}{1 + V(K_i - 1)} \tag{5.37}$$

$$y_i = \frac{K_i z_i}{1 + V(K_i - 1)} \tag{5.38}$$

Using Eqs. (5.36) to (5.38):

$$\sum_{i=1}^{N} (y_i - x_i) = \sum_{i=1}^{N} \frac{z_i(K_i - 1)}{V(K_i - 1) + 1} = 0 \tag{5.39}$$

This equation has a very important characteristic; with increasing V its value monotonically decreases. Eq. (5.39) could be solved through an iterative procedure. In the first step, the K-values at the system pressure and temperature should be found. Then, in the second step, a value of V should be assumed and solve the flash equation. If the objective is satisfied, the value of V is correct, if not, steps 2 and 3 should be repeated. At the final step, x_i and y_i are calculated.

This flash calculation procedure is the common case in which initial mixture composition, temperature, and pressure are the known values and the unknowns are x_i, y_i, and the molar amount of vapor phase. Evidently, the calculations could be modified for a case in which temperature or pressure is unknown and the molar amount of vapor phase is a known value. This will be discussed later in this chapter.

As can be seen, the key to flash calculation is the first step, finding the equilibrium ratio. In fact, the accuracy of flash calculation crucially depends on the equilibrium ratio (Riazi, 2005). There are numerous methods proposed in literature for determining K-values. In the following section, we will discuss the most common methods.

5.3 METHODS OF FINDING *K*-VALUE

5.3.1 Ideal Concept

An ideal solution is a solution in which all components have the same molecular diameter and intermolecular forces (attraction and repulsion) between unlike molecules the same as like molecules; upon mixing components, mutual solubility is achieved and no chemical interaction happens (McCain, 1990; Prausnitz et al., 1998).

5.3.1.1 Lewis Fugacity Rule

This rule is based on an assumption that states that, at constant temperature and pressure, mixture's molar volume is a linear function of the mole fraction (Prausnitz et al., 1998; Abbott et al., 2001). The direct result of this assumption is:

$$f_i(T, P) = z_i f_{i,\text{pure}}(T, P) \tag{5.40}$$

The simplicity of this rule has made it very interesting, particularly in special cases when restriction could be applied and simplify the problem. This rule is an excellent approximation of ideal gases or, in other words, gases at low pressure and liquids with ideal behavior. It also could be perfectly applied in any pressure to the mixtures in which their species have similar physical properties and gaseous mixtures that have a species that holds the majority of the mixture concentration. Comparing Eqs. (5.27) and (5.40), it could be known that applying the Lewis rule to the liquids will result in $\gamma_i = 1$.

5.3.1.2 Raoult's Law

Assume a VLE system at low pressure in which the vapor phase is nearly an ideal gas. If we apply Eqs. (5.23) and (5.40) to the system, we would have:

$$x_i f_{i,\text{pure}}^{\text{L}} = y_i f_{i,\text{pure}}^{\text{V}} \tag{5.41}$$

As the vapor phase is an ideal gas, the vapor fugacity would be equal to the total pressure. In the other hand, one can assume, at low pressures, the fugacity of pure liquids are equal to their fugacity at saturation pressure. According to Eq. (5.23), the fugacities of liquid and saturated vapor are equal. In addition, at low pressures it can be assumed the vapor fugacity is equal to the vapor pressure $\left(P_i^{\text{sa}}\right)$ so at a specified temperature:

$$f_{i,\text{pure}}^{\text{V}} = P \quad \text{and} \quad f_{i,\text{pure}}^{\text{L}} = P_i^{\text{sa}} \tag{5.42}$$

From Eqs. (5.41) and (5.42) it could be said that:

$$y_i P = x_i P_i^{\text{sa}} \tag{5.43}$$

Eq. (5.43) is known as Raoult's law and is used for ideal solutions at low pressures. From Raoult's law the equilibrium ratio would be:

$$K_i = \frac{P_i^{sa}}{P} \tag{5.44}$$

Raoult's law could also be derived from the concept of partial pressure. At low pressures in a VLE system, the vaporization rate of all species in equilibrium must be equal to the rate of condensation. Therefore, the vapor and liquid composition would not be changed. In other words, the driving force of each direction is equal. If the driving force is presented as partial pressures, then:

$$PP_i^V = PP_i^L, \quad PP_i^V = y_i P, \quad PP_i^L = x_i P_i^{sa} \Rightarrow y_i P = x_i P_i^{sa} \tag{5.45}$$

Raoult's law is applicable at pressures up to 400 kPa, in which the ideal gas concept is still valid (Campbell, 1979). If the vapor phase behaves as an ideal gas, but the liquid phase deviates from an ideal solution, Raoult's law, with an eye on Eq. (5.27), could be modified as:

$$y_i P = x_i \gamma_i P_i^{sa} \quad \text{and} \quad K_i = \frac{\gamma_i P_i^{sa}}{P} \tag{5.46}$$

Eq. (5.46) is known as modified Raoult's law.

5.3.1.3 Henry's Law

The Henry's law states that the solubility of gaseous species solubility in a liquid is proportional to its partial pressure in the vapor phase (Prausnitz et al., 1998; Abbott et al., 2001):

$$y_i P = x_i H_i \tag{5.47}$$

The symbol H_i is known as Henry's constant, which is temperature dependent and experimentally determined. This constant has a unit of pressure per weight or mole fraction and is independent of concentration but slightly changes with pressure change. Henry's law is best applied when the concentration of the solute is not exceeding 3 mol% and the pressure is low, not more than 5–10 bars (Danesh, 1998; Riazi, 2005). The temperature is also should be below the solvent's critical temperature. Henry's law will be very useful for determining the solubility of light hydrocarbons in heavy oils or solubility of hydrocarbons in water.

In fact, the right side of Eq. (5.47) assumes vapor phase is an ideal gas, whereas the left side implies that the fugacity of a component (at low concentrations) in a liquid mixture is proportional to its concentration:

$$f_i = x_i H_i \tag{5.48}$$

The equilibrium ratio from Eq. (5.47) could be found:

$$K_i = \frac{H_i}{P} \tag{5.49}$$

The pressure dependency of Henry's constant could be detected by:

$$H_i = H_i^0 \exp\left[\frac{v_i^\infty \left(P - P^0\right)}{RT}\right] \tag{5.50}$$

Here, the symbol of v_i^∞ shows the partial molar volume of species i in an infinite dilution (assuming constant in the pressure and composition ranges), H_i^0 shows Henry's constant at the reference pressure P^0.

5.3.2 Fugacity-Derived Equilibrium Ratio (ϕ–ϕ Approach)

With the use of Eq. (5.31) for a mixture, we have:

$$K_i = \frac{\widehat{\phi}_i^{\,L}(T, P, x_i)}{\widehat{\phi}_i^{\,V}(T, P, y_i)} \tag{5.51}$$

This equation shows that if one cloud finds the fugacity coefficient of both phases through a function that relates temperature, pressure, and composition of the system, the K-value would be found. Such a function is an equation of state. This method is may be the most common method for finding K-values of hydrocarbon mixtures. The fugacity coefficient could be found by following relations:

$$\begin{aligned}\ln\widehat{\phi}_i &= \int_0^P (Z_i - 1)\frac{dP}{P} = \frac{-1}{RT}\int_0^P \left(\frac{RT}{P} - \overline{V}_i\right) dP \\ &= \frac{1}{RT}\int_{V^t}^{\infty}\left[\left(\frac{\delta P}{\delta n_i}\right)_{T,V,n_{j\neq i}} - \frac{RT}{V^t}\right] dV^t - \ln Z\end{aligned} \tag{5.52}$$

in which $V^t = nV$ and represents the total volume. The $\overline{V}_i$ could be found from equations of state (EOSs); the combination of cubic EOSs and Eq. (5.52) is a common method for finding ϕ_i.

5.3.3 Activity-Derived Equilibrium Ratios (γ–ϕ Approach)

To find the fugacity in vapor phase, one always uses the EOSs because they give reasonably good results. However, in the liquid phase, especially when it consists of dissimilar molecules, the EOSs fail to give a proper result. In the petroleum industry, many mixtures show severe nonideality. In the hydrocarbon mixtures, polar substances like water, hydrogen sulfide, and glycol cause trouble for the EOSs (Campbell, 1979; Danesh, 1998; Orbey and Sandler, 1998). In this situation, it is better to calculate liquid fugacity from Eq. (5.27) and therefore the equilibrium ratio from Eq. (5.32).

The $\gamma-\phi$ approach could give dependable outcomes for systems with very different liquid and vapor-phase properties. However, at high pressures near the critical condition, this method is not recommended, because in this condition both phases approach similar properties (Orbey and Sandler, 1998).

5.3.4 Correlations for Finding Equilibrium Ratio

Despite the high development of theoretical models, empirical correlations are still a very common way of finding equilibrium ratios at low- and moderate-pressure conditions (in which for hydrocarbon mixtures the dependency of K-values to mixture composition is negligible). There are many empirical relations for predicting equilibrium ratio in the literature. Some of them are just simple mathematical relations, but some others are very complicated and have many composition-dependent variables. Here, we will introduce some of the empirical methods as examples.

5.3.4.1 Wilson's Correlation

This correlation provides reasonable predictions for determining K-values at low pressures:

$$K_i = \frac{P_{ci}}{P} \exp\left[5.37(1+\omega_i)\left(1-\frac{T_{ci}}{T}\right)\right] \tag{5.53}$$

here, the symbol P_{ci} shows critical pressure of component i in pounds per square inch absolute (psia), P represents system pressure in psia, T_{ci} is a sign of critical temperature in °R, T is the system temperature in °R and ω_i shows the acentric factor of species i. In fact, Wilson's equation uses Raoult's Law and relates vapor pressure to critical properties. This correlation overestimates the equilibrium ratios of supercritical components (Danesh, 1998).

5.3.4.2 Standing's Correlation

This correlation is derived from experimental K-values at temperatures less than 200°F and pressures below 1000 psia. The correlation is in the form of:

$$K_i = \frac{1}{P} 10^{(a+cF_i)} \tag{5.54}$$

in which F_i is called species characterization factor and defined as:

$$F_i = b_i\left[\frac{1}{T_{bi}} - \frac{1}{T}\right] \tag{5.55}$$

The T_{bi} represents species normal boiling point in R and b_i defines as follows:

$$b_i = \frac{\log\left(\frac{P_{ci}}{14.7}\right)}{\left[\frac{1}{T_{bi}} - \frac{1}{T_{ci}}\right]} \tag{5.56}$$

The symbol T_{ci} is the critical temperature of species i. The coefficients a and c are represented as a function of pressure.

$$a = 1.2 + 0.00045P + 15\left(10^{-8}\right)P^2 \tag{5.57}$$

$$c = 0.89 - 0.00017P - 3.5\left(10^{-8}\right)P^2 \tag{5.58}$$

Standing suggested that changing the values of b_i and boiling point of H_2S, N_2, CO_2 and C_1 to C_6 could considerably improve the prediction of K-values. Table 5.1 shows the suggested values by Ahmed (2006).

Table 5.1 Suggested Values for b_i and Boiling Point in Standing Correlation

Component	b_i	Boiling Point (°R)
H_2S	470	109
N_2	652	194
CO_2	1136	331
C_1	300	94
C_2	1145	303
C_3	1799	416
i-C_4	2037	471
n-C_4	2153	491
i-C_5	2368	542
n-C_5	2480	557
C_6 **(Lumped hexane fraction)**	2738	610
n-C_6	2780	616
n-C_7	3068	669
n-C_8	3335	718
n-C_9	3590	763
n-C_{10}	3828	805

To determine the equilibrium ratio for C_{7+}, first the F_i parameter should be found. For C_{7+} fraction the following relations are suggested:

$$n = 7.30 + 0.0075(T - 460) + 0.0016P \tag{5.59}$$

$$b = 1013 + 324n - 4.256n^2 \tag{5.60}$$

$$T_b = 301 + 59.85n - 0.971n^2 \tag{5.61}$$

With use of parameters found from Eqs. (5.59) to (5.61), the F_i for C_{7+} fraction could be found from Eq. (5.55).

5.3.4.3 Whitson and Torp Correlation

This equation is a modified form of Wilson's correlation, which is expected to give better results in higher pressures:

$$K_i = \left(\frac{P_{ci}}{P_{con}}\right)^{\left(1-\left(\frac{P}{P_{con}}\right)\right)}\left(\frac{P_{ci}}{P}\right)\exp\left[537\left(1-\left(\frac{P}{P_{con}}\right)\right)(1+\omega_i)\left(1-\frac{T_{ci}}{T}\right)\right] \tag{5.62}$$

The symbol P_{con} in Eq. (5.62) is called convergence pressure. It is known that in VLE systems when a fixed composition hydrocarbon mixture held at a constant temperature, while the pressure is increasing, at a specific pressure the equilibrium ratios of all components will reach a value of unity (McCain, 1990; Ahmed, 2006). This specific pressure is convergence pressure. Many graphs and correlations could be found in literature for finding convergence pressure. One simple correlation, which gives a rough estimate of the convergence pressure, is suggested by Standing:

$$P_{con} = -2381.8542 + 46.341487[M\gamma]_{C_{7+}} + \sum_{i=1}^{3}\left[\frac{[M\gamma]_{C_{7+}}}{T-460}\right]^i$$

$$a_1 = 6124.3049 \tag{5.63}$$

$$a_2 = -27532538$$

$$a_3 = 415.42049$$

Here, $M_{C_{7+}}$ is shown the molecular weight of the heptanes-plus fraction, $\gamma_{C_{7+}}$ denotes the specific gravity of the heptanes-plus fraction, T indicates temperature, °R, and a_i shows correlation coefficients.

5.4 BUBBLE AND DEW-POINT CALCULATIONS

In a liquid phase at constant temperature, the saturation pressure point is a pressure at which the first bubble of gas is formed in the liquid. This is why the saturation pressure for liquids is called bubble-point pressure. Analogous to this definition, bubble-point temperature is a temperature at which the first bubble of gas is formed in a liquid phase at constant pressure.

In a vapor phase, the saturation point is a pressure at which the first drop of liquid is formed in the vapor. This is why the saturation pressure for gases called dew point. Analogous to this definition, dew-point temperature is a temperature at which the first drop of liquid is formed in a vapor phase at constant pressure.

The following algorithms could be used for calculating bubble and dew points. The algorithms could be changed for a case in which the pressure is fixed and the temperature is unknown or vice versa.

Bubble-point pressure calculations algorithm.

1. Estimate a bubble-point pressure and find the equilibrium ratios. (Wilson correlation could be used for an initial approximation.)
2. Find the composition of vapor phase using Eq. (5.31) $y_i = z_i K_i$.
3. Calculate ϕ_i^{V} and ϕ_i^{L} with the estimated bubble-point pressure and vapor composition in step (2). Keep in mind that the liquid composition is in fact the feed composition.
4. With the use of fugacity coefficients from step (3) and Eq. (5.31), find the new equilibrium ratios.
5. Calculate $F = \sum_{i=1}^{N} z_i K_i - 1$ and $\frac{dF}{dP} = \sum_{i=1}^{N} z_i K_i \left(\frac{\partial \ln \phi_i^{\mathrm{L}}}{\partial P} - \frac{\partial \ln \phi_i^{\mathrm{V}}}{\partial P} \right)$
6. Calculate $P_{\mathrm{b}}^{j+1} = P_{\mathrm{b}}^{j} - \frac{F^j}{\frac{dF^j}{dP}}$ and check for convergence (j is an iteration counter). If the convergence did not happen, go back to the first step and calculate the equilibrium ratios with the new pressure.

Dew-point temperature calculation algorithm.

1. Estimate a dew-point temperature and find the equilibrium ratios. (Wilson correlation could be used for an initial approximation.)
2. Find the composition of liquid phase using Eq. (5.31) $x_i = z_i / K_i$.
3. Calculate ϕ_i^{V} and ϕ_i^{L} with the estimated dew-point temperature and liquid composition in step (2). Keep in mind that the vapor composition is in fact the feed composition.

4. With the use of fugacity coefficients from step (3) and Eq. (5.31), find the new equilibrium ratios.

5. Calculate $F = \sum_{i=1}^{N} \frac{z_i}{K_i} - 1$ and $\frac{dF}{dT} = \sum_{i=1}^{N} \frac{z_i}{K_i}\left(\frac{\partial \ln \phi_i^{L}}{\partial T} - \frac{\partial \ln \phi_i^{V}}{\partial T}\right)$

6. Calculate $T_d^{j+1} = T_d^{j} - \frac{F^j}{\frac{dF^j}{dT}}$ and check for convergence (j is an iteration counter). If the convergence did not happen, go back to the first step and calculate the equilibrium ratios with the new temperature.

Example 5.1

The molar composition of an oil sample is listed in Table 5.2 (Pedersen et al., 1992):

The molecular weight and specific gravity of C_{7+} fraction are 211.5 and 0.846, respectively. This oil flashes at 1000 psia and 650 R. Estimate the equilibrium ratio by Standing's correlation.

Solution

To estimate the equilibrium ratio by Standing's correlation the critical properties of the components are required. The required properties extracted from reference [Gas Processors Suppliers Association (GPSA, 2004)] for all components (excluded C_{7+}) are reported in below table.

Table 5.2 Molar Composition of an Oil Sample

Component	Mol%
N_2	0.69
CO_2	0.12
C_1	47.09
C_2	5.69
C_3	4.39
i-C_4	0.95
n-C_4	2.42
i-C_5	1.11
n-C_5	1.46
C_6	2.26
C_{7+}	33.82

(Continued)

Component	P_c (psi)	T_c (R)
CO_2	1070	547.43
N_2	492.5	227.14
C_1	667	343.01
C_2	706.6	549.59
C_3	615.5	665.59
i-C_4	527.9	734.08
n-C_4	550.9	765.22
i-C_5	490.4	828.67
n-C_5	488.8	845.47
C_6	436.9	913.47

The coefficients a and c in Eq. (5.54) determine by substituting $P = 1000$ psi in Eqs. (5.57) and (5.58), respectively.

$$a = 1.2 + 0.00045(1000) + 15(10^{-8})(1000)^2 = 1.80$$

$$c = 0.89 - 0.00017(1000) - 3.5(10^{-8})(1000)^2 = 0.685$$

The parameters b and T_b for C_{7+} calculate using Eqs. (5.59) through (5.61) as follows:

$$n = 7.30 + 0.00075(650 - 460) + 0.0016(1000) = 10.325$$

$$b = 1013 + 324(10.325) - 4.256(10.325)^2 = 3904.586$$

$$T_{bC_{7+}} = 301 + 59.85(10.325) - 0.971(10.325)^2 = 815.44°R$$

b and T_b for other components taken from Table (5.1). Substituting the calculated values in Eq. (5.54) gives the following results:

Component	P_c (psi)	T_c (R)	T_b (R) Table (5.1) and Eq. (5.61)	b Table (5.1) and Eq. (5.60)	F_i Eq. (5.55)	K_i Eq. (5.54)
CO_2	1070	547.43	194	652	2.358	2.601
N_2	492.5	227.14	109	470	3.589	18.129
C_1	667	343.01	94	300	2.730	4.678
C_2	706.6	549.59	303	1145	2.017	1.520
C_3	615.5	665.59	416	1799	1.557	0.735
i-C_4	527.9	734.08	471	2037	1.191	0.413
n-C_4	550.9	765.22	491	2153	1.073	0.343
i-C_5	490.4	828.67	542	2368	0.726	0.198
n-C_5	488.8	845.47	557	2480	0.637	0.172
C_6	436.9	913.47	610	2738	0.276	0.098
C_{7+}	320.3	1139.4	815.4372	3904.586	−1.219	0.0092

Example 5.2

Determine the composition of equilibrated phases in previous example.

Solution

Assuming the mole fraction of component "i" in liquid and vapor phase denoted by x_i and y_i, respectively. If number of moles of feed is $F = 1$, the overall mass balance and mass balance for component "i" gives the following relation:

$$1 = L + V$$

$$z_i = x_i L + y_i V = x_i(1 - V) + y_i V$$

in which L and V are the number of moles of liquid and vapor, respectively. Based on equilibrium ratio definition $y_i = K_i x_i$, hence:

$$z_i = x_i(1 - V) + K_i x_i L$$

Solving the previous equation for x_i gives:

$$x_i = \frac{z_i}{V(K_i - 1) + 1}$$

and:

$$\xrightarrow{y_i = K_i x_i} y_i = \frac{K_i z_i}{V(K_i - 1) + 1}$$

Therefore, the composition of liquid and vapor can be determined if the sum of the mole fractions of all components in each phase is equal to 1, which can be written in mathematical form as follows:

$$\sum_{i=1}^{N} x_i = 1, \; \sum_{i=1}^{N} y_i = 1 \Rightarrow \sum_{i=1}^{N} x_i - \sum_{i=1}^{N} y_i = 0$$

Here, N is the total number of components. The above equation can be rewritten as follows (by substituting definition of x_i and y_i in terms of z_i, K_i, and V):

$$\sum_{i=1}^{N} \frac{z_i(K_i - 1)}{V(K_i - 1) + 1} = 0$$

The previous equation can be solved for n^V by the Newton–Raphson method as follows:

$$h(V) = \sum_{i=1}^{N} \frac{z_i(K_i - 1)}{V(K_i - 1) + 1}$$

$$h'V = \frac{dh(V)}{dV} = -\sum_{i=1}^{N} \frac{z_i(K_i - 1)^2}{[V(K_i - 1) + 1]^2}$$

$$V_{k+1} = V_k - \frac{h(V_k)}{h'(V_k)}$$

in which k is iteration counter. The initial guess for n^V can be chosen as 0.5. The results for the first three iterations are reported in below table.

(Continued)

Component	K_i	z_i	$V = 0.5$		$V = 0.4380$		$V = 0.4356$	
			$\frac{z_i(K_i-1)}{V(K_i-1)+1}$	$-\frac{z_i(K_i-1)^2}{[V(K_i-1)+1]^2}$	$\frac{z_i(K_i-1)}{V(K_i-1)+1}$	$-\frac{z_i(K_i-1)^2}{[V(K_i-1)+1]^2}$	$\frac{z_i(K_i-1)}{V(K_i-1)+1}$	$-\frac{z_i(K_i-1)^2}{[V(K_i-1)+1]^2}$
CO_2	2.601	0.0069	0.00613	−0.00545	0.00649	−0.00611	0.00651	−0.00614
N_2	18.129	0.0012	0.00215	−0.00385	0.00242	−0.00487	0.00243	−0.00492
C_1	4.678	0.4709	0.61004	−0.79028	0.66328	−0.93425	0.66555	−0.94065
C_2	1.520	0.0569	0.02349	−0.00969	0.02410	−0.01021	0.02413	−0.01023
C_3	0.735	0.0439	−0.01340	−0.00409	−0.01315	−0.00394	−0.01314	−0.00393
i-C_4	0.413	0.0095	−0.00790	−0.00656	−0.00751	−0.00593	−0.00749	−0.00591
n-C_4	0.343	0.0242	−0.02370	−0.02321	−0.02234	−0.02063	−0.02229	−0.02054
i-C_5	0.198	0.0111	−0.01485	−0.01988	−0.01372	−0.01695	−0.01368	−0.01685
n-C_5	0.172	0.0146	−0.02062	−0.02911	−0.01896	−0.02461	−0.01890	−0.02446
C_6	0.098	0.0226	−0.03717	−0.06112	−0.03373	−0.05034	−0.03361	−0.04998
C_{7+}	0.0092	0.3382	−0.66403	−1.30377	−0.59201	−1.03629	−0.58951	−1.02756
Sum	—	—	−0.13985	−2.25701	−0.00512	−2.11412	−3.6E-06	−2.11116

Therefore, the number of moles of vapor and liquid are 0.4356 and 0.5644, respectively. The composition in vapor and liquid phases is given in following table.

Component	x_i	y_i
CO_2	0.004066	0.010572
N_2	0.000142	0.002571
C_1	0.180977	0.846526
C_2	0.046389	0.070518
C_3	0.049623	0.036485
i-C_4	0.012765	0.005270
n-C_4	0.033912	0.011617
i-C_5	0.017057	0.003382
n-C_5	0.022832	0.003935
C_6	0.037240	0.003633
C_{7+}	0.594997	0.005492
Sum	1	1

Example 5.3

Pressure–volume data for steam water at 773.15K reported in below table (Abbott et al., 2001).

P (Pa)	V (m^3/mol)
1,000	6.42258
10,000	0.64206
20,000	0.32094
30,000	0.21402
40,000	0.160517
50,000	0.128403
75,000	0.085585
100,000	0.064175
125,000	0.051331
150,000	0.042766
175,000	0.03665
200,000	0.032062
225,000	0.028494
250,000	0.025639
275,000	0.023305
300,000	0.021357
325,000	0.01971
350,000	0.018299
375,000	0.017076
400,000	0.016005
425,000	0.015061
450,000	0.014221
475,000	0.01347
500,000	0.012794

Determine the fugacity coefficient at 500,000 Pa and 773.15K.

(Continued)

Solution

According to Eq. (5.52), the fugacity coefficient for a pure component (water) is defined by following integral:

$$\ln \phi = \int_0^P (Z-1)\frac{dP}{P}$$

On the other hand, all gases at low pressure can be considered as an ideal gas, therefore the earlier integral can be rewritten as follows:

$$\ln \phi = \int_0^P (Z-1)\frac{dP}{P} \approx \int_{1000\ \mathrm{Pa}}^P (Z-1)\frac{dP}{P}$$

The fugacity coefficient is equal to the area under graph $\frac{(Z-1)}{P}$ versus P. The compressibility factor is defined as $Z = PV/RT$ in which R is equal to 8.315 J/mol K. The values of compressibility factor at different pressures are given in below table.

P (Pa)	V (m^3/mol)	Z	$\frac{(Z-1)}{P}$ (1/Pa)
1,000	6.42258	0.999162	−8.3838E−07
10,000	0.64206	0.998854	−1.1464E−07
20,000	0.32094	0.998574	−7.1322E−08
30,000	0.21402	0.998854	−3.8214E−08
40,000	0.160517	0.998865	−2.8380E−08
50,000	0.128403	0.998784	−2.4328E−08
75,000	0.085585	0.998581	−1.8926E−08
100,000	0.064175	0.998378	−1.6225E−08
125,000	0.051331	0.998189	−1.4492E−08
150,000	0.042766	0.997972	−1.3523E−08
175,000	0.03665	0.997782	−1.2671E−08
200,000	0.032062	0.997565	−1.2173E−08
225,000	0.028494	0.997383	−1.1629E−08
250,000	0.025639	0.997173	−1.1306E−08
275,000	0.023305	0.997012	−1.0864E−08
300,000	0.021357	0.996753	−1.0822E−08
325,000	0.01971	0.996543	−1.0636E−08
350,000	0.018299	0.996361	−1.0396E−08
375,000	0.017076	0.996186	−1.0170E−08
400,000	0.016005	0.995986	−1.0035E−08
425,000	0.015061	0.99579	−9.9056E−09
450,000	0.014221	0.995581	−9.8189E−09
475,000	0.01347	0.995385	−9.7148E−09
500,000	0.012794	0.995185	−9.6295E−09

Using trapezoidal integration formula, the value of $\int_{1000\ \mathrm{Pa}}^{500000\ \mathrm{Pa}} (Z-1)\frac{dP}{P}$ is −0.0118697. So the fugacity coefficient is:

$$\phi = \exp(-0.0118697) = 0.9882$$

Example 5.4

Given the composition that is reported in following table, what is the state of fluid at 600 R and 200 psi? Use Watson correlation.

Component	z_i
C_1	0.4
C_2	0.2
C_3	0.15
i-C_4	0.1
n-C_4	0.1
i-C_5	0.03
n-C_5	0.02

Solution

From example 5.2 this is known that at flash condition the number of moles of vapor phase is determined by solving the following equation for n^V.

$$\sum_{i=1}^{N} \frac{z_i(K_i - 1)}{V(K_i - 1) + 1} = 0$$

The previous equation gives a physically meaningful root for V if the following relation is satisfied:

$$\sum_{i=1}^{N} z_i K_i > 1$$

$$\sum_{i=1}^{N} z_i / K_i > 1$$

In other words, if the previous conditions satisfied a root between 0 and 1 found for V, the previous conditions simultaneously satisfy the fluid in the two-phase region.

If the fluid is at its bubble point, V is equal to 0 and the equation $\sum_{i=1}^{N} \frac{z_i(K_i - 1)}{V(K_i - 1) + 1} = 0$ is reduced to:

$$\sum_{i=1}^{N} z_i K_i = 1$$

If the fluid is at its dew point, V is equal to 1 and the equation $\sum_{i=1}^{N} \frac{z_i(K_i - 1)}{V(K_i - 1) + 1} = 0$ is reduced to:

$$\sum_{i=1}^{N} z_i / K_i = 1$$

Moreover, if the $\sum_{i=1}^{N} z_i K_i$ smaller than 1, the fluid is a compressed liquid, and if $\sum_{i=1}^{N} z_i / K_i$ is smaller than 1, the fluid is a superheated vapor.

(Continued)

All conditions are summarized in below table:

State of Fluid	Condition
Compressed liquid	$\sum_{i=1}^{N} z_i K_i < 1$
Saturated liquid (bubble point)	$\sum_{i=1}^{N} z_i K_i = 1$
Vapor + liquid	$\sum_{i=1}^{N} z_i K_i > 1$ and $\sum_{i=1}^{N} z_i K_i > 1$
Saturated vapor (dew point)	$\sum_{i=1}^{N} z_i / K_i = 1$
Superheated vapor	$\sum_{i=1}^{N} z_i / K_i < 1$

To estimate the equilibrium ratio by Watson's correlation, the critical properties and acentric factors of components required have been extracted from GPSA (2004) for all components and reported in below table.

Component	P_c (psi)	T_c (R)	Acentric factor
C_1	667	343.01	0.0115
C_2	706.6	549.59	0.0994
C_3	615.5	665.59	0.1529
i-C_4	527.9	734.08	0.1866
n-C_4	550.9	765.22	0.2003
i-C_5	490.4	828.67	0.2284
n-C_5	488.8	845.47	0.2515

The equilibrium for methane is calculated as follows:

$$K_{C_1} = \frac{667}{200} \exp\left[5.37(1 + 0.0115)\left(1 - \frac{343.01}{600}\right)\right] = 34.158$$

The results for other components are report in below table.

Component	z_i	K_i Eq. (5.53)	$z_i{}^*K_i$	z_i/K_i
C_1	0.4	34.1576	13.6630	0.0117
C_2	0.2	5.8018	1.1604	0.0345
C_3	0.15	1.5641	0.2346	0.0959
i-C_4	0.1	0.6355	0.0635	0.1574
n-C_4	0.1	0.4669	0.0467	0.2142
i-C_5	0.03	0.1985	0.0060	0.1512
n-C_5	0.02	0.1563	0.0031	0.1279
Sum	–	–	15.1773	0.7927

The $\sum_{i=1}^{N} z_i/K_i$ is smaller than 1, therefore the fluid is superheated vapor.

Example 5.5

Given a fluid with the composition.

Component	z_i	P_c (psi)	T_c (R)	w
C_1	0.2	667	343.01	0.0115
C_3	0.3	665.59	665.59	0.1529
n-C_5	0.5	488.8	845.47	0.2515

Estimate the bubble-point pressure at 620 R using Watson correlation.

Solution

The equilibrium ratio could be calculated similar to example 5.4. A value for pressure is estimated, and then the pressure is corrected using linear interpolation. The calculation for the first three iterations is reported in the table below.

		$P=1700$		$P=1300$		$\frac{1.2755-0.9754}{1300-1700}(1-1.2755)+1300 = 1667.168$	
Component	z_i	k	z^*K	k	z^*K	k	z^*K
C_1	0.2	4.4419	0.8884	5.8087	1.1617	4.5294	0.9059
C_3	0.3	0.2483	0.0745	0.3248	0.0974	0.2532	0.0760
n-C_5	0.5	0.0250	0.0125	0.0326	0.0163	0.0255	0.0127
Sum	–	–	0.9754	–	1.2755	–	0.9946

The converged bubble-point pressure is 1658.12 psi.

Example 5.6

Calculate the bubble-point pressure of an equimolar mixture of methane and normal decane using Peng–Robinson EOS (PR EOS) at 380K. Using the quadratic mixing rule. Set the binary interaction parameters to zero.

Solution

The Peng–Robinson EOS is presented by the following equations:

$$P = \frac{RT}{V-b} - \frac{a}{V(V+b)+b(V-b)} \tag{a}$$

$$a = \alpha a_c$$

$$a_c = 0.45724\frac{R^2T_c^2}{P_c}$$

(Continued)

$$b = 0.07780\frac{RT_c}{P_c}$$

$$\alpha = \left[1 + \kappa\left(1 - \left(\frac{T}{T_c}\right)^{0.5}\right)\right]^2$$

$$\kappa = 0.37464 + 1.54226\omega - 0.26992\omega^2$$

The fugacity coefficient may be calculated by the following equation.

$$\ln \phi_i = \frac{1}{RT}\int_V^\infty \left[\left(\frac{\partial P}{\partial n_i}\right)_{T,V,n_{j+I}} - RT/V\right] dV - \ln Z \quad \text{(b)}$$

Substituting Eq. (b) in Eq. (a) and applying the quadratic mixing rules to calculate parameters a and b for a mixture, gives the following equation for fugacity coefficient of component i in a mixture.

$$\ln\frac{f_i}{z_i P} = \ln \phi_i = \frac{b_i}{b}(Z-1) - \ln(Z-B) + \frac{A}{2\sqrt{2}B}\left(\frac{2}{a}\sum_{j=1}^{N} z_j a_{ij} - \frac{b_i}{b}\right)\ln\left[\frac{Z + (1-\sqrt{2})B}{Z + (1+\sqrt{2})B}\right] \quad \text{(c)}$$

in which z_i is the mole fraction of component i in the liquid or gas mixture, a_i and b_i are the parameters of PR EOS for component i in the mixture, a and b are the parameters of PR EOS for mixture and defined as below based on quadratic mixing rules:

$$a = \sum_i \sum_j z_i z_j (a_i a_j)^{0.5} \quad \text{(d)}$$

$$b = \sum_i z_i b_i \quad \text{(e)}$$

The A and B dimensionless parameters are determined by following relations:

$$A = \frac{aP}{R^2 T^2} \quad \text{(f)}$$

$$B = \frac{bP}{RT} \quad \text{(g)}$$

The critical temperature, critical pressure, and acentric factor for methane and ethane were taken from Danesh (1998).

Component	x_i	T_c (K)	P_c (MPa)	ω
C_1	0.5	190.56	4.60	0.0115
n-C_{10}	0.5	617.7	2.11	0.4923

For bubble-point calculation, consider $x_i = z_i$, Assuming $P = 20$ MPa as initial guess for bubble-point pressure. The parameters of PR EOS are calculated as follows:

Component	x_i	K_i	$y_i = K_i x_i$	Normalized y_i	a_{ci} (Pa m^6/mol^2)	b_i (m^3/mol)	α	$a_i = a_{ci}\alpha$ (Pa m^6/mol^2)
C_1	0.5	3.449	1.725	0.9998	0.2495	2.68E−05	0.703	0.175
n-C_{10}	0.5	7.018E−04	3.509E−04	2.0343E−04	5.1753	1.89E−04	1.491	8.519

The parameters a and b are determined using Eqs. (d) and (e) for both vapor and liquid phases.

For liquid phase:

$$a^L = \sum_i \sum_j x_i x_j (a_i a_j)^{0.5} = 2.7850 \text{ Pa m}^6/\text{mol}^2$$

$$b^L = \sum_i x_i b_i = 1.0808 \times 10^{-4} \text{ m}^3/\text{mol}$$

$$A^L = 5.5804, \quad B^L = 0.6842$$

For vapor phase:

$$a^V = \sum_i \sum_j y_i y_j (a_i a_j)^{0.5} = 0.176 \text{ Pa m}^6/\text{mol}^2$$

$$b^V = \sum_i y_i b_i = 2.6829 \times 10^{-5} \text{ m}^3/\text{mol}$$

$$A^V = 0.3525, \quad B^V = 0.1698$$

Z-form of PR EOS for compressibility is as follows.

$$Z^3 - (1 - B)Z^2 + \left(A - 3B^2 - 2B\right)Z - \left(AB - B^2 - B^3\right) = 0$$

(Continued)

Solving the Z-form of PR EOS for compressibility factor is performed using calculated parameter for each phase. Note that, if three roots are obtained by solving the equation prior to the last equation, for vapor phase select the biggest root and for liquid phase select the smallest root.

$$Z^L = 0.9383, \quad Z^V = 0.9063$$

The fugacity of each component for both phases is calculated by Eq. (c).

Component	f_i^L (MPa)	f_i^V (MPa)	$\phi_i^L = \frac{f_i^L}{z_i P}$	$\phi_i^V = \frac{f_i^V}{z_i P}$	$K_i = \frac{\phi_i^V}{\phi_i^L}$	$K_i x_i$
C_1	15.62	17.72	1.562	0.886	1.763	0.882
n-C_{10}	3.11E−02	3.73E−04	0.003	0.092	0.034	0.017

Checking the error by error $= \sum_{i=1}^{N} \left(1 - \frac{f_i^L}{f_i^V}\right)^2$ gives:

$$\text{error} = \sum_{i=1}^{N} \left(1 - \frac{f_i^L}{f_i^V}\right)^2 = 6.802 \times 10^3$$

Modifying the pressure for the next iteration as follows:

$$P_{new} = P_{old} \sum_i K_i x_i$$

$$P_{new} = 25 \times (0.882 + 0.017) = 17.972 \text{ MPa}$$

Now, with the P_{new} and adjusted equilibrium ratio, repeat previous steps until the error is greater than 10^{-12}. The converged bubble-point pressure is 15.902 MPa.

5.5 A DISCUSSION ON THE STABILITY

As pointed out before, an important obstacle in VLE problems is knowing whether a mixture actually forms two or more equilibrated phases or remains as a single stable phase. This is a very serious question and even if the results of the flash calculations seem physically consistent, this question should be answered to validate the outcomes.

The thermodynamic concept for phase stability states that a mixture will split in two or more phases if its total Gibbs free energy decreases after splitting, and remains as a single phase if splitting requires an increase in the total Gibbs energy (Prausnitz et al., 1998). This concept was developed by Baker and Michelsen (Baker et al., 1982; Michelsen, 1982) in the early 1980s. In this section, we briefly discuss stability based on these two papers.

Imagine a binary mixture with a given composition (known mole fraction, z_i, of each component). One could define the normalized Gibbs energy function for this mixture by Whitson and Brulé (2000):

$$g^* = G/RT = \sum_{i=1}^{N} z_i \ln f_i(z) \tag{5.64}$$

The g^* function could be shown as a curve in a two-dimensional plot versus one of the mole fractions as it could be seen in Fig. 5.1.

The graphic equilibrium examination requires a tangent plane to the Gibbs energy surface in a manner in which the tangent plane does not meet the surface except at the tangency points. The points of tangency are actually the compositions, which satisfy the condition of equal fugacity and are equilibrium phases. In fact, for this binary mixture, x and y, respectively, represent dew point and bubble point. If a mixture is split into two phases, it is a physical necessity that the mixture composition locates between bubble point and dew point.

$$n^{\mathrm{L}} + n^{\mathrm{V}} = n \tag{5.65}$$

$$n^{\mathrm{L}}(y - z) = n^{\mathrm{V}}(z - x) \tag{5.66}$$

The mixture Gibbs energy then will be:

$$g^*_{\mathrm{mix}} = F_{\mathrm{V}} g^*_{\mathrm{V}} + (1 - F_{\mathrm{V}}) g^*_{\mathrm{L}} \tag{5.67}$$

The F_{V} could be found by Eq. (5.68):

$$F_{\mathrm{V}} = \frac{z - y}{x - y} \tag{5.68}$$

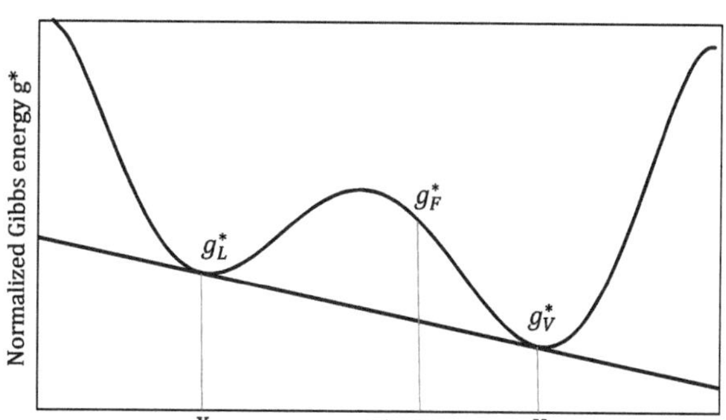

Figure 5.1 Reduced Gibbs energy surface of a hypothetical binary mixture.

If the mixture composition located at $z < x$ or $z > y$ the material balance emphasizes that the mixture could not form two phases. It can be shown if $z > y$ then:

$$n_L(y - z) = n_V(z - x) \tag{5.69}$$

Eq. (5.69) will be true if one of the mole numbers is negative, which is physically impossible.

Fig. 5.2 shows a normalized Gibbs energy curve for a hypothetical binary mixture. As can be seen from the figure, there are three valleys in the picture, which may suggest three tangents for the mixture. Nevertheless, only one of the tangents (AC) is valid and the other two (AB and BC) are invalid. This is because tangent AC does not intersect the g^* curve (other than at tangency points), whereas tangents AB and BC lie above the Gibbs energy curve. Despite satisfying equal-fugacity constraint, these false tangents only present local minima of the g^* curve and the AC tangent, which really indicates a valid two-phase solution. In other words, any solution with a composition in the range of $z_A < z < z_C$ will split in two equilibrated phases for which points A and C represent them. Despite invalidity of AB and AC, these are two potential two-phase solutions and could hardly be detected. Without previous data on actual equilibrium conditions, one could wrongly assume that these false tangents are valid solutions.

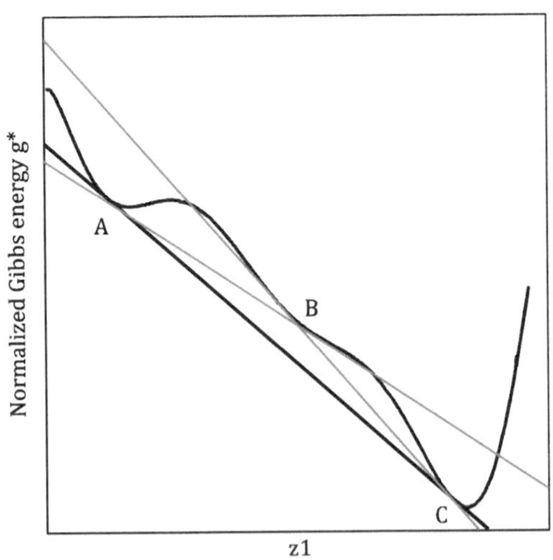

Figure 5.2 Reduced Gibbs energy surface of a hypothetical binary mixture with two invalid tangents (AB and BC).

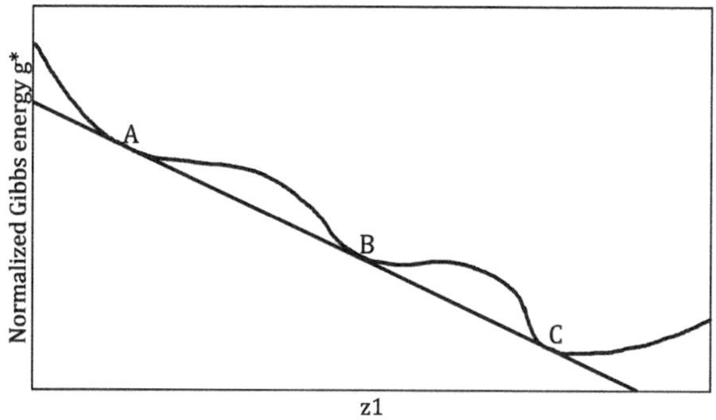

Figure 5.3 Reduced Gibbs energy surface of a hypothetical binary mixture with three valleys.

Fig. 5.3 shows a normalized Gibbs energy plot for a hypothetical binary mixture. As could be seen from the figure, the g^* curve has three valleys in a way that a single tangent line could pass through all of them. This is an indication of a three-phase equilibrium for any composition, which lies in the $z_A < z < z_C$ region. For $z < z_A$ and $z > z_C$, the mixture remains as a stable single phase.

Fig. 5.4 demonstrates a normalized Gibbs energy curve for a hypothetical binary mixture. As it is shown in the figure, this curve has two valid tangents AB and CD, each of which represents two different equilibrated two-phase systems. If the feed composition lies in the $z_A < z < z_B$ region, the mixture splits into Θ_1 and Θ_2 phases, whereas if it is in the $z_C < z < z_D$ region it splits into Θ_3 and Θ_4 phases. The mixture remains as a stable single phase outside these composition areas. The AD tangent line is not valid because it intersects the Gibbs energy curve.

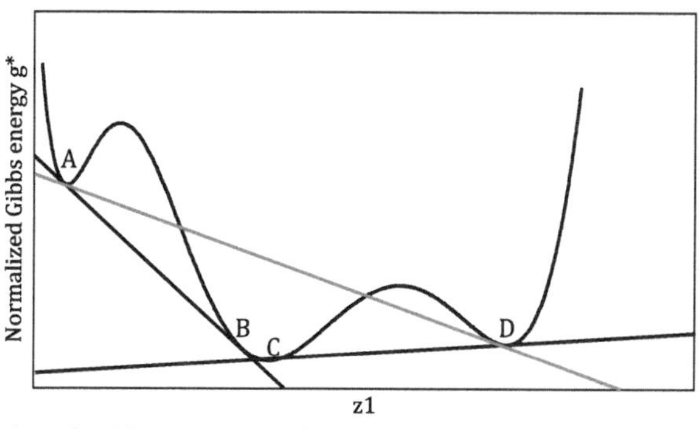

Figure 5.4 Reduced Gibbs energy surface of a hypothetical binary mixture with two valid tangents.

These examples point out that when the g^* curve is concave upward, the mixture remains a stable single phase. On the other hand, it is proven that when the Gibbs energy curve is convex $\left(\frac{\partial^2 g^*}{\partial z^2} < 0\right)$ downward, the mixture is intrinsically unstable (Danesh, 1998; Whitson and Brulé, 2000).

Despite that all the mentioned examples earlier are for binary mixtures, the basic rules remain valid for multicomponent mixtures, and the necessary and sufficient condition for equilibrium is curve-tangent citation. Nevertheless, for a multicomponent system the graphical presentation of the hyper surface g^* curve and hyper plane tangent is not practical. In this regard, many investigators tried to develop a numerical algorithm that could perform the calculation of phase stability.

The numerical algorithm for stability test, which is developed by Michelsen (1982) gives acceptable results for various multiphase mixtures. The detailed discussion of this stability test is beyond the scope of this book; nevertheless, its simplified algorithm is as follows. In the graphical presentation of this test, a tangent plane should be drawn on the g^* curve at the feed composition. The next step is to locate the other tangent planes to the Gibbs energy surface, which are parallel to the feed tangent plane. If any of the parallel, tangent planes are found below the feed tangent plane, the feed mixture would be unstable and will split into at least two equilibrated phases. If there is no other tangent plane parallel to the feed tangent plane or other tangent planes are all locating above the feed tangent plane, then the mixture will remain as a single stable phase. In addition, if a mixture composition locates on the very same feed tangent plane, the feed is in equilibrium state (bubble or dew point), and the second phase represents another equilibrium-stable phase (Firoozabadi, 1999; Whitson and Brulé, 2000).

Fig. 5.5 is a graphical representation of Michelsen test for a hypothetical binary mixture. Assume the feed stream composition is z_B. After drawing the tangent plane on the Gibbs free energy curve at point B, one must search for tangent planes parallel with the feed composition concentration. As can be seen from Fig. 5.5, the tangent planes at points C and E are parallel with the tangent plane at B. The tangent at point E lies under the feed tangent plane, and this means that the feed mixture is unstable and will split in two equilibrated phases. If we assume the feed composition is z_E, then, as could be seen from the figure, there will be no tangent planes parallel to the feed tangent plane, which locates below it. This is evidence for stability of the feed mixture as a stable single phase. If we assume a mixture with

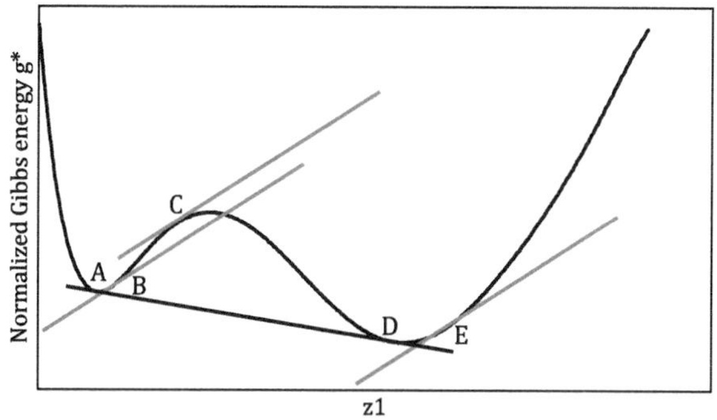

Figure 5.5 Graphical representation of Michelsen test for a hypothetical binary mixture.

composition of z_A, then observe the feed tangent plane with another point of tangency at point D. This means the feed composition at point A indicates an equilibrium state (bubble point or dew point), whereas the point D shows another equilibrium state.

This algorithm could be interpreted in mathematical relations. Michelsen indicated that finding a tangent plane parallel to the feed tangent plane equals finding a composition y, for which the fugacities of its components are equal to the fugacities of the feed components times a constant. This actually is the key to the rule of the Michelsen test and permits it to be used for multicomponent systems. This key concept could be mathematically shown as:

$$\frac{f_{z_i}}{f_{y_i}} = \text{cons.} \tag{5.70}$$

The stability test consists of two or more parts that should be calculated separately. In each part, one should find the second phase with a different assumption than in the other part. In one part, the second phase is assumed vapor like, and, in the other part, the second phase is considered liquid like. In principle, one can even assume N (the number of species in the mixture) parts for a stability test, starting each search with a pure component as the composition, but it is unnecessary. It would be sufficient if the assumptions were in a manner that covers a large composition range for searching. The following simplified algorithm could be used to perform the Michelsen stability test (Whitson and Brulé, 2000):

1. Find the fugacity of each component in the mixture.
2. Find the equilibrium ratio of each component with the use of Wilson correlation.

3. Find the mole number of each component in the second assumed phase.

$$\begin{aligned} Y_i &= z_i K_i \quad \text{if the 2nd assumed phase is vapor like} \\ X_i &= \frac{z_i}{K_i} \quad \text{if the 2nd assumed phase is liquid like} \end{aligned} \tag{5.71}$$

4. Find the mole fraction of each component in the second phase by normalizing the mole number.

$$\begin{aligned} y_i &= \frac{Y_i}{\sum_{j=1}^{N} Y_j} \quad \text{if the 2nd assumed phase is vapor like} \\ x_i &= \frac{X_i}{\sum_{j=1}^{N} X_j} \quad \text{if the 2nd assumed phase is liquid like} \end{aligned} \tag{5.72}$$

5. With the help of an EOS, find the fugacity of each component in the second assumed phase.
6. Find the fugacity ratio with the help of feed fugacity components f_{z_i} and the fugacities of the assumed phase, which has been found in step 5. This fugacity ratio will be used for sequential update of the equilibrium ratios.

$$\begin{aligned} (\mathrm{R}_i)_\mathrm{V} &= \frac{f_{z_i}}{f_{y_i}} \frac{1}{\sum_{j=1}^{N} Y_j} \quad \text{if the 2nd assumed phase is vapor like} \\ (\mathrm{R}_i)_\mathrm{L} &= \frac{f_{x_i}}{f_{z_i}} \sum_{j=1}^{N} X_j \quad \text{if the 2nd assumed phase is liquid like} \end{aligned} \tag{5.73}$$

7. Check for Convergence.

$$\sum_{i=1}^{N} (R_i - 1)^2 < \varepsilon \tag{5.74}$$

8. If the convergence is not achieved, the equilibrium ratios should be updated.

$$K_i^{(n+1)} = K_i^n R_i^n \tag{5.75}$$

9. with the use of the following criterion, check for the trivial solution.

$$\sum_{i=1}^{N} (\ln K_i)^2 < 1 \times 10^{-4} \tag{5.76}$$

10. If the trivial solution does not achieve, repeat the procedure from step 3. If both assumed phases reach convergence for a trivial solution, the feed composition is stable. If one of the assumed phases reaches a trivial solution whereas the sum of mole fractions of the other phase is less than or equal to unity $\left(\sum_{j=1}^{N} Y_j \text{ or } \sum_{j=1}^{N} X_j \le 1\right)$, the feed composition remains as a single stable phase. If the sum of mole fractions of both assumed phases became less than or equal to unity $\left(\sum_{j=1}^{N} Y_j \text{ and } \sum_{j=1}^{N} X_j \le 1\right)$ then the mixture would remain a stable single phase. In principle, it is difficult to give an unassailable guarantee for the mixture stability without checking all the compositions; however, this stability test usually confirms the stability. The stability test shows the feed mixture is unstable if the sum of the mole fractions of one of the assumed phases becomes more than unity $\left(\sum_{j=1}^{N} Y_j \text{ or } \sum_{j=1}^{N} X_j > 1\right)$. It is interesting to point out that, for the unstable phase, the equilibrium ratios from the stability test could be used as initial guesses for performing flash calculations. Note that if both $\sum_{j=1}^{N} Y_j$ and $\sum_{j=1}^{N} X_j$ become more than unity, then the following equation could be used for the initial guess.

$$K_i = (y_i/x_i) = K_{iV} K_{iL} \tag{5.77}$$

This initial guess would be very beneficial near the critical point for equilibrium calculations, as it will be crucial to have a close approximation of the K-values at that region. With the help of this initial guess and Eq. (5.39) one could solve for a vapor mole fraction V

and following Eqs. (5.78) and (5.79) the new vapor mole fraction could be found:

$$V_{j+1} = V_j - \frac{\sum_{i=1}^{N} \frac{z_i(K_i - 1)}{V(K_i - 1) + 1}}{\left(\frac{d}{dV} \sum_{i=1}^{N} \frac{z_i(K_i - 1)}{V(K_i - 1) + 1} \right)_j} \tag{5.78}$$

Note that the subscript j is a sign for an iteration counter. The derivative in the denominator is as follows:

$$\frac{d}{dV} \sum_{i=1}^{N} \frac{z_i(K_i - 1)}{V(K_i - 1) + 1} = - \sum_{i=1}^{N} \frac{z_i(K_i - 1)^2}{\left(V(K_i - 1) + 1\right)^2} \tag{5.79}$$

With the new vapor mole fraction, one could have new mole fractions with the help of Eqs. (5.37) and (5.38) and then use an EOS to calculate new equilibrium ratios. This procedure could successively repeated to the point of convergence.

Fig. 5.6 shows the normalized Gibbs energy for a hypothetical binary mixture at constant temperature and pressure. The composition M must be split into two phases of A and D to be stable. Nevertheless, there is the possibility that it remains as metastable phase. The metastable-phase region could develop by the increase of $z1$ until it reaches the inflation point I at which $\frac{\partial^2 g^*}{\partial z^2}$ turns negative and the mixture intrinsically could not be stable and will split into two phases of A and D. This point is called the limit of intrinsic stability and specifies the border of the metastable single-phase fluid.

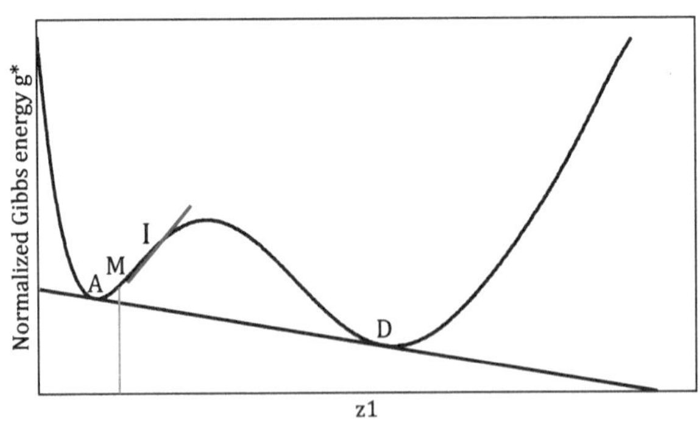

Figure 5.6 Reduced Gibbs energy for a hypothetical binary mixture at constant temperature and pressure.

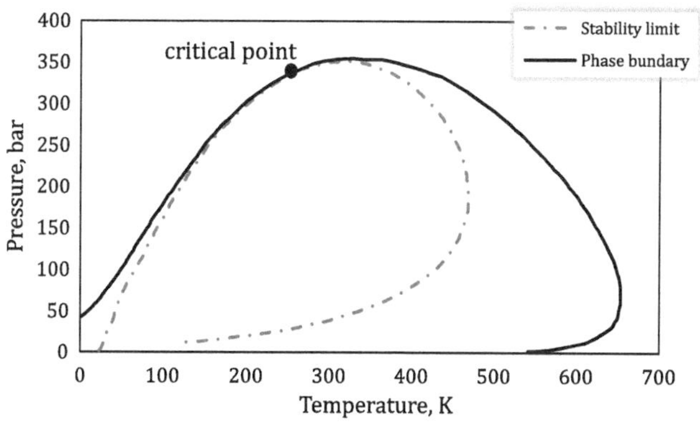

Figure 5.7 Phase diagram for a reservoir fluid (Nichita et al., 2007).

Fig. 5.7 shows the Nichita (Nichita et al., 2007) phase diagram for a reservoir fluid. The left part of the *dashed line* (starting from the critical point) in the figure shows the convergence of the stability test in which the second phase is assumed liquid like. The right part of the *dashed line* represents the convergence of the stability test in which the second phase is assumed vapor like. Inside the dashed curve, both $\sum_{j=1}^{N} Y_j$ and $\sum_{j=1}^{N} X_j$ are greater than unity and both tests have nontrivial solutions. The *dashed-line curve* is known as a spinodal curve, which distinguishes the metastable region from the unstable region (Prausnitz et al., 1998). This curve has an exciting feature. This curve meets the phase envelope at the critical point. This property comes very handy for determination of the critical point.

5.6 MULTIPHASE FLASH CALCULATIONS

When the systems under consideration are multicomponent, the complexity of stability analysis rises dramatically. Assume a single-phase system with vapor-like properties, which has a total Gibbs free energy of G_1. The stability of this system may be checked by considering a liquid-like phase. The system will split into two or more phases if and only if the total Gibbs energy of the system is reduced to G_2. If the stability analysis continued to search for a second liquid-like phase, it may be concluded that a further reduction of total Gibbs energy to G_3 is possible by forming the third phase. The stability analysis could be continued even for a fourth phase. This is a time-consuming and long process for ensuring a valid flash calculation. Therefore, a flash calculation with prior knowledge on the system behavior makes it much easier and saves much time.

Imagine a multicomponent system with with j phases and i components. For performing flash calculation on such a system, Eq. (5.39) could be modified as (Pedersen et al., 2014):

$$\sum_{i=1}^{N} \frac{z_i\left(K_i^m - 1\right)}{1 + \sum_{q=1}^{j-1} S^q\left(K_i^q - 1\right)} = 0 \quad q = 1, 2, \ldots, j - 1 \tag{5.80}$$

in which S^q indicates the molar fraction of phase q and K_i^q represents equilibrium ratios of species i in phases j and q. By initial estimation of the equilibrium ratios, one could find the molar fraction of each phase by Eq. (5.80), just like a two-phase flash calculation. The composition of each phase could be found by Eqs. (5.81) and (5.82).

$$y_i^q = \frac{z_i K_i^q}{1 + \sum_{q=1}^{j-1} S^q\left(K_i^q - 1\right)} \quad i = 1, 2, \ldots N \text{ and } q = 1, 2 \ldots, j - 1 \tag{5.81}$$

$$y_i^j = \frac{z_i}{1 + \sum_{q=1}^{j-1} S^q\left(K_i^q - 1\right)} \quad i = 1, 2, \ldots N \tag{5.82}$$

Here, y_i^q and y_i^j respectively represent, mole fractions of species i in phases q and j.

In the absence of water, oil and gas compositions are unlikely to form more than two phases and there is no need for a multiphase flash. Nevertheless, in the petroleum industry, multiphase flash will come in handy on many occasions. For example, when at low temperature CO_2 is displaced by reservoir oil, a three-phase system will form which has two liquid phases (one hydrocarbon rich and the other CO_2 rich) in equilibrium with a vapor phase. Presence of water as a separate phase is very common in the reservoirs and under proper temperature and pressure conditions; it also can take a solid-phase form as hydrate or ice. In addition, the formation of asphaltenes and waxes as precipitations are very common.

As mentioned earlier, water is very common in petroleum reservoirs and often forms a third phase during production. This water phase could be assumed pure water because the solubility of hydrocarbons in water is quite limited. This assumption would simplify the flash calculation by not considering other solubility of other species in the water phase. At first, it can be assumed that there are only two phases; a pure-water phase and a hydrocarbon phase, which contains an amount of dissolved water. The question is whether the dissolved water from the assumed hydrocarbon phase will separate. This may happen if dissolved water has a larger chemical potential than in its pure form.

$$(\mu_{\text{water}})_{\text{dissolved}} > (\mu_{\text{water}})_{\text{pure}} \tag{5.83}$$

Eq. (5.83) could be rewritten by expanding the chemical potentials.

$$\mu_{\text{w}}^{0} + RT(\ln \phi_{\text{dw}} + \ln P + \ln z) > \mu_{\text{w}}^{0} + RT(\ln \phi_{\text{pw}} + \ln P + \ln z) \tag{5.84}$$

in which subscripts dw and pw represent dissolved water and pure water, respectively, and the term z is the overall mole fraction of water in the mixture. By eliminating like terms on both sides, Eq. (5.84) could be simplified to Eq. (5.85).

$$\ln \phi_{\text{dw}} + \ln z > \ln \phi_{\text{pw}} \tag{5.85}$$

Therefore, if the dissolved water in the hydrocarbon phase precipitates, then its mole fraction x could be found by Eq. (5.86) and the rest of the water would be in the pure-water phase.

$$\ln \phi_{\text{dw}} + \ln x - \ln \phi_{\text{pw}} = 0 \tag{5.86}$$

Now that the amount of water in each phase is determined, the assumed hydrocarbon phase could be flashed independently. Note that if no pure water drifts from the feed mixture, Eq. (5.85) will no longer be valid and one should perform two-phase pressure–temperature (PT) flash on the mixture but check every iteration whether from any of the hydrocarbon phases a pure-water phase will be separated.

5.7 CALCULATION OF SATURATION PRESSURES WITH STABILITY ANALYSIS

Saturation pressure of a mixture at a fixed temperature refers to a pressure at which an equilibrium state with a minuscule quantity of an incipient phase is granted. In other words, in a PT flash calculation the vapor mole fraction has the value of zero or unity. This may could be interpreted in a simpler way of bubble and dew-point pressures. The conventional method for finding saturation pressures is to search for pressures that, with the value of vapor mole fractions equal to one or zero, the PT flash converges. This method has previously been discussed in this chapter. Despite the safety of this algorithm, it is time-consuming and has convergence problems at high pressure and near-critical point. There are other alternative procedures, which, with the use of stability analysis, calculate saturation pressures in a more resourceful way. The algorithm proposed by Michelsen is one of the most efficient methods for finding the saturation pressure. It starts with an estimated moderate pressure from either the bubble-point or dew-point curve.

There are two conditions for defining a saturation pressure. First is the equality of fugacities for all components in both phases:

$$f_{z_i} = f_{y_i} \tag{5.87}$$

And the second, for the incipient phase the mole fraction equals unity:

$$\sum_{i=1}^{N} y_i = 1 \quad \text{or} \quad \sum_{i=1}^{N} x_i = 1 \tag{5.88}$$

The conventional equations for solving bubble-point and dew-point calculations could be found by using the equilibrium ratio K.

$$\begin{aligned} &1 - \sum_{i=1}^{N} z_i K_i = 0 \quad \text{for Bubble calculation} \\ &1 - \sum_{i=1}^{N} z_i / K_i = 0 \quad \text{for Dew calculation} \end{aligned} \tag{5.89}$$

To use the concept of stability analysis for establishing the condition of saturation pressure, one should search for a secondary phase for which its tangent plane is parallel to the tangent plane of the mixture and has zero distance from it. In another word, the sum of all mole numbers in the incipient phase should be equal to unity.

$$\sum_{i=1}^{N} Y_i = 1 \tag{5.90}$$

Michelsen (1985) stated that for the determination of saturation pressure:

$$\chi(p_{\text{sat}}, y) = 1 - \sum_{i=1}^{N} z_i \left[\frac{\phi_i(z)}{\phi_i(y)}\right] = 0 = 1 - \sum_{i=1}^{N} y_i \left(\frac{f_{zi}}{f_{yi}}\right) = 1 - \sum_{i=1}^{N} Y_i \tag{5.91}$$

Here, the y_i is defined as:

$$y_i = \frac{Y_i}{\sum_{j=1}^{N} Y_j} \tag{5.92}$$

An algorithm proposed by Whitson and Brulé (2000) for solving this problem is proposed as follows:

1. Assume the vapor mole fraction V is equal one or zero. The convergence would not be affected if the validity of this assumption is not confirmed.
2. Consider a pressure and use that for the Michelsen Stability test. If the test results in the stability of the mixture, at the assumed pressure, indicate the higher limit of the saturation pressure search on the upper curve of the phase diagram. Go back to the first step and assume a lower pressure to find an unstable condition. The pressure at which the stability test results in an unstable system indicates the lower limit of the saturation pressure search on the upper-curve phase diagram.
3. With the unstable system at hand, the equilibrium ratios could be used for finding mole numbers of the incipient phase at dew and bubble points. It should be noted that if the stability test showed two unstable solutions, the equilibrium ratios of the one with the bigger $\sum_{i=1}^{N} Y_i$ should be used.

$$\begin{aligned} Y_i &= z_i K_i \quad \text{Bubble calculation} \\ Y_i &= z_i / K_i \quad \text{Dew calculation} \end{aligned} \tag{5.93}$$

4. Use Eq. (5.94) to find the normalized composition of the incipient phase. With the use of an EOS, and at the estimated saturation pressure, calculate Z factors and fugacities of the components for both z and y phases.

$$y_i = \frac{Y_i}{\sum_{j=1}^{N} Y_j} \tag{5.94}$$

5. Find the fugacity ratio

$$R_i = \frac{f_{zi}}{f_{yi}} \frac{1}{\sum_{j=1}^{N} Y_j} \tag{5.95}$$

6. With the use of the fugacity ratio, update the mole numbers of the incipient phase:

$$Y_i^{n+1} = Y_i^n \left[R_i^n\right]^\lambda \quad (5.96)$$

which here λ is defined as

$$\lambda = \left|\frac{b_{11}}{b_{11} - b_{01}}\right|$$

$$b_{01} = \sum_{i=1}^{N} \ln R_i^n \ln R_i^{n-1} \quad (5.97)$$

$$b_{11} = \sum_{i=1}^{N} \ln R_i^{n-1} \ln R_i^{n-1}$$

7. Using Newton–Raphson, the new saturation pressure could be estimated:

$$p_{\text{sat}}^{n+1} = p_{\text{sat}}^{n} - \frac{Q^n}{\left(\frac{\partial Q}{\partial p}\right)^n} \quad (5.98)$$

And the $\left(\frac{\partial Q}{\partial p}\right)^n$ is calculated in the nth iteration through Eq. (5.99):

$$\frac{\partial Q}{\partial P} = \sum_{i=1}^{N} Y_i R_i \left(\frac{\partial f_{yi}}{\partial p}\frac{1}{f_{yi}} - \frac{\partial f_{zi}}{\partial p}\frac{1}{f_{zi}}\right) \quad (5.99)$$

If the objective is finding an upper saturation pressure, the new estimate from Eq. (5.98) should be higher than the pressure which we guessed at the second step. If this condition was not fulfilled, go to step two and assume a new saturation pressure; if it did have a higher value than the estimated pressure, then check if the convergence criterion is satisfied.

$$\left|1 - \sum_{i=1}^{N} Y_i\right| < 10^{-13} \text{ and } \left[\sum_{i=1}^{N} \frac{\ln(R_i)}{\ln\left(\frac{Y_i}{z_i}\right)}\right]^2 < 10^{-8} \quad (5.100)$$

A trivial solution is achieved if the following criterion is satisfied:

$$\sum_{i=1}^{N}\left(\ln\frac{Y_i}{z_i}\right)^2 < 10^{-4} \tag{5.101}$$

8. If the algorithm did not converge to a solution, go back to step 4, but if it did converge to a solution, the determining the type of saturation could be easily performed by looking at the mole fraction of the mixture's heaviest species and comparing it with the mole fraction of that species in the incipient phase. If the mole fraction of the heaviest species in the incipient phase is greater than in the mixture, the saturation pressure is a dew point and the equilibrium ratio will be $K_i = z_i/y_i$. If the mole fraction of the heaviest species in the incipient phase was less than that in the mixture, then the saturation pressure is a bubble point and the equilibrium ratio would be $K_i = y_i/z_i$.

5.8 IDENTIFYING PHASES

One important and applicable aspect of flash calculation is to identify the phases that may be formed after performing the flash operation. For oil and gas mixtures, when the flash calculation results in splitting the mixture, it is usually convenient to say the one that has a lower density is the vapor phase and the other is the liquid phase. Nevertheless, in the case of a single phase, it may be hard to tell if the phase is really a liquid or a vapor. It is suggested that (Pedersen et al., 2014):

If $\frac{V}{b} <$ const the mixture is a liquid
If $\frac{V}{b} >$ const the mixture is a vapor

Here, V represents the molar volume, b is the parameter of the cubic EOS, and const depends on the EOS. The constant for the PR and SRK EOSs is suggested to have a value of 1.75.

Problems

5.1 Use SRK EOS for calculating the composition of equilibrated phases for the mixture with the following composition at 1500 psia and 150°F.

Component	z_i
C_1	0.5176
n-C_4	0.1593
C_{10}	0.3232

5.2 Calculate the equilibrium ratios with the Whitson and Torp Correlation for a mixture with the following composition at 750 psia and 15°F.

Component	z_i
C_1	0.429
C_2	0.178
C_3	0.108
i-C_4	0.102
n-C_4	0.082
i-C_5	0.049
n-C_5	0.042
C_6	0.001
CO_2	0.002
N_2	0.003
H_2O	0.004

5.3 Determine the bubble point temperature of the mixture with the following composition at 500 psia. Assume C_{7+} has the properties of C_{10}. Use Raoult's law for finding equilibrium ratios.

Component	z_i
C_1	0.2032
C_2	0.3483
C_3	0.0988
i-C_4	0.0618
n-C_4	0.0954
i-C_5	0.0255
n-C_5	0.0387
C_6	0.0377
C_{7+}	0.0303
CO_2	0.0120
N_2	0.0483

5.4 Repeat example 5.6 using SRK EOS.

5.5 Repeat example 5.5 using PR EOS.

REFERENCES

Abbott, M., Smith, J., Van Ness, H., 2001. Introduction to Chemical Engineering Thermodynamics. McGraw-Hill.

Ahmed, T., 2006. Reservoir Engineering Handbook. Gulf Professional Publishing.

Baker, L.E., Pierce, A.C., Luks, K.D., 1982. Gibbs energy analysis of phase equilibria. Society of Petroleum Engineers Journal 22 (05), 731–742.

Campbell, J.M., 1979. Gas Conditioning and Processing, vol. 1.

Danesh, A., 1998. PVT and Phase Behaviour of Petroleum Reservoir Fluids. Elsevier.

Firoozabadi, A., 1999. Thermodynamics of Hydrocarbon Reservoirs.

GPSA, F., 2004. Engineering Data Book. Gas Processors Suppliers Association, pp. 16–24.
McCain, W.D., 1990. The Properties of Petroleum Fluids. PennWell Books.
Michelsen, M.L., 1982. The isothermal flash problem. Part I. Stability. Fluid Phase Equilibria 9 (1), 1–19.
Michelsen, M.L., 1985. Saturation point calculations. Fluid Phase Equilibria 23 (2), 181–192.
Nichita, D.V., Broseta, D., Montel, F., 2007. Calculation of convergence pressure/temperature and stability test limit loci of mixtures with cubic equations of state. Fluid Phase Equilibria 261 (1), 176–184.
Orbey, H., Sandler, S.I., 1998. Modeling Vapor-Liquid Equilibria: Cubic Equations of State and Their Mixing Rules. Cambridge University Press.
Pedersen, K.S., Blilie, A.L., Meisingset, K.K., 1992. PVT calculations on petroleum reservoir fluids using measured and estimated compositional data for the plus fraction. Industrial & Engineering Chemistry Research 31 (5), 1378–1384.
Pedersen, K.S., Christensen, P.L., Shaikh, J.A., 2014. Phase Behavior of Petroleum Reservoir Fluids. CRC Press.
Prausnitz, J.M., Lichtenthaler, R.N., de Azevedo, E.G., 1998. Molecular Thermodynamics of Fluid-Phase Equilibria. Pearson Education.
Riazi, M., 2005. Characterization and Properties of Petroleum Fractions. ASTM International.
Whitson, C.H., Brulé, M.R., 2000. Phase Behavior. Henry L. Doherty Memorial Fund of AIME, Society of Petroleum Engineers.

CHAPTER SIX

Fluid Sampling

M.A. Ahmadi[1], A. Bahadori[2,3]
[1]Petroleum University of Technology (PUT), Ahwaz, Iran
[2]Southern Cross University, Lismore, NSW, Australia
[3]Australian Oil and Gas Services Pty Ltd, Lismore, NSW, Australia

6.1 INTRODUCTION

Appropriate production management from an oil and gas reservoir can increase the production of the hydrocarbons (oil and gas) originally in the reservoir. Building appropriate management policies needs precise understanding of oil and gas reservoir characterizations including reservoir rock and fluid characterization. In this chapter we try to explain the different sampling methods and experimental procedures for calculating reservoir fluid properties (Moffatt and Williams, 1998; Williams, 1994, 1998; Standing, 1951, 1952; API Recommended Practice 44, 2003).

The main aim of the fluid sampling in oil and gas reservoirs is to collect a sample that is demonstrative of the original oil and gas fluid. If the process of sampling is improper or if fluids are gathered from an inappropriately "conditioned" well, the subsequent samples may not be demonstrative of the original oil and gas fluid. A nondemonstrative sample may not show the same characters as the original oil and gas fluid. Using fluid property data gained from nondemonstrative samples, however precise the laboratory experiment approaches, may yield faults in the management of the oil and gas reservoir. Inadequate development can also result in insufficient data being taken throughout the sampling plan. Inadequate data can make it unfeasible or problematic for lab experts to conduct and deduce experiments that give precise and expressive fluid character info (API Recommended Practice 44, 2003).

The composition of the reservoir oil and gas samples gathered from different oil and gas fields in the world varies significantly. In some fields, the sample is in the liquid state and in others it is in the gaseous state; commonly, liquid and gas exist in a specified reservoir. Furthermore, the composition of the rocks that encompass these reservoir fluids also differs significantly, and in flow and physical characters. In specific cases, this can

Fluid Phase Behavior for Conventional and Unconventional Oil and Gas Reservoirs
http://dx.doi.org/10.1016/B978-0-12-803437-8.00006-3

oblige to complex the sampling process. It should be noted that different factors affect the choosing of fluid sampling method such as height of oil/gas column, fractured or homogeneous reservoir, and water coning (API Recommended Practice 44, 2003; Moffatt and Williams, 1998; Williams, 1994, 1998).

An appropriate sample taken from a sole well can be demonstrative of the original oil and gas reservoir fluid throughout the whole reservoir when a reservoir is fairly small. Samples collected from different wells and/or depths may be needed for reservoirs that are large or complicated. In reservoirs exposed to new tectonic disturbances and/or very thick formations in huge oil and gas reservoir, considerable changes in fluid composition often happen. Extra sampling throughout the later life of a reservoir is common for the reason that production knowledge can reveal that the reservoir is more complicated than shown by earlier info (API Recommended Practice 44, 2003; Moffatt and Williams, 1998; Williams, 1994, 1998).

Two main categories for reservoir fluid sampling are surface sampling and subsurface sampling. As clearly seen from the names of these methods, each group indicates the place at which the sampling method takes place. Subsurface sampling is also known as bottom-hole or downhole sampling. Choosing one specific approach over another is affected by the producing characteristics and the mechanical condition of the well, the reservoir fluid type, the design and mechanical situation of the surface production facilities, the comparative cost of the different approaches, and safety concerns. Explanation with details of the suggested sampling techniques is demonstrated in the following sections of this chapter. Parameters that should be taken into account in selection of a technique are also illustrated (API Recommended Practice 44, 2003; Moffatt and Williams, 1998; Williams, 1994, 1998).

The selection of either the bottom-hole or surface sampling technique cannot be reflected as a routine or straightforward issue. Each reservoir generally has specific limitations or conditions special to it. Wells that show rapid changes in rate of production exhibit particular complications in making the required experiments with satisfactory precision. Daily or seasonal weather variations can also affect the operation of fluid sampling. Therefore, the specifics of a prearranged sampling process regularly need amendment to avoid local difficulties (API Recommended Practice 44, 2003; Moffatt and Williams, 1998; Williams, 1994, 1998).

Well conditioning prior to fluid sampling is practically required. Normal production operations or initial well testing often yields the fluid near the wellbore that has a composition that is changed from the initial reservoir fluid. The main aim of the well conditioning is to eliminate this nonrepresentative fluid. Conditioning of the well comprises a production rate that will push the nonrepresentative fluid into the wellbore and let it to be substituted by representative fluid flowing from the reservoir. Conditioning of the well is particularly eminent when the fluid of the reservoir is at or near its bubble point/dew point pressure at the dominant reservoir circumstances for the reason that pressure reduction near the wellbore, which unavoidably happens from production well, will change the fluid composition that is flowing into the wellbore (API Recommended Practice 44, 2003; Moffatt and Williams, 1998; Williams, 1994, 1998).

6.2 SAMPLING METHOD

As mentioned in previous section, the choice of sampling method can be influenced by a number of important considerations. These include the volume of sample required by the laboratory, the type of reservoir fluid to be sampled, the degree of depletion of the reservoir, the mechanical condition of the wellbore, and the type of available gas–oil separation equipment (API Recommended Practice 44, 2003).

6.2.1 Subsurface Sampling

6.2.1.1 Bottom-Hole Samplers

The conventional subsurface method consists of lowering a sampling device, usually called a "bottom-hole sampler," down the well to a preselected depth. The bottom-hole samplers can be used in either open-hole or cased-hole wells and can be run in tubing. A sample of the fluid in the wellbore at that depth is trapped in a pressure-tight section of the sampler. The sampler is brought to the surface where the sample is repressured and restored to single-phase condition, then it may be transferred to a suitable pressure vessel for transporting to the laboratory. Bottom-hole samplers are available in a variety of configurations: design details and operating instructions can be obtained from the vendors of such equipment (Moffatt and Williams, 1998; Williams, 1994, 1998; API Recommended Practice 44, 2003; Danesh, 2003).

The subsurface sampling method is often used when the flowing bottom-hole pressure is greater than the reservoir oil saturation pressure. Some types of bottom-hole samplers function poorly with highly viscous and foaming oils. The operator should study the operation of the sampler and then decide if a representative sample can be collected with it.

Mechanical obstructions, such as a downhole choke or a bent or collapsed section of tubing, can prevent the sampler from reaching the desired sampling depth. Produced sand can also form an obstruction. When a large volume of sample is desired, the relatively small sample provided by a bottom-hole sampler requires repetition of the sampling operation. However, modern designs of subsurface samplers incorporate larger volume and/or multiple sample containers (API Recommended Practice 44, 2003; Proett et al., 1999; Smits et al., 1993; Danesh, 2003).

6.2.1.2 Formation Testers

The modern open-hole wire-line samplers (API Recommended Practice 44, 2003; Proett et al., 1999; Smits et al., 1993) consist of a probe and seal assembly that can be extended against the side of the wellbore to achieve a pressure-tight flow path between the reservoir layer and the tool flow line leading to one or several chambers that can be selectively opened and closed by control from the surface. A suitable pressure gauge enables accurate measurement of the flow line pressure.

Whereas the bottom-hole samplers collect a fraction of whatever fluid is inside the wellbore, the formation testers collect fluid samples directly from the formation. Modern formation tester tools can pump out drilling and completion fluids before collecting an uncontaminated sample of reservoir fluid. This ability to pump out unwanted fluids overcomes the serious disadvantages of earlier designs. However, increasing use of oil-based muds during drilling has resulted in common problems with contamination of fluid samples (API Recommended Practice 44, 2003; Proett et al., 1999; Williams, 1998; Danesh, 2003).

Procedures are identical to those used for conventional bottom-hole samplers. However, to minimize possible handling incidents that may affect their integrity, samples should be shipped to the pressure—volume—temperature (PVT) laboratory in the sampling chamber whenever possible rather than being transferred into shipping containers at the well site (API Recommended Practice 44, 2003).

6.2.1.3 Surface Sampling

The surface sampling method consists of taking samples of separator oil and gas with concurrent and accurate measurements of the rates of separator oil and gas flow. The reservoir fluid is reconstructed in the lab by recombining the gas and oil samples in appropriate proportion as determined from the producing gas—oil ratio (GOR). Large volumes of both oil and gas samples can be easily obtained with this method (Moffatt and Williams, 1998; Williams, 1994, 1998; API Recommended Practice 44, 2003; Danesh, 2003).

Before samples are taken, fluid flow into the wellbore, in the flow string, in the separators, and through the points where the oil and gas rates are measured must be stabilized. Also, the oil and gas flow rate determinations must be accurate. Therefore the facilities for making these determinations must be in excellent condition and operated by persons thoroughly instructed in their use. Metering equipment must be properly calibrated. The importance of these calibrations cannot be overemphasized. Any errors in the GOR measurement will be reflected in the recombination calculations and can prevent the laboratory personnel from properly reconstituting the reservoir fluid (Moffatt and Williams, 1998; Williams, 1994, 1998; API Recommended Practice 44, 2003; Danesh, 2003).

As an example, recording an incorrect orifice diameter for a gas orifice meter can easily result in a 50% error in bubble point pressure measured on a recombined oil sample. This is not trivial and helps to illustrate the importance of accurate flow measurement data for surface sampling.

On the other hand, other cases have shown excellent agreement in measured fluid properties between recombined surface samples and subsurface samples, thereby confirming that the surface sample method can provide good results, if good samples are collected and if flow measurement data are accurate and representative (Moffatt and Williams, 1998; Williams, 1994, 1998; Standing, 1951, 1952; API Recommended Practice 44, 2003; Danesh, 2003).

6.2.1.4 Wellhead Sampling

This is a less common, but potentially valuable, alternative to the previously mentioned approaches. If a fluid is known to be in the single-phase state at the wellhead conditions of temperature and pressure, this technique can produce the easiest and most reliable results. Typically, it is employed only for oils that are highly undersaturated at wellhead conditions or for dry gases. The problem in using wellhead sampling is knowing that the fluid is truly in the single phase at the sampling point (Moffatt and Williams, 1998; Williams, 1994, 1998; API Recommended Practice 44, 2003; Danesh, 2003).

6.2.1.5 Relative Advantages of Subsurface and Surface Sampling

The following summary of relative advantages of subsurface and surface sampling should be considered in selecting the more appropriate sampling technique for a given application (Moffatt and Williams, 1998; Williams, 1994, 1998; API Recommended Practice 44, 2003; Danesh, 2003).

1. Pros of subsurface sampling (Moffatt and Williams, 1998; Williams, 1994, 1998; API Recommended Practice 44, 2003; Danesh, 2003):
   a. Collects directly the preferred sample.
   b. With special sampling tool, can preserve full pressure on sample.
   c. Excludes using surface separators and the appropriate sizing of separators.
   d. Excludes the need for flow rate–metering devices and their proper sizing and calibration (for determination of producing GOR).
   e. Requires less sampling information be transmitted to testing laboratory.
   f. Eliminates potential errors in recombination of gas and oil samples required for surface samples.
   g. Fewer sample containers need to be transmitted to the field because three subsurface samples can supply an adequate quantity of sample for routine laboratory studies.
2. Pros of formation testers (Moffatt and Williams, 1998; Williams, 1994, 1998; API Recommended Practice 44, 2003; Danesh, 2003):
   a. Same advantages as subsurface sampling.
   b. Collects the desired sample directly from the formation.
   c. Sample represents reservoir fluid over a very narrow depth interval.
   d. Sample not affected by fluid segregation in the wellbore.
   e. Can sample reservoir fluid even if water is standing in wellbore.
   f. Before production from reservoir formation, testers can sample reservoir fluid at original circumstances.
   g. Controlled pressure drawdown during sample collection.
3. Pros of surface sampling (Moffatt and Williams, 1998; Williams, 1994, 1998; API Recommended Practice 44, 2003; Danesh, 2003):
   a. Relatively easy, convenient, and less expensive compared to subsurface sampling (e.g., no rig or wire-line unit is required on location).
   b. Avoids loss of production during required shut-in period for subsurface sampling (period of 1–4 days, or more for low deliverability wells).
   c. Avoids the potential for getting the subsurface sampling tool stuck or lost if the tubing is damaged or deviated, or if the sampling tool is lowered below tubing level.
   d. Applicable to cases where water is expected in tubing at the depth of the producing formation, where subsurface sampling cannot be used.
   e. Does not require that single-phase fluid be produced into the wellbore.

f. Preferred method for saturated gas-condensate reservoirs.
g. Applicable to gas condensates, wet and dry gases, where subsurface sampling is generally inappropriate.
h. Applicable to viscous and foamy oils, where obtaining satisfactory subsurface samples may be difficult.
i. Large volumes of samples and replicate samples are easier to obtain than by subsurface samplers.

In general, the pros stated for subsurface sampling recognize drawbacks in surface sampling and vice versa. One exception is the frequent problem of sample contamination by mud filtrate or drilling mud that happens for formation samples, but which is prevented when there is production to the surface and either subsurface "production" sampling or surface sampling (Moffatt and Williams, 1998; Williams, 1994, 1998; API Recommended Practice 44, 2003; Danesh, 2003).

6.3 RECOMBINATION

In this section we are trying to determine reservoir fluid composition or properties when the composition or properties of the aforementioned fluid are available at the surface facilities (separators and stock tank) (McCain, 1990). In this regard, we have four different types of case that are explained in detail in the following sections.

6.3.1 Case 1

In this case, we have separator gas composition, stock tank gas composition, stock tank liquid composition, GOR in separator and stock tank, and API (American Petroleum Institute) gravity of fluid. Consider the aforementioned input data and calculate the reservoir fluid composition at reservoir condition. To assess this goal, we should follow the following procedure:

1. Assume 1 lb_{mol} separator liquid and calculate molecular weight and density of the stock tank liquid
2. Calculate the number of lb_{mol} for stock tank gas ($n_{st.gas}$) and separator gas ($n_{sep.gas}$)
3. Calculate total number of lb_{mol} and reservoir fluid composition

Example 6.1

Consider a wet gas is produced through a separator that is operating at 350 Psi and 78°F to a stock tank. Separator produces 64,000 SCF/STB and stock tank vents 770 SCF/STB. Moreover, stock tank liquid gravity is 53.5°API. Composition of the surface streams is reported in the following table. Calculate the composition of the reservoir gas.

(Continued)

Component	Separator Gas Composition	Stock Tank Gas Composition	Stock Tank Liquid Composition
C_1	0.8809	0.2909	0.0015
C_2	0.0606	0.1749	0.0031
C_3	0.02955	0.2232	0.0395
i-C_4	0.016	0.0742	0.0097
n-C_4	0.0057	0.1101	0.0303
i-C_5	0.00208	0.0461	0.0519
n-C_5	0.003	0.0401	0.0336
C_6	0.002	0.039	0.1099
C_{7+}	0.00017	0.0015	0.7205

$\gamma_{C_{7+}}$ in Stock Tank $= 0.7825$

$MW_{C_{7+}}$ in Stock Tank $= 120$

Solution

At first we should assume $n_{st.liq} = 1\ lb_{mol}$ and then follow the abovementioned procedure.

Molecular weight of stock tank liquid is:

$$MW_{st.liq} = \sum_{i=1} x_i \times MW_i = 105.8411$$

$$API = 53.5 \rightarrow \rho_{st.liq} = 47.70\ \text{lbm/ft}^3$$

$$n_{sep.gas} = 64000\left(\frac{\text{SCF Separator Gas}}{\text{1 Stock Tank STB}}\right) \times \left(\frac{\text{1 lbmol Separator Gas}}{\text{380.7 SCF Separator Gas}}\right) \times \left(\frac{\text{1 STB Stock Tank Liquid}}{5.615\ \text{ft}^3\ \text{Stock Tank Liquid}}\right) \times \left(\frac{1\ \text{ft}^3\ \text{Stock Tank Liquid}}{\text{47.70 lbm Stock Tank Liquid}}\right) \times \left(\frac{\text{105.8411 lbm Stock Tank Liquid}}{\text{1 lbmol Stock Tank Liquid}}\right) = 66.432\left(\frac{\text{lbmol Separator Gas}}{\text{lbmol Stock Tank Liquid}}\right)$$

$$n_{st.gas} = 770\left(\frac{\text{SCF Stock Tank Gas}}{\text{1 Stock Tank STB}}\right) \times \left(\frac{\text{1 lbmol Stock Tank Gas}}{\text{380.7 SCF Stock Tank Gas}}\right) \times \left(\frac{\text{1 STB Stock Tank Liquid}}{5.615\ \text{ft}^3\ \text{Stock Tank Liquid}}\right) \times \left(\frac{1\ \text{ft}^3\ \text{Stock Tank Liquid}}{\text{47.70 lbm Stock Tank Liquid}}\right) \times \left(\frac{\text{105.8411 lbm Stock Tank Liquid}}{\text{1 lbmol Stock Tank Liquid}}\right) = 0.799\left(\frac{\text{lbmol Stock Tank Gas}}{\text{lbmol Stock Tank Liquid}}\right)$$

$$n_t = 66.432 + 0.799 + 1 = 68.231$$

$$n_i = y_{i\ sep.gas} \times 66.432 + y_{i\ st.gas} \times 0.799 + x_{i\ st.liq} \times 1$$

Consequently, the composition of the reservoir fluid is calculated and reported in the following table.

Component	Number of Moles (n_i)	Reservoir Fluid Composition (n_i/n_t)
C_1	58.75388	0.861102
C_2	4.168624	0.061096
C_3	2.180902	0.031964
i-C_4	1.131898	0.016589
n-C_4	0.496932	0.007283
i-C_5	0.226912	0.003326
n-C_5	0.264936	0.003883
C_6	0.273925	0.004015
C_{7+}	0.732992	0.010743

6.3.2 Case 2

In this case, we have separator gas composition, separator liquid composition, GOR in separator, and separator liquid volume factor (this parameter is defined as the ratio of separator liquid to stock tank liquid). Consider the aforementioned input data and calculate the reservoir fluid composition at reservoir condition. To assess this goal, we should follow the following procedure:

1. Assume 1 lb_{mol} separator liquid and calculate molecular weight and density of the separator liquid.
2. Calculate the number of lb_{mol} for separator gas ($n_{sep.gas}$).
3. Calculate total number of lb_{mol} and reservoir fluid composition.

Example 6.2

Consider a wet gas is produced through a separator that is operating at 365 Psi and 77°F to a stock tank. Separator produces 66,000 SCF/STB and separator liquid volume factor is 1.32 bbl/STB. Moreover, stock tank liquid gravity is 50°API. Composition of the surface streams is reported in the following table. Calculate the composition of the reservoir gas.

(Continued)

Component	Separator Gas Composition	Separator Liquid Composition
C_1	0.8809	0.0015
C_2	0.0606	0.0131
C_3	0.02955	0.0295
i-C_4	0.016	0.0197
n-C_4	0.0057	0.0203
i-C_5	0.00208	0.1519
n-C_5	0.003	0.1336
C_6	0.002	0.1099
C_{7+}	0.00017	0.5205

$\gamma_{C_{7+}}$ in Separator $= 0.7825$

$MW_{C_{7+}}$ in Separator $= 120$

Solution

At first we should assume $n_{\text{sep.liq}} = 1\ \text{lb}_{\text{mol}}$ and then follow the abovementioned procedure.

Molecular weight of separator liquid is:

$$MW_{\text{sep.liq}} = \sum_{i=1} x_i \times MW_i = 96.13(\text{lbm/lbmol})$$

$$\text{API} = 50 \rightarrow \rho_{\text{st.liq}} = 48.624\ \text{lbm/ft}^3$$

$$n_{\text{sep.gas}} = 66000\left(\frac{\text{SCF Separator Gas}}{\text{1 Stock Tank STB}}\right) \times \left(\frac{\text{1 lbmol Separator Gas}}{\text{380.7 SCF Separator Gas}}\right)$$
$$\times \left(\frac{\text{1 STB Stock Tank Liquid}}{\text{1.32 bb6 STB or liquid volume factor is 1.rocedirereported in below table.ents 366 SCF/STB.M}}\right)$$
$$\times \left(\frac{\text{1 STB Separator Liquid}}{5.615\ \text{ft}^3\ \text{Separator Liquid}}\right)$$
$$\times \left(\frac{1\ \text{ft}^3\ \text{Separator Liquid}}{\text{48.624 lbmol Separator Liquid}}\right)$$
$$\times \left(\frac{\text{96.13 lbm Separator Liquid}}{\text{1 lbmol Separator Liquid}}\right)$$
$$= 46.242\left(\frac{\text{lbmol Separator Gas}}{\text{lbmol Separator Liquid}}\right)$$

$$n_t = 46.242 + 1 = 47.242$$

$$n_i = 46.242 \times y_{i\ \text{sep.gas}} + 1 \times x_{i\ \text{sep.liq}}$$

Consequently, the composition of the reservoir fluid is calculated and reported in the following table.

Component	Number of Moles (n_i)	Reservoir Fluid Composition (n_i/n_t)
C_1	40.73608	0.862285
C_2	2.815365	0.059595
C_3	1.395951	0.029549
i-C_4	0.759572	0.016078
n-C_4	0.283879	0.006009
i-C_5	0.248083	0.005251
n-C_5	0.272326	0.005764
C_6	0.202384	0.004284
C_{7+}	0.528361	0.011184

6.3.3 Case 3

In this case, we have separator gas gravity, stock tank gas gravity, GOR in separator and stock tank, and API gravity of stock tank liquid. Consider the aforementioned input data and calculate the reservoir fluid gravity at reservoir condition. To assess this goal we should follow the following procedure:

1. Calculate the average gas specific gravity at surface using the following equation:

$$\gamma_{\text{g surf}} = \frac{\sum R_i \gamma_i}{\sum R_i} = \frac{R_{\text{sep}}\gamma_{\text{sep}} + R_{\text{st}}\gamma_{\text{st}}}{R_{\text{sep}} + R_{\text{st}}}$$

2. We have the following empirical equation for calculating molecular weight of stock tank liquid:

$$\text{MW}_{l\ \text{st}} = \frac{5954}{\text{API} - 8.81}$$

3. Calculate the mass of the gas in gas phase at surface condition using the following equation:

$$m_{\text{g surface}} = R\frac{\text{SCF Surface Gas}}{\text{STB}} \times \frac{1\ \text{lbmol Surface Gas}}{380.7\ \text{SCF Surface Gas}} \times 29\gamma_{\text{g surf}}$$
$$= 0.0762 \times R \times \gamma_{\text{g surf}}$$

4. Calculate the mass of the gas in liquid phase at surface condition using the following equation:

$$m_{l\ \mathrm{surface}} = 1\ \mathrm{STB} \times \frac{5.165\ \mathrm{ft}^3}{1\ \mathrm{STB}} \times 62.4\gamma_{l\ \mathrm{surf}} = 350.2\gamma_{l\ \mathrm{surf}}$$

5. We assume 1 STB stock tank liquid at surface and calculate based on this assumption. Moreover, we know that the produced gases from reservoir have both liquid and gas phases. Consequently, calculate the mass of the gas at reservoir condition using the following equation:

$$\text{mass of gas}_{\text{reservoir}} = \text{mass of gas}_{\text{surface}} + \text{mass of liquid}_{\text{surface}}$$

$$\text{mass of gas}_{\text{reservoir}} = 0.0762 \times R \times \gamma_{\mathrm{g\ surf}} + 350.2\gamma_{l\ \mathrm{surf}}$$

6. Calculate the number of gas moles at reservoir condition using the following equation:

$$n_{\mathrm{g\ reservoir}} = \frac{0.0762 \times R \times \gamma_{\mathrm{g\ surf}}}{29 \times \gamma_{\mathrm{g\ surf}}} + 350.2\frac{\gamma_{l\ \mathrm{surf}}}{\mathrm{MW}_{l\ \mathrm{surface}}}$$

$$= 0.00263R + 350.2\frac{\gamma_{l\ \mathrm{surf}}}{\mathrm{MW}_{l\ \mathrm{surface}}}$$

7. Determine the molecular weight of the reservoir gas using the following simple equation:

$$\mathrm{MW}_{\mathrm{g\ reservoir}} = \frac{m_{\mathrm{g\ reservoir}}}{n_{\mathrm{g\ reservoir}}}$$

8. Determine the specific gravity of the reservoir gas using the following simple equation:

$$\gamma_{\mathrm{g\ reservoir}} = \frac{\mathrm{MW}_{\mathrm{g\ reservoir}}}{29} = \frac{R\gamma_{\mathrm{g\ surface}} + 4600\gamma_{l\ \mathrm{surface}}}{R + 133300\frac{\gamma_{l\ \mathrm{surf}}}{\mathrm{MW}_{l\mathrm{surface}}}}$$

Example 6.3

Consider a wet gas is produced through a separator that is operating at 310 Psi and 76°F to a stock tank. Separator produces 70,500 SCF/STB and stock tank vents 500 SCF/STB and separator gas specific gravity is 0.683. Moreover, stock tank liquid gravity and specific gas gravity are 51.2°API and 1.119, respectively. Calculate the gas specific gravity of the reservoir gas.

Solution

At first we should calculate the total producing GOR as follows:

$$R_{total} = R_{sep} + R_{st} = 70500 + 500 = 71000\ (\text{SCF/STB})$$

Then we should determine the average specific gravity at surface using the following equation:

$$\gamma_{g\ surf} = \frac{\sum R_i \gamma_i}{\sum R_i} = \frac{R_{sep}\gamma_{sep} + R_{st}\gamma_{st}}{R_{sep} + R_{st}} = \frac{48151.5 + 559.5}{71000} = 0.686$$

We have

$$MW_{l\ st} = \frac{5954}{API - 8.81} = 140.45$$

Then

$$\gamma_{g\ reservoir} = \frac{R\gamma_{g\ surface} + 4600\gamma_{l\ surface}}{R + 133300\dfrac{\gamma_{l\ surf}}{MW_{l\ surface}}} = \frac{71000 \times 0.686 + 4600 \times 0.7744}{71000 + 133300\left(\dfrac{0.7744}{140.45}\right)}$$

$$= 0.72863$$

6.3.4 Case 4

In this case, we have separator gas gravity, GOR in separator, and API gravity of surface fluid. Consider the aforementioned input data and calculate the reservoir fluid gravity at reservoir condition. To assess this goal, we should follow the following procedure:

1. Calculate the average gas specific gravity at surface using the following equation:

$$\gamma_{g\ surf} = \frac{\sum R_i \gamma_i}{\sum R_i} = \frac{R_{sep}\gamma_{sep} + R_{st}\gamma_{st}}{R_{sep} + R_{st}}$$

2. We have the following empirical equation for calculating molecular weight of stock tank liquid:

$$MW_{l\ st} = \frac{5954}{API - 8.81}$$

3. Determine the mass of the gas in gas phase at surface condition using the following equation:

$$m_{g\ surface} = R\frac{\text{SCF Surface Gas}}{\text{STB}} \times \frac{1\ \text{lbmol Surface Gas}}{380.7\ \text{SCF Surface Gas}} \times 29\gamma_{g\ surf}$$
$$= 0.0762 \times R \times \gamma_{g\ surf}$$

4. Determine the mass of the gas in liquid phase at surface condition using the following equation:

$$m_{l\ \mathrm{surface}} = 1\ \mathrm{STB} \times \frac{5.165\ \mathrm{ft}^3}{1\ \mathrm{STB}} \times 62.4\gamma_{l\ \mathrm{surf}} = 350.2\gamma_{l\ \mathrm{surf}}$$

5. We assume 1 STB stock tank liquid at surface and calculate based on this assumption. Moreover, we know that the produced gases from reservoir have both liquid and gas phases. Consequently, calculate the mass of the gas at reservoir condition using the following equation:

$$\text{mass of gas}_{\mathrm{reservoir}} = \text{mass of gas}_{\mathrm{surface}} + \text{mass of liquid}_{\mathrm{surface}}$$

$$\text{mass of gas}_{\mathrm{reservoir}} = 0.0762 \times R \times \gamma_{\mathrm{g\ surf}} + 350.2\gamma_{l\ \mathrm{surf}}$$

6. Calculate the number of gas moles at reservoir condition using the following equation:

$$n_{\mathrm{g\ reservoir}} = \frac{0.0762 \times R \times \gamma_{\mathrm{g\ surf}}}{29 \times \gamma_{\mathrm{g\ surf}}} + 350.2\frac{\gamma_{l\ \mathrm{surf}}}{\mathrm{MW}_{l\ \mathrm{surface}}}$$

$$= 0.00263R + 350.2\frac{\gamma_{l\ \mathrm{surf}}}{\mathrm{MW}_{l\ \mathrm{surface}}}$$

7. Determine the molecular weight of the reservoir gas using the following simple expression:

$$\mathrm{MW}_{\mathrm{g\ reservoir}} = \frac{m_{\mathrm{g\ reservoir}}}{n_{\mathrm{g\ reservoir}}}$$

8. Determine the specific gravity of the reservoir gas using the following simple equation:

$$\gamma_{\mathrm{g\ reservoir}} = \frac{R_{\mathrm{sep1}}\gamma_{\mathrm{g\ sep1}} + R_{\mathrm{sep2}}\gamma_{\mathrm{g\ sep2}} + R_{\mathrm{st}}\gamma_{\mathrm{g\ st}} + 4600\gamma_{l\ \mathrm{surface}}}{R_{\mathrm{sep1}} + R_{\mathrm{sep2}} + R_{\mathrm{st}} + 133300\frac{\gamma_{l\ \mathrm{surf}}}{\mathrm{MW}_{l\ \mathrm{surface}}}}$$

9. In this case the information of stock tank is unknown and we should employ the following correlations for two- or three-stage separation units to determine the unknown parameters and replace the values into the abovementioned equation to determine the specific gravity of the reservoir gas.

For two-stage separators:

$$R_{st}\gamma_{st} = B_1(P_{sep} - 14.65)^{B_2}(\gamma_{sep})^{B_3}(\text{API})^{B_4}(T_{sep})^{B_5}$$

in which the constants are reported in the following table:

Constants	Values
B_1	1.45993
B_2	1.33940
B_3	7.09434
B_4	1.14356
B_5	−0.934460

$$R_{st} + 133300\frac{\gamma_{l\ surf}}{\text{MW}_{l\ surface}} = B_0 + B_1(P_{sep})^{B_2}(\gamma_{sep})^{B_3}(\text{API})^{B_4}(T_{sep})^{B_5}$$

in which the constants are reported in the following table:

Constants	Values
B_0	635.530
B_1	0.361821
B_2	1.05435
B_3	5.08305
B_4	1.58124
B_5	−0.791301

For three-stage separators:

$$R_{sep2}\gamma_{sep2} + R_{st}\gamma_{st} = B_1(P_{sep1} - 14.65)^{B_2}(\gamma_{sep1})^{B_3}(\text{API})^{B_4}(T_{sep1})^{B_5}(T_{sep2})^{B_6}$$

in which the constants are reported in the following table:

Constants	Values
B_1	2.99222
B_2	0.970497
B_3	6.80491
B_4	1.07916
B_5	−1.19605
B_6	0.553670

The abovementioned constants are valid for the following conditions:

$P_{sep1} = 100$ to 500 Psi
$\Upsilon_{sep1} = 0.6$ to 0.8
$\text{API} = 40°$ to $70°$
$T_{sep1} = 60°$ to 120°F
$T_{sep2} = 60°$ to 120°F

$$R_{sep2} + R_{st} + 133300\frac{\gamma_{l\ surf}}{MW_{l\ surface}} = B_0 + B_1(P_{sep1})^{B_2}(\gamma_{sep1})^{B_3}(API)^{B_4}(T_{sep1})^{B_5}(T_{sep2})^{B_6}$$

in which the constants are reported in the following table:

Constants	Values
B_0	535.916
B_1	2.62310
B_2	0.793183
B_3	4.66120
B_4	1.20940
B_5	−0.849115
B_6	0.269870

In abovementioned equations, T_{sep1} and T_{sep2} stand for the primary and secondary separator temperature (°F), P_{sep1} and P_{sep2} denote the primary and secondary pressure in Psi, and Υ_{sep1} represents the specific gravity of the primary separator gas.

Example 6.4

Consider a wet gas is produced through a separator that is operating at 380 Psi and 77°F to a stock tank. Separator produces 63,000 SCF/STB and separator gas specific gravity is 0.648. Moreover, stock tank liquid gravity is 49.9°API. Determine the gas specific gravity of the reservoir gas.

Solution

In this example we have the two-stage separator unit, and consequently we should use the equations for this system as follows:

$$R_{st}\gamma_{st} = B_1(P_{sep} - 14.65)^{B_2}(\gamma_{sep})^{B_3}(API)^{B_4}(T_{sep})^{B_5} = 274.8071$$

$$R_{st} + 133,300\frac{\gamma_{l\ surf}}{MW_{l\ surface}} = B_0 + B_1(P_{sep})^{B_2}(\gamma_{sep})^{B_3}(API)^{B_4}(T_{sep})^{B_5} = 961.3953$$

$$API = 49.9 \rightarrow \gamma_{l\ st} = 0.780044$$

$$\gamma_{g\ reservoir} = \frac{R_{sep1}\gamma_{g\ sep1} + R_{st}\gamma_{g\ st} + 4600\gamma_{l\ surface}}{R_{sep1} + R_{st} + 133,300\frac{\gamma_{l\ surf}}{MW_{l\ surface}}}$$

$$= \frac{63,000 \times 0.648 + 274.8071 + 4600 \times 0.780044}{63,000 + 961.3953} = 0.6986$$

Example 6.5

Consider a wet gas reservoir produced through a three-stage separator system. Derive an equation for calculating the specific gravity of the surface gas.

Solution

We know that

$$MW_g = \frac{m_g}{n_g}$$

$$\gamma_g = \frac{MW_g}{29}$$

$$\begin{aligned}
m_g &= m_{g\ sep1} + m_{g\ sep2} + m_{g\ st}\\
&= R_{sep1}\left(\frac{SCF}{STB}\right) \times \left(\frac{1\ lbmol}{380.7\ SCF}\right) \times \left(\frac{29 \times \gamma_{g\ sep1} lbm}{lbmol}\right) + R_{sep2}\left(\frac{SCF}{STB}\right)\\
&\quad \times \left(\frac{1\ lbmol}{380.7\ SCF}\right) \times \left(\frac{29 \times \gamma_{g\ sep2} lbm}{lbmol}\right) + R_{st}\left(\frac{SCF}{STB}\right)\\
&\quad \times \left(\frac{1\ lbmol}{380.7\ SCF}\right) \times \left(\frac{29 \times \gamma_{g\ st} lbm}{lbmol}\right)\\
&= 0.0762\left(R_{sep1}\gamma_{sep1} + R_{sep2}\gamma_{sep2} + R_{st}\gamma_{st}\right)
\end{aligned}$$

$$\begin{aligned}
n_g &= 0.0762 \times \left(\frac{R_{sep1}\gamma_{sep1}}{29 \times \gamma_{sep1}} + \frac{R_{sep2}\gamma_{sep2}}{29 \times \gamma_{sep2}} + \frac{R_{st}\gamma_{st}}{29 \times \gamma_{st}}\right)\\
&= \frac{0.0762}{29}\left(R_{sep1} + R_{sep2} + R_{st}\right)
\end{aligned}$$

$$MW_g = \frac{0.0762\left(R_{sep1}\gamma_{sep1} + R_{sep2}\gamma_{sep2} + R_{st}\gamma_{st}\right)}{\frac{0.0762}{29}\left(R_{sep1} + R_{sep2} + R_{st}\right)}$$

$$\gamma_{g\ reservoir} = \frac{MW_g}{29} = \frac{\left(R_{sep1}\gamma_{sep1} + R_{sep2}\gamma_{sep2} + R_{st}\gamma_{st}\right)}{\left(R_{sep1} + R_{sep2} + R_{st}\right)}$$

6.4 PVT TESTS

This section provides details of the PVT experiments employed to determining phase behavior of the oil and gas reservoir fluids including oil and gas samples.

6.4.1 Differential Test

PVT experiments are intended to investigate and determine the properties and phase behavior of an oil and gas sample at reservoir conditions that are simulated in the laboratory. It is worth stressing that in the most laboratory PVT experiments, the PVT measurements are carried out in the absence of water, and the influence of interstitial water on the phase behavior of petroleum samples is unseen. Depletion tests are the major part of PVT test, where the pressure of the single-phase fluid is declined in sequential stages either by removing part of fluid volume or by increasing the fluid volume. The decline of pressure eventuates in the formation of another phase, except in wet and dry gas hydrocarbons (Danesh, 2003; Ahmed, 2010).

Tests conducted in laboratories on liquid samples contained in a porous medium have resulted in some degree of supersaturation, with values as high as 5 MPa (Kennedy and Olson, 1952; Wieland and Kennedy, 1957; Ahmed, 2010). High supersaturation has been observed in tests where the pressure has been lowered rapidly. In a reservoir where the pressure decline is slow, significant supersaturation is not expected (Firoozabadi et al., 1992).

Surface forces can be significant in tight pores, affecting the phase behavior of fluids. Capillary condensation, where gas condenses in pores due to fluid–solid interaction, is a well-known phenomenon (Yeh et al., 1986; Yeh and Yeh, 1986). The effect would be of significance in pores typically less than 10^{-8} m. Gas-condensate reservoirs are generally assumed to be water-wet, with tight cavities filled with water. Hence, the capillary condensation effect may be ignored. Tests in a cell packed with 30–40 mesh beads have resulted in the same dew point as that measured conventionally in an equilibrium cell (Sigmund et al., 1973; Danesh, 2003; Ahmed, 2010).

The aforementioned review suggests that the assumption of equilibrium between the phases in reservoirs, and neglecting the surface effect on fluid equilibrium, is a reasonable engineering approach. This has greatly simplified experimental and theoretical studies of the phase behavior of reservoir fluids. In conventional PVT tests, the fluids are given ample time and agitation in equilibrium cells, to approach equilibrium. At certain conditions, such as in rapid pressure buildup near the wellbore or in high pressure gradient flow, the deviation from equilibrium may become significant. Should nonequilibrium information become important to field operation, such as bubble nucleation in water-invaded reservoirs during depletion (Kortekaas and Poelgeest, 1991; Moulu and Longeron, 1989; Danesh, 2003; Ahmed, 2010), special tests could be designed to generate the required data.

6.4.2 Swelling Test

The following steps should be followed in swelling experiments (Danesh, 2003; McCain, 1990):

1. The given volume of the fluid sample is injected into a PVT cell (high pressure cell) at reservoir pressure and heated up to reach the temperature of the reservoir (Danesh, 2003; McCain, 1990).
2. A constant composition experiment (this experiment is explained in Section 4.4.4) is conducted to calculate the relative volume, bubble point pressure, liquid shrinkage, and liquid density data (Danesh, 2003; McCain, 1990).
3. The cell pressure is declined until the volume of the sample is expanded to at least two times the sample volume at bubble point pressure. Moreover, the liquid shrinkage data are also recorded (Danesh, 2003; McCain, 1990).
4. A known volume of the gas (gas sample that is designed for injection) is added to the sample and the fluid is agitated until single-phase equilibrium is reached (Danesh, 2003; McCain, 1990).
5. The increase in the sample volume and the total sample volume is recorded. The lately produced sample is put through a constant composition test, as explained earlier, and the liquid shrinkage and bubble point pressure are rerecorded (Danesh, 2003; McCain, 1990; Ahmed, 2010).
6. The composition of each fluid mixture is determined from the recorded gas injection and the reservoir fluid composition, accompanied by the mole/mole recombination ratio (Danesh, 2003; McCain, 1990; Ahmed, 2010).
7. The swelling experiment comprises a number of gas injections, but it is fairly typical to conduct five constant composition tests and fluid mixtures, finishing in the production of a gas condensate fluid after the latter gas adding (Danesh, 2003; McCain, 1990; Ahmed, 2010).

At a given temperature, the lean gas to be mixed with reservoir fluid, and the amount of added gas along with the mole percentage of the gas and

GOR (volume of gas at STC/volume of oil at original saturation pressure) are calculated as following as:

$$\text{mol\%} = M = \frac{N_{\text{gas}}}{N_{\text{gas}} + N_{\text{res}}}$$

$$\text{GOR} = G = \frac{V_{\text{gas}}^{\text{st}}}{V_{\text{res}}\left(P_{\text{sat}}{}^{\text{o}}\right)}$$

From real gas law we have

$$V_{\text{gas}}^{\text{st}} = \frac{N_{\text{gas}} R T^{\text{st}}}{P^{\text{st}} V_{\text{res}}\left(P_{\text{sat}}{}^{\text{o}}\right)}$$

Therefore $G = G(N_{\text{gas}})$ and mixture

$$z_{\text{i}}^{\text{mix}} = (1 - M) z_{\text{i}}^{\text{res}} + M z_{\text{i}}^{\text{gas}}$$

6.4.3 Separator Test

Separators consist of a set of connected equilibrium flashed at user-prescribed pressures and temperatures. This test specifies (Danesh, 2003; Ahmed, 2010):

- composition of the feed-stream
- number of stages
- connection of vapor and liquid outputs of each stage

The separator test consists of the following steps:

Step 1. The separator test comprises insertion of a reservoir fluid sample at the temperature of the reservoir and bubble point pressure of the reservoir fluid sample in a PVT cell.

Step 2. The sample volume is recorded as V_{sat}.

Step 3. The reservoir fluid sample is then flashed via a research laboratory multistep separator system in usually one to three steps.

Step 4. The temperature and pressure of these steps are tuned to denote the real or preferred surface separation amenities.

Step 5. The gas released from each step is eliminated and its volume and specific gravity at standard circumstances are calculated.

Step 6. The last stage is represented the stock tank condition, and consequently the volume of the residual oil in the latter step is calculated and documented as $(V_{\text{o}})_{\text{st}}$.

Step 7. Laboratory data points recorded from the abovementioned steps are employed to calculate the solubility of gas and the formation volume factor of oil at the saturation pressure as follows (Danesh, 2003; McCain, 1990):

$$B_{\mathrm{ofb}} = \frac{V_{\mathrm{sat}}}{(V_{\mathrm{o}})_{\mathrm{st}}} \quad R_{\mathrm{sfb}} = \frac{(V_{\mathrm{g}})_{\mathrm{sc}}}{(V_{\mathrm{o}})_{\mathrm{st}}}$$

where $(V_g)_{sc}$ denotes the total volume of gas released from separators (scf), R_{sfb} represents the solution GOR at the saturation pressure as calculated by flash process (scf/STB), and B_{ofb} stands for the formation volume factor of oil at saturation pressure, as determined by flash process, oil volume at the saturation pressure (bbl)/STB.

Step 8. The aforementioned experimental framework is continued at a sequence of various separator pressures and at a constant temperature.

To calculate the optimum separator pressure, it is generally suggested that four of these tests should be employed. It is typically assumed the separator pressure that yields minimum oil formation volume factor, maximum oil gravity in the stock tank, and the minimum total evolved gas (summation of separator gas and stock tank gas). As noted previously the differential experiment is conducted at the given reservoir temperature and multiple steps of flashes although the separator experiment is usually a one- or two-step flash at low temperature and low pressure. It is worth mentioning that both quality and quantity of the gas released in the two aforementioned experiments are totally different (Danesh, 2003; McCain, 1990; Ahmed, 2010). Fig. 6.1 depicts the schematic of the separator test.

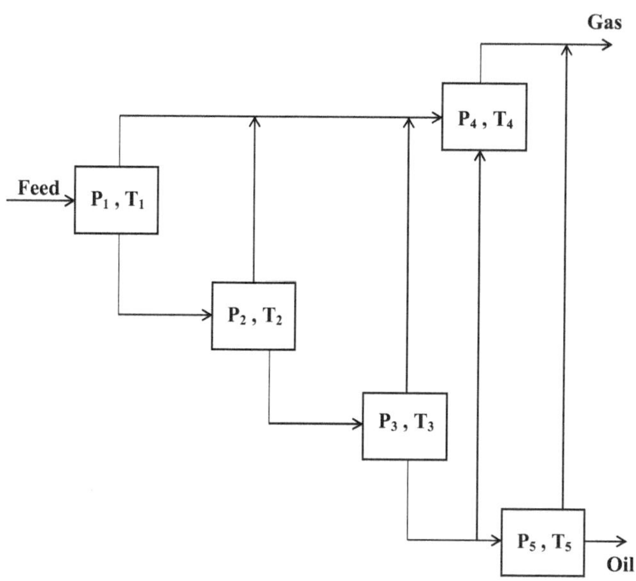

Figure 6.1 Schematic of the separator test.

Example 6.6

The following table shows the PVT properties from the example mixture at its bubble point and at a separator pressure and temperature of 200 psia and 90°F. Calculate the oil formation volume factor, gas volume at standard condition, and other PVT properties.

P (Psi)	T (°F)	f_o	f_g	z_o	z_g	V_o (cm³)	V_g (cm³)	ρ_o (lb/ft³)	ρ_g (lb/ft³)
2200	190	1.000	0.000	0.9987	0.9011	32.265	0.00099	43.0874	6.2345
200	90	0.6345	40.124	0.1234	0.9803	27.874	128.156	46.6524	0.5042
14.7	60	0.9226	0.0763	0.01534	0.9922	26.763	130.123	47.435	0.0534

Solution

The corresponding PVT properties from the separator test are calculated as follows:

$$B_{ofb} = \frac{V_{sat}}{(V_o)_{st}} = \frac{32.265}{26.763} = 1.205582$$

The volume of gas from the separator at standard conditions is

$$V_{sc} = \frac{P_{sep} V_{sep}}{Z_{sep} T_{sep}} \frac{T_{sc}}{P_{sc}} = \frac{200 \times 128.156}{0.9803 \times (90 + 460)} \frac{(60 + 460)}{14.7} = 1681.64 \text{ cm}^3$$

The solution-GOR of the separator is then

$$R_{sfb} = \frac{(V_g)_{sc}}{(V_o)_{st}} = \frac{1681.64 + 130.123}{26.763} \times 5.615 \frac{\text{SCF}}{\text{STB}} = 380.1161 \frac{\text{SCF}}{\text{STB}}$$

$$\rho_o(P_{sc}, T_{sc}) = 47.435 \frac{\text{lbm}}{\text{ft}^3} \rightarrow \text{API} = 54.18$$

6.4.4 Constant Composition Test

The constant composition test consists of the following steps (Danesh, 2003; Ahmed, 2010):

Step 1. The constant composition test comprises placing a sample of reservoir fluid (gas or oil) in a visual PVT cell at a pressure greater than the reservoir pressure and at the temperature of reservoir (Danesh, 2003; Ahmed, 2010).

Step 2. The pressure is declined in stages at fixed temperature by exiting mercury from the cell, and the variation in the volume of total hydrocarbon V_t is recorded versus each pressure increase (Danesh, 2003; Ahmed, 2010).

Step 3. The dew point/bubble point pressure (P_d or P_b) and the conforming volume (as a reference volume V_{sat}) are monitored and documented (Danesh, 2003; Ahmed, 2010).

Step 4. The ratio of total hydrocarbon volume and reference volume is called the relative volume and is formulated by the following expression (Danesh, 2003; Ahmed, 2010):

$$V_{rel} = \frac{V_t}{V_{sat}}$$

where V_{rel} stands for the relative volume, V_t represents the volume of total hydrocarbon, and V_{sat} denotes the volume at the bubble point/dew point pressure.

It is worth mentioning that hydrocarbons are not released from the PVT cell; thus the composition of the total hydrocarbons in the PVT cell remains constant at the initial composition.

Step 5. Above the saturation pressure, the oil density can be determined by employing the measured relative volume:

$$\rho = \frac{\rho_{sat}}{V_{rel}}$$

where ρ stands for the density at given pressure above the dew point/bubble point pressure, ρ_{sat} stands for the density at the dew point/bubble point pressure, and V_{rel} represents the relative volume at the given pressure.

Fig. 6.2 illustrates the schematic of the constant composition experiment procedure.

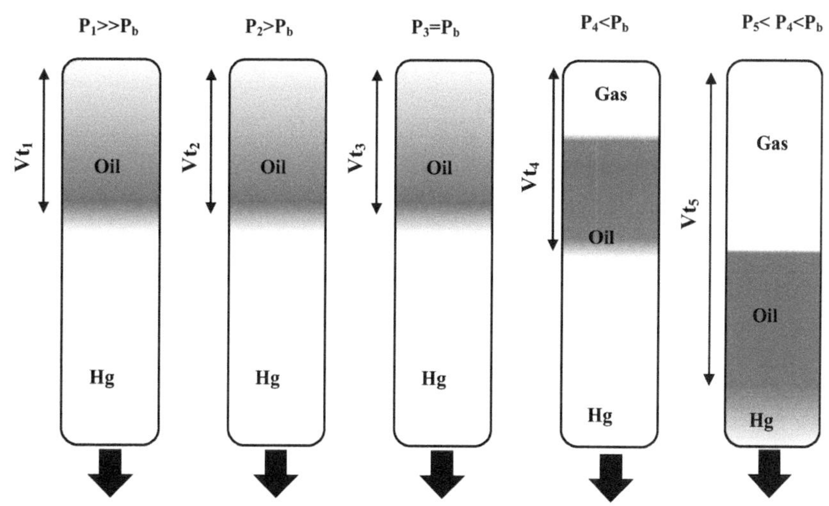

Figure 6.2 Schematic of the constant composition experiment (CCE).

6.4.5 Constant Volume Depletion

Specify a temperature (below cricondentherm) and a series of pressures. This test can be applied to both oil and condensate systems. Vapor is removed to restore the cell to original volume. Relative volume reported is the fraction of the cell filled with liquid after the gas is removed (Danesh, 2003; McCain, 1990; Ahmed, 2010). Fig. 6.3 demonstrates the schematic of the constant volume depletion (CVD) experiment. The CVD test consists of the following steps:

Step 1. As shown in Fig. 6.3 (part A), a determined volume of a demonstrative fluid sample of the reservoir oil and gas fluid with a known overall composition of z_i is placed into a visual PVT cell. The pressure of the PVT cell is equal to the dew point pressure P_d of the fluid sample. Moreover, the PVT cell temperature is equal to the temperature of reservoir (T) during the CVD test. The reference volume throughout this test is equal to the initial volume V_i of the saturated fluid.

Step 2. Via real gas equation the initial gas compressibility factor is determined as following as:

$$Z_d = \frac{P_d V_i}{n_i R T}$$

where P_d stands for the dew point pressure (Psi), V_i denotes the initial gas volume (ft^3), n_i represents the initial number of the gas moles (m/MW$_a$), R denotes the universal gas constant (10.73), T stands for the temperature (°R), and Z_d represents the compressibility factor at dew point pressure.

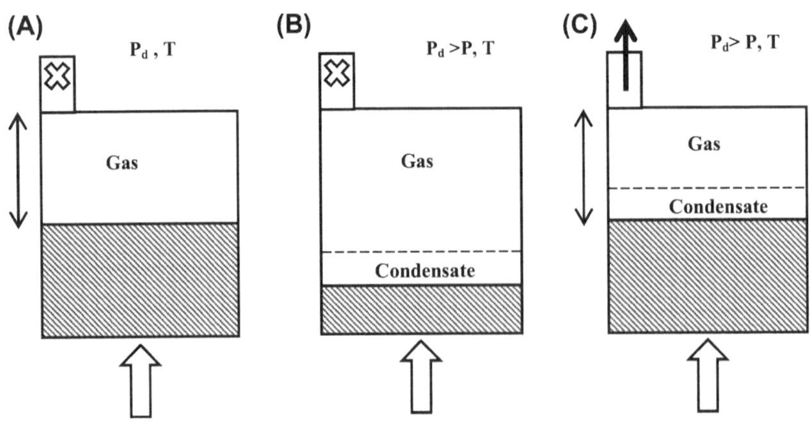

Figure 6.3 Schematic of the constant volume depletion (CVD) experiment.

Step 3. The pressure of the PVT cell pressure is declined from the dew point pressure to a prearranged level pressure. As demonstrated in Fig. 6.3 (Part B), this can be done by removing the mercury from the PVT cell. In this stage, a retrograde liquid as a second phase is created. The fluid in the PVT cell is brought to equilibrium, and retrograde liquid volume V_L and the volume of gas V_g are recorded. The volume of the retrograde liquid is recorded as a percent of the initial volume V_i that principally stands for the retrograde liquid saturation S_L:

$$S_L = \left(\frac{V_L}{V_i}\right) \times 100$$

Step 4. Mercury is reinjected into the PVT cell at fixed pressure (P), whereas a corresponding volume of gas is concurrently released. As depicted in Fig. 6.3 (Part C), injection of mercury is stopped when the initial volume V_i is achieved. This stage simulates a reservoir with only gas production and immobile retrograde liquid remained in the reservoir.

Step 5. The released gas is placed into analytical tool where its volume is recorded at standard circumstances and reported as $(V_{gp})_{sc}$ and its composition y_i is calculated. The equivalent moles of the produced gas can be determined using the following equation:

$$n_p = \frac{P_{sc}(V_{gp})_{sc}}{RT_{sc}}$$

where n_p stands for the moles of the produced gas, $(V_{gp})_{sc}$ represents volume of the produced gas recorded at standard circumstances (scf), T_{sc} denotes standard temperature (°R), P_{sc} stands for the standard pressure (Psi), and R represents universal gas constant (10.73).

Step 6. Using the real gas equation of state, the gas compressibility factor at cell temperature and pressure is determined as follows:

$$Z = \frac{P(V_g)}{n_p RT}$$

The two-phase compressibility factor stands for the total compressibility of all the residual retrograde liquid and gas in the cell and is calculated as follows:

$$Z_{two-phase} = \frac{PV_i}{(n_i - n_p)RT}$$

where $(n_i - n_p)$ stands for the residual moles of fluid in the cell, n_i represents the initial moles in the cell, and n_p denotes the cumulative moles of gas removed.

The two-phase Z-factor is an important parameter for the reason that it is employed when the P/Z against cumulative gas production plot is created for assessing production of gas condensate. Former expression can be formulated in a more appropriate shape by substituting moles of gas, i.e., n_i and n_p, with their equivalent volumes of gas, as follows:

$$Z_{\text{two-phase}} = \left(\frac{Z_d}{P_d}\right)\left[\frac{P}{1-(G_p/\text{GIIP})}\right]$$

where Z_d stands for the gas deviation factor at the dew point pressure, P_d represents the dew point pressure (Psi), P denotes the reservoir pressure (Psi), GIIP stands for initial gas in place (Scf), and G_p represents the cumulative produced gas at given pressure (Scf).

Step 7. By dividing the cumulative volume of the gas produced by the initial gas in place, the volume of the produced gas as a percentage of initial gas in place is determined.

$$\%G_p = \left[\frac{\sum (V_{gp})_{sc}}{\text{GIIP}}\right] \times 100$$

$$\%G_p = \left[\frac{\sum n_p}{(n_i)_{\text{original}}}\right] \times 100$$

The abovementioned experimental protocol is continued until a lowest pressure of the test is achieved, after which the composition and quantity of the gas and retrograde liquid residual in the cell are calculated. The experiment protocol can also be carried out on a volatile oil sample. In this case, instead of gas, the PVT cell initially comprises liquid at its saturation pressure. It should be carried out on all volatile oils and condensates as these are the fluids that are going to experience the significant compositional variations if the pressure of the reservoir is permitted to decline under the dew point/bubble point pressure. As the pressure declines under the bubble point/dew point pressure, the following calculations and procedures are undertaken (Danesh, 2003; McCain, 1990; Ahmed, 2010). The volume occupied by 1 mol of the sample fluid at P_{sat} is given by

$$V_{cell} = V_{sat} = V(P_{sat})$$

The total volume of the liquid and vapor phases is determined and then compared with the control volume, V_{sat}. The excess of the new total volume compared with the control volume, $V_{del} = V_{tot} - V_{sat}$, is then removed from the gas volume:

$$V_{gas}^{after} = V_{gas}^{before} - V_{del}$$

Oil volume is left unchanged. The gas and oil saturations are calculated using the new volume:

$$S_{gas} = \frac{V_{gas}^{after}}{V_{sat}}$$

The total mole composition that will be the feed-stream for the next pressure depletion must be calculated.

$$z_i = x_i(1 - v_f) + y_i v_f \frac{V_{gas}^{after}}{V_{total}^{before}}$$

where x_i and y_i are the liquid and vapor mole compositions from the flash prior to the gas removal. V_f is the vapor fraction from the flash = total fluid volume before the gas removal. This procedure continues down to the lowest specified pressure.

6.4.6 Differential Liberation Test

Specify a temperature and a series of pressures. The liberation test can be applied to liquid/oil systems only. All gas is removed at each pressure step. Last pressure step will be a reduction to standard conditions automatically (Danesh, 2003; McCain, 1990; Ahmed, 2010). The liberation test consists of the following steps:

Step 1. The test comprises placing a reservoir fluid sample into a visual PVT cell at the temperature of the reservoir and bubble point pressure.

Step 2. The pressure is declined in stages, typically 10 to 15 levels of pressure, and all the released gas is eliminated and its volume is recorded at standard circumstances.

Step 3. The volume of the remaining oil V_L is also recorded at each level of pressure.

It is worth highlighting that the remaining oil is put through repeated compositional variations as it turns into gradually richer in the heavier component.

Step 4. The aforementioned framework is repeated at atmospheric pressure where the residual oil volume is recorded and transformed to a volume at 60°F, V_{sc}.

Step 5. The differential formation volume factors of oil B_{od} at all the different levels of pressure are determined by dividing the measured volumes of oil V_L by the residual oil volume V_{sc}, or:

$$B_{od} = \frac{V_L}{V_{sc}} \quad \text{or} \quad B_o = \frac{V_o}{V_{stosc}}$$

Step 6. By dividing the volume of solution gas by the residual oil volume, the differential solution GOR R_{sd} is determined as follows:

$$R_s = \frac{V_{gsc}}{V_{stosc}}$$

Step 7. Relative total volume B_{td} from differential liberation experiment is determined from the following equation:

$$B_{td} = B_{od} + (R_{sdd} - R_{sd})B_g \quad \text{or} \quad B_t = B_o + R_P B_g$$

where B_{td} stands for the relative total volume (bbl/STB) and B_g represents the gas formation volume factor (bbl/scf).

Step 8. The gas compressibility factor (Z) denotes the Z-factor of the released solution gas at the given pressure, and these values are determined from the experimental gas volume measured as follows:

$$Z = \left(\frac{T_{sc}}{V_{sc}P_{sc}}\right)\left(\frac{VP}{T}\right)$$

where V_{sc} stands for the volume of the released gas at standard situation and V represents the liberated gas volume in the PVT cell at a given temperature and pressure.

Step 9. The gas formation volume factor B_g is formulated by the following expression:

$$B_g = \left(\frac{P_{sc}}{T_{sc}}\right)\frac{zT}{P}$$

or

$$B_g = \frac{V_g}{V_{gsc}}$$

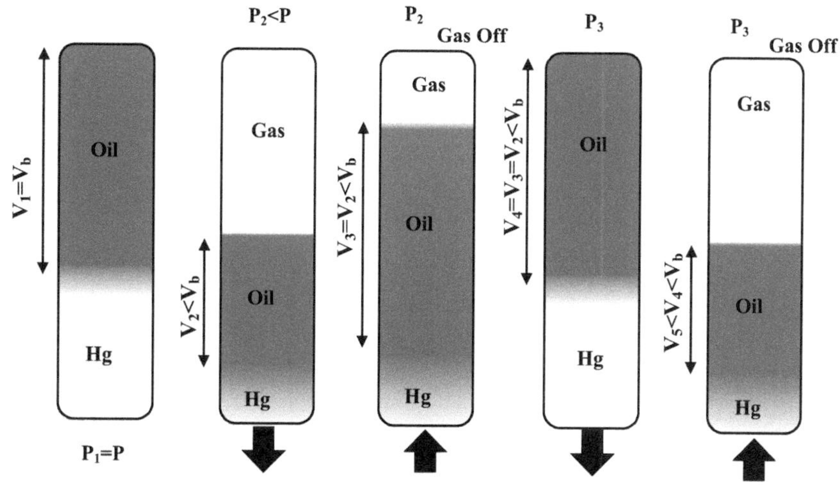

Figure 6.4 Schematic of the differential liberation (DL) test.

where B_g stands for the gas formation volume factor in terms of ft^3/scf, P denotes the pressure of the cell in terms of Psi, T represents the temperature in terms of °R, P_{sc} denotes the standard pressure in terms of Psi, and T_{sc} stands for the standard temperature in terms of °R.

The schematic of the liberation test is depicted through Fig. 6.4.

The PVT data that can be achieved from the differential experiment comprise (Ahmed, 2010):

- The variation in amount of solution gas versus corresponding pressure
- The variation of oil volume shrinkage versus corresponding pressure
- The released gas composition
- The compressibility factor of gas
- The specific gravity of gas
- Variation of the remaining oil density versus corresponding pressure

6.5 FLASH CALCULATION

In flash process, a liquid mixture is partially separated and the gas is allowed to come to equilibrium with the liquid. The graphical demonstration of the flash process is illustrated in Fig. 6.5. The gas and liquid phases are then separated.

Making a component i balance gives

$$Fx_{iF} = Vy_i + Lx_i = Vy_i + (F - V)x_i$$

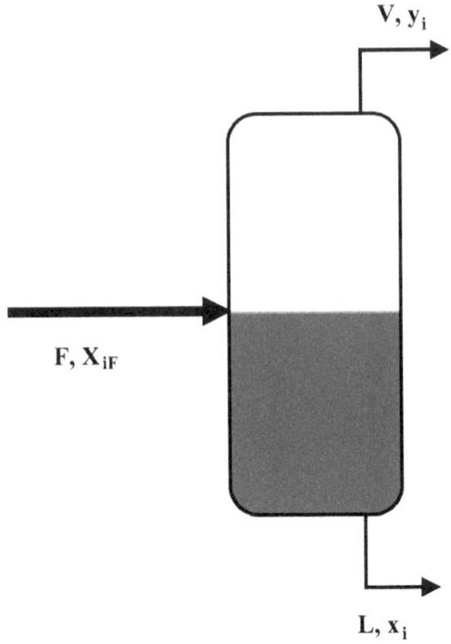

Figure 6.5 Graphical illustration of flash process.

Defining $f = V/F$, the above equation becomes

$$x_{iF} = f y_i + (1 - f) x_i$$

The previous equation can be solved for y_i:

$$y_i = K_i x_i = \frac{f-1}{f} x_i + \frac{x_{iF}}{f}$$

or for x_i:

$$x_i = \frac{x_{iF}}{f(K_i - 1) + 1}$$

We will discuss the solution of isothermal flash calculation. If the temperature T, feed composition x_{iF}, and pressure P of separator are given, then the compositions x_i and y_i and fraction of the feed vaporized V/F can be determined. The above equations can be arranged so that $f = V/F$ is the only unknown.

$$\sum y_i - \sum x_i = 0$$

$$\sum \frac{K_i x_{iF}}{f(K_i - 1) + 1} - \sum \frac{x_{iF}}{f(K_i - 1) + 1} = 0$$

$$F = \sum \frac{(K_i - 1)x_{iF}}{f(K_i - 1) + 1} = 0$$

The above equation, which is known as the Rachford-Rice equation, has excellent convergent properties and can be solved by iterative or Newton–Raphson method. Taking the derivative of the function F with respect to V/F (or f):

$$\frac{dF}{df} = -\sum \frac{(K_i - 1)^2 x_{iF}}{[f(K_i - 1) + 1]^2}$$

The following procedure can be used to solve V/F:

1. Evaluate $K_i = K_i\ (T, P)$
2. Check to see if T is between T_b and T_d.
   If all K-values are less than 1, the feed is a subcooled liquid below the bubble point. If all K-values are greater than 1, the feed is a superheated vapor above the dew point. If one or more K-values are greater than 1 and one or more K-values are less than 1, we need to evaluate the Rachford-Rice equation at $f = 0$ and at $f = 1$.
   a. If $\sum(K_i - 1)x_{iF} < 0$, the feed is below its bubble point pressure
   b. If $\sum\frac{(K_i-1)x_{iF}}{(K_i)} > 0$, the feed is above its dew point pressure
3. Assume $f = 0.5$
4. Evaluate $F = \sum\frac{(K_i-1)x_{iF}}{f(K_i-1)+1}$
5. Evaluate $\frac{dF}{df} = -\sum\frac{(K_i-1)^2 x_{iF}}{[f(K_i-1)+1]^2}$
6. We know error (E) $E = F/\frac{dF}{df}$ and $f = f - E$
7. If $|E| > 0.001$ go to step 4, otherwise

$$x_i = \frac{x_{iF}}{f(K_i - 1) + 1} \quad \text{and} \quad y_i = K_i x_i$$

Example 6.7

Consider a gas with the following composition. This mixture is flashed at 1000 Psi. Determine the fraction of the feed-vaporized and composition of gas and liquid streams leaving the separator if the temperature of the separator is 150°F.

(Continued)

Component	Mole Fraction
C_1	0.70
C_2	0.07
C_3	0.03
C_4	0.05
C_5	0.05
C_6	0.1

Solution

In this case we should determine the equilibrium ratio of each component at $T = 150°F$ and $P = 1000$ Psi as reported in the following table.

Component	Mole Fraction	P_c	T_c	ω_i	K_i
C_1	0.70	666.4	343.33	0.0104	3.1692
C_2	0.07	706.5	549.92	0.0979	1.1520
C_3	0.03	616.0	666.06	0.1522	0.5616
C_4	0.05	527.9	765.62	0.1852	0.2830
C_5	0.05	488.6	845.8	0.2280	0.1421
C_6	0.1	453	923	0.2500	0.0686

Then assume the vapor fraction $f = 0.5$. The results after three iterations are reported in the following table and the vapor fraction is equal to 0.78.

Component	Gas Mole Fraction	Liquid Mole Fraction
C_1	0.8235	0.2598
C_2	0.0721	0.0626
C_3	0.0256	0.0456
C_4	0.0322	0.1136
C_5	0.0215	0.1515
C_6	0.0251	0.3669

Example 6.8

Consider a gas with the following composition. This mixture is flashed at 300 Psi. Determine the fraction of the feed vaporized and composition of gas and liquid streams leaving the separator if the temperature of the separator is 90°F.

Component	Mole Fraction
CO_2	0.005
N_2	0.007
C_1	0.635
C_2	0.015
C_3	0.098
$i\text{-}C_4$	0.013
$n\text{-}C_4$	0.013
$i\text{-}C_5$	0.025
$n\text{-}C_5$	0.065
C_6	0.065
C_{7+}	0.059

$MW_{C_{7+}} = 160$

$\gamma_{C_{7+}} = 0.794$

Solution

In this case, at first we should determine the equilibrium ratio of the components.

Component	Mole Fraction	K_i
CO_2	0.005	3.9234
N_2	0.007	29.3494
C_1	0.635	9.9912
C_2	0.015	1.9991
C_3	0.098	0.6484
$i\text{-}C_4$	0.013	0.2749
$n\text{-}C_4$	0.013	0.1988
$i\text{-}C_5$	0.025	0.0878
$n\text{-}C_5$	0.065	0.0680
C_6	0.065	0.0241
C_{7+}	0.059	0.001

In the second step we should assume $f = 0.5$ and then start the flash calculations. The results after five iterations are reported in the following table and the vapor fraction is equal to 0.7267.

Component	Liquid Mole Fraction	Gas Mole Fraction
CO_2	0.0016	0.0063
N_2	0.0003	0.0095
C_1	0.0843	0.8421
C_2	0.0087	0.0174
C_3	0.1316	0.0854
$i\text{-}C_4$	0.0275	0.0076
$n\text{-}C_4$	0.0311	0.0062
$i\text{-}C_5$	0.0742	0.0065
$n\text{-}C_5$	0.2014	0.0137
C_6	0.2235	0.0053
C_{7+}	0.2158	0.00002

Problems

6.1 Consider a retrograde gas is produced through a separator which is operating at 350 Psi and 77°F to a stock tank. Separator produces 55,672 SCF/STB and separator gas specific gravity is 0.612. Moreover, stock tank liquid gravity is 49°API. Calculate the gas specific gravity of the reservoir gas.

6.2 Consider a gas reservoir is produced through a separator, which is operating at 400 Psi and 90°F to a stock tank. Separator produces 60,015 SCF/STB and separator liquid volume factor is 1.421 bbl/STB. Moreover, stock tank liquid gravity is 53°API. The composition of the surface streams are reported in the following table. Calculate the composition of the reservoir gas.

Component	Separator Gas Composition	Separator Liquid Composition
C_1	0.8609	0.0013
C_2	0.0806	0.0133
C_3	0.02855	0.0291
$i\text{-}C_4$	0.017	0.0195
$n\text{-}C_4$	0.0047	0.0204
$i\text{-}C_5$	0.00308	0.1519
$n\text{-}C_5$	0.002	0.1339
C_6	0.003	0.1099
C_{7+}	0.00017	0.5207

$\gamma_{C_{7+}}$ in separator $= 0.7731$

$MW_{C_{7+}}$ in separator $= 118$

6.3 The following table shows the PVT properties from the example mixture at its bubble point and at a separator pressure and temperature of 180 psia and 100°F, respectively. Calculate the oil formation volume factor, gas volume at standard condition, and other PVT properties.

P (Psi)	T (°F)	f_o	f_g	z_o	z_g	V_o (cm^3)	V_g (cm^3)	ρ_o (lb/ft^3)	ρ_g (lb/ft^3)
2200	180	1.000	0.000	0.9977	0.9008	34.665	0.00093	44.1274	7.12345
180	100	0.6345	40.124	0.1104	0.9706	29.474	129.356	47.2424	0.7042
14.7	60	0.9226	0.0763	0.01334	0.9922	27.863	132.423	48.1405	0.0634

6.4 Consider a gas with the following composition. This mixture is fed to a separator at 350 Psi. Determine the composition of gas and liquid streams leaving the separator if the separator temperature is 100°F.

Component	Mole Fraction (%)
C_1	75
C_2	5
C_3	5
i-C_4	5
n-C_4	1
i-C_5	4
n-C_5	4
C_6	1

6.5 Consider a gas with the following composition. This mixture is fed to a separator at 500 Psi. Determine the fraction of the feed vaporized and composition of gas and liquid streams leaving the separator if the temperature of the separator is 100°F.

Component	Separator Gas Composition (%)
C_1	80
C_2	3
C_3	3
i-C_4	4
n-C_4	2
i-C_5	3
n-C_5	3
C_6	2

6.6 Consider a wet gas reservoir produced through a four-stage separator system. Derive an equation for calculating the specific gravity of the surface gas. Moreover, derive an equation for calculating the specific gravity of the reservoir fluid.

6.7 Derive an equation for flash calculation in a single separator with pressure P_{sep} and temperature T_{sep} when the amount of feed vaporization and amount of liquid drainage from the separator is equal. Hint: Consider the feed composition z_i, liquid composition x_i, and gas composition y_i.

6.8 Consider a gas reservoir is produced through a separator that is operating at 340 Psi and 78°F to a stock tank. Separator produces

55,021 SCF/STB and stock tank vents 500 SCF/STB and separator gas specific gravity is 0.709. Moreover, stock tank liquid gravity and specific gas gravity are 50°API and 1.145, respectively. Calculate the gas specific gravity of the reservoir gas.

6.9 The following table shows the PVT properties from the mixture at its bubble point and at two-stage separators. The first separator pressure and temperature are 480 psia and 110°F and the second separator pressure and temperature are 120 psia and 90°F, respectively. Calculate the oil formation volume factor, gas volume at standard condition, and other PVT properties.

P (Psi)	T (°F)	f_o	f_g	z_o	z_g	V_o (cm³)	V_g (cm³)	ρ_o (lb/ft³)	ρ_g (lb/ft³)
2200	180	1.000	0.000	0.9977	0.9008	34.665	0.00093	44.1274	7.12345
480	110	0.7321	0.3064	0.4012	0.9364	30.235	78.7246	46.2346	1.6724
120	90	0.6345	40.124	0.1104	0.9706	29.474	129.356	47.2424	0.7042
14.7	60	0.9312	0.07551	0.01221	0.9933	26.763	133.176	49.2405	0.0514

6.10 Consider a retrograde gas reservoir is produced through a separator that is operating at 315 Psi and 76°F to a stock tank. Separator produces 72,000 SCF/STB and stock tank vents 487 SCF/STB. Moreover, stock tank liquid gravity is 58.7°API. The composition of the surface streams is reported in the following table. Calculate the composition of the reservoir gas.

Component	Separator Gas Composition	Stock Tank Gas Composition	Stock Tank Liquid Composition
C_1	0.8109	0.2609	0.0011
C_2	0.0806	0.1949	0.0133
C_3	0.04955	0.2332	0.0297
i-C_4	0.046	0.0642	0.0192
n-C_4	0.0037	0.1201	0.0208
i-C_5	0.00406	0.0361	0.0313
n-C_5	0.001	0.0301	0.0539
C_6	0.004	0.057	0.0199
C_{7+}	0.00019	0.0035	0.8108

$\gamma_{C_{7+}}$ in stock tank = 0.8112

$MW_{C_{7+}}$ in stock tank = 128

6.11 A feed is flashed through a flash drum with the conditions 700 Psi and 110°F. The composition of the feed is as follows. Calculate the composition at the top and the bottom streams.

Component	Composition
CO_2	0.03
C_1	0.7909
C_2	0.0706
C_3	0.03955
i-C_4	0.036
n-C_4	0.0137
i-C_5	0.01406
n-C_5	0.001
C_6	0.004
C_{7+}	0.00019

$\gamma_{C_{7+}}$ in stock tank = 0.8019

$MW_{C_{7+}}$ in stock tank = 123

6.12 Consider a gas reservoir is produced through a separator that is operating at 300 Psi and 80°F to a stock tank. Separator produces 45,000 SCF/STB and separator liquid volume factor is 1.111 bbl/STB. Moreover, stock tank liquid gravity is 46°API. The composition of the surface streams is reported in the following table. Calculate the composition of the reservoir gas.

Component	Separator Gas Composition	Separator Liquid Composition
C_1	0.7606	0.0025
C_2	0.0909	0.0021
C_3	0.04858	0.0394
i-C_4	0.027	0.0292
n-C_4	0.0347	0.0109
i-C_5	0.03305	0.0419
n-C_5	0.002	0.0434
C_6	0.003	0.1097
C_{7+}	0.01017	0.7209

$\gamma_{C_{7+}}$ in separator = 0.8131

$MW_{C_{7+}}$ in separator = 144

6.13 Consider a wet gas is produced through a separator that is operating at 475 Psi and 93°F to a stock tank. Separator produces 39,236 SCF/STB and the separator gas specific gravity is 0.7045. Moreover, stock tank liquid gravity is 53°API. Calculate the gas specific gravity of the reservoir gas.

6.14 Derive an equation for calculating the temperature of a single separator with pressure P_{sep} when the amount of feed vaporization and amount of liquid drainage from the separator is equal. Hint: Consider the feed composition z_i, liquid composition x_i, and gas composition y_i.

6.15 Consider a wet gas reservoir is produced through a three-stage separator that is operating at 200 Psi and 80°F to a stock tank. Primary separator produces 60,000 SCF/STB, secondary separator produces 14,000 SCF/STB, and stock tank vents 200 SCF/STB. Moreover, stock tank liquid gravity is 53°API. The composition of the surface streams is reported in the following table. Calculate the composition of the reservoir gas.

Component	Primary Separator Gas Composition	Primary Separator Liquid Composition	Secondary Separator Gas Composition	Stock Tank Gas Composition	Stock Tank Liquid Composition
C_1	0.7009	0.0031	0.9106	0.2302	0.0114
C_2	0.1906	0.0113	0.0209	0.2149	0.0030
C_3	0.02955	0.0207	0.07955	0.2439	0.0199
i-C_4	0.0248	0.0292	0.016	0.0442	0.0390
n-C_4	0.0035	0.0198	0.0057	0.1401	0.0105
i-C_5	0.00410	0.0233	0.00509	0.0261	0.0416
n-C_5	0.003	0.0419	0.001	0.0411	0.0338
C_6	0.002	0.0109	0.001	0.057	0.0199
C_{7+}	0.00015	0.8398	0.00016	0.0025	0.8309

$\gamma_{C_{7+}}$ in stock tank = 0.8305

$MW_{C_{7+}}$ in stock tank = 136

6.16 Consider a sour gas with the following composition. This mixture is fed to a separator at 225 Psi. Determine the fraction of the feed vaporized and composition of gas and liquid streams leaving the separator if the temperature of the separator is 88°F.

Component	Separator Gas Composition (Mol%)
N_2	0.5
H_2S	2
CO_2	4.5
C_1	70
C_2	6
C_3	4
i-C_4	3
n-C_4	3
i-C_5	2
n-C_5	2
C_6	3

6.17 The following table reports data of constant composition experiment performed on an oil sample. Calculate the compressibility factor of oil sample and the formation volume factor at bubble point pressure.

Pressure (Psi)	Relative Volume
5049	0.9875
4050	0.9909
3549	0.9927
3047	0.9946
2543	0.9967
2442	0.9971
2341	0.9975
2241	0.9980
2140	0.9984
2039	0.9988
1939	0.9993
1777	1.0000
1745	1.0043
1725	1.0071
1705	1.0097
1686	1.0123
1667	1.0149
1648	1.0175
1625	1.0215
1591	1.0276
1540	1.0367
1477	1.0497
1392	1.0706
1290	1.0992
1169	1.1429
1021	1.2187
873	1.3253
726	1.4838
585	1.7316
475	2.0457
367	2.5723
280	3.2656

REFERENCES

Ahmed, T., 2010. Reservoir Engineering Handbook, fourth ed. Gulf Publication, United States of America.

API Recommended Practice 44, April 2003. Sampling Petroleum Reservoir Fluid, second ed.

Danesh, A., 2003. PVT and phase behavior of petroleum reservoir fluids. In: Developments in Petroleum Science, third ed., vol. 47, Netherlands.

Firoozabadi, A., Ottesen, B., Mikkelsen, M., Dec., 1992. Measurement of Supersaturation and Critical Gas Saturation. SPE Formation Evaluation, pp. 337–344.

Kennedy, H.T., Olson, R., 1952. Bubble formation in supersaturated hydrocarbon mixtures. Transactions of the American Institutetute of Mining, Metallurgical and Petroleum Engineers 195, 271–278.

Kortekaas, T.F.M., van Poelgeest, F., Aug., 1991. Liberation of solution gas during pressure depletion of virgin and watered-out oil reservoirs. Transactions of the Society of Petroleum Engineers 291, 329–335.

McCain, W.D., 1990. Properties of Petroleum Fluids, second ed. PennWell Corporation.

Moffatt, B.J., Williams, J.M., 1998. Identifying and meeting the key needs for reservoir fluid properties a multi-disciplinary approach. In: Presented at the SPE Annual Technical Conference and Exhibition, New Orleans, 27–30 September. SPE-49067-MS. http://dx.doi.org/10.2118/49067-MS.

Moulu, J.C., Longeron, D., April 1989. Solution gas drive, experiments and simulation. In: Proc. of 5th Europ. IOR Symp., Budapest, pp. 145–154.

Proett, M.A., Gilbert, G.N., Chin, W.C., Monroe, M.L., 1999. New wireline formation testing tool with advanced sampling technology. In: Paper 56711 Presented at the Annual Technical Conference and Exhibition of the Society of Petroleum Engineers of AIME, Houston, October 3–6, 1999.

Sigmund, P.M., Dranchuk, P.M., Morrow, N.R., Purvis, R.A., April 1973. Retrograde condensation in porous media. The Society of Petroleum Engineers Journals 93–104.

Smits, A.R., Fincher, D.V., Nishida, K., Mullins, O.C., Schroeder, R.J., Yamate, T., 1993. In-situ optical fluid analysis as an aid to wireline formation sampling. In: Paper 26496 Presented at the Annual Technical Conference and Exhibition of the Society of Petroleum Engineers of AIME, Houston, October 3–6, 1993.

Standing, M.B., 1951. Volumetric and Phase Behavior of Oil Field Hydrocarbon Systems. SPE, Richardson, Texas, pp. 10–19.

Standing, M.B., 1952. Volumetric and Phase Behavior of Oil Field Hydrocarbon Systems. Reinhold Publishing Co., p. 10

Wieland, D.R., Kennedy, H.T., 1957. Measurement of bubble frequency in cores. Transactions of the American Institutetute of Mining, Metallurgical and Petroleum Engineers 210, 122–125.

Williams, J.M., 1994. Getting the best out of fluid samples. Journal of Petroleum Technology 46 (9), 752. http://dx.doi.org/10.2118/29227-PA. SPE-29227-PA.

Williams, J.M., May–June 1998. Fluid sampling under adverse conditions. Oil & Gas Science and Technology – Revue d'IFP Energies nouvelles 53 (3), 355–365. http://dx.doi.org/10.2516/ogst:1998031.

Yeh, G.C., Yeh, B.V., June 1986. Vapour-liquid equilibria of non electrolyte solutions in small capillaries. 2. Theoretical calculations of equilibrium compositions. In: 60th Colloid and Surface Science Symposium. ACS.

Yeh, G.C., Shah, M.S., Yeh, B.V., 1986. Vapour-liquid equilibria of non electrolyte solutions in small capillaries. 1. Experimental determination of equilibrium compositions. ACS Langmuir 2, 90.

CHAPTER SEVEN

Retrograde Gas Condensate

M.A. Ahmadi[1], A. Bahadori[2,3]
[1]Petroleum University of Technology (PUT), Ahwaz, Iran
[2]Southern Cross University, Lismore, NSW, Australia
[3]Australian Oil and Gas Services Pty Ltd, Lismore, NSW, Australia

7.1 INTRODUCTION

There are five main groups of reservoir fluids, namely: volatile oil, black oil, wet gas, retrograde condensate, and dry gas. Gas condensate as an important part of natural gas resources is usually located in the deep strata under high-temperature, high-pressure conditions (Ungerer et al., 1995; Sun et al., 2012). It is mainly composed of methane and derives its high molecular weight from the quantity of heavy hydrocarbon fractions (Sutton, 1985). The retrograde condensate fluid is very complex due to fluid behavior and properties. This reservoir is usually located between the critical temperature and the cricondentherm on the phase diagram of the reservoir fluid (Fig. 7.1) (Thomas et al., 2009). Fluid flow in gas-condensate reservoir is very complex and involves phase changes, phase redistribution in and around the wellbore, retrograde condensation, multiphase flow of the fluid (oil and gas), and possibly water (Kool et al., 2001). Gas-condensate fluid usually emerges as a single gas phase in the reservoir at exploration time. Gas condensation to liquid phase is a result of pressure reductions from the reservoir to the producing well and production facilities. Retrograde condensation is defined as the isothermal condensation owing to pressure reduction lower than the dew-point pressure of the primary hydrocarbon fluid (Fasesan et al., 2003; Bozorgzadeh and Gringarten, 2006; Fevang, 1995; Moses and Donohoe, 1962; Thomas et al., 2009; Zendehboudi et al., 2012).

The range of liquid production in gas-condensate reservoirs is 30−300 bbl/MMSCF (barrels of liquid per million standard cubic feet of gas). Moreover, the ranges of temperature and pressure for the gas-condensate reservoirs typically are 200−400°F and 3000−8000 psi, correspondingly. The temperature and pressure values accompanied by the broad range of compositions result in the gas-condensate mixtures, which demonstrate complicated and different thermodynamic trends

Fluid Phase Behavior for Conventional and Unconventional Oil and Gas Reservoirs
http://dx.doi.org/10.1016/B978-0-12-803437-8.00007-5

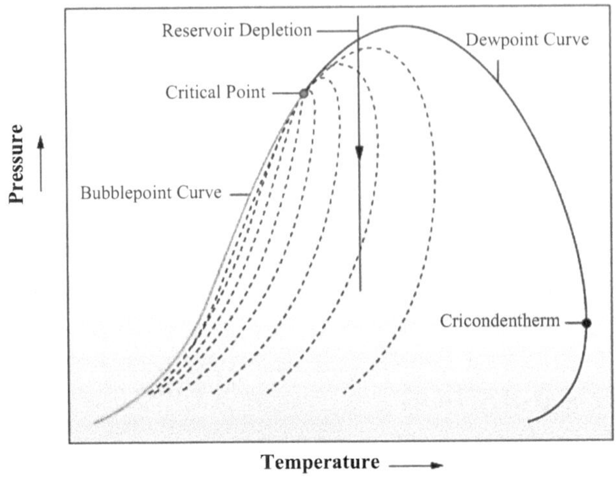

Figure 7.1 Typical gas-condensate phase envelope. *From Zendehboudi, S., Ahmadi, M.A., James, L., Chatzis, I., 2012. Prediction of condensate-to-gas ratio for retrograde gas condensate reservoirs using artificial neural network with particle swarm optimization. Energy & Fuels 26, 3432–3447.*

(Fasesan et al., 2003; Bozorgzadeh and Gringarten, 2006; Fevang, 1995; Moses and Donohoe, 1962; Mott, 2002; Thomas et al., 2009; Zendehboudi et al., 2012).

The main part of the condensed liquid in the reservoir is unrecoverable and considered as condensate loss because the ratio of liquid viscosity to gas viscosity is fairly high and also the formation has lower permeability to liquid in the gas-condensate reservoirs. Condensate loss is one of the most economical concerns because the liquid condensate holds valuable intermediate and heavier constituents of the original hydrocarbons that are trapped in the porous medium (Fasesan et al., 2003; Bozorgzadeh and Gringarten, 2006; Fevang, 1995; Moses and Donohoe, 1962; Hosein and Dawe, 2011; Babalola et al., 2009; Chowdhury et al., 2008; Vo et al., 1989; Mott, 2002; Thomas et al., 2009; Li et al., 2010; Zendehboudi et al., 2012).

Characterization of gas-condensate systems is a complex task for petroleum experts and researchers, because variation of the fluid composition and multiphase flow in the formation significantly obscures the analysis of well tests. In the research area of gas-condensate reservoirs, the topics, for instance, well-test interpretation, pressure–volume–temperature (PVT) analysis, well deliverability, and multiphase flow, have been the common challenges for a long time (Fasesan et al., 2003; Bozorgzadeh and Gringarten, 2006; Fevang, 1995; Moses and Donohoe, 1962; Hosein and Dawe, 2011;

Babalola et al., 2009; Chowdhury et al., 2008; Vo et al., 1989; Mott, 2002; Thomas et al., 2009; Li et al., 2010; Hazim, 2008; Zendehboudi et al., 2012).

7.2 GAS-CONDENSATE FLOW REGIONS

As stated by Fevang (1995), in condensate gas reservoirs there are three different flow regimes for reservoir fluid flow toward a producing well during the production scenario, as illustrated through Fig. 7.2 (Fevang, 1995; Fevang and Whitson, 1996; Zendehboudi et al., 2012):

- Near wellbore (Region 1): This region (1) is an inner near-wellbore part where reservoir pressure lowers further below the dew point. The saturation of the liquid condensate is greater than the critical value, and the condensate buildup turns out to be moveable. Due to the presence of the liquid condensate phase, the gas-phase mobility remains significantly low.
- Condensate buildup (Region 2): The reservoir pressure throughout this region is below the dew-point pressure. In this region, liquid condensate production is observed in the reservoir. However, due to low saturation degree of condensate, the liquid condensate phase will not flow. Accordingly, the flowing phase in this region still comprises just the single gas phase. As the reservoir pressure decreases during the production process the flowing gas loses the heavier fractions in Region 2.

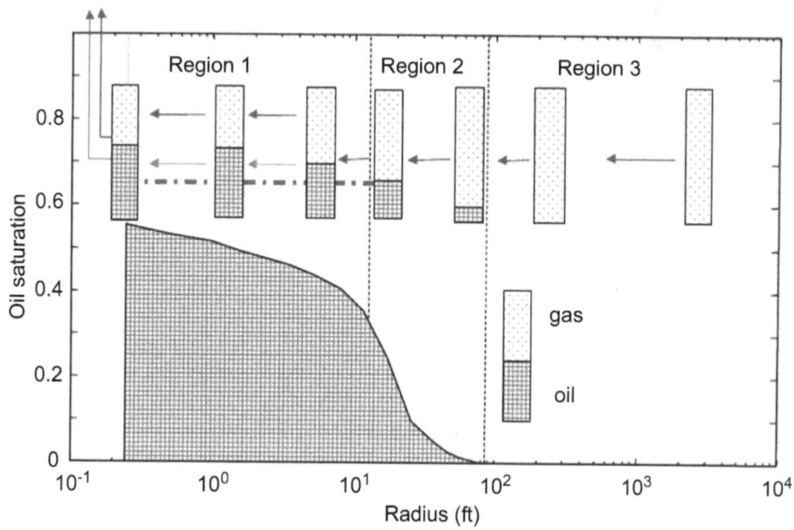

Figure 7.2 Flow regimes in gas-condensate reservoirs. *After Roussennac, B., 2001. Gas Condensate Well Test Analysis (M.Sc. thesis). Stanford University, p. 121.*

- Single-phase gas (Region 3): A region which is far from the producing well and a single-phase gas exists in the region due to pressure higher than the dew-point value.

7.2.1 Condensate Blockage

As the fluid flows toward the production well (Fig. 7.2), the gas mobility increases somewhat in Region 3, but a significant decline is observed in the mobility value as the condensate accumulates in Region 3 and remains at a lower magnitude in Region 1 where the liquid condensate commences to move. A challenging phenomenon called "condensate blockage" that occurs in the near-wellbore region in condensate gas reservoirs is due to negligible production of liquid condensate at the well, and the gas mobility and, accordingly, the well productivity remain considerably low (Roussennac, 2001; Fevang, 1995; Fevang and Whitson, 1996; Zendehboudi et al., 2012).

7.2.2 Composition Change and Hydrocarbon Recovery

As the initial gas moves toward Region 2 in the porous medium, its composition varies and the flowing gas becomes leaner of heavy and intermediate elements (e.g., C_4^+) in the reservoir (Roussennac, 2001; Fevang, 1995; Fevang and Whitson, 1996; Zendehboudi et al., 2012). Therefore, the heavier oil builds up in Regions 1 and 2 as the reservoir pressure declines during depletion. This shows that the overall composition of the mixture everywhere in Region 1 or 2 comprises heavier elements with decrease of the reservoir pressure below the dew point. At the scale of production times, Region 1 will rapidly develop at the wellbore, and thus the production well flow will keep a constant composition, which is leaner than the initial gas fluid. Moreover, the intermediate and heavier elements will be left in Regions 1 and 2 (Roussennac, 2001; Fevang, 1995; Fevang and Whitson, 1996; Zendehboudi et al., 2012).

To adequately handle this fluid, an Equation of State (EOS) model is required. This is an analytical formulation that correlates volume of a fluid to the temperature and pressure which is employed to describe reservoir fluids. The PVT relationship for real hydrocarbon fluids needs to be properly described to ascertain the volumetric and phase behavior of petroleum reservoir fluids. Reservoir and production engineers usually require PVT measurements for effective operations, and one major issue is the use of an EOS for the description of phase behavior of fluids for development of compositional simulators (Wang and Pope, 2001; Nagarajan et al., 2007). Different types of equations of state (EOSs) include Van der Waals, Peng–Robinson (PR), Redlich–Kwong (RK), Patel and Teja (PT), Soave–Redlich–Kwong (SRK), etc. These EOSs were developed in the literature to characterize the phase behavior of the fluids in gas-condensate reservoirs (Sarkar et al., 1991; Khan et al., 2012). It is important to know the gas-condensate phase behavior with the purpose of estimating the future processing requirements and performance of the reservoir. The experimentally measured data are usually matched (by linear regression) with the simulated data to increase the degree of confidence of the EOS model.

7.3 EQUATIONS OF STATE

An EOS defined as an analytical expression relates the thermodynamic variables including temperature T, the pressure P, and the volume V of the system. Finding an accurate PVT relationship for real hydrocarbon mixtures has great importance for calculating petroleum reservoir fluid phase behavior and consequently estimating the performance of surface production and separation amenities. Generally, most EOSs need only the acentric factor and the critical properties of pure substances. The major benefit of employing an EOS is that a similar equation can be employed to demonstrate the phase behavior of other phases, thus confirming reliability when implementing phase-equilibrium computations.

The ideal gas equation, which is formulated by the following expression, is the simplest form of an EOS

$$P = \frac{RT}{V} \tag{7.1}$$

in which V stands for the gas volume, ft^3/mol.

Eq. (7.1) is employed at atmospheric pressure to model the volumetric behavior of hydrocarbon gases for which it was experimentally developed.

Based on the pressure and temperature restrictions of the capability of Eq. (7.1), various research studies have been performed to propose an EOS proper for modeling the phase behavior of real hydrocarbons at reservoir temperatures and pressures. The following sections demonstrate different EOSs which can be useful for modeling phase behavior of petroleum reservoir fluids.

7.3.1 Van der Waals's Equation of State

Two following assumptions have been made for developing Eq. (7.1) to model ideal gas PVT behavior.

1. There are no repulsive or attractive forces between the molecules or the walls of the cell.
2. The volume of the gas molecules is unimportant in comparison with both the distance between the molecules and the volume of the cell.

Van der Waals (1873) attempted to exclude the aforementioned presumptions in proposing an empirical EOS for real gases. For the first one, Van der Waals mentioned that the molecules of gas occupy a major portion of the volume in high-pressure conditions and suggested that the molecules' volume, represented by the factor b, be subtracted from the actual molar volume v in Eq. (7.1), to give

$$P = \frac{RT}{v - b} \tag{7.2}$$

in which the factor b is called the covolume and reflects the molecules' volume. The parameter v stands for the molar volume in ft^3/mol.

For elimination of the second one, Van der Waals proposed a new term represented by a/V^2 to consider the attractive forces between the gas molecules. Considering the aforementioned correction, the Van der Waals EOS can be written as follows:

$$P = \frac{RT}{v - b} - \frac{a}{v^2} \tag{7.3}$$

in which v denotes the molar volume of system (ft^3/mol), P represents pressure of the system (psi), R stands for the gas constant (10.73 psi ft^3/lb mol °R), T denotes temperature of the system (°R), a stands for the "attraction" parameter, and b represents "repulsion" parameter.

The parameters a and b are constants describing the molecular features of the specific components. The parameter a represents a value of the attractive forces between the gas molecules.

A more generalized form of the Van der Waals or any other EOS can be written as follows:

$$P = P_{\text{repulsion}} - P_{\text{attraction}}$$

in which the term $RT/(v - b)$ stands for the repulsion pressure term, $P_{\text{repulsion}}$, and a/v^2 stands for the attraction pressure term, $P_{\text{attraction}}$.

Van der Waals pointed out that for calculating the values of the two constants, *a* and *b*, for any pure component, at the critical point the first and second derivatives of pressure with respect to volume are equal to 0. The mathematical expression of this observation can be expressed as follows:

$$\left[\frac{\partial P}{\partial v}\right]_{T_c,P_c} = 0 \quad \left[\frac{\partial^2 P}{\partial v^2}\right]_{T_c,P_c} = 0 \tag{7.4}$$

To determine the parameters *a* and *b*, the previous equations should be solved at the same time. The solutions for parameters *a* and *b* are as follows:

$$a = \left(\frac{8}{9}\right) RT_c v_c \quad b = \left(\frac{1}{3}\right) v_c \tag{7.5}$$

Experimental investigations reveal that the value of the covolume parameter, *b*, may vary from 0.24 to 0.28 of the critical volume in pure substances; however, as shown in Eq. (7.5), the Van der Waals EOS suggests that the value of covolume is about 0.333 of the critical volume of the component.

Set critical pressure and temperature into Eq. (7.3) and use the calculated values for *a* and *b* by Eq. (7.5) yields

$$P_c v_c = (0.375) RT_c \tag{7.6}$$

As can be seen from Eq. (7.36), the Van der Waals EOS produces a unique critical compressibility factor, Z_c, of 0.375 for any substances regardless of their types. However, based on experimental investigations the value of critical compressibility factor for substances may vary from 0.23 to 0.31. Owing to this point, the Van der Waals EOS has significant drawbacks because it assumes that the critical compressibility factor of different substances is 0.375.

To calculate the two parameters of the Van der Waals EOS (*a* and *b*), the critical molar volume in Eq. (7.5) should be replaced with $v_c = \frac{(0.375)RT_c}{P_c}$ as follows

$$a = \Omega_a \left(\frac{R^2 T_c^2}{P_c}\right) \quad b = \Omega_b \left(\frac{RT_c}{P_c}\right) \tag{7.7}$$

in which P_c denotes the critical pressure (psi); *R* stands for the gas constant, 10.73 (psi ft^3/lb mol °R); T_c represents the critical temperature (°R); and values of Ω_a and Ω_b are as follows:

$$\Omega_a = 0.421875 \quad \Omega_b = 0.12$$

By substituting ZRT/P instead of molar volume in Eq. (7.3) and rearranging into cubic form, a more practical EOS in terms of the compressibility factor Z is achieved as follows:

$$Z^3 - (1 + B)Z^2 + AZ - AB = 0 \tag{7.8}$$

with

$$A = \frac{aP}{R^2T^2} \text{ and } B = \frac{bP}{RT} \tag{7.9}$$

in which P stands for the pressure of the system (psi), Z denotes the compressibility factor, and T represents the temperature of the system (°R).

If we have a one-phase system, solving Eq. (7.8) results in one real root and two imaginary roots (we do not consider these imaginary roots in our calculations). On the other hand, if we have a two-phase system, solving Eq. (7.8) results in three real roots. In the two-phase system, the smallest positive root corresponds to that of the liquid phase, Z_L, whereas the largest positive root corresponds to the compressibility factor of the gas phase, Z_g.

Despite the fact that the Van der Waals EOS is simple easy to use, and predicts some thermodynamic properties for both liquid and gaseous substances, at least qualitatively, it is not adequately precise for use in design of thermodynamic cycles.

7.3.2 Soave–Redlich–Kwong Equation of State

The Redlich–Kwong (RK) (1949) EOS effectively relates the PVT of gases; however, it poorly estimates the liquid density and vapor pressure of pure substances. Soave (1972) introduced the temperature dependence (α) for the attractive term of the Redlich–Kwong (RK) (1949) EOS as explained in a further section. This parameter meaningfully enhances the precision of the EOS to estimate vapor pressure, although the accuracy of the EOS to estimate liquid density was not improved (Nasrifar and Moshfeghian, 1999). Because of the aforementioned amendment, the Soave–Redlich–Kwong can effectively be employed in fluid-phase equilibrium in a system of hydrocarbon mixtures. Although, the α-function causes the Soave–Redlich–Kwong to estimate inconsistent behaviors at high pressures (Segura et al., 2003).

$$P = \frac{RT}{v - b} - \frac{a_c\alpha(T_r)}{v(v + b)} \tag{7.10}$$

in which

$$\alpha = \left[1 + m\left(1 - \sqrt{T_r}\right)\right]^2 \quad (7.11)$$

$$m = 0.480 + 1.574\omega - 0.176\omega^2 \quad (7.12)$$

$$a_c = 0.42747\left(\frac{R^2 T_c^2}{P_c}\right) \quad (7.13)$$

$$b = 0.08664\left(\frac{RT_c}{P_c}\right) \quad (7.14)$$

in which T denotes the temperature of the system (°R), ω stands for the acentric factor of the component, and T_r represents the reduced temperature (T/T_c).

Replacing (ZRT/P) instead of the molar volume, v, in Eq. (5.33) and rearranging results in the following equation

$$Z^3 - Z^2 + \left(A - B - B^2\right)Z - AB = 0 \quad (7.15)$$

$$A = \frac{(\alpha a)P}{(RT)^2} \quad (7.16)$$

$$B = \frac{bP}{RT} \quad (7.17)$$

in which R represents the gas constant (10.730 psi ft^3/lb mol °R), T denotes the temperature of the system (°R), and P stands for the pressure of the system (psi).

7.3.3 The Soave–Redlich–Kwong–Square Well Equation of State

Nasrifar and Bolland (2004) took the advantage of the square-well (SW) potential to account for the supercritical behavior of fluids in the Soave–Redlich–Kwong EOS. The SRK-SW can be employed for predicting the liquid density of gas-condensate mixtures. A brief description of SRK-SW EOS is expressed as follows:

$$P = \frac{RT}{v - b} - \frac{a_c\alpha(T_r)}{v(v + b)} \quad (7.18)$$

$$a_c = 0.42747\left(\frac{R^2 T_c^2}{P_c}\right) \tag{7.19}$$

$$b = 0.08664\left(\frac{R T_c}{P_c}\right) \tag{7.20}$$

$$\alpha(T_r) = \left\{\begin{array}{ll} \left[1 + m\left(1 - \sqrt{T_r}\right)\right]^2 & T_r \leq 1 \ \text{(Subcritical Condition)} \\ \dfrac{b_1}{T_r} + \dfrac{b_2}{T_r^2} + \dfrac{b_3}{T_r^3} & T_r > 1 \ \text{(Supercritical Condition)} \end{array}\right\} \tag{7.21}$$

in which

$$b_1 = 0.25\left(12 - 11m + m^2\right) \tag{7.22}$$

$$b_2 = 0.5\left(-6 + 9m - m^2\right) \tag{7.23}$$

$$b_3 = 0.25\left(4 - 7m + m^2\right) \tag{7.24}$$

$$m = 0.480 + 1.574\omega - 0.175\omega^2 \tag{7.25}$$

7.3.4 Peng–Robinson Equation of State

Another successful PVT relation among EOSs is the Peng–Robinson EOS. In comparison with Redlich–Kwong (1949) family EOSs, the Peng–Robinson family EOSs generally estimate more precisely the liquid density of mixtures (Nasrifar and Moshfeghian, 1999); however, the precision is not good enough for industrial purposes. The Peng–Robinson EOS employs the pros of Soave-type α-function, therefore show comparable excellence with temperature. However, Peng and Robinson (Nasrifar and Moshfeghian, 1998) employed a reduced temperature range from 0.7 to 1 to correlate the Peng–Robinson α-function. The α-function was first correlated to the vapor pressure of pure compounds with acentric factor less than 0.5, and, later in 1978, Peng and Robinson developed the α-function for mixtures with larger acentric factor.

$$P = \frac{RT}{v - b} - \frac{\alpha a}{v(v + b) + b(v - b)} \tag{7.26}$$

By applying boundary conditions at critical point we have

$$a = \Omega_a \frac{R^2 T_c^2}{P_c} \tag{7.27}$$

$$b = \Omega_b \frac{R T_c}{P_c} \tag{7.28}$$

in which

$$\Omega_a = 0.45724 \quad \Omega_b = 0.07780$$

A universal critical gas-compressibility factor, Z_c, for the Peng–Robinson EOS is equal to 0.307; however, the value of Z_c for the Soave–Redlich–Kwong EOS is equal to 0.333. As noted previously, Peng and Robinson employed the α-function proposed by Soave as follows:

$$\alpha = \left[1 + m(1 - \sqrt{T_R})\right]^2 \tag{7.29}$$

$$m = 0.3796 + 1.54226\omega - 0.2699\omega^2 \quad \text{for } \omega < 0.49 \tag{7.30}$$

For heavier elements with acentric values $\omega > 0.49$, Robinson and Peng (1978) recommended the following improved formulation for m as follows:

$$m = 0.379642 + 1.48503\omega - 0.1644\omega^2 + 0.016667\omega^3 \tag{7.31}$$

Rearranging the original Peng–Robinson EOS into the compressibility factor form gives

$$Z^3 + (B - 1)Z^2 + \left(A - 3B^2 - 2B\right)Z - \left(AB - B^2 - B^3\right) = 0 \tag{7.32}$$

in which the parameters A and B can be calculated as follows

$$A = \frac{\alpha a p}{(RT)^2} \tag{7.33}$$

$$B = \frac{bp}{RT} \tag{7.34}$$

It is worth mentioning that, for calculating the aforementioned parameters for the hydrocarbon mixtures, we should use a mixing rule, which is explained in the next sections.

7.3.5 Peng–Robinson–Gasem Equation of State

Gasem et al. (2001) proposed a new α-function in exponential formulation having determined that the Soave-type α-function employed by the

Peng–Robinson (PR) EOS does not decline uniformly to 0 by increasing the temperature. Their EOS also tries to enhance the estimation potential of the PR EOS for vapor pressure.

$$P = \frac{RT}{v - b} - \frac{\alpha a}{v(v + b) + b(v - b)} \tag{7.35}$$

in which

$$b = \frac{0.0778RT_c}{P_c} \tag{7.36}$$

$$a_c = \frac{0.45724R^2T_c^2}{P_c} \tag{7.37}$$

$$\alpha(T)^{1/2} = 1 + k\left(1 - T_r^{1/2}\right) \tag{7.38}$$

$$k = 0.480 + 1.574\omega - 0.176\omega^2 \tag{7.39}$$

in which P stands for the pressure, T denotes the temperature, R represents the gas constant, v stands for the molar volume, and a and b are the constants of the EOS. T_c represents the critical temperature, T_r stands for the reduced temperature, P_c denotes the critical pressure, ω stands for the acentric factor, and $\alpha(T)$ represents the temperature dependence in the parameter α.

7.3.6 Nasrifar and Moshfeghian (NM) Equation of State

Nasrifar and Moshfeghian (2001) employed a linear temperature dependence for molar covolume and a modified Soave's temperature dependence for the attractive parameter of the PVT relation proposed by Twu et al. (1995) to obtain an accurate EOS for simple pure compounds and their mixtures. The NM EOS (Nasrifar and Moshfeghian, 2001) quite accurately predicts the liquid density of liquefied natural gas (LNG) mixtures.

Nasrifar and Moshfeghian proposed the following EOS:

$$P = \frac{RT}{v - b} - \frac{a}{v^2 + 2bv - 2b^2} \tag{7.40}$$

Considering the critical-point limitations, the parameters a and b at the critical point can be calculated as follows:

$$a_c = 0.497926\left(\frac{R^2T_c^2}{P_c}\right) \tag{7.41}$$

$$b_c = 0.094451\left(\frac{RT_c}{P_c}\right) \tag{7.42}$$

with $Z_c = 0.302$.

It should be noted that the temperature is scaled according to the following expression to determine a temperature dependency for the parameter a:

$$\theta = \frac{T - T_{pt}}{T_c - T_{pt}} \tag{7.43}$$

in which T_{pt} stands for the pseudo triple-point temperature and its value could be larger or smaller than the triple-point temperature of the element. The variable T_{pt} differs from the EOS and component. It is worth to mention that the EOS can be performed to a broad range of temperatures, i.e., from the critical point to the pseudo triple-point temperature.

Soave (1972) and Peng and Robinson (1976) considered a linear relationship between the parameters $\sqrt{a}$ and $\sqrt{T_r}$. The same relationship between $\sqrt{a}$ and $\sqrt{\theta}$ can be expressed as follows:

$$\sqrt{a} = l_a + m_a\left(1 - \sqrt{\theta}\right) \tag{7.44}$$

in which l_a and m_a are unknown parameters and using the following boundary conditions helps us to calculate the aforementioned unknown parameters.

$$a \rightarrow a_c \text{ as } \theta \rightarrow 1$$

$$a \rightarrow a_{pt} \text{ as } \theta \rightarrow 0$$

in which the parameter a_{pt} stands for the value of parameter a at the pseudo triple-point temperature. Performing boundary conditions to Eq. (7.44) results in

$$\alpha = a_c\left[1 + m_a\left(1 - \sqrt{\theta}\right)\right]^2 \tag{7.45}$$

in which

$$m_a = \sqrt{\frac{a_{pt}}{a_c}} - 1 \tag{7.46}$$

Moreover,

$$b = l_b + m_b(1 - \theta) \tag{7.47}$$

By considering the boundary conditions, we have

$$b \to b_c \text{ as } \theta \to 1$$

$$b \to b_{pt} \text{ as } \theta \to 0$$

Consequently, the final equation for calculation parameter b is as follows:

$$b = b_c[1 + m_b(1 - \theta)] \quad (7.48)$$

with

$$m_b = \frac{b_{pt}}{b_c} - 1 \quad (7.49)$$

It should be noted that Nasrifar and Moshfeghian (2001) assumed the following expression to determine three parameters including a_{pt}, b_{pt}, and T_{pt}

$$\frac{a_{pt}}{b_{pt}RT_{pt}} = 29.7056 \quad (7.50)$$

$$\frac{T_{pt}}{T_c} = 0.2498 + 0.3359\omega - 0.1037\omega^2 \quad (7.51)$$

$$\frac{b_{pt}}{b_c} = 1 - 0.1519\omega - 3.9462\omega^2 + 7.0538\omega^3 \quad (7.52)$$

7.3.7 Schmidt and Wenzel Equation of State

Schmidt and Wenzel (1980) recognized that the Soave–Redlich–Kwong EOS precisely estimates the thermodynamic parameters of compounds with acentric factor near 0 whereas the Peng–Robinson EOS precisely estimates the thermodynamic parameters of compounds with acentric factor near 0.3. Considering this point, they proposed a new EOS that reduces to the Redlich–Kwong (1949) EOS at acentric factor of 0 and to the Peng–Robinson EOS at acentric factor of 1/3. In the Schmidt–Wenzel EOS, acentric factor is a third parameter. The Schmidt–Wenzel EOS precisely estimates the vapor pressure and liquid density of moderate and light fluids.

$$P = \frac{RT}{v - b} - \frac{a_c\alpha(T_r)}{v^2 + (1 + 3\omega)bv - 3\omega b^2} \quad (7.53)$$

in which a_c, b as follows

$$a_c = \Omega_{ac}\frac{R^2T_c^2}{P_c} \tag{7.54}$$

$$b = \Omega_b\frac{RT_c}{P_c} \tag{7.55}$$

$$\Omega_{ac} = [1 - \eta(1 - q)]^3 \tag{7.56}$$

$$\Omega_b = \eta q \tag{7.57}$$

in which η stands for the critical compressibility factor and is related to the correlating factor q (defined as b/v_c) as follows:

$$\eta = \frac{1}{[3(1 + q\omega)]} \tag{7.58}$$

The smallest positive value of the following equation is q

$$(6\omega + 1)q^3 + 3q^2 + 3q - 1 = 0 \tag{7.59}$$

7.3.8 The Patel–Teja Equation of State and Modifications

The Patel–Teja (PT) (1982) and modified PT EOSs have the same PVT relationship and a-function. The difference is in calculating the EOS parameters, i.e., a, b, and c. In the modified PT EOS, the actual compressibility factor is used, whereas the critical compressibility factor in the PT EOS is a conventional parameter. Consequently, near the critical point the modified PT EOS estimates liquid densities more precisely than the PT EOS.

$$P = \frac{RT}{v - b} - \frac{a}{v(v + b) + c(v - b)} \tag{7.60}$$

In which a, b, and c are calculated as follows

$$a = \Omega_a\alpha(RT_c)^2/P_c \tag{7.61}$$

$$b = \Omega_b RT_c/P_c \tag{7.62}$$

$$c = \Omega_c RT_c/P_c \tag{7.63}$$

$$\Omega_a = 3\xi_c^2 + 3(1 - 2\xi_c)\Omega_b + \Omega_b^2 + (1 - 3\xi_c) \tag{7.64}$$

and Ω_b is the smallest positive root of its cubic form

$$\Omega_b^3 + (2 - 3\xi_c)\Omega_b + 3\xi_c^2\Omega_b - \xi_c^3 = 0 \tag{7.65}$$

α has the same expression as the recommended by Soave—Redlich—Kwong EOS.

$$\alpha = \left[1 + F\left(1 - \sqrt{T_R}\right)\right]^2 \tag{7.66}$$

By fitting in both vapor pressure and liquid density of pure substances, generalized expressions for parameter F and ξ_c can be gained as follows:

$$F = 0.452413 + 1.30982\omega - 0.295937\omega^2 \tag{7.67}$$

$$\xi_c = 0.329032 - 0.076799\omega + 0.0211947\omega^2 \tag{7.68}$$

Mixing parameters a_M, b_M, and c_M are calculated by Van der Waals mixing rule which is explained in further sections.

7.3.9 Mohsen-Nia—Modarress—Mansoori Equation of State

Mohsen-Nia et al. (2003) did not employ a repulsive term proposed by Van der Waals for their EOS called MMM EOS. Instead, they employed a more precise practical repulsive formulation which included the molecular simulation data of hard spheres. The MMM EOS is precise for estimating liquid density and vapor pressure of moderate and light pure fluids. The a-function of the MMM EOS is a Soave type. They employed a temperature-dependence term for the molecular covolume parameter. As illustrated by Mohsen-Nia et al. (2003), this enhances the potential of the EOS for predicting liquid density; however, as pointed out by Salim and Trebble (1991), it may cause abnormal trends at high pressures.

$$P = \frac{RT}{v}\left(\frac{v + ab}{v - b}\right) - \frac{\alpha_c \alpha(T_r)/\sqrt{T}}{v(v + N\alpha b)} \tag{7.69}$$

in which

$$\alpha = \alpha_c\left[1 + m\left(1 - \sqrt{T_R}\right)\right]^2 \tag{7.70}$$

$$b = b_c\left[1 + n_1\left(1 - \sqrt{T_R}\right) + n_2\left(1 - T_R^{0.75}\right)\right]^2 \quad \text{for } \mathrm{T_R} < 1 \tag{7.71}$$

$$m = 0.32 + 0.64\omega \tag{7.72}$$

$$n_1 = 3.270572 - 6.4127\omega + 10.6821\omega^2 \tag{7.73}$$

$$n_2 = -1.72192 + 3.85288 - 7.202286\omega^2 \tag{7.74}$$

$$a_c = 0.47312\left(\frac{R^2 T_c^{2.5}}{P_c}\right) \tag{7.75}$$

$$b_c = 0.04616\left(\frac{RT_c}{P_c}\right) \tag{7.76}$$

7.3.10 Adachi–Lu–Sugie Equation of State

Adachi et al. (1983) (ALS) developed a four-parameter EOS. Jensen (1987) modified the Adachi–Lu–Sugie EOS for application in the oil and gas industries. The Adachi–Lu–Sugie EOS is quite precise for determining the properties of pure fluids.

$$P = \frac{RT}{v - b_1} - \frac{a(T)}{(v - b_2)(v + b_3)} \tag{7.77}$$

It is cubic in volume with three temperature-independent parameters b_1, b_2, and b_3 and one temperature-dependent parameter $a(T)$.

The behavior of real fluid indicates that at $T = T_c$ there are three equal roots for v at the critical point, at $T > T_c$ there is only one root for v, and at $T < T_c$ there are three roots for v. Rearrangement of Adachi–Lu–Sugie EOS in terms of reduced volume v_r, and matching coefficients to the equation $(v_r - 1)^3 = 0$ at the critical points results in:

$$B_1 + B_2 - B_3 - 3Z_c = -1 \tag{7.78}$$

$$B_1B_2 - B_2B_3 - B_3B_1 - 3Z_c^2 = -A - B_2 + B_3 \tag{7.79}$$

$$B_1B_2B_3 + Z_c^3 = AB_1 - B_2B_3 \tag{7.80}$$

in which

$$A = \frac{aP_c}{R^2T_c^2} \tag{7.81}$$

$$B_i = \frac{b_iP_c}{RT_c}, \quad i = 1, 2, 3 \tag{7.82}$$

As the number of limitations is not adequate to calculate the values of the aforementioned constants, a straightforward search-optimization approach was employed to determine the optimum values of B_1 and Z_c along the critical isotherm. The values illustrated in Pitzer et al. (1955) were employed in the computation. As a result, the values of Z_c, B_1, B_2, B_3, and A were specified and generalized as functions of the acentric factor, ω, as follows:

$$Z_c = 0.3242 - 0.0576\omega \tag{7.83}$$

$$B_1 = 0.08974 - 0.03452\omega + 0.00330\omega^2 \tag{7.84}$$

$$B_2 = 0.03686 + 0.00405\omega - 0.01073\omega^2 + 0.00157\omega^3 \tag{7.85}$$

$$B_3 = 0.15400 + 0.00405\omega - 0.00272\omega^2 - 0.00484\omega^3 \tag{7.86}$$

$$A = 0.44869 + 0.04024\omega + 0.01111\omega^2 - 0.00576\omega^3 \tag{7.87}$$

At temperatures other than the critical, the formulation proposed by Soave (1972) was employed to calculate the parameter $a(T)$ as follows

$$\alpha = a\left[1 + \alpha\left(1 - \sqrt{T_R}\right)\right]^2 \tag{7.88}$$

The constant α in the previous equation has been formulated with the acentric factor ω as follows:

$$\alpha = 0.4070 + 1.3787\omega - 0.2933\omega^2 \tag{7.89}$$

7.4 MIXING RULES

The Van der Waals quadratic mixing rule with geometric merging rule is employed to calculate the attractive parameter of the EOSs as expressed as follows. It is worth mentioning that the Van der Waals mixing rules have been confirmed worthwhile in calculating the thermodynamic parameters in hydrocarbon mixtures.

$$\alpha = \sum_{i=1}^{n} \sum_{j=1}^{m} x_i x_j \alpha_{ij} \tag{7.90}$$

in which

$$\alpha_{ij} = \sqrt{\alpha_{ii}\alpha_{jj}}\left(1 - K_{ij}\right) \tag{7.91}$$

in which K_{ij} denotes the binary interaction parameter. For the other parameters of the EOSs, including second, third, and fourth parameters, the following mixing rule is employed:

$$w = \sum_{j} x_j w_j \tag{7.92}$$

in which w stands for the constants including b, c, d, and ω in different EOSs described in the previous sections.

7.5 HEAVY FRACTIONS

The heavy components of a real gas are typically lumped and reported as C_{7+}. The C_{7+} fraction is usually identified by specific gravity and molecular weight. The C_{7+} fractions may comprise various families of hydrocarbons and many components. However, it is promising to analyze the C_{7+} fractions precisely by employing novel methods like gas capillary chromatography. Pedersen et al. (1992) pointed out that the specification of hydrocarbon fluids up to C_{20} may be adequate for a precise determination of thermodynamic parameters. Nevertheless, usually the composition up to C_6 is available with a heavy end that must be characterized. Splitting a C_{7+} fraction into a number of single-carbon number (SCN) groups and then determine the critical properties of each group by employing available correlations is a routine job for determining phase behavior of reservoir fluid (Cavett, 1962; Riazi and Daubert, 1987; Twu, 1984; Lee and Kesler, 1980). Katz (1983) proposed a straightforward decline exponential function to formulate the distribution of SCN groups. Pedersen et al. (1992) demonstrated that an exponential function for the compositional distribution of SCNs in North Sea petroleum fractions had the best description. Starling (2003) has also proposed an exponential decline function for splitting heavy fractions. The decay functions are different in form and accuracy. However, the decline function proposed by Pedersen et al. (1992) seems simple but still useful; hence, it will be used in this study. The decay function reads

$$Z_{\mathrm{n}} = \exp(\mathrm{A} + B\mathrm{MW}_{\mathrm{n}}) \tag{7.93}$$

in which Z_{n} stands for the SCN-group mole fraction and MW_{n} represents the SCN-group molecular weight. Using the following expressions helps us determine the unknowns A and B for a C_{7+} fraction:

$$Z_{C_{7+}} = \sum_{C_7}^{C_n} Z_{C_n} \tag{7.94}$$

and

$$Z_{C_{7+}} = \sum_{C_7}^{C_N} Z_{C_n} \tag{7.95}$$

in which C_N stands for the heaviest SCN to be considered in a C_{7+} fraction and C_n is a model parameter. For determining A and B and consequently the SCN distribution, the SCN volume and molecular weight are required. Whitson (1983) generalized the properties of SCN, and these generalized properties can be employed in calculations.

7.6 GAS PROPERTIES

7.6.1 Viscosity

Fluid flow through gas reservoirs depends on the viscosity of natural gas mixtures, which plays a noteworthy role in analytical calculations for gas production and numerical simulations for long-term predictions. Same as oil reservoirs, the lab studies and measurements are time-consuming, expensive, and have more limitations compared with analytical and empirical correlations. The laboratory limitations for oil/gas property measurements including recombining the fluid sample, simulating reservoir condition in the lab, and preserving equilibrium circumstances. Consequently, upstream experts are very interested in using empirical correlations (Chen and Ruth, 1993; Naseri et al., 2014) for calculating the reservoir fluid properties. The major advantage of the empirical correlation and EOSs is highlighted when the required PVT data in laboratory conditions are not available. Owing to time and money consumption of laboratory investigations, the required oil/gas properties are usually predicted by EOSs along with empirical correlations (Hemmati-Sarapardeh et al., 2013). It is worth highlighting that the aforementioned empirical correlation should be capable of determining the effect of dynamic reservoir parameters including temperature and pressure, in the case of gas viscosities. Thoughtful inconsistencies in estimation viscosity of gas in reservoir conditions have been explained with details in the previous research work. Supporting approaches are illustrated to boost prediction of viscosities from given specific gravity of the gaseous phase, pressure, and temperature of a hydrocarbon mixture (Carr et al., 1954; Dranchuk and Abou-Kassem, 1975). Furthermore, most empirical correlations attempted to develop correlations for estimation of gas viscosity at reservoir conditions as a function of ambient viscosity, which are known as two-step correlations (McCain, 1990).

7.6.1.1 Empirical Correlations

Owing to the complications of laboratory measurements of viscosity, this variable can be predicted by empirical methods with reasonable accuracy.

Moreover, viscosity of a natural gas is expressed by the following function (Ahmed, 1989):

$$\mu_g = f(P, T, y_i)$$

μ_g stands for the gas-phase viscosity, P represents the pressure of the system, T denotes the temperature of the system, and y_i stands for the composition of the gas phase. As can clearly be seen from this simple relation, the gas viscosity strongly depends on the temperature, pressure, and composition of the gas phase in the hydrocarbon mixture. In the following sections in this chapter, we attempt to highlight the advantages and disadvantages of the empirical correlations developed for estimation of gas viscosity in the hydrocarbon mixtures along with the brief description of them.

7.6.1.1.1 Lee—Gonzalez—Eakin Method (1966)

Lee—Gonzalez—Eakin in 1966 developed a semiempirical correlation for estimating the natural gas viscosity at the reservoir condition. They attempted to correlate the gas viscosity as an output molecular weight of gas, gas density, and reservoir temperature as input variables. Their developed correlation is expressed as follows:

$$\mu_g = 10^{-4} K \exp\left[X\left(\frac{\rho_g}{62.4}\right)^Y\right] \tag{7.96}$$

$$K = \frac{(9.4 + 0.02\text{MW})T^{1.5}}{209 + 19\text{MW} + T} \tag{7.97}$$

$$X = 3.5 + \left(\frac{986}{T}\right) + 0.01\text{MW} \tag{7.98}$$

$$Y = 2.4 - 0.2X \tag{7.99}$$

in which ρ stands for the density (g/cc), T denotes temperature (°R), and MW represents molecular weight of the gas.

7.6.1.1.2 Dempsey's Standing Method (1965)

Carr et al. (1954) proposed charts for calculating natural gas viscosity at the reservoir temperature and ambient pressure. Later, Standing made an attempt to propose a correlation representing Carr et al. (1954) charts at the aforementioned conditions. Moreover, his correlation considers the impact of nonhydrocarbon elements (including H_2S, N_2, and CO_2) on the viscosity of the natural gas (Standing, 1977; Carr et al., 1954). The validity ranges of his correlation for temperature and gas specific gravity are 100–300°F and 0.55–1.55, correspondingly. Furthermore, Standing tried to determine natural gas viscosity at any given pressure, in this regard, he employed the correlation proposed by Dempsey to estimate the viscosity of natural gas at the given pressure (Standing, 1951, 1977; Standing and Katz, 1942). The following correlations demonstrate the mathematical expression of Dempsey's Standing method:

$$\begin{aligned}\mu_g &= \left(\mu_g \text{uncorrected}\right) + (N_2\,\text{correction}) \\ &\quad + (CO_2\,\text{correction}) + (H_2S\,\text{correction})\mu_g(\text{uncorrected}) \\ &= \left[1.709\left(10^{-3}\right) - 2.062\left(10^{-3}\right)\gamma_g\right]T_R + 8.188\left(10^{-3}\right) \\ &\quad - 6.15\left(10^{-3}\right)\log\gamma_g\end{aligned} \tag{7.100}$$

$$N_2\,\text{correction} = y_{N_2}\left[8.43\left(10^{-3}\right)\log\gamma_g + 9.59\left(10^{-3}\right)\right] \tag{7.101}$$

$$CO_2\,\text{correction} = y_{CO_2}\left[9.08\left(10^{-3}\right)\log\gamma_g + 6.24\left(10^{-3}\right)\right] \tag{7.102}$$

$$H_2S\,\text{correction} = y_{H_2S}\left[8.49\left(10^{-3}\right)\log\gamma_g + 3.73\left(10^{-3}\right)\right] \tag{7.103}$$

$$\begin{aligned}\text{Ln}\left[\left(\frac{\mu}{\mu_g}\right)T_r\right] &= a_0 + a_1P_r + a_2P_r^2 + a_3P_r^3 + T_r\left(a_4 + a_5P_r + a_6P_r^2\right. \\ &\quad \left. + a_7P_r^3\right) + T_r^2\left(a_8 + a_9P_r + a_{10}P_r^2 + a_{11}P_r^3\right) \\ &\quad + T_r^3\left(a_{12} + a_{13}P_r + a_{14}P_r^2 + a_{15}P_r^3\right)\end{aligned} \tag{7.104}$$

in which T_R stands for the temperature of the system (°F), γ_g represents the specific gravity of the natural gas, and P_r, and T_r are reduced pressure and temperature, correspondingly. The coefficient of the aforementioned equation is reported in Table 7.1.

7.6.1.1.3 Chen and Ruth Method (1993)

Dadash-zade and Gurbanov employed a multivariable regression approach to develop a correlation from laboratory measurements with the ranges of 16–100 for molecular weight of the gas and 310.8–477.4K for temperature. Chen and Ruth (1993) made attempt to optimize the correlation

Table 7.1 Coefficients of Eq. (7.104)

Coefficient	Value	Coefficient	Value
a_0	-2.46211820	a_9	1.39643306
a_1	2.970547414	a_{10}	$-1.49144925 \times (10^{-1})$
a_2	$-2.86264054 \times (10^{-1})$	a_{11}	$4.41015512 \times (10^{-3})$
a_3	$8.05420522 \times (10^{-3})$	a_{12}	$8.39387178 \times (10^{-2})$
a_4	2.80860949	a_{13}	$-1.86408848 \times (10^{-1})$
a_5	-3.49803305	a_{14}	$2.03367881 \times (10^{-2})$
a_6	$3.60373020 \times (10^{-1})$	a_{15}	$-6.09579263 \times (10^{-4})$
a_7	$-1.044324 \times (10^{-2})$		
a_8	$-7.93385648 \times (10^{-1})$		

proposed by Dadash-zade and Gurbanov and the following correlation is final form of Chen and Ruth (1993):

$$\mu_g = (a_1 + a_2 T_K) - (a_3 + a_4 T_K)\sqrt{MW} \tag{7.105}$$

$$\frac{\mu}{\mu_g} = \left(b_1 + b_2 P_r + b_3 P_r^2\right)\frac{1}{T_r} + \left(b_4 + b_5 P_r + b_6 P_r^2\right)\frac{1}{T_r^4} + \left(b_7 + b_8 P_r + b_9 P_r^2\right) \tag{7.106}$$

in which T_K represents the temperature (K), MW stands for the molecular weight of the gas, and P_r denotes the reduced pressure. The constants of the Chen and Ruth (1993) correlation are illustrated in Table 7.2.

Table 7.2 Coefficients of Equations (7.106) and (7.107)

Coefficient	Tuned Coefficient
a_1	0.38539×10^{-2}
a_2	0.00356×10^{-2}
a_3	0.04131×10^{-2}
a_4	0.00016×10^{-2}
b_1	-0.4888439
b_2	-0.0943952
b_3	0.0199591
b_4	0.8266923
b_5	1.7124100
b_6	-0.0700968
b_7	1.2076900
b_8	0.0301188
b_9	-0.0048318

7.6.1.1.4 Elsharkawy Method (2004)

Elsharkawy in (2004) made attempt to modify the correlation for natural gas viscosity estimation proposed by the Lee–Gonzalez–Eakin to consider the effect of the nonhydrocarbon gases and presence of heptane-plus fraction. His proposed correlation is expressed as follows:

$$\mu_g = D_1 \times 10^{-4} \exp\left(D_2 \rho_g^{D_3}\right) \tag{7.107}$$

in which

$$D_1 = \frac{(9.379 + 0.01607 M_g) T^{1.5}}{209.2 + 19.26 M_g + T} \tag{7.108}$$

$$D_2 = 3.448 + \left(\frac{986.4}{T}\right) + 0.01009 M_g \tag{7.109}$$

$$D_3 = 2.447 - 0.224 D_2 \tag{7.110}$$

in which M_g stands for the gas molecular weight, ρ_g represents the gas density (g/cc), and T denotes the temperature (°R). The modifications considered in the Elsharkawy method (Eqs. 7.111–7.113) try to improve the potential of the original correlation by taking into account the effects of hydrogen sulfide, the presence of high content of heptane-plus fraction, and carbon dioxide in natural gases. These corrections are considered via the following correlations:

$$\Delta\mu_g = y_{H_2S}\left[-3.2268 \times \left(10^{-3}\right)\log\gamma_g + 2.1479 \times \left(10^{-3}\right)\right] \tag{7.111}$$

$$\Delta\mu_g = y_{CO_2}\left[6.4366 \times \left(10^{-3}\right)\log\gamma_g + 6.7255 \times \left(10^{-3}\right)\right] \tag{7.112}$$

$$\Delta\mu_g = y_{C_{7+}}\left[-3.2875 \times \left(10^{-1}\right)\log\gamma_g + 1.2885 \times \left(10^{-1}\right)\right] \tag{7.113}$$

in which γ_g stands for the gas gravity and y denotes the mole fraction of the component.

7.6.1.1.5 Sutton Method (2007)

Sutton (1985) developed a math-based equation for two gas samples in which a high-gravity associated gas is characteristically rich in C_1 to C_5; however, retrograded condensate gases are rich in heptane-plus fractions. Consequently, his correlation is appropriate for all light natural gases and the condensate gases/heavier gas; however, this correlation cannot be employed for high-gravity hydrocarbon gases that do not comprise considerable heptane-plus fractions. Sutton studied condensate gases/high-gravity gas and proposed approaches for predicting pseudocritical properties which yielded in more-precise Z factors. Sutton improved his method based on the data from both condensate gas and associated gas (Sutton, 2007). His correlation expressed mathematically as follows:

$$\mu_g \zeta = 10^{-4}\Big[0.807 T_{pr}^{0.618} - 0.357\exp(-0.449 T_{pr}) + 0.340\exp(-4.058 T_{pr}) + 0.018\Big] \tag{7.114}$$

$$\mu = \mu_g \exp\left(X\rho^Y\right) \tag{7.115}$$

$$\zeta = 0.949\left(\frac{T_{pc}}{\mathrm{MW}^3 P_{pc}^4}\right)^{1/6} \tag{7.116}$$

$$X = 3.47 + \left[\frac{1.588}{T}\right] + 0.0009\mathrm{MW} \tag{7.117}$$

$$Y = 1.66378 - 0.04679X \tag{7.118}$$

in which P_{pr} represents pseudoreduced pressure, T_{pr} stands for the pseudoreduced temperature, ζ denotes normalizing parameter for the viscosity, T_{pc} denotes the pseudocritical temperature (°R), and P_{pc} stands for the pseudocritical pressure (psi). Sutton attempted to evaluate appropriate approaches for predicting PVT properties of the condensate gases/high-gravity gas by handling extensive data samples of associated gas compositions comprising more than 3200 compositions. Most of the investigations concentrated on proposing an appropriate correlation between natural gas viscosities and corresponding input variables; however, they cannot be generalized to other gases with various compositions produced from various gas fields throughout the world and suffer from localized validity (Dempsey, 1965; Londono Galindo et al., 2005; Jossi et al., 1962; Diehl et al., 1970; Golubev, 1959).

7.6.1.1.6 Shokir and Dmour Method (2009)

Shokir and Dmour (2009) proposed a correlation for predicting viscosity of hydrocarbon gases via a genetic programming approach. They presented a viscosity model for both gas mixture and pure hydrocarbon gas over a broad range of pressures and temperatures as a function of pseudoreduced temperature, gas density, molecular weight, and the pseudoreduced pressure of hydrocarbon gas mixtures and pure gases. Their model was straightforward and excluded the various complications included in any EOS computation (Shokir and Dmour, 2009).

$$\mu_g = A_1 + A_2 + A_3 \tag{7.119}$$

$$A_1 = -0.003338\left((\mathrm{MW} - P_{pr})\rho_g\right) - 0.745356\left(\rho_g\left(\frac{\rho_g}{\left(T_{pr} - \left(\frac{T_{pr}}{\rho_g}\right)\right) - \left(\frac{\rho_g}{P_{pr} - \mathrm{MW}}\right)}\right)\right) \tag{7.120}$$

$$A_2 = -0.590871\left(\rho_g\left(\frac{\frac{\frac{\mathrm{MW}}{T_{pr}}}{T_{pr}}}{\mathrm{MW}}\right)\right) + 0.004602(T_{pr}P_{pr}) - 0.007935P_{pr} + 1.063654\rho_g \tag{7.121}$$

$$A_3 = -0.392638\left(\rho_g T_{pr}\right) - 0.004755\left(\frac{P_{pr}}{T_{pr}}\right) + 0.000463\mathrm{MW} + 0.011707T_{pr} - 0.017994 \tag{7.122}$$

μ stands for the viscosity of the hydrocarbon gas mixtures or pure gases, P_{pr} represents the pseudoreduced pressure, T_{pr} denotes the pseudoreduced temperature, ρ_g stands for the density of the hydrocarbon gas mixtures and/or pure gases, and MW stands for the molecular weight of the hydrocarbon gas mixtures and/or pure gases.

7.6.1.1.7 Sanjari–Nemati Lay–Peymani Method (2011)

Sanjari et al. (2011) suggested another correlation to estimate viscosity of natural gases. This method is usable for estimating gas viscosity in the range of $0.01 \le P_{pr} \le 21$ and $1.01 \le T_{pr} \le 3$. The mathematical expression of their correlation is as follows:

$$\mu = \frac{A(P_{pr}) + B(T_{pr})}{C(P_{pr}) + D(T_{pr})} \tag{7.123}$$

$$A = \alpha_1 + \alpha_2 P_r + \alpha_3 P_r^2 + \alpha_4 \mathrm{Ln}(P_r) + \alpha_5 \mathrm{Ln}^2(P_r) \tag{7.124}$$

$$B = \frac{\alpha_6}{T_r} + \alpha_7 \mathrm{Ln}^2(T_r) \tag{7.125}$$

Table 7.3 Coefficients of Eqs. (7.124)–(7.127)

Coefficient	Tuned Coefficient
α_1	−0.141645
α_2	0.018076
α_3	0.00214
α_4	−0.004192
α_5	−0.000386
α_6	0.187138
α_7	0.569211
α_8	0.000387
α_9	−2.857176
α_{10}	2.925776
α_{11}	−1.062425

$$C = 1 - \alpha_8 P_r^2 \tag{7.126}$$

$$D = \frac{\alpha_9}{T_r} + \frac{\alpha_{10}}{T_r^2} + \frac{\alpha_{11}}{T_r^2} \tag{7.127}$$

in which μ stands for the gas viscosity (cP), and α_1 to α_{11} are tuned coefficients which were calculated based on the minimizing the summation of square errors of the correlation, as reported in Table 7.3.

Example 7.1

Consider a condensate gas fluid with the following composition. Determine the gas viscosity at reservoir conditions via Sanjari–Nemati Lay–Peymani method. The reservoir pressure and temperature are 4300 Psi and 380K, correspondingly.

Component	Mol%	Critical Pressure (Psi)	Critical Temperature (K)
N_2	0.03496	493.12	126.4
CO_2	0.02331	1070.814	304.4
C_1	0.84990	666.4	190.7
C_2	0.05529	706.5	305.5
C_3	0.02008	616	370
i-C_4	0.00401	529	408.3
n-C_4	0.00585	550.56	425.3
i-C_5	0.00169	491	460.6
n-C_5	0.00147	488.777	469.9
C_{7+}	0.00344	310	700

Solution

In this case, we should calculate the pseudocritical pressure and temperature using the critical pressure and temperature of each component. Then, we should calculate the pseudoreduced critical pressure and temperature.

(Continued)

Component	Mol%	P_c (Psi)	T_c (K)	y_i*P_c	y_i*T_c
N_2	0.03496	493.12	126.4	17.2394752	4.418944
CO_2	0.02331	1070.814	304.4	24.96067434	7.095564
C_1	0.84990	666.4	190.7	566.37336	162.07593
C_2	0.05529	706.5	305.5	39.062385	16.891095
C_3	0.02008	616	370	12.36928	7.4296
i-C_4	0.00401	529	408.3	2.12129	1.637283
n-C_4	0.00585	550.56	425.3	3.220776	2.488005
i-C_5	0.00169	491	460.6	0.82979	0.778414
n-C_5	0.00147	488.777	469.9	0.71850219	0.690753
C_{7+}	0.00344	310	700	1.0664	2.408

$$P_{pc} = \sum y_i P_{ci} = 668.1 \text{ psia}$$

$$T_{pc} = \sum y_i T_{ci} = 205.91\text{K}$$

$$6P_{pr} = \frac{P}{P_{pc}} = \frac{4300}{668.1} = 6.436$$

$$T_{pr} = \frac{T}{T_{pc}} = \frac{380}{205.91} = 1.845$$

Using Eqs. (7.125)–(7.129), we have:

$$\mu = \frac{A(P_{Pr}) + B(T_{Pr})}{C(P_{Pr}) + D(T_{Pr})} = 0.0032 \text{ cP}$$

$$A = a_1 + a_2 P_r + a_3 P_r^2 + a_4 \, Ln\,(P_r) + a_5 \, Ln^2(P_r) = 0.0542$$

$$B = \frac{a_6}{T_r} + a_7 \, Ln^2(T_r) = 0.315$$

$$C = 1 - a_8 P_r^2 = 119.35$$

$$D = \frac{a_9}{T_r} + \frac{a_{10}}{T_r^2} + \frac{a_{11}}{T_r^2} = -1.0012$$

7.6.2 *Z* Factor

Compressibility factor of the real gas can be expressed as a function of pressure, volume, and temperature as follows:

$$Z = \frac{PV}{nRT} \tag{7.128}$$

in which Z stands for the compressibility factor, V denotes the volume, n represents the number of gas moles, T represents the system temperature, R stands for the gas constant, and P stands for the system pressure.

Based on the theory of corresponding states, Z also can be defined as a function of pseudo-reduced pressure (P_{pr}) and pseudo-reduced temperature (T_{pr}) as follows:

$$P_{pr} = \frac{P}{P_{pc}} \tag{7.129}$$

$$T_{pr} = \frac{T}{T_{pc}} \tag{7.130}$$

in which T_{pc} and P_{pc} represent the pseudocritical temperature and pressure, correspondingly.

The pseudocritical temperature and pressure can be determined by some mixing rules (Stewart et al., 1959; Sutton, 1985; Corredor et al., 1992; Piper et al., 1993; Elsharkawy et al., 2001; Elsharkawy, 2004). Along with these works, several correlations were developed to calculate the pseudocritical parameters through using gas specific gravity (Standing, 1981; Elsharkawy, 2002; Elsharkawy and Elkamel, 2001; Londono Galindo et al., 2005; Sutton, 2007).

7.6.2.1 Empirical Correlations

For the sake of calculating compressibility factors, more than 20 empirical correlations have been proposed. This kind of compressibility factor calculating method is classified into two categories: indirect and direct methods (Chamkalani et al., 2013). The empirical correlations adopted in this study are presented in the following sections.

7.6.2.1.1 Papay (1968)

Papay (1968) proposed a simple relationship to calculate the compressibility factor as follows:

$$Z = 1 - \frac{P_{pr}}{T_{pr}}\left[0.3648758 - 0.04188423\left(\frac{P_{pr}}{T_E}\right)\right] \tag{7.131}$$

7.6.2.1.2 Beggs and Brill (1973)

Beggs and Brill (1973) proposed a correlation which was generated from the Standing–Katz chart to predict compressibility factor:

$$Z = A + \frac{1 - A}{e^B} + CP_{pr}^D \tag{7.132}$$

$$A = 1.39(T_{pr} - 0.92)^{0.5} - 0.36T_{pr} - 0.101 \tag{7.133}$$

$$B = (0.62 - 0.23T_{pr})P_{pr} + \left[\frac{0.066}{(T_{pr} - 0.86)} - 0.037\right]P_{pr}^2 + \left(\frac{0.32}{10^{9(T_{pr}-1)}}\right)P_{pr}^6 \quad (7.134)$$

$$C = 0.132 - 0.32\log(T_{pr}) \quad (7.135)$$

$$D = 10^{\left(0.3016 - 0.49T_{pr} + 0.1824T_{pr}^2\right)} \quad (7.136)$$

7.6.2.1.3 Shell Oil Company

Kumar (2004) referenced the Shell Oil Company model for calculation of compressibility factor as:

$$Z = A + BP_{pr} + (1 - A)\exp(-C) - D\left(\frac{P_{pr}}{10}\right)^4 \quad (7.137)$$

$$A = -0.101 - 0.36T_{pr} + 1.3868\sqrt{T_{pr} - 0.919} \quad (7.138)$$

$$B = 0.021 + \frac{0.04275}{T_{pr} - 0.65} \quad (7.139)$$

$$C = P_{pr}\left(E + FP_{pr} + GP_{pr}^4\right) \quad (7.140)$$

$$D = 0.122\exp[-11.3(T_{pr} - 1)] \quad (7.141)$$

$$E = 0.6222 - 0.224T_{pr} \quad (7.142)$$

$$F = \frac{0.0657}{T_{pr} - 0.85} - 0.037 \quad (7.143)$$

$$G = 0.32\exp[-19.53(T_{pr} - 1)] \quad (7.144)$$

7.6.2.1.4 Bahadori et al. (2007)

Bahadori et al. (2007) presented an equation which correlated compressibility factor to reduced temperature and pressure. The application ranges of this correlation are $0.2 < P_{pr} < 16$ and $1.05 < T_{pr} < 2.4$. This equation is given by:

$$Z = a + bP_{pr} + cP_{pr}^2 + dP_{pr}^3 \quad (7.145)$$

$$a = A_a + B_a T_{pr} + C_a T_{pr}^2 + D_a T_{pr}^3 \quad (7.146)$$

$$b = A_b + B_b T_{pr} + C_b T_{pr}^2 + D_b T_{pr}^3 \quad (7.147)$$

$$c = A_c + B_c T_{pr} + C_c T_{pr}^2 + D_c T_{pr}^3 \quad (7.148)$$

$$d = A_d + B_d T_{pr} + C_d T_{pr}^2 + D_d T_{pr}^3 \quad (7.149)$$

Constants of the previous equations are reported in Table 7.4.

It is worth mentioning that Bahadori used the following correlations to predict pseudocritical properties:

$$P_{pc} = 756.8 - 131.07\gamma_g - 3.6\gamma_g^2 \quad (7.150)$$

$$T_{pc} = 169.2 + 349.5\gamma_g - 74.0\gamma_g^2 \quad (7.151)$$

in which T_{pc} and P_{pc} are pseudocritical temperature (K) and pseudocritical pressure (kPa), correspondingly, and γ_g stands for the specific gravity of gas. It is worth mentioning that these equations are applicable in range of $0.55 < \gamma_g < 1.75$.

Bahadori proposed a cubic expression to take into account the effect of the hydrogen sulfide (H_2S) and carbon dioxide (CO_2) known as acid gases on the value of compressibility factor. Once the modified pseudocritical properties are calculated, they are employed to determine pseudoreduced properties and the Z factor is calculated from Eq. (7.145). The new equation first determines a deviation variable ε as follows:

$$\varepsilon = \frac{a + b y_{H_2S} + c y_{H_2S}^2 + d y_{H_2S}^3}{1.8} \quad (7.152)$$

$$a = A_a + B_a y_{CO_2} + C_a y_{y_{CO_2}}^2 + D_a y_{y_{CO_2}}^3 \quad (7.153)$$

Table 7.4 Coefficients of Eqs. (7.146)–(7.149)

Coefficient	Tuned Coefficient	Coefficient	Tuned Coefficient
A_a	0.969469	A_c	0.0184810
B_a	−1.349238	B_c	0.0523405
C_a	1.443959	C_c	−0.050688
D_a	−0.36860	D_c	0.010870
A_b	−0.107783	A_d	−0.000584
B_b	−0.127013	B_d	−0.002146
C_b	0.100828	C_d	0.0020961
D_b	−0.012319	D_d	−0.000459

$$b = A_b + B_b y_{CO_2} + C_b y^2_{y_{CO_2}} + D_b y^3_{y_{CO_2}} \quad (7.154)$$

$$c = A_c + B_c y_{CO_2} + C_c y^2_{y_{CO_2}} + D_c y^3_{y_{CO_2}} \quad (7.155)$$

$$d = A_d + B_d y_{CO_2} + C_d y^2_{y_{CO_2}} + D_d y^3_{y_{CO_2}} \quad (7.156)$$

in which y_{H_2S} and y_{CO_2} are the mole fractions of H_2S and CO_2 in the gas mixture, respectively. The tuned coefficients used in Eqs. (7.153)–(7.156) are also given in Table 7.5. These tuned coefficients help to cover sour natural gases up to as much as 90% total acid gas.

Then, ε is employed to calculate the adjusted pseudocritical properties as follows:

$$T_{pc}^{correct} = T_{pc} - \varepsilon \quad (7.157)$$

$$P_{pc}^{correct} = \frac{P_{pc}(T_{pc} - \varepsilon)}{T_{pc} + \varepsilon y_{H_2S}(1 - y_{H_2S})} \quad (7.158)$$

Eq. (7.152) is appropriate when the acid gas concentrations of $H_2S < 75$ mol% and $CO_2 < 55$ mol%, and has an average absolute error of 1% over the following ranges of data: 1000 kPa $< P <$ 45,000 kPa, and 275K $< T <$ 425K.

7.6.2.1.5 Azizi et al. (2010)

Based on the Standing–Katz chart with 3038 points, Azizi et al. (2010) presented an empirical correlation to estimate the sweet-gas compressibility factor values over the range of $0.2 < P_{pr} < 11$ and $1.1 < T_{pr} < 2$. The form is:

$$Z = A + \frac{B + C}{D + E} \quad (7.159)$$

Table 7.5 Coefficients of Eqs. (7.153)–(7.156)

Coefficient	Tuned Coefficient	Coefficient	Tuned Coefficient
A_a	4.094086	A_c	−1.95766763E2
B_a	1.15680575E2	B_c	3.835331543E2
C_a	−1.6991417E2	C_c	−6.08818159E2
D_a	5.62209803E1	D_c	3.704173461E2
A_b	1.45517461E2	A_d	5.24425341E1
B_b	−3.9672762E2	B_d	−2.0133960E2
C_b	3.93741592E2	C_d	3.51359351E2
D_b	−2.17915813E2	D_d	−2.20884255E2

in which

$$A = aT_{pr}^{2.16} + bP_{pr}^{1.028} + cT_{pr}^{-2.1}P_{pr}^{1.58} + d\,\mathrm{Ln}(T_{pr})^{-0.5} \quad (7.160)$$

$$B = e + fT_{pr}^{2.4} + gP_{pr}^{1.56} + hT_{pr}^{3.033}P_{pr}^{0.124} \quad (7.161)$$

$$C = i\mathrm{Ln}(T_{pr})^{-1.28} + j\mathrm{Ln}(T_{pr})^{1.37} + k\mathrm{Ln}(P_{pr}) + l\mathrm{Ln}(P_{pr})^2 + m\mathrm{Ln}(P_{pr})\mathrm{Ln}(T_{pr}) \quad (7.162)$$

$$D = 1 + nT_{pr}^{5.55} + cT_{pr}^{0.33}P_{pr}^{0.68} \quad (7.163)$$

$$E = p\,\mathrm{Ln}(T_{pr})^{1.18} + q\mathrm{Ln}(T_{pr})^{2.1} + r\,\mathrm{Ln}(P_{pr}) + s\,\mathrm{Ln}(P_{pr})^2 + t\,\mathrm{Ln}(P_{pr})\mathrm{Ln}(T_{pr}) \quad (7.164)$$

in which

$a = 0.0373142485385592$	$k = -24449114791.1531$
$b = -0.0140807151485369$	$l = 19357955749.3274$
$c = 0.0163263245387186$	$m = -126354717916.607$
$d = -0.0307776478819813$	$n = 623705678.385784$
$e = 13843575480.943800$	$o = 17997651104.3330$
$f = -16799138540.763700$	$p = 151211393445.064$
$g = 1624178942.6497600$	$q = 139474437997.172$
$h = 13702270281.086900$	$r = -24233012984.0950$
$i = -41645509.896474600$	$s = 18938047327.5205$
$j = 237249967625.01300$	$t = -141401620722.689$

7.6.2.1.6 Sanjari and Nemati Lay (2012)

The model developed by Sanjari and Nemati Lay (2012) was derived from 5844 experimental data of compressibility factors for a range of $0.01 < P_{pr} < 15$ and $1 < T_{pr} < 3$. The results of their study indicate the superiority of this empirical correlation over the other methods, such as Dranchuk and Abou-Kassem (1975), Azizi et al. (2010) correlations, and PR EOS with average absolute relative deviation percent of 0.6535. This correlation is written by:

$$Z = 1 + A_1P_{pr} + A_2P_{pr}^2 + \frac{A_3P_{pr}^{A_4}}{T_{pr}^{A_5}} + \frac{A_6P_{pr}^{(A_4+1)}}{T_{pr}^{A_7}} + \frac{A_8P_{pr}^{(A_4+2)}}{T_{pr}^{(A_7+1)}} \quad (7.165)$$

The constants of the Sanjari and Nemati Lay correlation are illustrated in Table 7.6.

Table 7.6 Coefficients of Eq. (7.165)

Coefficient	$0.01 < P_{pr} < 3$	$3 < P_{pr} < 15$
A_1	0.007698	0.015642
A_2	0.003839	0.000701
A_3	−0.467212	2.341511
A_4	1.018801	−0.657903
A_5	3.805723	8.902112
A_6	−0.087361	−1.136000
A_7	7.138305	3.543614
A_8	0.083440	0.134041

7.6.2.1.7 Shokir et al. (2012)

Shokir et al. (2012) employed genetic programming to present a novel correlation as a function of pseudoreduced pressure and pseudo-reduced temperature for predicting compressibility factors for sour gases, sweet gases, and gas condensates based on data samples reported in open literature. The correlation proposed by Shokir and colleagues is expressed as follows:

$$Z = A + B + C + D + E \tag{7.166}$$

$$A = 2.679562\left(\frac{2T_{pr} - P_{pr} - 1}{\left(P_{pr}^2 + T_{pr}^3\right)/P_{pr}}\right) \tag{7.167}$$

$$B = -7.686825\left(\frac{T_{pr}P_{pr} + P_{pr}^2}{T_{pr}P_{pr} + 2T_{pr}^2 + T_{pr}^3}\right) \tag{7.168}$$

$$C = -0.000624\left(T_{pr}^2P_{pr} - T_{pr}P_{pr}^2 + T_{pr}P_{pr}^3 + 2T_{pr}P_{pr} - 2P_{pr}^2 + 2P_{pr}^3\right) \tag{7.169}$$

$$D = 3.067747\left(\frac{T_{pr} - P_{pr}}{P_{pr}^2 + T_{pr} + P_{pr}}\right) \tag{7.170}$$

$$E = \left(\frac{0.068059}{T_{pr}P_{pr}}\right) + 0.139489T_{pr}^2 + 0.081873P_{pr}^2 - \left(\frac{0.041098T_{pr}}{P_{pr}}\right) + \left(\frac{8.152325P_{pr}}{T_{pr}}\right) - 1.63028P_{pr} + 0.24287T_{pr} - 2.64988 \tag{7.171}$$

7.6.2.1.8 Mahmoud (2014)

Mahmoud (2014) developed an empirical equation for predicting the gas-compressibility factor for high-temperature and high-pressure gas reservoirs by employing more than 300 data samples of measured compressibility factor. This correlation is a function of pseudo-reduced pressure and temperature and has simpler functional form than the conventional complex correlations. This correlation can be expressed as follows:

$$Z = \left(0.702e^{-2.5T_{\text{pr}}}\right)P_{pr}^2 - \left(5.524e^{-2.5T_{\text{pr}}}\right)P_{\text{pr}} + \left(0.044T_{\text{pr}}^2 - 0.164T_{\text{pr}} + 1.15\right) \quad (7.172)$$

Example 7.2

Consider a gas with the following properties (Bahadori et al., 2007):

$\gamma_g = 0.7$

$H_2S = 7\%$

$CO_2 = 10\%$

The reservoir temperature and pressure are 297K and 13,860 kPa, correspondingly. Determine compressibility factor using Bahadori method.

Solution

The following procedure should be followed:

At first, we should calculate the pseudocritical properties by using correlations proposed by Sutton (1985) [after unit conversion to International System of Units (SI)]:

$$P_{pc} = 756.8 - 131.07\gamma_g - 3.6\gamma_g^2 = 4572.28 \text{ kPa}$$

$$T_{pc} = 169.2 + 349.5\gamma_g - 74.0\gamma_g^2 = 209.59\text{K}$$

Then, we should determine the adjustment parameter for modifying the pseudocritical properties using Eqs. (7.152) and (7.157)–(7.158):

$$\varepsilon = 11.612$$

$$P'_{pc} = 4287.73 \text{ kPa}$$

$$T'_{pc} = 197.98\text{K}$$

Finally, using Eq. (7.145) with calculated pseudoreduced properties results the Z factor will be $Z = 0.7689$.

7.6.2.2 Equations of State

Among virial type, cubic and complex or molecular-based principle EOSs, cubic EOSs are more widely recommended and used (Forero and Velásquez, 2012). They are simple expressions and have the ability to describe quickly and reliably the phase behavior of vapor and liquid over a wide range of pressures, temperatures, and thermodynamic properties of fluids (Farrokh-Niae et al., 2008; Guria and Pathak, 2012). According to the number of variables which emerge in the attractive and repulsive terms, cubic EOSs can be divided into two-parameter, three-parameter, four- and five-parameter cubic EOSs (Forero and Velásquez, 2012). The SRK (Soave, 1972) and PR (Peng and Robinson, 1976) equations belong to two-parameter cubic EOSs, and the PT (Patel and Teja, 1982) equation is a three-parameter cubic EOS. They are the commonly used volumetric property-predicating methods for gas condensates and sour gases, and the application of modified and other EOSs has been popular in recent years.

Example 7.3

Two-phase butane exists in a sealed container at 200°F. Using PR EOS, calculate compressibility factor of each phase which are in equilibrium in the container.

Solution

Vapor pressure of this system is 195.1 psi.

The parameters of PR EOS are calculated as follows:

$$a = \Omega_a \frac{R^2 T_c^2}{P_c} = 56044.628$$

$$b = \Omega_b \frac{RT_c}{P_c} = 1.1607972$$

$$\alpha = \left[1 + m\left(1 - \sqrt{T_R}\right)\right]^2 = 1.0984264$$

$$a_T = a_c \times \alpha = 61560.902$$

$$m = 0.37464 + 1.54226\omega - 0.2699\omega^2 = 0.6716$$

$$A = \frac{\alpha_T p}{(RT)^2} = \frac{61560.92 \times 195.1}{(660 \times 10.73)^2} = 0.2394831$$

$$B = \frac{bp}{RT} = \frac{1.1607972 \times 195.1}{10.73 \times 660} = 0.0319793$$

$Z_1 = 0.05343 \rightarrow$ Liquid compressibility factor
$Z_2 = 0.1648455 \rightarrow$ Rejected
$Z_3 = 0.749748 \rightarrow$ Gas-compressibility factor

Example 7.4

A PVT cell with pressure of 1200 psi contains a mixture with the following compositions. The equilibrium temperature is 80°F. Calculate the compressibility factor of the mixture by using Schmidt–Wenzel and SRK EOSs.

Component	Mole Fraction	P_c	T_c	ω_i
C_1	0.85	666.4	343.33	0.0104
C_2	0.10	706.5	549.92	0.0979
C_3	0.05	616.0	666.06	0.1522

Solution

At first, we should determine the parameters of the EOS by using Eqs. (7.54)–(7.59). Results obtained for each component are reported in the following tables.

For Schmidt–Wenzel EOS:

Component	q	η	Ω_{ac}	a_c	Ω_b	b
C_1	0.2596647	0.332435	0.428467	8725.8	0.08632	0.47719
C_2	0.2578015	0.3231275	0.4367114	21,521.911	0.083818	0.70004
C_3	0.2566741	0.3208009	0.4416504	36,613.893	0.08234	0.95523

Component	m_o	T_r	m	α	$\alpha \times a_c$
C_1	0.47895	1.5728	0.89987	0.594934	5191.28
C_2	0.5918107	0.98196	0.6569	1.0119393	21,778.87
C_3	0.637782	0.81081	0.67447	1.1387951	41,695.72

Then, we should determine the parameters of EOS for the mixture by using mixing rule as follows:

$$b = \sum_{i=1}^{3} y_i b_i = 0.523377$$

Assuming $K_{ij} = 0$,

$$a_T = \sum_{i=1} \sum_{j=1} y_i y_j (a_{Ti} a_{Tj})^{0.5} (1 - K_{ij}) = 11{,}363.86$$

$$\omega_m = \frac{\sum_i y_i \omega_i b_i^{0.7}}{y_i b_i^{0.7}} = 0.032$$

(Continued)

Z-form of Schmidt–Wenzel EOS is as follows:

$$Z^3 - [1 + (1 - (1+3\omega))B]Z^2 + [A - (1+3\omega)B - (1+6\omega)B^2]Z - [AB - 3\omega(B^2 + B^3)] = 0$$

$$A = \frac{\alpha_T p}{(RT)^2} = \frac{11{,}363.86 \times 1200}{(540 \times 10.73)^2} = 0.406182$$

$$B = \frac{bp}{RT} = \frac{0.523377 \times 1200}{10.73 \times 540} = 0.108393$$

$$Z^3 - 0.989594Z^2 + 0.273378Z - 0.042777 = 0.$$

Solving this equation yields
$Z_1 = 0.680115 \rightarrow$ Accepted compressibility factor
$Z_2 = 0.154739 + 0.197364i \rightarrow$ Rejected
$Z_2 = 0.154739 - 0.197364i \rightarrow$ Rejected

For SRK EOS:
Using Eqs. (7.11)–(7.14) for calculating parameters of the SRK EOS yields:

Component	b_i	a_{ci}	m_i	α_i	a_{Ti}
C_1	0.4789	8703.98	0.4964	0.7636	6646.29
C_2	0.7236	21,064.95	0.6324	1.0115	21,306.56
C_3	1.0053	35,448.94	0.7155	1.1476	40,682.25

Using mixing rule we have:

$$b = \sum_{i=1}^{3} y_i b_i = 0.529698$$

Assuming $K_{ij} = 0$,

$$a_T = \sum_{i=1} \sum_{j=1} y_i y_j (a_{Ti} a_{Tj})^{0.5} (1 - K_{ij}) = 8831.8$$

$$A = \frac{\alpha_T p}{(RT)^2} = \frac{8831.8 \times 1200}{(540 \times 10.73)^2} = 0.315677$$

$$B = \frac{bp}{RT} = \frac{0.529698 \times 1200}{10.73 \times 540} = 0.109702$$

Solving this equation yields:
$Z_1 = 0.814 \rightarrow$ Accepted compressibility factor
$Z_2 = 0.093 + 0.184i \rightarrow$ Rejected
$Z_2 = 0.093 - 0.184i \rightarrow$ Rejected

Example 7.5

Consider gas-condensate fluid with the following composition. The equilibrium pressure and temperature are 3720 psi and 188°F, respectively. The composition of gas-condensate fluid is reported in the following table.

Component	Mole Fraction	P_c	T_c	ω_i
C_1	0.86	666.4	343.33	0.0104
C_2	0.05	706.5	549.92	0.0979
C_3	0.05	616.0	666.06	0.1522
C_4	0.02	527.9	765.62	0.1852
C_5	0.01	488.6	845.8	0.2280
C_6	0.005	453	923	0.2500
C_{7+}	0.0005	255	1180	0.5400

The heptane-plus fraction has the following characteristics:

$MW_{C_{7+}} = 211$

$P_c = 255$

$T_c = 720°F$

$\omega = 0.54$

Use PR EOS to determine compressibility factor of the aforementioned mixture.

Solution

At first, we should determine the parameters of the PR EOS using Eqs. (7.26)–(7.31) as shown in the following table.

Component	a_i	b_i	m	α_i	a_T
C_1	9311.769	0.430087	0.39561	1.226884	11,424.46
C_2	22,533.6	0.64978	0.528	1.084925	24,447.25
C_3	37,913.13	0.902635	0.60808	0.98324	37,277.7
C_4	58,454.58	1.210712	0.655969	0.88915	51,974.92
C_5	77,077.13	1.445085	0.717205	0.806075	62,129.97
C_6	99,003.12	1.700916	0.748296	0.731407	72,411.57
C_{7+}	287,453.4	3.862968	1.136244	0.36355	104,503.8

Using mixing rule we have:

$$b = \sum_{i=1}^{3} y_i b_i = 0.496364$$

Assuming $K_{ij} = 0$,

$$a_T = \sum_{i=1} \sum_{j=1} y_i y_j (a_{Ti} a_{Tj})^{0.5} (1 - K_{ij}) = 13912.71$$

(Continued)

$$A = \frac{\alpha a p}{(RT)^2} = \frac{13{,}912.71 \times 3720}{(10.73 \times (188 + 460))^2} = 1.070546$$

$$B = \frac{bp}{RT} = \frac{(0.496364 \times 3720)}{10.73 \times (188 + 460)} = 0.265564$$

$$Z^3 - 0.73444Z^2 + 0.327847Z - 0.195045 = 0.$$

Solving this equation yields:
$Z_1 = 0.676179$ → Accepted compressibility factor
$Z_2 = 0.0291307 + 0.536287i$ → Rejected
$Z_2 = 0.0291307 - 0.536287i$ → Rejected

7.6.3 Density

The density is defined as the mass per unit volume of the substance at any pressure and temperature. Two main categories can be used to predict density of gas mixtures, specifically gas condensates including empirical correlations and EOSs which are explained in the next sections.

7.6.3.1 Empirical Correlations

7.6.3.1.1 Nasrifar and Moshfeghian

Nasrifar and Moshfeghian (1998) proposed a saturated-liquid density correlation in conjunction with EOSs. The correlation in its generalized form is formulated as follows:

$$\frac{\rho}{\rho_c} = 1 + d_1\varphi^{1/3} + d_2\varphi^{2/3} + d_3\varphi + d_4\varphi^{4/3} \tag{7.173}$$

in which

$$\varphi = 1 - \frac{T_r}{\alpha(T_r)} \tag{7.174}$$

in which $d_1 = 1.1688$, $d_2 = 1.8177$, $d_3 = -2.6581$, $d_4 = 2.1613$. The parameter T_r is the reduced temperature and α denotes the α function from any EOS. Eq. (7.178) is extended to mixtures by the following mixing rules:

$$T_c = \sum_j^n x_j T_{c,j} \tag{7.175}$$

$$\alpha = \sum_{i=1}^{n}\sum_{j=1}^{n} x_i x_j \sqrt{\alpha_i \alpha_j} \tag{7.176}$$

$$\rho_c = \left[\sum_{j}^{n} x_j \rho_{c,j}^{-3/4}\right]^{-4/3} \tag{7.177}$$

7.6.3.2 Equation of State

To determine the gas-mixture density including gas phase and liquid phase, the following steps should be followed:

Step 1: Determine critical properties of the gas mixture, calculating reduced pressure and temperature of the mixture, and using random mixing rule to determine EOS parameters.
Step 2: Rearrange the EOS in terms of compressibility factor (Z)
Step 3: Determine the roots of cubic form of EOS (Z_v and Z_l)
Step 4: Calculate the apparent molecular weight for both gas and liquid phases as follows:

$$\text{For gas:} \quad M_a = \sum y_i M_i \tag{7.178}$$

$$\text{For liquid phase:} \quad M_a = \sum x_i M_i \tag{7.179}$$

Step 5: Use the following equation for determining both liquid and gas density

$$\rho = \frac{PM_a}{RTZ} \tag{7.180}$$

It should be noted that, based on the formulation of each EOS, the mixing rules and parameters may vary.

Example 7.6

Consider Example 7.3, using PR EOS, calculate density of liquid and gas phases which are at equilibrium in the container.

(Continued)

Solution

By solving the cubic form of PR EOS in terms of compressibility factor, we have:

$Z_1 = 0.05343 \rightarrow$ Liquid-compressibility factor
$Z_2 = 0.1648455 \rightarrow$ Rejected
$Z_3 = 0.749748 \rightarrow$ Gas-compressibility factor

Using Eq. (7.185) for both liquid and gas phases, we have:

$$\rho_L = \frac{PM_a}{RTZ_L} = \frac{195.1 \times 58.123}{0.05443 \times 10.73 \times 660} = 29.97\ \text{lbm/ft}^3$$

$$\rho_g = \frac{PM_a}{RTZ_g} = \frac{195.1 \times 58.123}{0.749748 \times 10.73 \times 660} = 2.1357\ \text{lbm/ft}^3$$

Example 7.7

Consider gas-condensate fluid with the following composition. The equilibrium pressure and temperature are 3500 psi and 180°F, respectively. The composition of gas-condensate fluid is reported in the following table:

Component	Mole Fraction	P_c	T_c	ω_i
C_1	0.67	666.4	343.33	0.0104
C_2	0.13	706.5	549.92	0.0979
C_3	0.03	616.0	666.06	0.1522
C_4	0.01	527.9	765.62	0.1852
C_5	0.02	488.6	845.8	0.2280
C_6	0.05	453	923	0.2500
C_{7+}	0.09	285	1210	0.5700

The heptane-plus fraction has the following properties:

$MW_{C_{7+}} = 227$
$P_c = 285$
$T_c = 750°F$
$\omega = 0.57$

Use PR EOS to determine gas density of the aforementioned mixture.

Solution

At first, we should determine the parameters of PR EOS using Eqs. (7.26)–(7.31) as shown in the following table:

Component	a_i	b_i	m	α_i	a_T
C_1	9311.769	0.430087	0.39561	1.222912	11,387.48
C_2	22,533.6	0.64978	0.528	1.07862	24,305.2
C_3	37,913.13	0.902635	0.60808	0.975637	36,989.45
C_4	58,454.58	1.210712	0.655969	0.880792	51,486.33
C_5	77,077.13	1.445085	0.717205	0.796934	61,425.4
C_6	99,003.12	1.700916	0.748296	0.72192	71,472.36
C_{7+}	270,439.1	3.544213	1.136244	0.329371	89,074.84

Using mixing rule we have:

$$b = \sum_{i=1}^{3} y_i b_i = 0.844742$$

Assuming $K_{ij} = 0$,

$$a_T = \sum_{i=1} \sum_{j=1} y_i y_j (a_{Ti} a_{Tj})^{0.5} (1 - K_{ij}) = 21021.52$$

$$A = \frac{\alpha a p}{(RT)^2} = \frac{13{,}912.71 \times 3500}{(10.73 \times (180 + 460))^2} = 1.032572$$

$$B = \frac{bp}{RT} = \frac{(0.496364 \times 3500)}{10.73 \times (180 + 460)} = 0.252981$$

$$Z^3 - 0.74702Z^2 + 0.33461Z - 0.181031 = 0.$$

Solving this equation yields:

$Z_1 = 0.657076 \rightarrow$ Accepted compressibility factor

$Z_2 = 0.044972 + 0.52296i \rightarrow$ Rejected

$Z_2 = 0.044972 - 0.52296i \rightarrow$ Rejected

Then we should determine the apparent molecular of gas phase as follows:

$$M_a = \sum y_i M_i = 42.72920$$

Finally, using Eq. (7.180) results in:

$$\rho_g = \frac{PM_a}{RTZ_g} = \frac{3500 \times 42.7292}{0.657076 \times 10.73 \times 640} = 33.14343 \text{ lbm/ft}^3$$

7.6.4 Formation Volume Factor

Gas-volume factor, B_g, is defined as the ratio of gas volume at reservoir conditions ($P_{reservoir}$ and $T_{reservoir}$) to the ideal-gas volume at standard conditions,

$$B_g = \left(\frac{P_{sc}}{T_{sc}}\right)\frac{ZT}{P} \tag{7.181}$$

Substituting values of standard pressure and temperature ($P_{sc} = 14.7$ psia and $T_{sc} = 520°R$) results:

$$B_g = 0.02827\frac{ZT}{P} \tag{7.182}$$

in which T stands for the reservoir temperature (°R) and P denotes the reservoir pressure (psi). This definition of B_g presumes that the gas volume at P and T remains as a gas in standard circumstances. For gas condensates and wet gases, the surface gas will not comprise all the original gas mixture because liquid is produced after separation. For these mixtures, the conventional definition of B_g may still be helpful; however, we denote to this quantity as a theoretical wet-gas volume factor, B_{gw}, which is determined by Eq. (7.181).

Example 7.8

Consider the gas mixture in Example 7.7; determine the formation volume factor at reservoir condition. Reservoir pressure and temperature are 3660 psi and 190°F, respectively.

Solution

As explained in Example 7.7, solving EOS in terms of compressibility factor yields:

$Z_1 = 0.657076 \rightarrow$ Accepted compressibility factor

$Z_2 = 0.044972 + 0.52296i \rightarrow$ Rejected

$Z_2 = 0.044972 - 0.52296i \rightarrow$ Rejected

Finally, using Eq. (7.182) results in:

$$B_g = 0.02827\frac{ZT}{P} = 0.02827 \times \frac{0.657076 \times 650}{3660} = 0.003299$$

7.6.5 Equilibrium Ratio

For real hydrocarbon mixtures, the equilibrium ratios are a function of the composition of the hydrocarbon mixture, temperature, and pressure of the system. This phenomenon can be expressed as follows:

$$K_i = K(P, T, Z_i)$$

Various approaches were suggested for estimating the equilibrium ratios of hydrocarbon mixtures. These approaches range from a straightforward mathematical equation to a complex equation including different compositional dependent parameters. Several useful approaches for predicting equilibrium ratio for both hydrocarbon and nonhydrocarbon mixtures are explained in the following sections.

7.6.5.1 Equilibrium Ratio for Hydrocarbon Mixtures

7.6.5.1.1 Wilson's Correlation

Wilson (1968) developed a straightforward thermodynamic equation for predicting *K*-values. The mathematical formulation of Wilson correlation is as follows:

$$K_i = \frac{P_{ci}}{P}\exp\left[5.37(1+\omega_i)\left(1-\frac{T_{ci}}{T}\right)\right] \tag{7.183}$$

in which P denotes the pressure of the system (psi), P_{ci} stands for the critical pressure of component i, (psi), T stands for the temperature of the system (°R), and T_{ci} represents the critical temperature of component i (°R). At low system pressure, this correlation produces logical values for the equilibrium ratio.

7.6.5.1.2 Standing's Correlation

Several scholars (Hoffmann et al., 1953; Brinkman and Sicking, 1960; Kehn, 1964; Dykstra and Mueller, 1965) pointed out that any nonhydrocarbon and/or pure hydrocarbon fluid could be exclusively described by merging its critical temperature, critical pressure, and boiling-point temperature into a representative variable, which is determined via the following equation:

$$F_i = b_i\left[\frac{1}{T_{bi}}-\frac{1}{T}\right] \tag{7.184}$$

in which

$$b_i = \frac{\log(P_{ci}/14.7)}{\left[\dfrac{1}{T_{bi}}-\dfrac{1}{T}\right]} \tag{7.185}$$

in which T_{bi} represents the normal boiling point of component i (°R) and F_i stands for the component characterization factor.

Standing (1979) proposed a series of formulations which fit the equilibrium ratio data reported by Katz and Hachmuth (1937) at temperatures below 200°F and pressures less than 1000 psi, which are mainly proper for surface-separator circumstances. The suggested formulation of the correlation is based on the plots of log (K_iP) versus F_i at a certain pressure often form straight lines with an intercept of a and slope of c. The basic expression of the straightline equation is as follows:

$$\log(K_iP) = a + cF_i \tag{7.186}$$

Rearranging for calculating equilibrium ratio K_i results in

$$K_i = \frac{1}{P}10^{(a+cF_i)} \tag{7.187}$$

Standing made an attempt to correlate the coefficients a and c with the pressure from six isobar plots of log (K_iP) versus F_i for 18 equilibrium ratio values as follows:

$$a = 1.2 + 0.00045P + 15\left(10^{-8}\right)P^2 \tag{7.188}$$

$$c = 0.89 - 0.00017P - 3.5\left(10^{-8}\right)P^2 \tag{7.189}$$

Standing mentioned that the estimated values of the equilibrium ratios of CO_2, N_2, H_2S, and C_1 through C_6 can be enhanced significantly by modifying the boiling point of these elements and the correlating parameter, b_i. Standing reported the modified values for boiling point and b_i in Table 7.7.

Katz and Hachmuth (1937) proposed a rule of thumb for calculating the equilibrium ratio for C_{7+} in which the K-value for C_{7+} is equal to 15% of the K of C_7 as follows:

$$K_{C_{7+}} = 0.15K_{C_7} \tag{7.190}$$

Standing proposed a substitute method for calculating the K-value of the heptanes and heavier fractions. Standing proposed the flowchart for calculating the parameters b and T_b of the heptane-plus fraction and consequently $F_{C_{7+}}$ as shown in Fig. 7.3.

Table 7.7 Modified Values of Boiling Point and Parameter b_i Proposed by Standing

Component	b_i	T_{bi} (°R)	Component	b_i	T_{bi} (°R)
N_2	470	109	i-C_5	2368	542
CO_2	652	194	n-C_5	2480	557
H_2S	1136	331	C_6	2738	610
C_1	300	94	n-C_6	2780	616
C_2	1145	303	n-C_7	3068	616
C_3	1799	416	n-C_8	3335	718
i-C_4	2037	471	n-C_9	3590	763
n-C_4	2153	491	n-C_{10}	3828	805

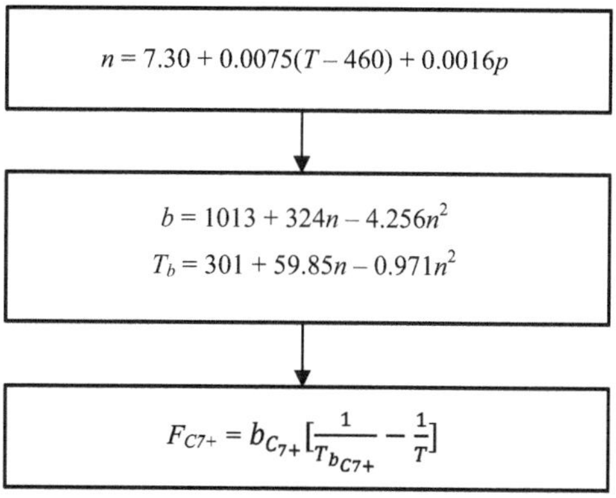

Figure 7.3 Flowchart proposed by Standing for calculation of $F_{C_{7+}}$.

It should be noted that based on experimental data obtained for the equilibrium ratio of carbon dioxide, K_{CO_2} can be roughly estimated by the following equation (Ahmed, 2007):

$$K_{CO_2} = \sqrt{K_{C_1} K_{C_2}} \tag{7.191}$$

in which K_{CO_2} stands for the equilibrium ratio of CO_2 at desired T and P, K_{C_1} represents the equilibrium ratio of methane at desired T and P, and K_{C_2} denotes the equilibrium ratio of ethane at desired T and P.

7.6.5.1.3 Whitson and Torp's Method

Whitson and Torp (1981) modified Wilson's correlation (Eq. 7.183) to predict precisely the equilibrium ratio at higher pressures. Wilson's correlation was revised by including the convergence pressure into Eq. (7.183) as follows:

$$K_i = \left(\frac{P_{ci}}{P_K}\right)^{A-1} \frac{P_{ci}}{P} \exp\left[5.37A(1+\omega_i)\left(1-\frac{T_{ci}}{T}\right)\right] \tag{7.192}$$

with

$$A = 1 - \left(\frac{P}{P_K}\right)^{0.7} \tag{7.193}$$

in which P_K represents the convergence pressure (psi), P stands for the system pressure (psi), ω_i stands for acentric factor of component i, and T denotes the system temperature (°R).

To calculate convergence pressure, two main methods can be used. These methods including Rzasa et al. (1952) and Standing (1977) are explained in the next paragraph.

Rzasa et al. (1952) presented a straightforward graphical method for estimating the convergence pressure of light hydrocarbons. They employed the product of the specific gravity and molecular weight of the heptane-plus fraction and the temperature as input variables. Their graphical method is formulated via the following correlation:

$$P_K = -2381.8542 + 46.31487\left[(M\gamma)_{C_{7+}}\right] + \sum_{i=1}^{3} a_i \left[\frac{(M\gamma)_{C_{7+}}}{T-460}\right]^i \tag{7.194}$$

in which T denotes the desired temperature (°R), $\gamma_{C_{7+}}$ represents the specific gravity of C_{7+} and $(M)_{C_{7+}}$ stands for the molecular weight of C_{7+}. Values of the constant coefficients of Eq. (7.194) are as follows:

$a_1 = 6124.3049$

$a_2 = -2753.2538$

$a_3 = 415.42049$

Standing (1977) recommended that the convergence pressure be approximately calculated via linear correlation with the molecular weight of the heptane-plus fraction. Whitson and Torp (1981) proposed the following equation for predicting the convergence pressure:

$$P_K = 60MW_{C_{7+}} - 4200 \tag{7.195}$$

in which $MW_{C_{7+}}$ stands for the molecular weight of the heptane-plus fraction.

7.6.5.2 Equilibrium Ratio for Nonhydrocarbon Mixtures

Lohrenze et al. (1963) proposed the following equations which estimate the values of the equilibrium ratio of the H_2S, N_2, and CO_2 as a function of temperature, pressure, and the convergence pressure, P_K.

For CO_2

$$\mathrm{Ln}(K_{CO_2}) = \left(1-\frac{P}{P_K}\right)^{0.6}\left[7.0201913 - \frac{152.7291}{T} - \mathrm{Ln}(P)\left(1.8896974 - \frac{1719.2956}{T} + \frac{644740.69}{T^2}\right)\right] \tag{7.196}$$

For H_2S

$$\mathrm{Ln}(K_{H_2S}) = \left(1-\frac{P}{P_K}\right)^{0.8}\left[6.3992127 + \frac{1399.2204}{T} - \mathrm{Ln}(P)\left(0.76885112 + \frac{18.215052}{T}\right) - \frac{1112446.2}{T^2}\right] \tag{7.197}$$

For N_2

$$\mathrm{Ln}(K_{N_2}) = \left(1 - \frac{P}{P_K}\right)^{0.4}\left[11.294748 - \frac{1184.2409}{T} - 0.90459907\,\mathrm{Ln}(P)\right] \tag{7.198}$$

in which P denotes the pressure of the system (psi), T stands for the temperature of the system (°R), and P_K stands for the convergence pressure (psi).

7.6.6 Dew-Point Pressure

To determine dew-point pressure of gas condensate different methods can be employed. Two main categories for predicting dew-point pressure of gas-condensate fluids are empirical correlations and EOSs. The next sections provide description of each methodology.

7.6.6.1 Empirical Correlations

This section presents empirical correlations for estimating dew-point pressure of the retrograded gas condensate.

7.6.6.1.1 Nemeth and Kennedy (1967)

Nemeth and Kennedy (1967) proposed a mathematical formulation relating dew-point pressure, composition, and temperature—which is formulated in terms of mole fraction of methane through C_{7+}, the molecular weight, non-hydrocarbon components, and specific gravity of the heptane-plus fraction. Their expression is as follows:

$$\begin{aligned}\mathrm{Ln}(P_d) = {} & A_1(Z_{C_2} + Z_{CO_2} + Z_{H_2S} + Z_{C_6} + 2(Z_{C_3} + Z_{C_4}) + Z_{C_5} \\ & + 0.4Z_{C_1} + 0.2Z_{N_2}) + A_2\mathrm{SG}_{C_{7+}} + A_3\left(\frac{Z_{C_1}}{Z_{C_{7+}} + 0.002}\right) + A_4 T \\ & + A_5(Z_{C_{7+}}\mathrm{MW}_{C_{7+}}) + A_6(Z_{C_{7+}}\mathrm{MW}_{C_{7+}})^2 + A_7(Z_{C_{7+}}\mathrm{MW}_{C_{7+}})^3 \\ & + A_8\left[\frac{\mathrm{MW}_{C_{7+}}}{\mathrm{SG}_{C_{7+}} + 0.001}\right] + A_9\left[\frac{\mathrm{MW}_{C_{7+}}}{\mathrm{SG}_{C_{7+}} + 0.001}\right]^2 \\ & + A_{10}\left[\frac{\mathrm{MW}_{C_{7+}}}{\mathrm{SG}_{C_{7+}} + 0.001}\right]^3 + A_{11}\end{aligned} \tag{7.199}$$

Table 7.8 Values of the Nemeth and Kennedy (1967) Correlation Constants

Coefficient	Value
A_1	-2.0623054
A_2	6.6259728
A_3	$-4.4670559 \times 10^{-3}$
A_4	$-1.0448346 \times 10^{-4}$
A_5	3.2673714×10^{-2}
A_6	$-3.6453277 \times 10^{-3}$
A_7	$-7.4299951 \times 10^{-5}$
A_8	$-1.1381195 \times 10^{-1}$
A_9	$-6.2476497 \times 10^{-4}$
A_{10}	$-1.1381195 \times 10^{-1}$
A_{11}	-1.0746622×10

in which Z_i represents mole fraction of gas components ($i = C_1$ through C_{7+}, nonhydrocarbons including CO_2, N_2, and H_2S), $MW_{C_{7+}}$ stands for molecular weight of C_{7+}, and $SG_{C_{7+}}$ denotes specific gravity of C_{7+}. Values of Nemeth and Kennedy (1967) approach parameters are reported in Table 7.8.

Example 7.9

Consider a gas condensate with composition given as follows (Sage and Olds, 1947). Predict the dew-point pressure of this condensate fluid by Nemeth and Kennedy method. The reservoir temperature is 40°F.

Component	Mol%
N_2	0.0
H_2S	0.0
CO_2	0.0
C_1	0.8238
C_2	0.0428
C_3	0.0351
i-C_4	0.0161
n-C_4	0.0303
i-C_5	0.0060
n-C_5	0.0068
C_6	0.009
C_{7+}	0.0292

$MW_{C_{7+}} = 125$

$\gamma_{C_{7+}} = 0.74$

Solution

Input the mole fractions of hydrocarbon and nonhydrocarbon components along with heptane-plus molecular weight and specific gravity in Eq. (7.199) results:

$P_d = 2823$ psi

7.6.6.1.2 Elsharkawy (2002)

A model for gas-condensate dew-point pressure estimation was presented by Elsharkawy (2002). The correlation was proposed employing laboratory measurements from 340 data samples covering a broad range of properties and was a function of reservoir temperature and routinely measured gas analysis. It contains 19 terms, correlating dew-point pressure with reservoir composition of nonhydrocarbons, and hydrocarbons expressed as mole fraction, reservoir temperature, molecular weight of C_{7+}, and specific gravity of C_{7+}.

$$\begin{aligned} P_d = {} & A_0 + A_1 T + A_2 Z_{H_2S} + A_3 Z_{CO_2} + A_4 Z_{N_2} \\ & + A_5 Z_{C_1} + A_6 Z_{C_2} + A_7 Z_{C_3} + A_8 Z_{C_4} + A_9 Z_{C_5} \\ & + A_{10} Z_{C_6} + A_{11} Z_{C_{7+}} + A_{12} \mathrm{MW}_{C_{7+}} + A_{13} \mathrm{SG}_{C_{7+}} \\ & + A_{14}\left(Z_{C_{7+}} \mathrm{MW}_{C_{7+}}\right) + A_{15}\left(\frac{\mathrm{MW}_{C_{7+}}}{\mathrm{SG}_{C_{7+}}}\right) \\ & + A_{16}\left(\frac{Z_{C_{7+}} \mathrm{MW}_{C_{7+}}}{\mathrm{SG}_{C_{7+}}}\right) + A_{17}\left(\frac{Z_{C_{7+}}}{Z_{C_1} + Z_{C_2}}\right) \\ & + A_{18}\left(\frac{Z_{C_{7+}}}{Z_{C_3} + Z_{C_4} + Z_{C_5} + Z_{C_6}}\right) \end{aligned} \tag{7.200}$$

in which Z_i represents the mole fraction of gas components ($i = C_1$ through C_{7+}, nonhydrocarbons including CO_2, N_2, and H_2S); T stands for the reservoir temperature; $\mathrm{MW}_{C_{7+}}$ represents molecular weight of C_{7+}; and $\mathrm{SG}_{C_{7+}}$ denotes specific gravity of C_{7+}. Values of Elsharkawy (2002) approach parameters are reported in Table 7.9.

Table 7.9 Values of the Parameters in Elsharkawy (2002) Correlation

Coefficient	Value	Coefficient	Value
A_0	4268.85	A_{10}	691.5298
A_1	0.094056	A_{11}	40660.36
A_2	−7157.87	A_{12}	205.26
A_3	−4540.58	A_{13}	−7260.32
A_4	−4663.55	A_{14}	−352.413
A_5	−1357.56	A_{15}	−114.519
A_6	−7776.10	A_{16}	8.13300
A_7	−9967.99	A_{17}	94.916
A_8	−4257.10	A_{18}	238.252
A_9	−1417.10		

Example 7.10

Consider a gas condensate with composition given as follows (Al-Mahroos and Tejoa, 1987). Predict the dew-point pressure of this condensate fluid by Elsharkawy method. The reservoir temperature is 337°F.

Component	Mol%
N_2	0.1171
H_2S	0.0005
CO_2	0.0650
C_1	0.7906
C_2	0.0162
C_3	0.0035
$i\text{-}C_4$	0.0008
$n\text{-}C_4$	0.0010
$i\text{-}C_5$	0.0004
$n\text{-}C_5$	0.0004
C_6	0.0006
C_{7+}	0.0039
$MW_{C_{7+}} = 161.9$	
$\gamma_{C_{7+}} = 0.8$	

Solution

Input the reservoir temperature, mole fractions of hydrocarbon and nonhydrocarbon components, heptane-plus molecular weight, and specific gravity in Eq. (7.200). Elsharkawy (2002) gives dew-point pressure as follows:

$$P_d = 6541 \text{ psi}$$

7.6.6.1.3 Humoud and Al-Marhoun (2001)

Based on field and laboratory PVT data of several gas-condensate fluid samples from different Middle Eastern reservoirs, Humoud and Al-Marhoun (2001) developed a correlation for dew-point pressure of gas-condensate fluids. They assumed a direct relationship between dew-point pressure and pseudoreduced temperature and pressure, reservoir temperature, primary separator gas—oil ratio, the primary separator temperature and pressure, heptane-plus fraction, and relative densities of separator gas. Their proposed correlation is as follows:

$$\mathrm{Ln}(P_d) = \beta_0 + \beta_1 \,\mathrm{Ln}(T) + \beta_2 \,\mathrm{Ln}(R_m) + \beta_3 \,\mathrm{Ln}\left(P_{sp} T_{sp}\right) + \frac{\beta_4}{T_{pr}} + \frac{\beta_5}{P_{pr}} + \frac{\beta_6}{\gamma_{C_{7+}}} \tag{7.201}$$

$$R_m = \frac{R_{sp}\gamma_{gsp}}{\gamma_{C_{7+}}} \tag{7.202}$$

$$\beta_0 = 43.777183, \quad \beta_1 = -3.594131, \quad \beta_2 = -0.247436,$$
$$\beta_3 = -0.053527, \quad \beta_4 = -4.291404, \quad \beta_5 = -3.698703,$$
$$\beta_6 = -4.590091$$

in which T, reservoir temperature (°F), T_{sp}, primary separator temperature (°F), P_{sp}, primary separator pressure (psi), T_{pr}, pseudo-reduced temperature, P_{pr}, pseudo-reduced pressure, $\gamma_{C_{7+}}$, specific gravity of C_{7+}, γ_{gsp}, separator gas specific gravity, R_{sp}, gas—oil ratio (Scf/STB), $MW_{C_{7+}}$, molecular weight of C_{7+}, $SG_{C_{7+}}$, specific gravity of C_{7+}.

It is worth mentioning that when the compositions of the fluid are not available, the following expressions in terms of gas specific gravity should be used to predict the pseudocritical pressure and temperature.

$$P_{pc} = 694.5 - 55.3\gamma_g \tag{7.203}$$

$$T_{pc} = 208.5 + 213.7\lambda_g \tag{7.204}$$

7.6.6.1.4 Alternating Conditional Expectations

Al-Dhamen in (2010) developed a new correlation based on nonparametric model called Alternating Conditional Expectations (ACE) to estimate dew-point pressure in a retrograde gas-condensate reservoir. The ACE method produces new transformation functions from the independent and dependent parameters. In general, the ACE method better estimates than do the classical methods. Al-Dhamen's developed correlation is as follows:

$$P_d = e^{C_1^{T(P_d)^2} + C_2^{T(P_d)^2} + C_3^{T(P_d)} + C_4} \quad (7.205)$$

in which

$$T(P_d) = \mathrm{Ln}\left[T(T_R) + T(\mathrm{GOR}) + T\left(\gamma_g\right) + T(\gamma_{\mathrm{cond}}) + 10\right] \quad (7.206)$$

$$T(T_R) = p_1 T_R^3 + p_2 T_R^2 + p_3 T_R + p4 \quad (7.207)$$

$$T(\mathrm{GOR}) = r_1 \,\mathrm{Ln}(\mathrm{GOR}) + r_2 \quad (7.208)$$

$$T\left(\gamma_g\right) = q_1 \gamma_g^2 + q_2 \gamma_g + q_3 \quad (7.209)$$

$$T(\gamma_{\mathrm{cond}}) = s_1 \gamma_{\mathrm{cond}}^3 + s_2 \gamma_{\mathrm{cond}}^2 + s_3 \gamma_{\mathrm{cond}} + s_4 \quad (7.210)$$

where

$C_1 = 49.1377,\ C_2 = -336.5699,\ C_3 = 770.0995,\ C_4 = -580.0322,$
$p_1 = -0.35014 \times 10^{-6},\ p_2 = 0.18048 \times 10^{-3},\ p_3 = -0.32315 \times 10^{-1},$
$p_4 = 1.2058, r_1 = -0.3990,\ r_2 = 5.1377,\ q_1 = -23.8741,$
$q_2 = 36.9448, q_3 = -12.0398, s_1 = -30120.78,$
$s_2 = 69{,}559, s_3 = -53484.21,\ s_4 = 13689.39.$

in which γ_{cond} denotes the condensate specific gravity, GOR represents the gas–oil ratio (Scf/STB), T_R denotes the reservoir temperature (°F), γ_g stands for the gas specific gravity, and P_d stands for the dew-point pressure (psi).

7.6.6.1.4.1 Marruffo–Maita–Him–Rojas (2002) Marruffo et al. (2002) developed an equation to estimate the dew-point pressure. Their correlation correlates dew-point pressure to C_{7+} content as mole fraction, gas–condensate ratio, and reservoir temperature. Furthermore, a model was proposed to predict C_{7+} content from specific separator gas gravity and gas–condensate ratio.

$$P_d = K_1 \left[\frac{\mathrm{CGR}^{K2}}{\%\mathrm{C}_{7+}^{K3}} \times K_8 \times \mathrm{API}^{\left(K_4 \times T_R^{K_5} - K_6 \times C_{7+}^{K_7}\right)}\right] \quad (7.211)$$

in which

$$K_1 = 346.7764689, \quad K_2 = 0.0974139, \quad K_3 = -0.294782419,$$
$$K_4 = -0.047833243, \quad K_5 = 0.281255219, \quad K_6 = 0.00068358,$$
$$K_7 = 1.906328237, \quad K_8 = 8.4176216$$

$$\%C_{7+} = \left(\frac{\text{GCR}}{70,680}\right)^{-0.8207} \tag{7.212}$$

in which P_d denotes the dew-point pressure (psi), GCR stands for the gas-to-condensate ratio [standard cubic feet per stock tank barrel (Scf/STB)], T_R represents the reservoir temperature (°R), and API denotes American Petroleum Institute condensate gravity.

Example 7.11

Consider a gas-condensate reservoir with the following composition (Humoud and Al-Marhoun, 2001). Determine the dew-point pressure of this reservoir via the Humoud and Al-Marhoun method.

Reservoir Pressure = 7630 psi
Reservoir Temperature = 282°F
Separator Pressure = 795 psi
Separator Temperature = 110°F
R_{sp} = 13,000 Scf/STB
$MW_{C_{7+}}$ = 144
$\gamma_{C_{7+}}$ = 0.7923
γ_{gsp} = 0.7399

Component	Mol%
N_2	10.51
H_2S	2.28
CO_2	1.71
C_1	68.44
C_2	6.68
C_3	3.01
i-C_4	0.58
n-C_4	1.21
i-C_5	0.46
n-C_5	0.52
C_6	0.67
C_{7+}	3.93

(Continued)

Solution 1

In this solution, we assume that the composition of the condensate gas is available as reported in the previous table. At first, we should calculate the pseudocritical pressure and temperature. Using the following equations, we have:

$$P_{pc} = \sum y_i P_{pci} = 639.8 \text{ psi}$$
$$T_{pc} = \sum y_i P_{pci} = 404.9°\text{R}$$

Then, using Eqs. (7.129) and (7.130) results in the pseudoreduced temperature and pressure are as follows:

$$P_{pr} = 11.926 \quad T_{pr} = 1.833$$

Using Eq. (7.201) for predicting dew-point pressure results in:

$$P_d = 5188 \text{ psi}$$

Solution 2

In this step, we assume that the composition of gas condensate is not available; however, specific gas gravity is available.

The reservoir gas specific gravity for this example is:

$$\gamma_{gR} = 0.9218$$

Using the correlation proposed by Humoud and Al-Marhoun for predicting pseudocritical pressure and temperature based on the reservoir specific gravity results in:

$$T_{pc} = 405.5°\text{R} \quad P_{pc} = 643.5 \text{ psi}$$

Using Eqs. (7.129) and (7.130) for calculating pseudoreduced temperature and pressure results in:

$$T_{pr} = 1.83 \quad P_{pr} = 11.857$$

Using the equation proposed by Humoud and Al-Marhoun for prediction of dew-point pressure results in:

$$P_d = 5158 \text{ psi}$$

7.6.6.2 *Iterative Method*

The pressure at which a large quantity of gas is in equilibrium with a negligible quantity of liquid is named the dew-point pressure (P_d) of a

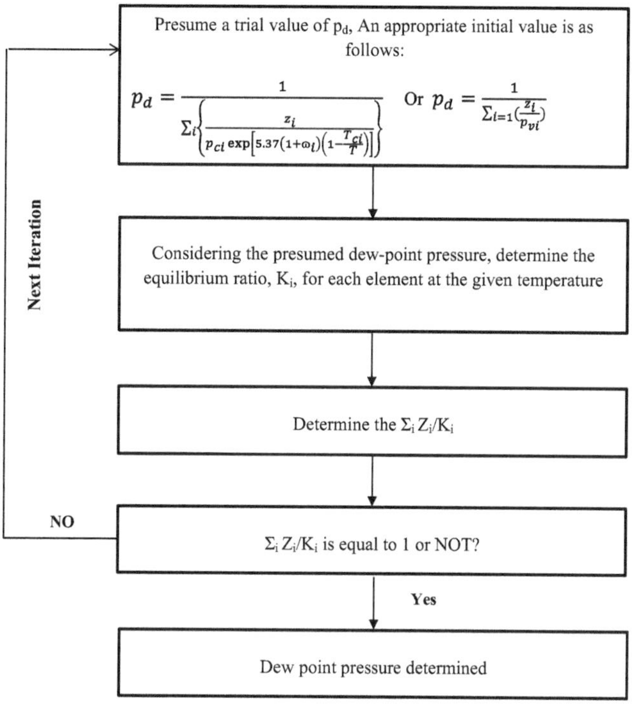

Figure 7.4 Flowchart for predicting dew-point pressure by iterative equilibrium ratio method.

hydrocarbon system. For 1 lb mol of a hydrocarbon mixture, i.e., $n = 1$, at the dew-point pressure, we have the following conditions:

$n_l \approx 0$

$n_v \approx 1$

At aforementioned circumstances, the overall composition, z_i, and the composition of the vapor phase, y_i, are the same. Performing these limitations to Eqn (5.14) results

$$\sum_i \frac{z_i}{n_v K_i + n_l} = \sum_i \frac{z_i}{K_i} = 1 \tag{7.213}$$

in which z_i = total composition of the system under consideration.

A trial and-error method should use to determine the dew-point pressure, P_d. Following flow chart (Fig. 7.4) demonstrates the process of iterative method for predicting dew-point pressure.

Example 7.12

Consider a gas condensate with composition given as follows. Calculate the dew-point pressure of the gas-condensate fluid with the iterative equilibrium ratio method. Reservoir temperature is 220°F.

The heptane-plus fraction has the following properties:

$MW_{C_{7+}} = 153.33$

$P_c = 290$

$T_c = 790°F$

$\omega = 0.53$

Component	Mole Fraction	P_c	T_c	ω_i
C_1	0.60	666.4	343.33	0.0104
C_2	0.04	706.5	549.92	0.0979
C_3	0.04	616.0	666.06	0.1522
C_4	0.02	527.9	765.62	0.1852
C_5	0.04	488.6	845.8	0.2280
C_6	0.06	453	923	0.2500
C_{7+}	0.20	290	1250	0.5300

Solution

At first, we should determine the convergence pressure. Using Eq. (7.195) results in the convergence pressure:

$$P_K = 5000 \text{ psi}$$

Next, we should assume P_d and calculate K_i at the assumed pressure using Eqs. (7.183) and (7.190). Then, calculate Z_i/K_i until the summation is equal to 1.

Component	z_i	K_i at $P = 2800$ psi	Z_i/K_i	K_i at $P = 4000$ psi	Z_i/K_i	K_i at $P = 4875$ psi	Z_i/K_i
C_1	0.9532	2.23375564	0.426725	1.377383	0.692037	1.037803	0.918479
C_2	0.0168	1.354242388	0.012405	1.108771	0.015152	1.010812	0.01662
C_3	0.0091	0.926428569	0.009823	0.940518	0.009676	0.990805	0.009184
C_4	0.0059	0.645608431	0.009139	0.804227	0.007336	0.972142	0.006069
C_5	0.0027	0.480743941	0.005616	0.707732	0.003815	0.957165	0.002821
C_6	0.0025	0.360055452	0.006943	0.624377	0.004004	0.942706	0.002652
C_{7+}	0.0068	0.072273416[a]	0.094087	0.311272[a]	0.021846	0.137482[a]	0.049461
Σ			0.564739		0.753865		1.005287

[a] Eq. (7.191).

After three iterations, we have $P_d = 4875$ psi.

Example 7.13

Consider a sour-gas condensate with composition given as follows. Calculate the dew-point pressure of the gas-condensate fluid with the iterative equilibrium ratio method. Reservoir temperature is 240°F.

Component	z_i	P_c	T_c	ω_i
CO_2	0.0031	1071	547.9	0.225
N_2	0.0005	493	227.6	0.040
C_1	0.9302	666.4	343.33	0.0104
C_2	0.0068	706.5	549.92	0.0979
C_3	0.0098	616.0	666.06	0.1522
C_4	0.0169	527.9	765.62	0.1852
C_5	0.0139	488.6	845.8	0.2280
C_6	0.0137	453	923	0.2500
C_{7+}	0.0051	290	1250	0.5300

The heptane-plus fraction has following properties:

$MW_{C_{7+}} = 153.33$

$P_c = 290$

$T_c = 790°F$

$\omega = 0.53$

Solution

At first, we should determine the convergence pressure. Using Eq. (7.195) results in the convergence pressure:

$P_K = 5000$ psi

Next, we should assume P_d and calculate K_i at the assumed pressure using Eqs. (7.183) and (7.190). Then, calculate the Z_i/K_i until the summation is equal to 1. It should be noted that for calculating equilibrium ratio of nonhydrocarbon gases (CO_2 and N_2) we should use Eqs. (7.196) and (7.197).

(Continued)

Component	z_i	K_i at $P = 3000$ psi	Z_i/K_i	K_i at $P = 3750$ psi	Z_i/K_i	K_i at $P = 4475$ psi	Z_i/K_i
CO_2	0.0031	0.211808101[a]	0.014636	0.100694[a]	0.030786	0.03237[a]	0.095767
N_2	0.0005	0.436222366[b]	0.001146	0.234634[b]	0.002131	0.086748[b]	0.005764
C_1	0.9302	2.087767882	0.445548	1.528605	0.608529	1.181655	0.787201
C_2	0.0068	1.35329153	0.005025	1.175053	0.005787	1.060953	0.006409
C_3	0.0098	0.971909314	0.010083	0.961243	0.010195	0.977164	0.010029
C_4	0.0169	0.708612592	0.023849	0.793565	0.021296	0.90337	0.018708
C_5	0.0139	0.548120818	0.025359	0.67907	0.020469	0.847515	0.016401
C_6	0.0137	0.425727357	0.03218	0.582548	0.023517	0.795928	0.017213
C_{7+}	0.0051	0.039873437[c]	0.127905	0.065664[c]	0.077668	0.106202[c]	0.048022
Σ			0.685731		0.800379		1.005513

[a] Eq. (7.197).
[b] Eq. (7.199).
[c] Eq. (7.191).

After three iterations, we have $P_d = 4475$ psi.

Problems

7.1 A retrograde gas condensate with following composition is existed in a reservoir. However, the discovery pressure in the reservoir is 6998 psi, and higher than its dew-point pressure, 5990 psi. Reservoir temperature is 256°F. Determine compressibility factor via:

a) Sanjari–Nemati Lay method
b) Peng–Robinson EOS

Component	Mole Fraction
C_1	0.5904
C_2	0.0864
C_3	0.0534
n-C_4	0.0338
n-C_5	0.0178
C_6	0.0173
C_{7+}	0.2009

$\gamma_{C_{7+}} = 0.8325$
$MW_{C_{7+}} = 167$

7.2 Consider following gas composition. Determine the value of formation volume factor at the reservoir conditions of 4800 psi and 268°F.

Component	Mol%
H_2S	10.9
CO_2	24.69
C_1	31.80
C_2	4.90
C_3	2.05
n-C_4	1.67
n-C_5	1.02
C_6	1.72
C_{7+}	21.25

$\gamma_{C_{7+}} = 0.8475$
$MW_{C_{7+}} = 147$

7.3 A gas condensate has composition at given as follows.

a) Determine the value of compressibility factor at reservoir conditions of 5645 psi and 278°F via Patel–Teja EOS.

b) Determine the value of compressibility factor for this gas at 5500 psi and 256°F via Peng–Robinson EOS.

c) Determine the value of viscosity for this gas at 5324 psi and 263°F via:

c-1) Shokir and Dmour Method

c-2) Lee–Gonzalez–Eakin Method

Component	Mol%
H_2S	8.90
CO_2	14.69
C_1	31.80
C_2	9.90
C_3	3.05
n-C_4	2.67
n-C_5	1.02
C_6	1.72
C_{7+}	26.25

$\gamma_{C_{7+}} = 0.8337$
$MW_{C_{7+}} = 136.5$

7.4 Consider retrograde gas condensate with following composition. Determine the values of gas-compressibility factor, formation volume factor, and density of gas at pressure = 4200 psi, and temperature = 196°F.

Component	Mole Fraction
N_2	0.0046
H_2S	0.0
CO_2	0.0061
C_1	0.6864
C_2	0.139
C_3	0.0689
$i\text{-}C_4$	0.0066
$n\text{-}C_4$	0.0266
$i\text{-}C_5$	0.0062
$n\text{-}C_5$	0.0094
C_6	0.0114
C_{7+}	0.0348
$\gamma_{C_{7+}} = 0.7763$	
$MW_{C_{7+}} = 152.3$	

7.5 Consider a gas-condensate fluid with following composition. Determine the values of gas viscosity, and gas-compressibility factor at pressure = 4673 psi, and temperature = 205°F.
Gas viscosity should determine by:
a) Elsharkawy Method
b) Chen and Ruth Method
c) Dempsey's standing Method
d) Sutton Method
Gas-compressibility factor should calculate by:
a) Papay method
b) Bahadori et al. method
c) Shokir et al. method
d) SRK-SW EOS
e) Mohsen-Nia et al. (MMM) EOS

7.6 Consider a retrograde gas-condensate fluid with composition given as follows. Determine the dew-point pressure of this fluid by:
a) Nemeth and Kennedy
b) Alternating Conditional Expectations (ACE)
c) Iterative Equilibrium Ratio Method

Component	Mole Fraction
N_2	0.0054
H_2S	0.0053
CO_2	0.0
C_1	0.689
C_2	0.1364
C_3	0.0689
n-C_4	0.0332
n-C_5	0.0156
C_6	0.0112
C_{7+}	0.035
$\gamma_{C_{7+}} = 0.7763$	
$MW_{C_{7+}} = 152.3$	

Component	Mole Fraction
CO_2	0.0101
C_1	0.5556
C_2	0.2194
C_3	0.0699
n-C_4	0.0312
n-C_5	0.0136
C_6	0.0142
C_{7+}	0.086
$\gamma_{C_{7+}} = 0.8123$	
$MW_{C_{7+}} = 155.9$	

7.7 Consider the gas-condensate reservoir with following composition. Determine the values of compressibility factor and gas density via Peng–Robinson EOS. Moreover, predict the value of dew-point pressure via iterative equilibrium ratio. Reservoir pressure and temperature are 5110 psi and 238°F, respectively.

Component	Mole Fraction
H_2S	0.0107
C_1	0.659
C_2	0.1654
C_3	0.0699
n-C_4	0.0312
n-C_5	0.0176
C_6	0.0102
C_{7+}	0.036
$\gamma_{C_{7+}} = 0.7988$	
$MW_{C_{7+}} = 150.3$	

7.8 A retrograde gas condensate with following composition is existed in a reservoir. Reservoir temperature and pressure are 256°F and 6764 psi, respectively. Determine density of this retrograde gas via:

a) Nasrifar–Moshfeghian method
b) Nasrifar and Moshfeghian (NM) EOS
c) Schmidt and Wenzel EOS

Component	Mole Fraction
C_1	0.5904
C_2	0.0864
C_3	0.0534
$n\text{-}C_4$	0.0338
$n\text{-}C_5$	0.0178
C_6	0.0173
C_{7+}	0.2009
$\gamma_{C_{7+}} = 0.8325$	
$MW_{C_{7+}} = 167$	

7.9 Consider a gas-condensate reservoir with following properties. Determine the dew-point pressure of this reservoir via Humoud and Al-Marhoun method.

Reservoir Pressure = 6840 psi
Reservoir Temperature = 271.2°F
Separator Pressure = 685 psi
Separator Temperature = 132°F
$R_{sp} = 11{,}230$ Scf/STB
$MW_{C_{7+}} = 150.34$
$\gamma_{C_{7+}} = 0.7893$
$\gamma_{gsp} = 0.7221$

7.10 Consider a gas-condensate reservoir with following composition. Determine the dew-point pressure of this reservoir via

a) Alternating Conditional Expectations (ACE) method.
b) Soave–Redlich–Kwong EOS

Reservoir Pressure = 5982 psi
Reservoir Temperature = 252°F
Separator Pressure = 546 psi
Separator Temperature = 128°F

Component	Mol%
N_2	8.51
H_2S	4.28
CO_2	5.71
C_1	58.44
C_2	10.68
C_3	5.01
i-C_4	0.58
n-C_4	1.21
i-C_5	0.96
n-C_5	1.02
C_6	0.67
C_{7+}	2.93
$MW_{C_{7+}} = 140$	
$\gamma_{C_{7+}} = 0.7713$	

7.11 Consider a gas-condensate reservoir with following properties. Determine the dew-point pressure of this reservoir via Marruffo–Maita–Him–Rojas method.

Reservoir Pressure = 6654 psi
Reservoir Temperature = 272°F
Separator Pressure = 675 psi
Separator Temperature = 119°F
Condensate–Gas Ratio (CGR) = 10,000 Scf/STB
$\gamma_{gR} = 0.9218$
$\gamma_{gsp} = 0.7399$

7.12 Consider following gas composition. Determine the value of formation volume factor at the reservoir conditions of 3925 psi and 274°F.

Component	Mol%
N_2	14.69
H_2S	20.09
C_1	26.90
C_2	9.80
C_3	3.05
n-C_4	0.67
n-C_5	0.02
C_6	7.92
C_{7+}	16.05
$\gamma_{C_{7+}} = 0.7975$	
$MW_{C_{7+}} = 137$	

7.13 A wet gas with following composition is existed in a reservoir. Reservoir temperature and pressure are 270°F and 5900 psi, respectively. Determine density of this retrograde gas via Patel–Teja EOS and Schmidt and Wenzel EOS.

Component	Mole Fraction
C_1	0.3901
C_2	0.1865
C_3	0.0136
n-C_4	0.0537
n-C_5	0.0379
C_6	0.0176
C_{7+}	0.3006
$\gamma_{C_{7+}} = 0.8571$	
$MW_{C_{7+}} = 187$	

7.14 Consider a retrograde gas-condensate fluid with composition given as follows. Determine the dew-point pressure of this fluid by Peng–Robinson EOS and Patel–Teja EOS. Moreover, compare the results obtain from two EOSs.

Component	Mole Fraction
N_2	0.0101
C_1	0.4656
C_2	0.1094
C_3	0.0699
n-C_4	0.0212
n-C_5	0.0236
C_6	0.0142
C_{7+}	0.286
$\gamma_{C_{7+}} = 0.8327$	
$MW_{C_{7+}} = 166.6$	

7.15 A gas condensate has composition at given as follows.

a) Determine the value of compressibility factor at reservoir conditions of 6000 psi and 288°F via MMM EOS.

b) Determine the value of compressibility factor for this gas at 4950 psi and 240°F via Patel–Teja EOS.

c) Determine the value of viscosity for this gas at 4800 psi and 235°F via:

c-1) Lee–Gonzalez–Eakin Method

c-2) Sanjari–Nemati Lay–Peymani Method

Component	Mol%
H_2S	12.70
CO_2	10.89
C_1	35.90
C_2	5.80
C_3	1.35
n-C_4	1.37
n-C_5	1.02
C_6	1.72
C_{7+}	29.25
$\gamma_{C_{7+}} = 0.8411$	
$MW_{C_{7+}} = 141.25$	

7.16 Consider a retrograded gas condensate with following composition. The discovery pressure in the reservoir is 5502 psi and the reservoir temperature is 220°F. Calculate compressibility factor and density of the gas via Schmidt and Wenzel EOS.

Component	Mole Fraction
C_1	0.5304
C_2	0.1464
C_3	0.0334
n-C_4	0.0538
n-C_5	0.0078
C_6	0.0073
C_{7+}	0.2209
$\gamma_{C_{7+}} = 0.8501$	
$MW_{C_{7+}} = 174$	

REFERENCES

Adachi, Y., Lu, B.C.Y., Sugie, H., 1983. A four-parameter equation of state. Fluid Phase Equilibria 11, 29–84.

Ahmed, T., 1989. Hydrocarbon phase behavior. In: Contributions in Petroleum Geology & Engineering, vol. 7. Gulf Publishing Company, Houston, Texas.

Ahmed, T., 2007. Equations of State and PVT Analysis: Applications for Improved Reservoir Modeling. Gulf Publishing Company, Houston, Texas.

Al-Dhamen, M.A., 2010. New Models for Estimating the Dew-point Pressure of Gas Condensate Reservoir (Master thesis). King Fahd University of Petroleum & Minerals.

Al-Mahroos, F.M., Tejoa, G.H., 1987. Analysis and Phase Behavior of Khuff Gas System. In: Bahrain Field, Middle East Oil Show, Manama, Bahrain, March 7–10, 1987. SPE Paper No. 15766.

Azizi, N., Behbahani, R., Isazadeh, M.A., 2010. An efficient correlation for calculating compressibility factor of natural gases. Journal of Natural Gas Chemistry 19, 642–645.

Babalola, F.U., Susu, A.A., Olaberinjo, F.A., 2009. Gas condensate reservoir modeling, part I: from deterministics to stochastics. Petroleum Science and Technology 27 (10), 1007–1013.

Bahadori, A., Mokhatab, S., Towler, B.F., 2007. Rapidly estimating natural gas compressibility factor. Journal of Natural Gas Chemistry 16, 349–353.

Beggs, H.D., Brill, J.P., 1973. A study of two-phase flow in inclined pipes. Transactions of the AIME 255, 607–617.

Bozorgzadeh, M., Gringarten, A.C., 2006. Condensate-bank characterization from well-test data and fluid PVT properties. SPE Reservoir Evaluation and Engineering 9 (5), 596–611.

Brinkman, F.H., Sicking, J.N., 1960. Equilibrium ratios for reservoir studies. Transactions of the AIME 219, 313–319.

Carr, N.L., Kobayashi, R., Burrows, D.B., 1954. Viscosity of hydrocarbon gases under pressure. Journal of Petroleum Technology 6, 47–55.

Cavett, R.H., 1962. Physical data for distillation calculations, vapor-liquid equilibria. In: Proceedings of 27th API Meeting, San Francisco, pp. 351–366.

Chamkalani, A., Mae'soumi, A., Sameni, A., 2013. An intelligent approach for optimal prediction of gas deviation factor using particle swarm optimization and genetic algorithm. Journal of Natural Gas Science and Engineering 14, 132–143.

Chen, Z.A., Ruth, D.W., 1993. On the viscosity correlation of natural gas. In: Annual Technical Meeting. Petroleum Society of Canada, Calgary, Canada.

Chowdhury, N., Sharma, R., Pope, G.A., Sepehrnoori, K., 2008. A semianalytical method to predict well deliverability in gas-condensate reservoirs. SPE Reservoir Evaluation and Engineering 11 (1), 175–185.

Corredor, J.H., Piper, L.D., McCain Jr., W.D., 1992. Compressibility factors for naturally occurring petroleum gases. In: Paper SPE 24864, Presented at the SPE Annual Technical Meeting and Exhibition, Washington, DC, October 4–7, 1992.

Dempsey, J.R., 1965. Computer routine treats gas viscosity as a variable. Oil Gas Journal 63, 141–146.

Diehl, J., Gondouin, M., Houpeurt, A., Neoschil, J., Thelliez, M., Verrien, J.P., Zurawsky, R., 1970. Viscosity and density of light paraffins, nitrogen and carbon dioxide. CREPS/Geopetrole.

Dranchuk, P.M., Abou-Kassem, J.H., 1975. Calculation of z-factors for natural gases using equations of state. Journal of Canadian Petroleum Technology 14, 34–36.

Dykstra, H., Mueller, T.D., 1965. Calculation of phase composition and properties for lean- or enriched-gas drive. Society of Petroleum Engineers Journal 239–246.

Elsharkawy, A.M., Elkamel, A., 2001. The accuracy of predicting compressibility factor for sour natural gases. Petroleum Science and Technology 19 (5&6), 711–731.

Elsharkawy, A.M., Yousef, S.Kh, Hashem, S., Alikhan, A.A., 2001. Compressibility factor for gas condensates. Energy Fuels 15, 807–816.

Elsharkawy, A., 2002. Predicting the dew point pressure for gas condensate reservoirs: empirical models and equations of state. Fluid Phase Equilibria 193 (1–2), 147–165.

Elsharkawy, A.M., 2004. Efficient methods for calculations of compressibility, density and viscosity of natural gases. Fluid Phase Equilibria 218, 1–13.

Farrokh-Niae, A.H., Moddarress, H., Mohsen-Nia, M., 2008. A three-parameter cubic equation of state for prediction of thermodynamic properties of fluids. Journal of Chemical Thermodynamics 40, 84–95.

Fasesan, S.O., Olukink, O.O., Adewumi, O.O., 2003. Characteristics of gas condensate. Petroleum Science and Technology 21 (1–2), 81–90.
Fevang, Ø., Whitson, C.H., 1996. Modeling gas-condensate well deliverability. SPE Reservoir Engineering 11 (4), 221–230.
Fevang, O., 1995. Gas Condensate Flow Behaviour and Sampling (Ph.D. thesis). University of Trondheim.
Forero, L.A., Velásquez J, J.A., 2012. The Patel–Teja and the Peng–Robinson EoSs performance when Soave alpha function is replaced by an exponential function. Fluid Phase Equilibria 332, 55–76.
Gasem, K.A.M., Gao, W., Pan, Z., Robinson, R.L., 2001. A modified temperature dependence for the Peng-Robinson equation of state. Fluid Phase Equilibria 181, 113–125.
Golubev, I.F., 1959. Viscosity of Gases and Gas Mixtures: A Handbook Israel Program for Scientific Translations. US Department of Commerce, Clearinghouse for Federal Scientific and Technical Information, Springfield, VA.
Guria, C., Pathak, A.K., 2012. An improved generalized three-parameter cubic equation of state for pure fluids. Journal of Petroleum Science and Engineering 96–97, 79–92.
Hazim, A., 2008. Performance of wellhead chokes during sub-critical flow of gas condensates. Journal of Petroleum Science and Engineering 60 (3–4), 205–212.
Hemmati-Sarapardeh, A., Khishvand, M., Naseri, A., Mohammadi, A.H., 2013. Toward reservoir oil viscosity correlation. Chemical Engineering Science 90 (7), 53–68.
Hoffmann, A.E., Crump, J.S., Hocott, R.C., 1953. Equilibrium constants for a gas-condensate system. Transactions of the AIME 198, 1–10.
Hosein, R., Dawe, R., 2011. Characterization of trinidad gas condensates. Petroleum Science and Technology 29 (8), 817–823.
Humoud, A.A., Al-Marhoun, M.A., 2001. A new correlation for gas-condensate dewpoint pressure prediction. In: Proceedings of the Presentation at the SPE Middle East Oil Show Held in Bahrain, March 17–20, 2001. SPE Paper No. 68230.
Jensen, B.H., 1987. Densities, Viscosities and Phase Behavior in Enhanced Oil Recovery (Ph.D. thesis). Department of Chemical Engineering, The Technical University of Denmark, Denmark.
Jossi, J.A., Stiel, L.I., Thodos, G., 1962. The viscosity of pure substances in the dense gaseous and liquid phases. AIChE Journal 8 (1), 59–62.
Katz, D.L., Hachmuth, K.H., 1937. Vaporization equilibrium constants in a crude oil–natural gas system. Industrial and Engineering Chemistry 29 (9), 1072–1077.
Katz, D., 1983. An overview of phase behavior of oil and gas production. Journal of Petroleum Technology 1205–1214.
Kehn, D.M., April 1964. Rapid analysis of condensate systems by chromatography. Journal of Petroleum Technology 435–440.
Khan, N.M., Bilal, H.M., Shoaib, M., Manzoor, A., Shaukat, W., Shakil, T., 2012. Fluid characterization of a retrograde gas condensate reservoir for the simulation study. In: SPE 163134 Presented at the SPE/PAPG Annual Technical Conference Held in Islamabad, Pakistan, December 3–5.
Kool, H., Azari, M., Soliman, M.Y., Proett, M.A., Irani, C.A., 2001. Testing of gas condensate reservoirs-sampling, test design and analysis. In: Paper SPE 68668-MS, Presented at the SPE. Asia Pacific Oil and Gas Conference and Exhibition, Jakarta, Indonesia, April 17–19.
Kumar, N., 2004. Compressibility Factor for Natural and Sour Reservoir Gases by Correlations and Cubic Equations of State (M.S. thesis). Texas Tech University, Lubbock, Tex, USA.
Lee, B.I., Kesler, M.G., 1980. Improve vapor pressure prediction. Hydrocarbon Processing 163–167.

Lee, A.L., Gonzalez, M.H., Eakin, B.E., 1966. The viscosity of natural gases. Journal of Petroleum Technology 18, 997–1000.

Li, Y., Li, B., Hu, Y., Jiao, Y., Zhu, W., Xiao, X., Niu, Y., 2010. Water production analysis and reservoir simulation of the Jilake gas condensate field. Petroleum Exploration and Development 37 (1), 89–93.

Lohrenze, J., Clark, G., Francis, R., 1963. A compositional material balance for combination drive reservoirs. Journal of Petroleum Technology 15.

Londono Galindo, F.E., Archer, R.A., Blasingame, T.A., 2005. Correlations for hydrocarbon-gas viscosity and gas density-validation and correlation of behavior using a large-scale database. SPE Reservoir Evaluation & Engineering 8 (6), 561–572.

Mahmoud, M., 2014. Development of a new correlation of gas compressibility factor (Z-Factor) for high pressure gas reservoirs. Journal of Energy Resources Technology 136 (1), 012903.

Marruffo, I., Maita, J., Him, J., Rojas, G., 2002. Correlation to determine retrograde dew pressure and C_{7+} percentage of gas condensate reservoirs on basis of production test data of Eastern Venezuelan fields. In: Presented at the SPE Gas Technology Symposium Held in Calgary, Alberta, Canada, April 30.

McCain Jr., W.D., 1990. The Properties of Petroleum Fluids, second ed. Pennwell Publishing Company, Tulsa, Oklahoma, pp. 178–180.

Mohsen-Nia, M., Modarress, H., Mansoori, G.A., 2003. A cubic hard-core equation of state. Fluid Phase Equilibria 206, 27–39.

Moses, P.L., Donohoe, C.W., 1962. Gas Condensate Reservoirs. In: Petroleum Engineering Handbook, vol. 39. SPE Publications, pp. 1–28.

Mott, R., 2002. Engineering calculation of gas condensate well productivity. In: SPE 77551, SPE Annual Technical Conference and Exhibition, San Antonio, USA, October.

Nagarajan, N.R., Honarpour, M.M., Sampath, K., 2007. Reservoir fluid sampling and characterization—key to efficient reservoir management. SPE 103501 Distinguished Author Series Journal of Petroleum Technology 59 (08), 80–91.

Naseri, A., Khishvand, M., Sheikhloo, A.A., 2014. A correlations approach for prediction of PVT properties of reservoir oils. Petroleum Science and Technology 32, 2123–2136.

Nasrifar, Kh, Bolland, O., 2004. Square-well potential and a new α function for the Soave-Redlich-Kwong equation of state. Industrial and Engineering Chemistry Research 43, 6901–6909.

Nasrifar, Kh, Moshfeghian, M., 1999. Evaluation of saturated liquid density prediction methods for pure refrigerants. Fluid Phase Equilibria 158–160, 437–445.

Nasrifar, Kh, Moshfeghian, M., 1998. A saturated liquid density equation in conjunction with the predictive-Soave-Redlich-Kwong equation of state for pure refrigerants and LNG multicomponent systems. Fluid Phase Equilibria 153, 231–242.

Nasrifar, Kh, Moshfeghian, M., 2001. A new cubic equation of state for simple fluids: pure and mixture. Fluid Phase Equilibria 190, 73–88.

Nemeth, L.K., Kennedy, H.T., 1967. A correlation of dewpoint pressure with fluid composition and temperature. Society of Petroleum Engineers Journal 7 (2), 99–104.

Papay, J., 1968. A Termelestechologiai Parametrrek Valtozasa a Gazlelepk Muvelese Soran. Ogil Musz, Tud, Kuzl., Budapest, pp. 267–273.

Patel, N.C., Teja, A.S., 1982. A new equation of state for fluids and fluid mixtures. Chemical Engineering Science 37, 463–473.

Pedersen, K.S., Blilie, A.L., Meisingset, K.K., 1992. PVT calculations on petroleum reservoir fluids using measured and estimated compositional data for the plus fraction. Industrial and Engineering Chemistry Research 3, 1378–1384.

Peng, D.Y., Robinson, D.B., 1976. A new two-constant equation of state. Industrial and Engineering Chemistry Fundamental 15, 59–64.

Piper, L.D., McCain Jr., Corredor, J.H., 1993. Compressibility factors for naturally occurring petroleum gases. In: Paper SPE 26668, Presented at the Society of Petroleum Engineer International Petroleum Conferences and Exhibition, Houston, TX, October 3–6.
Pitzer, K.S., Lippman, D.Z., Curl, R.F., Huggins, C.M., Peterson, D.E., 1955. The volumetric and thermodynamic properties of fluids. II. Compressibility factor, vapor pressure and entropy of vaporization. Journal of American Chemical Society 77, 3433–3440.
Redlich, O., Kwong, J.N.S., 1949. On the thermodynamics of solutions: V. An equation of state: fugacity of gaseous solutions. Chemical Reviews 44, 233–244.
Riazi, M.R., Daubert, T.E., 1987. Characterization parameters for petroleum fractions. Industrial and Engineering Chemistry Research 26, 755–759.
Robinson, D.B., Peng, D.Y., 1978. The Characterization of the Heptanes and Heavier Fractions for the GPA Peng–Robinson Programs. Gas Processors Association. Research Report RR-28. (Booklet only sold by the GPA, Gas Processors Association).
Roussennac, B., 2001. Gas Condensate Well Test Analysis (M.Sc. thesis). Stanford University, p. 121.
Rzasa, M.J., Glass, E.D., Opfell, J.B., 1952. Prediction of critical properties and equilibrium vaporization constants for complex hydrocarbon systems. Chemical Engineering Progress, Symposium Series 48 (2), 28.
Sage, B.H., Olds, R.H., 1947. Volumetric behavior of oil and gas from several San Joaquin valley fields. Transactions of the AIME 170, 156–173.
Salim, P.H., Trebble, M.A., 1991. A modified Trebble-Bishnoi equation of state. Fluid Phase Equilibria 65, 59–71.
Sanjari, E., Nemati Lay, E., 2012. An accurate empirical correlation for predicting natural gas compressibility factors. Journal of Natural Gas Chemistry 21 (2), 184–188.
Sanjari, E., Nemati Lay, E., Peymani, M., 2011. An Accurate empirical correlation for predicting natural gas viscosity. Journal of Natural Gas Chemistry 20, 654–658.
Sarkar, R., Danesh, A.S., Todd, A.C., 1991. Phase behavior modeling of gas-condensate fluids using an equation of state. In: Paper SPE 22714, Presented at the 66th Annual Technical Conference and Exhibition of the Society of Petroleum Engineers Held in Dallas, TX, October 6–9.
Schmidt, G., Wenzel, H., 1980. A modified Van der Waals type equation of state. Chemical Engineering Science 35, 1503–1512.
Segura, H., Kraska, T., Mejía, A., Wisniak, J., Polishuk, I., 2003. Unnoticed pitfalls of soave-type alpha functions in cubic equations of state. Industrial and Engineering Chemistry Research 42, 5662–5673.
Shokir, E.M.El.-M., Dmour, H.N., 2009. Genetic programming (GP)-based model for the viscosity of pure and hydrocarbon gas mixtures. Energy Fuels 23 (7), 3632–3636.
Shokir, E.M.El.-M., El-Awad, M.N., Al-Quraishi, A.A., Al-Mahdy, O.A., 2012. Compressibility factor model of sweet, sour, and condensate gases using genetic programming. Chemical Engineering Research and Design 90 (6), 785–792.
Soave, G., 1972. Equilibrium constants from a modified Redlich-Kwong equation of state. Chemical Engineering Science 27, 1197–1203.
Standing, M.B., Katz, D.L., 1942. Density of natural gases. AIME Transactions 146, 140–149.
Standing, M.B., 1951. Volumetric and Phase Behavior of Oil Field Hydrocarbon Systems: PVT for Engineers. California Research Corp., San Francisco, CA.
Standing, M.B., 1977. Volumetric and Phase Behavior of Oil Field Hydrocarbon Systems. Society of Petroleum Engineers of AIME, Dallas.
Standing, M.B., 1979. A set of equations for computing equilibrium ratios of a crude oil/natural gas system at pressures below 1,000 psia. Journal of Petroleum Technology 1193–1195.

Standing, M.B., 1981. Volumetric and Phase Behavior of Oil Field Hydrocarbon Systems. Society of Petroleum Engineers, Dallas, TX.
Starling, K.E., 2003. Peng-Robinson equation of state and calculating natural gas dew points. In: American Gas Association Operations Conference, Orlando, April 27–30.
Stewart, W.F., Burkhardt, S.F., Voo, D., 1959. Prediction of pseudo-critical parameters for mixtures. In: Presented at the AIChE Meeting, Kansas City, Missouri, USA, May 18, 1959.
Sun, C.Y., Liu, H., Yan, K.L., Ma, Q.L., Liu, B., Chen, G.J., Xiao, X.J., Wang, H.Y., Zheng, X.T., Li, S., 2012. Experiments and modeling of volumetric properties and phase behavior for condensate gas under ultra-high-pressure conditions. Industrial and Engineering Chemistry Research 51 (19), 6916–6925.
Sutton, R.P., 1985. Compressibility factors for high molecular weight reservoir gases. In: Paper SPE 14265, Presented at the SPE Annual Technical Meeting and Exhibition, Las Vegas, September, pp. 22–25.
Sutton, R.P., 2007. Fundamental PVT calculations for associated and gas/condensate natural-gas systems. SPE Formation Evaluation 10 (3), 270–284.
Thomas, F.B., Andersen, G., Bennion, D.B., 2009. Gas condensate reservoir performance. Journal of Canadian Petroleum Technology 48 (7), 18–24.
Twu, C.H., Coon, J.E., Cunningham, J.R., 1995. A new generalized alpha function for a cubic equation of state: part 2. Redlich-Kwong equation. Fluid Phase Equlibria 105, 61–69.
Twu, C.H., 1984. An internally consistent correlation for predicting the critical properties and molecular weights of petroleum and coal tar liquids. Fluid Phase Equilibria 16, 137–150.
Ungerer, P., Faissat, B., Leibovici, C., Zhou, H., Behar, E., Moracchini, G., Courcy, J., 1995. High pressure-high temperature reservoir fluids: investigation of synthetic condensate gases containing a solid hydrocarbon. Fluid Phase Equilibria 111 (2), 287–311.
Vo, D.T., Jones, J.R., Raghavan, R., 1989. Performance predictions for gas-condensate reservoirs. SPE Formation Evaluation 4 (4), 576–584.
Van der Waals, J.D., June 14, 1873. On the Continuity of the Gas and Liquid State (Over de Continuïteit van den Gas- en Vloeistoftoestand) (Ph.D. Thesis). Leiden University.
Wang, P., Pope, G.A., 2001. Proper use of equations of state for compositional reservoir simulation. SPE Journal 53 (7), 74–81.
Whitson, C.H., Torp, S.B., 1981. Evaluating constant volume depletion data. In: Paper SPE 10067, Presented at the SPE 56th Annual Fall Technical Conference, San Antonio, October 5–7.
Whitson, C.H., 1983. Characterizing hydrocarbon plus fractions. SPE Journal 683–694.
Wilson, G., 1968. A modified Redlich-Kwong EOS, application to general physcial data calculations. In: Paper 15C, Presented at the Annual AIChE National Meeting, Cleveland, May 4–7.
Zendehboudi, S., Ahmadi, M.A., James, L., Chatzis, I., 2012. Prediction of condensate-to-gas ratio for retrograde gas condensate reservoirs using artificial neural network with particle swarm optimization. Energy & Fuels 26, 3432–3447.

CHAPTER EIGHT

Gas Hydrates

M.A. Ahmadi[1], A. Bahadori[2,3]
[1]Petroleum University of Technology (PUT), Ahwaz, Iran
[2]Southern Cross University, Lismore, NSW, Australia
[3]Australian Oil and Gas Services Pty Ltd, Lismore, NSW, Australia

8.1 INTRODUCTION

Clathrate hydrates are ice-like inclusion compounds which form at high pressures (*P*) and low temperatures (*T*) with nonpolar guest molecules surrounded by hydrogen-bonded water cages. Hydrates are applicable to wide ranges of industrial and scientific environments comprising modeling of climate change, CO_2 sequestration, hydrocarbon extraction, natural gas and hydrogen storage, refrigeration and separation technologies, planetary surface chemistry, and marine biology. Clathrate hydrates created from small gas molecules are normally denoted as gas hydrates and are influenced by the type of gas molecule and the thermodynamic circumstances to adopt various configurations (Sloan and Koh, 2008).

8.2 TYPES AND PROPERTIES OF HYDRATES

The three best-known structures formed from gas molecules are the structure I (sI), structure II (sII), and structure H (sH) hydrates, which are shown in Fig. 8.1.

sI hydrate consists of two types of cages: a small cage consisting of 12 pentagonal rings (5^{12}) of water and a larger cage consisting of 12 pentagonal and 2 hexagonal rings ($5^{12}6^2$). sII hydrate also consists of two types of cavities: the small 5^{12} cage and a larger cage consisting of 12 pentagonal and four hexagonal rings ($5^{12}6^4$) of water. sH hydrate consists of three types of cages: the 5^{12} cage, a larger $5^{12}6^8$ cage, and an intermediate cage consisting of three square, six pentagonal and three hexagonal rings ($4^35^66^3$)of water. Although hydrates formed in nature seem to favor formation of sI, those found in artificial systems, like oil and gas pipelines, most often form sII. sH is only favored when a heavy hydrocarbon such as methylcyclohexane is present with methane and water (Sloan and Koh, 2008).

Fluid Phase Behavior for Conventional and Unconventional Oil and Gas Reservoirs
http://dx.doi.org/10.1016/B978-0-12-803437-8.00008-7

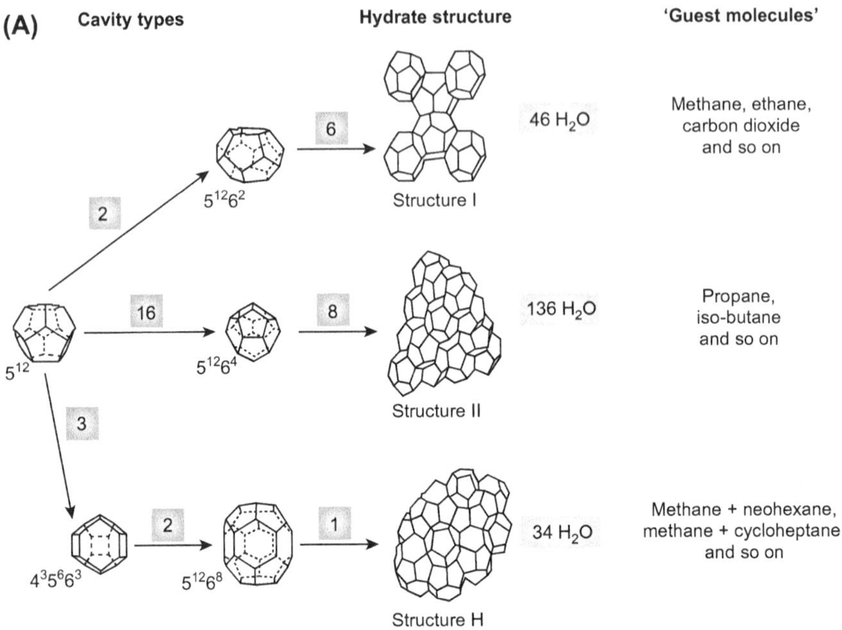

(B)

Hydrate crystal structure	I		II		H		
Cavity	Small	Large	Small	Large	Small	Medium	Large
Description	5^{12}	$5^{12}6^{2}$	5^{12}	$5^{12}6^{4}$	5^{12}	$4^{3}5^{6}6^{3}$	$5^{12}6^{8}$
Number of cavities per unit cell	2	6	16	8	3	2	1
Average cavity radius (Å)	3.95	4.33	3.91	4.73	3.91†	4.06†	5.71†
Coordination number*	20	24	20	28	20	20	36
Number of waters per unit cell	46		136		34		

*Number of oxygens at the periphery of each cavity.
†Estimates of structure H cavities from geometric models.

Figure 8.1 Graphical illustration for hydrate structures (http://www.nature.com/nature/journal/v426/n6964/full/nature02135.html).

Because it is not possible for all cages to be occupied by a guest molecule, hydrates always have more water than the stoichiometric composition. The ratio between the guest molecule and water bound in the cage lattice usually ranges from 6 to 19 moles of water for each mole of hydrate formed, with typical fractional occupancies of the smaller cages between 0.3 and 0.9, whereas the large-cage occupancy is close to unity. This variation causes clathrate hydrates to be called "non-stoichiometric hydrates" to distinguish them from stoichiometric salt hydrates (Sloan and Koh, 2008).

The upstream industry is very susceptible to gas hydrate formation: transmission lines of oil and gas, tiebacks, and offshore process equipment are likely to being choked by hydrates, producing potential risks and/or economic loss [a number of case studies related to this can be found in the literature (Sloan and Koh, 2008; Sloan, 2000)]. Hydrates are the leading (compared to wax, asphaltenes, and scale) deepwater flow assurance problem by an order of magnitude (Sloan, 2005) and in a survey among 110 energy companies, flow assurance was recorded as the main technical difficulty in offshore energy development (Welling and Associates, 1999).

8.3 THERMODYNAMIC CONDITIONS FOR HYDRATE FORMATION

Hydrate formation is supported by low temperature and high pressure. For each gas, it is possible to generate a hydrate curve that maps the region in the pressure−temperature plane in which hydrates can form. Much of the rest of this book is dedicated to the tools used to predict this locus. Again, without getting too far ahead of ourselves, some preliminary discussion of hydrate curves is appropriate (Carroll, 2014).

Fig. 8.2 shows a typical hydrate curve (labeled "hydrate curve"). The region to the left and above this curve (high pressure, low temperature) is where hydrates can form. In the region to the right and below the hydrate curve, hydrates can never form in this region, because the first criterion is violated. Therefore, if your process, pipeline, well, etc. operates in the region labeled "no hydrates," then hydrates are not a problem. On the other hand, if it is in the region labeled "hydrates region," then some remedial action is required to avoid hydrates (Carroll, 2014).

It might seem as though we can treat the temperature and pressure as separate variables, but when discussing hydrates they are linked. For example, you cannot say, "A hydrate will not form at 10°C for the gas mixture shown in Fig. 8.2." You must qualify this with a pressure. So at 10°C and 5 MPa the process is in the "hydrate region," whereas at 10°C and 1 MPa the process is in the region where a hydrate will not form. Thus, we must talk about a combination of temperature and pressure, and not each variable on its own (Carroll, 2014).

Furthermore, it is common to add a margin of safety even to the best hydrate prediction methods. This margin can be 3−5°C (5−10°F), but typically 3°C is used. The author typically uses 3°C, but the reader may have

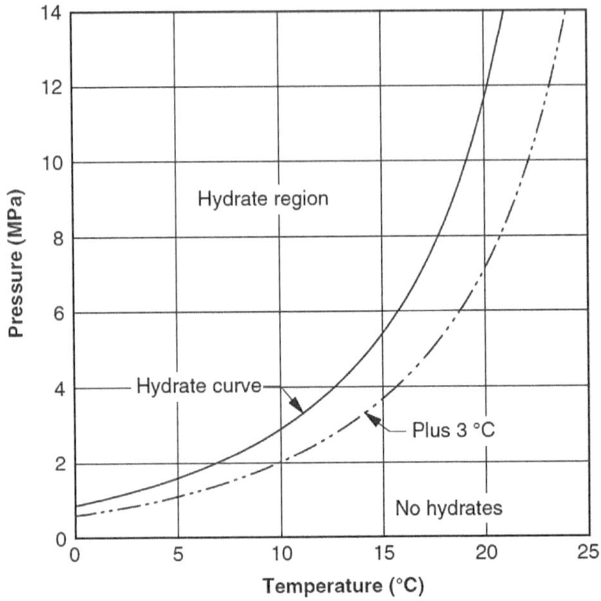

Figure 8.2 Pressure–temperature diagram for hydrate region and safety margins (Carroll, 2014).

their own margin or perhaps there is one specified by their company (Carroll, 2014).

A margin of safety is shown in Fig. 8.2 ("plus 3°C") and the buffer zone between the estimated hydrate curve and the +3°C curve is noted.

8.3.1 Calculating Hydrate Formation Condition

This section describes different models for estimating hydrate formation conditions. These methods including empirical correlations, equation of states are illustrated in the next sections.

8.3.1.1 Correlations

The first issue when designing processes containing hydrates is to estimate the situations of temperature and pressure at which hydrates will form. To begin the discussion of this topic, a series of approaches can be employed without a computer. Unfortunately, the shortcomings to these approaches are that they are not highly precise, and, overall, the less info needed as input, the less precise the outcomes of the estimation.

8.3.1.1.1 Makogon (1981)

Makogon (1981) developed a straightforward correlation for predicting the hydrate formation pressure as a function of gas gravity and temperature for paraffin hydrocarbons. The mathematical correlation is follows:

$$\log P = \beta + 0.0497\left(t + kt^2\right) - 1 \tag{8.1}$$

in which *P* stands for the pressure in terms of MPa and *t* denotes the temperature in terms of C. Makogon developed graphic correlations for *k* and β, but Elgibaly and Elkamel (1998) provided the following straightforward expressions:

$$\beta = 2.681 - 3.811\,\gamma + 1.679\,\gamma^2 \tag{8.2}$$

$$k = -0.006 + 0.011\,\gamma + 0.011\,\gamma^2 \tag{8.3}$$

It is worth mentioning that the aforementioned correlation by Elgibaly and Elkamel (1998) has deviations but their equations of β and *k* are correct.

Example 8.1

Calculate the hydrate formation pressure of ethane at 15°C by the Makogon method. It should be noted that molecular weight of ethane is about 30. Hint : $\gamma = \frac{MW}{28.96}$

Solution

At first we should determine ethane specific gravity. In this regard we have

$$\gamma = \frac{MW}{28.96} = \frac{30}{28.96} = 1.035$$

Then we should determine *k* and β

$$\beta = 2.681 - 3.811\,\gamma + 1.679\,\gamma^2 = 0.5352$$

$$k = -0.006 + 0.011\,\gamma + 0.011\,\gamma^2 = 0.0171$$

Finally, we have

$$\log P = \beta + 0.0497\left(t + kt^2\right) - 1 = 0.472688$$

$$P = 1.604\ \text{MPa}$$

8.3.1.1.2 Kobayashi et al. (1987)

Kobayashi et al. (1987) proposed the following, rather complex, correlation for predicting hydrate formation circumstances as a function of the gas gravity:

$$\begin{aligned}\frac{1}{T} = {} & 2.7707715 \times 10^{-3} - 2.782238 \times 10^{-3} \ln P - 5.649288 \times 10^{-4} \ln \gamma - 1.298593 \\ & \times 10^{-3} \ln P^2 + 1.407119 \times 10^{-3} \ln(P)\ln(\gamma) + 1.785744 \\ & \times 10^{-4} \ln(\gamma)^2 + 1.130284 \times 10^{-3}(P)^3 + 5.9728235 \\ & \times 10^{-4} \ln(P)^2 \ln(\gamma) - 2.3279181 \times 10^{-4} \ln(P)\ln(\gamma)^2 - 2.6840758 \\ & \times 10^{-5} \ln(\gamma)^3 + 4.6610555 \times 10^{-3} \ln(P)^4 + 5.5542412 \\ & \times 10^{-4} \ln(P)^3 \ln(\gamma) - 1.4727765 \times 10^{-5} \ln(P)^2 \ln(\gamma)^2 \\ & + 1.393808 \times 10^{-5} \ln(P)\ln(\gamma)^3 + 1.4885010 \times 10^{-5} \ln(\gamma)^4 \end{aligned} \tag{8.4}$$

T stands for the temperature in terms of Rankine, P denotes the pressure in terms of psi, and Υ represents the gas specific gravity.

Unfortunately, this correlation and the constants reported seem incorrect. No matter what value is input for the pressure, the resultant temperature is constantly near 0°R (−460°F). Enormous attempts were made to identify this big mistake, but the difficulty could not be entirely isolated.

8.3.1.1.3 Motiee (1991)

Motiee (1991) developed the following mathematical expression for predicting the hydrate formation temperature as a function of the gas gravity and the pressure:

$$\begin{aligned} T = {} & -283.24469 + 78.99667 \log(P) - 5.352544(\log(P))^2 \\ & + 349.473877\gamma - 150.854675\gamma^2 - 27.604065 \log(P)\gamma \end{aligned} \tag{8.5}$$

in which T stands for the hydrate temperature in terms of °F, P represents the pressure in terms of psi, and Υ denotes the gas gravity.

Example 8.2

Consider following mixture and calculate the hydrate formation temperature by Motiee method at $P = 5500$ psi.

Component	Mole Fraction
CO_2	0.08
H_2S	0.05
C_1	0.87

Solution

At first, we should determine ethane specific gravity. In this regard, we determine molecular weight of the gas mixture and then calculate gas specific gravity

$$\text{MW} = 0.08 \times 44 + 0.05 \times 34.08 + 0.87 \times 16 = 19.144$$

$$\gamma = \frac{\text{MW}}{28.96} = \frac{19.144}{28.96} = 0.66105$$

$$\begin{aligned} T &= -283.24469 + 78.99667 \log(P) - 5.352544 \log(P)^2 + 349.473877\gamma \\ &\quad - 150.854675\gamma^2 - 27.604065 \log(P)\gamma \\ &= 34.19^\circ\text{F} \end{aligned}$$

8.3.1.1.4 Østergaard et al. (2000)

Østergaard et al. (2000) developed another correlation for predicting hydrate formation pressure. Their proposed correlation was developed based on the gas gravity. It is worth highlighting that their proposed equation is applicable for sweet gases.

$$\begin{aligned} \ln(P) = {} & \left(c_1(\gamma + c_2)^{-3} + c_3 F_\text{m} + c_4 F_\text{m}^2 + c_5\right) T + c_3(\gamma + c_7)^{-3} \\ & + c_8 F_\text{m} + c_9 F_\text{m}^2 + c_{10} \end{aligned} \tag{8.6}$$

In which P stands for the pressure in terms of kPa, Υ represents the gas specific gravity, T denotes the temperature in terms of K, and F_m stands for the mole ratio between formers and nonformers throughout the mixture. The coefficients of their correlation are reported through Table 8.1.

It is worth highlighting that this correlation due to gas specific limitations is not valid to pure ethane or to pure methane. Østergaard et al. (2000) made enormous efforts to include corrections for H_2S in their approach; however, they could not achieve this.

Table 8.1 Coefficients of the Østergaard et al. Correlation

Coefficient	Value	Coefficient	Value
C_1	4.5134×10^{-3}	C_6	3.6625×10^{-4}
C_2	0.46852	C_7	−0.485054
C_3	2.18636×10^{-2}	C_8	−5.44376
C_4	-8.417×10^{-4}	C_9	3.89×10^{-3}
C_5	0.129622	C_{10}	−29.9351

8.3.1.1.5 Sun et al. (2003)

Sun et al. (2003) took a set of measurements for sour gas mixtures; remember that sour gas is a mixture containing H_2S. These data are from 1 to 26.5°C and 0.58−8.68 MPa. The specific gravity of these mixtures ranges from 0.656 to 0.787. The data set is approximately 60 points in total. It was noted earlier that the simple gas gravity method is not applicable to sour gas mixtures, thus, this set of data provides a severe test for our simplified local models. Using least squares regression to fit the set of data, one obtains the following correlation:

$$\frac{1000}{T} = 4.343295 + 1.07340 \times 10^{-3} P - 9.19840 \times 10^{-2} \ln P - 1.071989\,\gamma \tag{8.7}$$

Example 8.3

Consider following mixture and calculate the hydrate formation temperature by Sun et al. method at $P = 4$ MPa.

Component	Mole Fraction
CO_2	0.15
H_2S	0.15
C_1	0.70

Solution

At first, we should determine ethane specific gravity. In this regard we determine molecular weight of gas mixture and then calculate gas specific gravity

$$MW = 0.15 \times 44 + 0.15 \times 34.08 + 0.70 \times 16 = 22.912$$

$$\gamma = \frac{MW}{28.96} = \frac{22.912}{28.96} = 0.79176$$

$$T(k) = \frac{1000}{4.343295 + 1.07340 \times 10^{-3}P - 9.19840 \times 10^{-2}\ln P - 1.071989\gamma}$$
$$= 296.56K$$

8.3.1.1.6 Towler and Mokhatab (2005)

Towler and Mokhatab (2005) developed a somewhat straightforward correlation for predicting hydrate formation temperatures as a function of the gas gravity and the pressure as follows:

$$T = 13.47\ln(P) + 34.27\ln(\gamma) - 1.675\ln(P)\ln(\gamma) - 20.35 \quad (8.8)$$

Example 8.4

Consider following mixture and calculate the hydrate formation temperature by Towler and Mokhatab method at $P = 2000$ psi.

Component	Mole Fraction
CO_2	0.02
H_2S	0.02
C_1	0.70
C_2	0.16
C_3	0.10

Solution

At first we should determine ethane specific gravity. In this regard, we determine molecular weight of gas mixture and then calculate gas specific gravity

$$MW = 0.02 \times 44 + 0.02 \times 34.08 + 0.70 \times 16 + 0.16 \times 30 + 0.10 \times 44$$
$$= 21.9616$$

$$\gamma = \frac{MW}{28.96} = \frac{21.9616}{28.96} = 0.7583$$

Then we have

$$T = 13.47\ln(P) + 34.27\ln(\gamma) - 1.675\ln(P)\ln(\gamma) - 20.35 = 76.0761°F$$

8.3.1.1.7 Bahadori and Vuthaluru (2009)

Bahadori and Vuthaluru (2009) developed robust correlations for predicting hydrate formation pressure and temperature. They employed data of hydrate formation condition reported in previous literature to develop their correlation. They proposed two different correlations for estimating hydrate formation conditions. The first equation was used for predicting hydrate formation temperature as a function of pressure as formulated by Eq. (8.9). Moreover, the coefficient of this equation was correlated to molecular weight of gas mixture. The second equation was used for estimating hydrate formation pressure as a function of temperature as formulated by Eq. (8.10). It is worth mentioning that before predicting hydrate formation condition we should determine the appropriate coefficients of Eqs. (8.9) and (8.10), depending on our case. For predicting hydrate formation temperature, we should use the coefficients reported through Table 8.2 and for estimating hydrate formation pressure we should use the coefficients illustrated in Table 8.3.

The mathematical expressions of hydrate formation temperature and pressure are expressed by Eqs. (8.9) and (8.10), correspondingly, as follows:

$$\ln(T) = a + b\left(\frac{1}{P}\right) + c\left(\frac{1}{P}\right)^2 + d\left(\frac{1}{P}\right)^3 \tag{8.9}$$

Table 8.2 Coefficients of the Bahadori and Vuthaluru Correlation for Estimation of Hydrate Formation Pressure (kPa) via Eq. (8.9) (Bahadori and Vuthaluru, 2009)

Coefficient	Molecular weight < 23, 265K < Temperature < 298K	Molecular weight > 23, 265K < Temperature < 298K
A_1	$-2.8375555003183 \times 10^5$	$9.6485148281011 \times 10^4$
B_1	$4.188723721533 \times 10^4$	$1.2987255223562 \times 10^4$
C_1	$-2.0426785680161 \times 10^3$	$5.6943123183493 \times 10^2$
D_1	$3.2999427860007 \times 10^1$	-8.0291736544591
A_2	$2.3518577113598 \times 10^8$	$-8.3851942305767 \times 10^7$
B_2	$-3.470311070979 \times 10^7$	$1.1292443545403 \times 10^7$
C_2	$1.6921307674758 \times 10^6$	$-4.9481203210497 \times 10^5$
D_2	$-2.7331526571044 \times 10^4$	$6.9743729419639 \times 10^3$
A_3	$-6.4899035506028 \times 10^{10}$	$2.4283950487232 \times 10^{10}$
B_3	$9.5728921505256 \times 10^9$	$-3.2713325876178 \times 10^9$
C_3	$-4.667233443707 \times 10^8$	$1.4325969896394 \times 10^8$
D_3	$7.5373257072387 \times 10^6$	$-2.018536147544 \times 10^6$
A_4	$5.9653477415552 \times 10^{12}$	$-2.3430538061379 \times 10^{12}$
B_4	$-8.796372864875 \times 10^{11}$	$3.1570181175788 \times 10^{11}$
C_4	$4.2881972248701 \times 10^{10}$	$-1.38180509474908 \times 10^{10}$
D_4	$-6.9241414046235 \times 10^8$	$1.9463506733398 \times 10^8$

Table 8.3 Coefficients of the Bahadori and Vuthaluru Correlation to Estimate Hydrate Formation Temperature (K) via Eq. (8.10) (Bahadori and Vuthaluru, 2009)

Coefficient	Molecular Weight > 23 and Pressure 1200 kPa < P < 40,000 kPa	Molecular Weight < 23 and Pressure 1200 kPa < P < 5000 kPa	Molecular Weight < 23 and Pressure Range 5000 kPa < P < 40,000 kPa
A_1	6.4185071105353	−4.1812132784232	7.0959703947586
B_1	$-8.8017107875666 \times 10^{-2}$	1.472639349108	$-2.1806030070795 \times 10^{-1}$
C_1	$3.5573429357137x\ 10^{-3}$	$-7.2745386271251 \times 10^{-2}$	$1.1305933439794 \times 10^{-2}$
D_1	$-4.7499843881244 \times 10^{-5}$	$1.1897795879884 \times 10^{-3}$	$-1.927203195626 \times 10^{-4}$
A_2	$-8.6426289139868 \times 10^{3}$	$4.5284975000181 \times 10^{4}$	$-1.2584649421592 \times 10^{5}$
B_2	$1.0243307852297 \times 10^{3}$	$-6.8628124449813 \times 10^{3}$	$1.8993111766336 \times 10^{4}$
C_2	$-4.09663925465509 \times 10^{1}$	$3.4240721860406 \times 10^{2}$	$-9.5260058127234 \times 10^{2}$
D_2	$5.4450050757729 \times 10^{-1}$	−5.642533019	$1.5806820089029 \times 10^{1}$
A_3	$1.159643030462 \times 10^{7}$	$-8.317075073225 \times 10^{7}$	$9.2190382283151 \times 10^{8}$
B_3	$-1.3859027774109 \times 10^{6}$	$1.2604810249225 \times 10^{7}$	$-1.4030410567488 \times 10^{8}$
C_3	$5.5353148270822 \times 10^{4}$	$-6.3018579466138 \times 10^{5}$	$7.0820417989994 \times 10^{6}$
D_3	$-7.339994547645 \times 10^{2}$	$1.0408848430973 \times 10^{4}$	$-1.1818763471949 \times 10^{5}$
A_4	$-4.0200951475377 \times 10^{9}$	$5.8589773993386 \times 10^{9}$	$-2.1053548626211 \times 10^{12}$
B_4	$4.791331833062 \times 10^{8}$	$-9.6634962535354 \times 10^{8}$	$3.213992597219 \times 10^{11}$
C_4	$-1.9036325296009 \times 10^{7}$	$5.13473142241307 \times 10^{7}$	$-1.6274767262739 \times 10^{10}$
D_4	$2.5113297404156 \times 10^{5}$	$-8.87818586492 \times 10^{5}$	$2.724884324573 \times 10^{8}$

$$\ln(P) = a + b\left(\frac{1}{T}\right) + c\left(\frac{1}{T}\right)^2 + d\left(\frac{1}{T}\right)^3 \tag{8.10}$$

in which,

$$a = A_1 + B_1\text{MW} + C_1\text{MW}^2 + D_1\text{MW}^3 \tag{8.11}$$

$$b = A_2 + B_2\text{MW} + C_2\text{MW}^2 + D_2\text{MW}^3 \tag{8.12}$$

$$c = A_3 + B_3\text{MW} + C_3\text{MW}^2 + D_3\text{MW}^3 \tag{8.13}$$

$$d = A_4 + B_4\text{MW} + C_4\text{MW}^2 + D_4\text{MW}^3 \tag{8.14}$$

It is worth stressing that the coefficients of Eqs. (8.9) and (8.10) cover the data points of Katz (1945) gravity chart in which temperature varies from 260 to 298K and gas molecular weight changes from 16 to 29.

8.3.1.2 Equation of States

A system is in thermodynamic equilibrium when it is in thermal, mechanical, and chemical equilibrium. For a system at constant pressure and temperature, thermodynamic equilibrium can be characterized by minimum Gibbs energy. For a transfer of dn_i moles of a substance between two phases 1 and 2, in equilibrium at constant temperature and pressure, the change in Gibbs energy (G) is

$$dG = \left(\mu_i^2 - \mu_i^1\right) dn_i \tag{8.15}$$

in which μ_i stands for the chemical potential of a substance i, and n_i represents the number of moles of i. At equilibrium G is minimum, thus:

$$\left(\frac{\partial G}{\partial n_i}\right)_{T,P,n_j} = 0 \tag{8.16}$$

yielding

$$\mu_i^2 = \mu_i^1 \tag{8.17}$$

For multiphase–multicomponent equilibrium, Eq. (8.17) can be extended to

$$\mu_i^1 = \mu_i^2 = \ldots = \mu_i^k \quad i = 1, 2, \ldots., N \tag{8.18}$$

and k is the number of coexisting phases.

Based on the previous equations, the equilibrium condition may be calculated either by direct minimization of the Gibbs energy or by using the principle of equality of chemical potentials (Walas, 1985). The chemical potential can be expressed in terms of the fugacity of a component by the following equation:

$$\mu = \mu^0 + RT \ln \frac{f(p)}{p^0} \tag{8.19}$$

in which μ^0 is the chemical potential at reference state, T is the temperature, R is the universal gas constant, P^0 is the pressure at the reference state, and $f(P)$ is the fugacity as a function of pressure.

Combination of Eqs. (8.18) and (8.19) results in the equality of fugacities for the thermodynamic equilibrium under consideration:

$$f_A^1 = f_A^2, \quad f_B^1 = f_B^2 \tag{8.20}$$

in which f stands for the fugacity of component A or B in phase 1 or 2.

In the present work, the hydrate phase equilibrium is modeled by using the fugacity approach as proposed by Klauda and Sandler (2000, 2002, 2003). This approach is based on solving the condition of equal fugacities of water in the hydrate phase and the fluid phases:

$$f_W^H(T,P) = f_W^{\pi}(T,P,x) \tag{8.21}$$

For solving these conditions, the fugacity of water in the fluid phase is calculated with an equation of state (EOS), whereas the fugacity of water in the hydrate phase is calculated from Eq. (8.22).

$$f_W^H(T,P) = f_W^{\beta}(T,P)\exp\left(\frac{-\Delta\mu_W^H(T,P)}{RT}\right) \tag{8.22}$$

in which

$$f_W^{\beta}(T,P) = f_W^{\text{sat},\beta}(T,P)\exp\left(\frac{V_W^{\beta}(T,P)\left(P - P_w^{\text{sat},\ \beta}(T)\right)}{RT}\right) \tag{8.23}$$

In Eq. (8.23), f_W^{β} stands for the fugacity of the hypothetical, empty hydrate lattice. This fugacity is influenced by the guest molecule(s) of the clathrate hydrate cavities, which take into consideration the various lattice distortion degrees caused by various guests (Klauda and Sandler, 2000). The chemical potential difference between the empty and occupied cage of the hydrate $\Delta\mu_W^H$ is calculated according to the van der Waals and Platteeuw (VdWP) (1959) statistical thermodynamic theory.

8.3.1.2.1 The Cubic-Plus-Association Equation of State

The Cubic-Plus-Association (CPA) EOS demonstrated by Kontogeorgis et al. (1996) combines an association term like that found in Statistical Associating Fluid Theory (SAFT) approaches with the physical term from the cubic Soave–Redlich–Kwong (SRK) equation of state. This approach has been proven to provide accurate descriptions of complex systems involving water and other complex chemicals of hydrogen-bonding character (Kontogeorgis and Folas, 2010).

In pressure-explicit form, the CPA equation of state can be formulated as follows (Michelsen and Hendriks, 2001; Kontogeorgis et al., 2006):

$$P = \frac{RT}{V_m - b} - \frac{\alpha(T)}{V_m(V_m + b)} - \frac{RT}{2V_m}\left[1 + \frac{1}{V_m}\frac{\partial \ln g}{\partial(1/V_m)}\right] \times \sum_i X_i \sum_{A_i}(1 - X_{A_i}) \tag{8.24}$$

in which R stands for the gas constant and T represents temperature. V_m stands for the molar volume, $\alpha(T)$ denotes the temperature-dependent SRK energy parameter and b represents the SRK covolume parameter. g is the hard sphere radial distribution function. A_i denotes association site A on component i; x_i is the mole fraction of component i; X_{A_i} is the fraction of sites, type A on component i, not bonded to other sites. CPA simplifies to the SRK equation of state for nonassociating systems.

The fraction of nonbonded sites, X_{A_i}, is predicted by unraveling Eqs. (8.25) and (8.26) as follows:

$$X_{A_i} = \frac{1}{\left[1 + \frac{1}{V_m}\sum_j X_j \sum_{B_j} X_{B_j}\Delta^{A_iB_j}\right]} \tag{8.25}$$

Eq. (8.25) is examined for all site types on all associating elements. The summation over B_j in Eq. (8.25) specifies summation over all association sites.

$\Delta^{A_iB_j}$ stands for the association strength between site A on molecule i and site B on molecule j. It may be estimated by

$$\Delta^{A_iB_j} = g(V_m)^{\text{ref}} \times \left[\exp\left(\frac{\varepsilon^{A_iB_j}}{RT}\right) - 1\right] \times b_{ij} \times \beta^{A_iB_j} \tag{8.26}$$

$\varepsilon^{A_iB_j}$ and $\beta^{A_iB_j}$ are the association energy and volume, respectively, between site A on molecule i and site B on molecule j. $g(V_m)_{ref}$ stands for the contact value of the radial distribution function for the reference hard-sphere fluid system.

The radial distribution function, $g(V_m)$, was demonstrated in a simplified CPA (SCPA) formula by Kontogeorgis et al. (1999). Although previous forms of CPA utilized the Carnahan–Starling expression for the hard-sphere radial distribution function, SCPA employs the formula illustrated through Eq. (5.3.4) for the simplified hard-sphere radial distribution function.

$$g(V_m) = \frac{1}{\left[1 - 1.9 \times b \times \frac{1}{4V_m}\right]} \tag{8.27}$$

The temperature-dependent energy parameter, $\alpha_i(T)$, for pure component i, in the SRK term is determined through Eq. (8.28).

$$\alpha_i(T) = \alpha_{0,i}\left[1 + C_{i,i}\left(1 - \sqrt{T_R}\right)\right]^2 \tag{8.28}$$

in which $a_{0,i}$ and $c_{1,i}$ stand for pure-component parameters and T_R represents the reduced temperature for component i. For associating components, the CPA EOS employs five pure-element parameters, $a_{0,i}$, b_i, $c_{1,i}$, $\varepsilon^{A_iB_i}$, and $\beta^{A_iB_i}$. Nonassociating elements are depicted through three pure-element parameters, $a_{0,i}$, b_i, and $c_{1,i}$ in a manner like that of the "standard" SRK EOS. Via fitting the model to the saturated-liquid densities and experimental vapor pressures of the pure component, pure-element parameters for associating components can be determined. Moreover, via critical temperature, $T_{c,i}$, critical pressure, $P_{c,i}$, and the acentric factor, ω_i, the three pure-element parameters for nonassociating compounds can also be calculated.

In binary systems, the van der Waals one-fluid mixing rules are employed for calculating the SRK parameters, $\alpha(T)$ and b. This is done according to Eqs. (8.29) and (8.30) (Kontogeorgis et al., 2006).

$$\alpha = \sum_{i=1}^{n}\sum_{j=1}^{m} x_i x_j \alpha_{ij} \tag{8.29}$$

$$w = \sum_i x_i b_i \tag{8.30}$$

In which the "classical" combining rules are applied for the binary $\alpha_{ij}(T)$ in the SRK term and the binary b_{ij} in the association term.

$$\alpha_{ij} = \sqrt{\alpha_{ii}\alpha_{jj}}(1 - K_{ij}) \tag{8.31}$$

$$b_{ij} = \frac{(b_{ii} + b_{jj})}{2} \tag{8.32}$$

k_{ij} in Eq. (8.31) stands for the binary interaction parameter (BIP) between component i, and component j. k_{ij} may be temperature dependent, e.g., according to Eq. (8.33)

$$k_{ij} = a_{kij} + \frac{b_{kij}}{T} \tag{8.33}$$

For the association parameters of CPA, no mixing rules are required. Only for cross-associating systems, combining rules is necessary to obtain the two association parameters $\varepsilon^{A_iB_j}$ and $\beta^{A_iB_j}$. This work utilizes the CR1 combining rules (CRs) according to Eqs. (8.34) and (8.35):

$$\varepsilon^{A_iB_j} = \frac{(\varepsilon^{A_jB_j} + \varepsilon^{A_iB_i})}{2} \tag{8.34}$$

$$\beta^{A_jB_j} = \sqrt{\beta^{A_jB_j} \times \beta^{A_iB_i}} + \gamma^{A_iB_j} \tag{8.35}$$

The CR for $\beta^{A_iB_j}$, Eq. (8.35), has been formulated in a general form, which handles both cross-association between two self-associating molecules as well as cross-association between one self-associating and one non-self-associating molecule (solvation). In the case of cross-association between two self-associating molecules, $\gamma^{A_iB_j}$ may either be set to zero, to allow approach estimation according to the standard CR1 combining rule, or it can be used as an adjustable parameter on the cross-association interactions. In cases with cross-association involving one non-self-associating molecule, a nonzero $\gamma^{A_iB_j}$ is required to give cross-association interactions.

Because only binary interactions may be accounted for (directly) in the process of parameter prediction, CPA becomes predictive for systems containing three or more elements.

8.3.1.2.2 Peng–Robinson Equation of State

The Peng–Robinson (PR) EOS is a common model between both industrial engineers and scientific scholars as it is relatively precise for the estimation of density, vapor pressure, and other thermodynamic properties of slightly polar and nonpolar fluids. However, it is not as accurate for estimating properties of strongly polar and the compounds associated with water (Peng and Robinson, 1976). The equation is a pressure-explicit expression formulated as follows:

$$P = \frac{RT}{v - b} - \frac{a(T)}{v(v + b) + b(v - b)} \tag{8.36}$$

in which T stands for the temperature of the system, R represents the gas constant, v is the molar volume, and a and b are constants specific to each component. For a fluid, b is proportional to the size of the molecule or the molecular volume and is calculated based only on the critical temperature, T_c, and the critical pressure, P_c, according to the expression:

$$b = 0.07780\frac{RT_c}{P_c} \tag{8.37}$$

The term $a(T)$ is an expression that characterizes the intermolecular attractive interactions as a product of a temperature-dependent term, $\alpha(T)$, and a constant, $a(T_c)$ as indicated by Eq. (8.38):

$$a(T) = a(T_c)\alpha(T, \omega) \tag{8.38}$$

in which each of these terms is expressed by Eqs. (8.39)–(8.41) and depend on an additional parameter, ω, the acentric factor, which represents the deviations of the intermolecular potential from that of a perfectly spherical molecule (Prausnitz et al., 1999):

$$a(T_c) = 0.45724\frac{(RT_c)^2}{P_c} \tag{8.39}$$

$$\alpha(T) = \left[1 + \beta\left(1 - \sqrt{\frac{T}{T_c}}\right)\right]^2 \tag{8.40}$$

$$\beta = 0.37464 + 1.54226\omega - 0.26992\omega^2 \tag{8.41}$$

From this formulation, for a system to be fully defined the critical temperature, critical pressure, and the acentric factor for each component are required. In Table 8.4, these properties are given for the various components examined here.

The PR EOS is extended to mixtures using appropriate mixing rules. The most common mixing rules are the one-fluid van der Waals mixing rules:

$$a = \sum_i \sum_j x_i x_j a_{ij} \tag{8.42}$$

Table 8.4 Molar Mass, Critical Properties, and Acentric Factor of Pure Components (Prausnitz et al., 1999)

Chemical Species	MM (g/mol)	T_c (K)	P_c (MPa)	ω
O_2	32.00	154.6	5.04	0.025
Ar	39.95	150.8	4.87	0.001
N_2	28.00	126.2	3.39	0.039
CO	28.01	132.9	3.50	0.066
CO_2	44.01	304.1	7.38	0.239
CH_4	16.04	190.4	4.60	0.011
C_2H_6	30.07	305.3	4.87	0.099
C_3H_8	44.10	369.8	4.25	0.153
i-C_4H_{10}	58.12	408.2	3.65	0.183
H_2O	18.01	647.3	22.12	0.344
H_2S	34.08	373.2	8.94	0.081

in which a_{ij} is the cross-interaction parameter and is mathematically defined using Eq. (8.43):

$$a_{ij} = \sqrt{a_i a_j} \tag{8.43}$$

$$b = \sum_i x_i b_i \tag{8.44}$$

8.3.1.2.3 Perturbed Chain-Statistical Associating Fluid Theory

Unlike cubic EOSs, which are based on the van der Waals EOS, the SAFT family of EOSs is based on statistical mechanics principles. These molecular theories have become popular in the recent years due to their improved accuracy compared to more classical methods. Chapman and co-workers developed SAFT EOS (Chapman et al., 1990, 1989) based on Wertheim's first-order thermodynamic perturbation theory that defined a relationship between the Helmholtz energy and the association interactions of a molecule (Wertheim, 1984a,b, 1986a,b). In particular, this theory models the behavior of real fluids by describing different interactions as a series of perturbations to a reference fluid. SAFT is based on a reference fluid composed of hard spheres in which the attractive and repulsive interactions are based on the modified-square well potential model suggested by Chen and Kreglewski (Chen and Kreglewski, 1977).

After the development of the original SAFT, many other versions of the EOS were derived and vary from the original one in the type of the reference fluid or potential model used. SAFT-Variable Range (VR) (Gil-Villegas et al., 1997; Galindo et al., 1998) implemented the use of a variable-range square well intermolecular potential; soft-SAFT (Blas and Vega, 1997, 1998) used a Lennard–Jones reference fluid, whereas Perturbed Chain (PC)-SAFT is based on a hard-chain reference fluid. There are many other variations of the SAFT EOSs, and several comprehensive reviews (Müller and Gubbins, 2001; Economou, 2002) have covered their differences in detail.

The formulation of the PC-SAFT EOS is based on the calculation of the residual Helmholtz energy, a^{res}, in terms of the summation the Helmholtz contributions of different intermolecular interactions according to the expression:

$$\frac{a^{res}}{RT} = \frac{a}{RT} - \frac{a^{ideal}}{RT} = \frac{a^{hc}}{RT} + \frac{a^{disp}}{RT} + \frac{a^{assoc}}{RT} \tag{8.45}$$

The reference fluid is composed of a hard-chain fluid, in which its segments are freely jointed and defined exclusively by their hard-core repulsive interactions. The Helmholtz energy contribution of the reference fluid, a^{hc}, is a mathematical combination of the Helmholtz free energy of a hard-sphere reference fluid used in the SAFT EOS [the Carnahan–Starling expression (Mansoori et al., 1971)] and the energy of chain formation.

The addition of the dispersion perturbation, a^{disp}, to the reference fluid is used to calculate attractive interactions in the fluid. This potential model is described by the chain segment diameter, σ, and the energy of dispersion interactions between segments, ε. For simple nonassociating molecules, PC-SAFT utilizes an additional parameter, the number of segments in the nonspherical molecule, m, for a full description of the molecular shape and size.

Dispersion interactions in PC-SAFT are modeled using a two-term perturbation expansion. Both terms in the expansion are dependent on the integral of the radial distribution function. Gross and Sadowski (2001) simplified these integrals to a density power series (Eqs. (8.60) and (8.62)). After this simplification, these power series depend only on constant coefficients fitted to pure alkane data and the number of segments.

In this formulation, the final perturbation to the system that was considered is the associating interactions of a molecule, such as the ability of water to form hydrogen bonds. The contribution of the association interactions to the Helmholtz energy, a^{assoc}, have been derived based on Wertheim's perturbation theory. The central conclusion of Wertheim's work was the derivation of the fraction of associating sites of a component, X.

Chapman et al. later extended this theory and introduced the strength of association between unlike sites, Δ^{AB}. To be able to calculate this quantity, two additional parameters are used: the association energy between sites A and B of molecule i, $\varepsilon^{A_iB_i}$, and the volume of associating interactions, $\kappa^{A_iB_i}$.

All five parameters described here were fitted to pure-component saturation data from temperatures near the triple point to slightly below the critical point. These parameters were fitted by minimizing the difference between the equilibrium pressures and saturated-liquid density calculated using the EOS and the experimental values. The fundamental basis of the EOS remains unchanged when multicomponent mixtures are studied by fitting the parameters this way.

A number of other intermediate- and long-range intermolecular forces such as polarizability effects and ionic interactions can be accounted for by the inclusion of additional terms to the expression. Such forces are not accounted explicitly in this work.

To extend PC-SAFT to mixtures, the mixing rules in Eqs. (8.46) and (8.47), derived based on the van der Waals mixing rules are used to describe the dispersion interactions between different molecules (Gross and Sadowski, 2001):

$$\overline{m^2\varepsilon\sigma^3} = \sum_i \sum_j x_i x_j m_i m_j \left(\frac{\varepsilon_{ij}}{kT}\right) \sigma_{ij}^3 \tag{8.46}$$

$$\overline{m^2\varepsilon^2\sigma^3} = \sum_i \sum_j x_i x_j m_i m_j \left(\frac{\varepsilon_{ij}}{kT}\right)^2 \sigma_{ij}^3 \tag{8.47}$$

In these equations, the binary cross-interaction parameters, σ_{ij} and ε_{ij}, are calculated using the classical Lorentz–Berthelot-combining rules:

$$\sigma_{ij} = \frac{1}{2}(\sigma_i + \sigma_j) \tag{8.48}$$

$$\varepsilon_{ij} = \sqrt{\varepsilon_i \varepsilon_j} \tag{8.49}$$

To calculate thermodynamic properties of fluids in different phases, the pure-component PC-SAFT parameters m, σ, ε, ε^{AB}, and κ^{AB} need to be fitted to vapor pressure and saturated-liquid density data. In this work, experimental data from the National Institute of Standards and Technology (NIST) (Lemmon et al., n.d.) database were used for this purpose. All molecules studied within this work are modeled as nonassociating molecules with the exception of water, which is modeled as a molecule with two

Table 8.5 PC-SAFT Pure-Component Parameters

Component	MM (g/mol)	m	σ (Å)	ε/k (K)	ε^{AB}/κ (K)	κ^{AB}
CO_2	44.01	2.6037	2.555	151.04		
CH_4	16.04	1.0000	3.704	150.03		
O_2	32.00	1.1217	3.210	114.96		
Ar	39.95	0.9285	3.478	122.23		
N_2	28.00	1.2053	3.313	90.96		
CO	28.01	1.3195	3.231	91.41		
H_2O	18.02	1.9599	2.362	279.42	2059.28	0.1750
H_2S	34.08	1.7129	3.053	224.01		
C_2H_6	30.07	1.6040	3.532	191.47		
C_3H_8	44.10	2.0011	3.630	207.90		
i-C_4H_{10}	58.12	2.2599	3.774	216.25		

associating sites. The optimization of these parameters was achieved through the minimization of the deviation of the EOS prediction of both the saturation pressure and liquid density of the components from the experimental value. The parameters for all of the components studied in this work are displayed in Table 8.5. These parameters were refitted here and are consistent with parameters previously reported in the literature (Diamantonis et al., 2013; Diamantonis, 2013).

8.3.1.2.3.1 Mathematical Formulation of PC-SAFT In the rest of the appendix, the mathematical description of each Helmholtz free-energy term in the PC-SAFT EOS (Gross and Sadowski, 2001, 2000, 2002) is provided. The starting expression is:

$$\frac{a^{\text{res}}}{RT} = \frac{a}{RT} - \frac{a^{\text{ideal}}}{RT} = \frac{a^{\text{hc}}}{RT} + \frac{a^{\text{disp}}}{RT} + \frac{a^{\text{assoc}}}{RT} \tag{8.50}$$

The ideal Helmholtz free energy is:

$$\frac{a^{\text{ideal}}}{RT} = \ln \rho - 1 \tag{8.51}$$

8.3.1.2.3.1.1 Hard-chain Reference Fluid This section explains the terms needed to calculate the Helmholtz free energy of the hard-chain reference fluid, a^{hc}:

$$\frac{a^{\text{hc}}}{RT} = \overline{m}\frac{a^{\text{hs}}}{RT} - \sum_i x_i(m_i - 1)\ln g_{ii}^{\text{hs}}(d_{ii}) \tag{8.52}$$

The only input parameters that are directly required by Eq. (8.52) are m, the number of segments in the nonspherical molecule, and x_i, the mole fraction of component i. $\overline{m}$ is defined as the average number of segments:

$$\overline{m} = \sum_i x_i m_i \tag{8.53}$$

This equations also requires the Helmholtz energy of hard spheres that constitute the chain, a^{hs}:

$$\frac{a^{\text{hs}}}{RT} = \frac{1}{\zeta_0}\left(\frac{3\zeta_1\zeta_2}{(1-\zeta_3)} + \frac{\zeta_2^3}{\zeta_3(1-\zeta_3)^2} + \left(\frac{\zeta_2^3}{\zeta_3^2} - \zeta_0\right)\ln(1-\zeta_3)\right) \tag{8.54}$$

The final parameter in Eq. (8.52) is the hard-sphere radial distribution function, g_{ij}^{hs}:

$$g_{ij}^{\text{hs}} = \frac{1}{(1-\zeta_3)} + \left(\frac{d_i d_j}{d_i + d_j}\right)\frac{3\zeta_2}{(1-\zeta_3)^2} + \left(\frac{d_i d_j}{d_i + d_j}\right)^2 \frac{3\zeta_2^2}{(1-\zeta_3)^3} \tag{8.55}$$

Both the radial distribution function and the Helmholtz free energy of the hard-sphere reference fluid require the following expression:

$$\zeta_n = \frac{\rho\pi}{6}\sum_i x_i m_i d_i^n \quad n \in \{0, 1, 2, 3\} \tag{8.56}$$

The number density of the fluid is defined as the partial volume fraction at $n = 3$. The number density can then be converted to the density of the fluid:

$$\eta = \zeta_3 \tag{8.57}$$

In Eq. (8.56), d is the segment diameter that is a function of temperature and two parameters, the chain-segment diameter, σ_i, and the energy of dispersion, ε_i:

$$d_i = \sigma_i\left[1 - 0.12\exp\left(-3\frac{\varepsilon_i}{kT}\right)\right] \tag{8.58}$$

8.3.1.2.3.1.2 Dispersion Interactions The dispersion contribution to the residual Helmholtz free energy is based on the second-order perturbation theory that models a^{disp} as a function of two terms:

$$\frac{a^{\text{disp}}}{RT} = -2\pi\rho I_1(\eta, \overline{m})\overline{m^2\varepsilon\sigma^3} - \pi\rho\overline{m}C_1 I_2(\eta, \overline{m})\overline{m^2\varepsilon^2\sigma^3} \tag{8.59}$$

I_1 and I_2 are power series expansions that represent simplified integrals of the radial distribution function of the hard chain and depend only on the average number of segments and the number density:

$$I_1(\eta, \overline{m}) = \sum_{i=0}^{6} a_i(\overline{m})\eta^i \tag{8.60}$$

$$a_i(\overline{m}) = a_{0i} + \frac{\overline{m}-1}{\overline{m}}a_{1i} + \frac{\overline{m}-1}{\overline{m}}\frac{\overline{m}-2}{\overline{m}}a_{2i} \tag{8.61}$$

$$I_2(\eta, \overline{m}) = \sum_{i=0}^{6} b_i(\overline{m})\eta^i \tag{8.62}$$

$$b_i(\overline{m}) = b_{0i} + \frac{\overline{m}-1}{\overline{m}}b_{1i} + \frac{\overline{m}-1}{\overline{m}}\frac{\overline{m}-2}{\overline{m}}b_{2i} \tag{8.63}$$

The parameters, a_{0i} to a_{2i} and b_{0i} to b_{2i}, are constants presented by Gross and Sadowski (2001). In Eq. (8.59), C_1 is given from the expression:

$$C_1 = 1 + \overline{m}\frac{8\eta - 2\eta^2}{(1-\eta)^4} + (1-\overline{m})\frac{20\eta - 27\eta^2 + 12\eta^3 - 2\eta^4}{(2-3\eta+\eta^2)^2} \tag{8.64}$$

These equations are extended to mixtures using the van der Waal mixing rules.

8.3.1.2.3.1.3 Association Interactions The association contribution to Helmholtz free energy is defined as follows:

$$\frac{a^{\text{assoc}}}{RT} = \sum_i x_i\left[\sum_{A_i}\left(\ln X^{A_i} - \frac{X^{A_i}}{2}\right) + \frac{M_i}{2}\right] \tag{8.65}$$

in which X^{A_i} is the fraction of associating sites in a fluid:

$$X^{A_i} = \left[1 + \sum_j \sum_{B_j} \rho_j X^{B_j} \Delta^{A_i B_j}\right]^{-1} \tag{8.66}$$

The strength of association between unlike sites, $\Delta^{A_i B_j}$, is:

$$\Delta^{A_i B_j} = d_{ij}^3 g_{ij}(d_{ij})^{\text{seg}} \kappa^{A_i B_j}\left[\exp\left(\frac{\varepsilon^{A_i B_j}}{kT}\right) - 1\right] \tag{8.67}$$

in which,

$$d_{ij} = \frac{1}{2}(d_i + d_j) \tag{8.68}$$

8.3.1.3 Iterative Method (K-value Method)

The K-factor is defined as the distribution of the component between the hydrate and the gas (Carroll, 2014):

$$K_i = \frac{y_i}{x_i} \tag{8.69}$$

in which y_i and x_i stand for the mole fractions of component i in the vapor and hydrate, correspondingly. These mole fractions are on a water-free base and water is not involved in the computations. It is presumed that adequate water exists to create a hydrate (Carroll, 2014).

A chart is available for each of the hydrate-forming components commonly encountered in natural gaser: methane, ethane, propane, isobutane, n-butane, hydrogen sulfide, and carbon dioxide (Carroll, 2014).

The vapor—liquid K-factors can be gained from the K-factor charts in the Gas Processors Suppliers Association (GPSA) Engineering Data Book or one of the other straightforward or complicated methods accessible in the literature (Carroll, 2014).

All nonformers are simply assigned a value of infinity, because by definition $x_i = 0$ for nonformers; there is no nonformer in the hydrate phase. This is true for both light nonformers, such as hydrogen, and heavy ones, such as n-hexane and n-pentane (Carroll, 2014).

For a total of 1 lb-mole of a hydrocarbon mixture, i.e., $n = 1$, at the hydrate formation pressure we have following conditions (Carroll, 2014):

$$n_h \approx 0$$

$$n_v \approx 1$$

In the aforementioned circumstances, the overall composition, z_i, and the composition of the vapor phase, y_i, are the same. Performing these limitations to Eq. (8.69) results in (Carroll, 2014)

$$\sum_i \frac{z_i}{n_v K_i + n_l} = \sum_i \frac{z_i}{K_i} = 1 \tag{8.70}$$

in which $z_i =$ total composition of the system under consideration.

A trial-and-error method should be used to determine the hydrate-formation pressure, p_{hyd}. The following flowchart (Fig. 8.3) demonstrates the process of iterative method for predicting hydrate-formation pressure (Carroll, 2014).

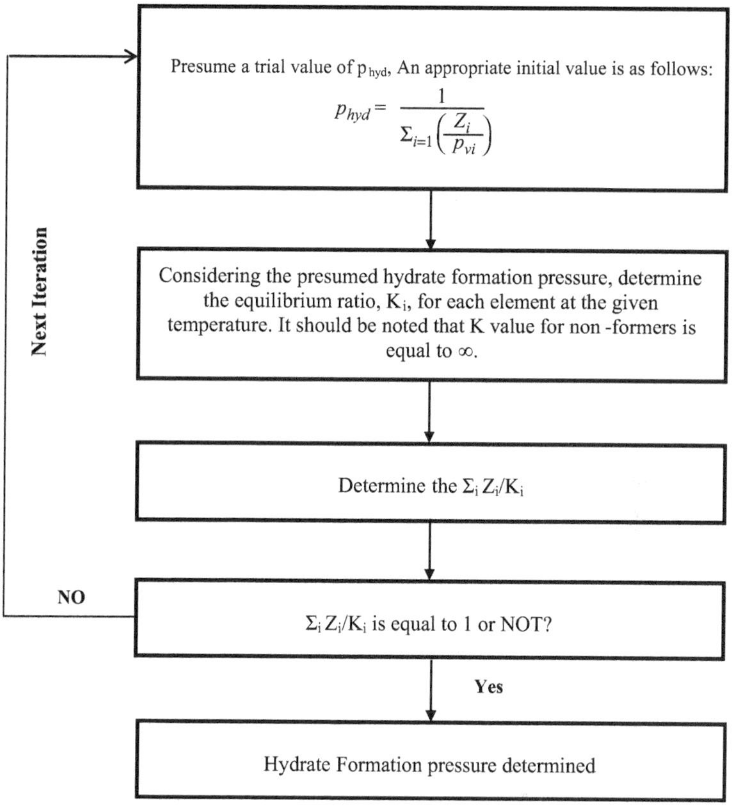

Figure 8.3 Flowchart for predicting hydrate-formation pressure by iterative equilibrium ratio method.

8.4 HYDRATE DEPOSITION

The main area in which hydrate can deposit is from water-saturated fluids in gas-export pipelines creating possible plugs owing to dehydrator failure (Kane et al., 2008). Although a few hydrate formation flow-loop investigations have been conducted for gas-dominated systems (Sloan et al., 2011; Matthews et al., 2000), hydrate deposition has not been precisely detected and/or fully investigated on a lab scale. In annular flow of water-saturated natural gas, it is possible that water condenses out of the vapor phase on the walls. In the area of hydrate formation, an important uncertainty in gas pipelines is the mechanism for hydrate deposition on the wall of the pipe (Rao et al., 2013).

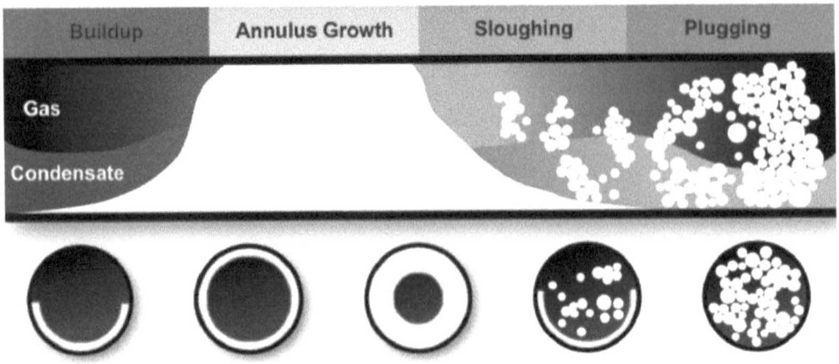

Figure 8.4 Graphical illustration for hydrate formation, deposition, and plugging in gas dominated/condensate systems (Sloan et al., 2011; Rao et al., 2013). *Adapted from Lingelem, M.N., Majeed, A.I., Stange, E., 1994. Industrial experience in evaluation of hydrate formation, inhibition, and dissociation in pipeline design and operation. In: Proceedings of International Conference on Natural Gas Hydrates. Wiley, New York.*

Different scholars (Turner et al., 2009; Camargo et al., 2000; Hernandez et al., 2004) have studied hydrate formation and plugging experiments throughout oil pipelines. Aspenes et al. (2010) studied one mechanism for hydrate deposition for hydrate adhesion on water-wetted surfaces. They concluded that higher surface free energy caused larger adhesion forces. It was found that hydrates formed throughout the bulk fluid phase would adhere to the wetted wall surface. Nicholas et al. (2009) illustrated that if the hydrates form on a cold carbon-steel surface, they are likely to adhere to the surface, and a considerable larger fracture force $\gg 100$ mN/m is needed to remove them. The process of growing hydrate film on surfaces is very slow which may need days to be detected, a timescale too lengthy to be experienced in flow loops (Sloan et al., 2011; Rao et al., 2013; Lachance et al., 2012).

The probability of condensation of the dissolved water in the vapor on the surfaces of the wall is considerable in annular flow of water-saturated hydrocarbon fluid. A hydrate film creates and the thickness of the hydrate film grows from the wall in a stenosis buildup, as illustrated in Fig. 8.4 (Rao et al., 2013).

8.5 HYDRATE INHIBITIONS

For hydrate to be stable, necessary conditions are presence of water, suitably sized gas–liquid molecules, and suitable temperature and pressure conditions. To avoid hydrate problems, injecting inhibitors has been utilized as the most economical method. These chemicals based on their operational concentration are classified to: (1) thermodynamic inhibitors, e.g., methanol, ethanol, glycols; (2) Low Dosage Hydrate Inhibitors which are in turn classified into (1) Kinetic hydrate inhibitors (KHI), and (2) Antiagglomerants (AA) (Sloan, 2003).

A broadly used thermodynamic approach is based on methanol injection. Thermodynamic methods using methanol and glycol are costly in offshore developments and onshore processing facilities because of the high treatment amounts required (10–50% of the water phase). Thermodynamic inhibitors prevent hydrate formation by shifting the equilibrium conditions so hydrates form at lower temperatures and higher pressures. Although there are opportunities to optimize thermodynamic inhibitor requirements (McIntyre et al., 2004; Bullin and Bullin, 2004), the high cost of thermodynamic inhibitors has stimulated the search for kinetic inhibitors. The flow-assurance industry is increasingly moving away from such prevention of hydrate formation toward risk management. The risk management viewpoint lets hydrates form, but avoids hydrates agglomerating and creating a plug, or delays hydrate creation within the period of the water residence in the hydrate-prone section of the flow line. Kinetic inhibition approaches are based onpolymer-based chemical injection at low dosages throughout the aquatic phase. As a result, these chemicals are named low-dosage hydrate inhibitors (LDHIs). LDHIs inhibit hydrate nucleation, growth, and agglomeration of hydrate particles. Consequently, they are split up into so-called antiagglomerates (AAs) and kinetic inhibitors (KIs) (Lee and Englezos, 2005). For a successful LDHI design, many parameters must be considered. The most important issues are as follows: hydrate stability zone and maximum degree of subcooling, water cut and other important fluid parameters, salinity and composition, whether to use KIs or AAs, fluid residence times, inhibitor limitations (low temperatures, high pressures, etc.), economical evaluations, safety, operational and environmental issues, initial laboratory testing, corrosion, scale, inhibitor dosage optimization at lab conditions, field tests, monitoring, and reevaluation.

8.5.1 Calculating the Amount of Hydrate Inhibitors

To calculate the amount of the hydrate inhibitors we need to understand the amount of depression in the freezing point. In other words, the freezing point depression let us know the dosage of the hydrate inhibitors should be used. This method is commonly employed to calculate the molar mass of the inhibitors.

The derivation begins with the fundamental relationship for the equilibrium between a solid and a liquid, and, after some simplifying assumptions, the resulting equation is (Carroll, 2003, 2009)

$$x_i = \frac{h_{sl}\Delta T}{RT_m^2} \tag{8.71}$$

in which x_i is the mole fraction of the solute (inhibitor), ΔT is the temperature depression in °C, R stands for the universal gas constant (8.314 J/mol K), and T_m represents the melting point of the pure solvent in K. Rearranging this equation slightly and converting from mole fraction to mass fraction gives (Carroll, 2003, 2009, 2014):

$$\Delta T = \frac{M_s R T_m^2}{h_{sl}} \times \frac{W_i}{(100 - W_i)M_i} = K_S \frac{W_i}{(100 - W_i)M_i} \tag{8.72}$$

in which M_s is the molar mass of the solvent, W_i is the weight percent solute (inhibitor), and M_i is the molar mass of the inhibitor. For water, it is $K_s = 1861$, when International System of Units (SI) units are used. The leading term in this equation contains only constants, so the freezing-point depression is a function of the concentration of the inhibitor and its molar mass (Carroll, 2003, 2009, 2014).

It is worth noting that this equation is not applicable to ionic solutions, such as salt.

Example 8.5

Estimate the freezing point of a 14% solution of methanol in water. Consider following information.

$M_s = 18.015$ g/mol
R = 8.314 J/mol K
$T_m = 273.15$K
$h_{sl} = 6006$ J/mol
$M_i = 32.042$ g/mol

Solution

Using Eq. (8.72) we have

$$\Delta T = \frac{M_s R T_m^2}{h_{sl}} \times \frac{W_i}{(100 - W_i)M_i} = 9.45°\text{C}$$

Thus, the freezing point of the mixture is predicted to be −9.45°C.

8.5.1.1 The Hammerschmidt Method

A comparatively straightforward and broadly employed correlation to estimate the effect of chemicals on the hydrate-forming temperature is the Hammerschmidt expression (Hammerschmidt, 1934, Carroll, 2003, 2009, 2014):

$$\Delta T = \mathrm{K_H}\frac{W}{(100-\ W)M} \tag{8.73}$$

in which ΔT stands for the temperature depression in terms of C, W denotes the concentration of the inhibitor in weight percent throughout the aqueous phase, M represents the molar mass of the inhibitor in terms of g/mol, and $\mathrm{K_H}$ stands for a constant with a value of 1297. If we like to employ this equation in American engineering units, the value of $\mathrm{K_H}$ is equal to 2355 and ΔT should be present in terms of °F. It is worth mentioning that the units on the other two terms remain unchanged (Hammerschmidt, 1934; Carroll, 2003, 2009, 2014).

The concentration in this correlation is on an inhibitorplus-water basis.

Eq. (8.73) can be reorganized to estimate the concentration of the inhibitor needed to produce the anticipated temperature depression (Hammerschmidt, 1934; Carroll, 2003, 2009, 2014), as:

$$W = \frac{100M\Delta T}{\mathrm{K_H} + M\Delta T} \tag{8.74}$$

Example 8.6

The methane hydrate creates at 5°C and 4.26 MPa. Estimate the amount of methanol needed to reduce this temperature by 20°C via the Hammerschmidt method.

Solution

We know that the molar mass of methanol is 32.042 g/mol. from the Hammerschmidt method we have

$$W = \frac{100M\Delta T}{\mathrm{K_H} + M\Delta T} = \frac{100 \times 32.042 \times 20}{1297 + 32.042 \times 20} = 33.069\ Wt\%$$

Consequently, we need 33 wt% of methanol to reduce hydrate-formation temperature by 20°C.

8.5.1.2 The Nielsen–Bucklin Method

Nielsen and Bucklin (1983) employed theories to propose new method for predicting hydrate inhibition of methanol solutions. The mathematical expression of their correlation is as follows:

$$\Delta T = -72\ \ln(1 - x_M) \tag{8.75}$$

in which ΔT stands for the temperature depression in terms of C and x_M denotes the mole fraction of methanol. They noted that the aforementioned correlation is precise up to 88 wt% (mole fraction of 0.8). Their correlation can be reorganized to predict the methanol concentration at the desired temperature depression as follows:

$$x_M = 1 - \exp(-\Delta T/72) \tag{8.76}$$

and then to determine the weight percent from this mole fraction, the following expression is employed:

$$X_M = \left(\frac{x_M M_M}{18.015 + x_M(M_M - 18.015)}\right) \tag{8.77}$$

in which X_M stands for the weight fraction of methanol and M_M represents the molar mass of methanol.

The Nielsen–Bucklin equation has been proposed for employ with methanol; however, the correlation is really free from the choosing of inhibitor. The correlation includes only the concentration of the inhibitor and the characteristics of water. Consequently, supposedly it can be employed for any inhibitor, for which the molecular weight of the solvent is substituted for M_M in Eq. (8.77).

8.5.1.3 McCain Method

In much the same way that they inhibit the formation of ice, ionic solids also prevent the hydrate formation. Several rapid rules of thumb can be found in the literature based on experimental records (Carroll, 2014).

McCain (1990) developed the following equation for predicting the effect of brine on the temperature of hydrate formation:

$$\Delta T = AS + BS^2 + CS^3 \tag{8.78}$$

in which ΔT stands for the temperature depression in terms of °F; S denotes the salinity of ionic liquid in terms of weight percent; and the coefficients A,

B, and C are functions of the gas gravity, γ, and can be calculated via the following equations:

$$A = 2.20919 - 10.5746\gamma + 12.1601\gamma^2 \quad (8.79)$$

$$B = -0.106056 + 0.722692\gamma - 0.85093\gamma^2 \quad (8.80)$$

$$C = 0.00347221 - 0.0165564\gamma + 0.049764\gamma^2 \quad (8.81)$$

Eq. (8.78) is restricted to gas gravities in the range $0.55 < g < 0.68$ and salt concentrations of 20 wt%.

8.5.1.4 Østergaard et al. (2005)

Østergaard et al. (2005) developed an equation that was meaningfully dissimilar from its prototypes. First, they constructed a correlation valid to both organic compounds (such as glycols and alcohols) and inorganic salts (such as $CaCl_2$). Second, their correlation considers the effect of pressure. The mathematical expression of their equation is as follows:

$$\Delta T = \left(c_1 W + c_2 W^2 + c_3 W^3\right)\left(c_4 \ln P + c_5\right)\left(c_6\left(P_0 - 1000\right) + 1\right) \quad (8.82)$$

in which ΔT stands for the temperature depression in terms of K or C; P represents the system pressure in terms of kPa; P_0 stands for the dissociation pressure of hydrocarbon fluid in pure water at 0°C in terms of kPa; W denotes the inhibitor concentration in liquid-water phase in terms of mass percent; and the c_1, c_2, and c_3 are constants which vary for each inhibitor.

The correlation is straightforward to employ if you know the inhibitor concentration and you need to predict the depression. It is more complicated to employ if the temperature is given, and we like to predict the needed concentration of inhibitor because it involves an iterative answer.

8.5.2 Calculating Inhibitor Loss in Hydrocarbon Phase

A simple estimation of the inhibitor losses to the vapor can be predicted supposing that the nonidealities in the vapor phase can be ignored and that Raoult's Law employs. This results in the straightforward expression (Carroll, 2014):

$$y_i = x_i\left(\frac{P_i^{sat}}{P}\right) \quad (8.83)$$

in which x_i is the inhibitor's mole fraction in the aqueous phase, y_i is the mole fraction in the vapor phase, P_{sat} is the inhibitor's vapor pressure, and P is the total pressure. Rewriting the aforementioned equation and employing conversion factors results in this equation presented in more familiar units (SI units) as follows (Carroll, 2014):

$$y_i = \left(\frac{760.4x_iM_i}{100M_i - (M_i - 18.015)x_i}\right)\left(\frac{P_i^{sat}}{P}\right) \tag{8.84}$$

in which x_i is the weight percent inhibitor in the aqueous phase, y_i is the inhibitor in the vapor phase in terms of kilograms per thousand standard cubic meter (kg/MSm^3), and M_i is the molar mass of the inhibitor. Via conversion factor Eq. (8.84) can be written in American engineering units as follows:

$$y_i = \left(\frac{47484x_iM_i}{100M_i - (M_i - 18.015)x_i}\right)\left(\frac{P_i^{sat}}{P}\right) \tag{8.85}$$

in which y_i is the inhibitor in the vapor phase in pounds per million standard cubic feet (lb/MMCF), x_i is the weight percent inhibitor in the aqueous phase, and M_i is the molar mass of the inhibitor. It is worth mentioning that through suitable properties, this equation can be employed for any nonionic inhibitor (Carroll, 2014).

In addition to the loss of inhibitor to the gas, if a liquid hydrocarbon is available, some of the inhibitor will enter that phase too (Carroll, 2014). The GPSA Engineering Data Book gives a graph for the distribution of methanol between an aqueous solution and a liquid hydrocarbon reprinted as Fig. 8.5. Fig. 8.6 is an analogous graph with some leveling. These graphs are appropriate for rough engineering calculations; however, the predicted values from the two graphs are not very similar.

The graph is a chart of the mole fraction in the hydrocarbon liquid as a function of the concentration of methanol in the water-rich phase and the temperature. Using this graph for a certain purpose needs the molar mass of the hydrocarbon liquid. For heavier oils, it may be as large as 1000 g/mol and for light condensate, it could be as low as 125 g/mol (Carroll, 2014).

For weight fractions between 20 and 70 wt%, it is adequately precise to "eye ball" your estimation. For methanol concentrations less than 20 wt%, a linear estimate can be employed, given the fact that at 0 wt% in the water the concentration in the hydrocarbon liquid is also 0 (Carroll, 2014). The resultant expression is:

$$x = \frac{x(20\text{ wt\%})}{20}X \tag{8.86}$$

Parameter x stands for the given weight percent methanol in the aqueous phase, x (20 wt%) represents the mole percent methanol in the condensate at 20 wt% in the water, and x denotes the mole fraction in the hydrocarbon liquid for the specified x. If the hydrocarbon liquid is aromatic, the methanol

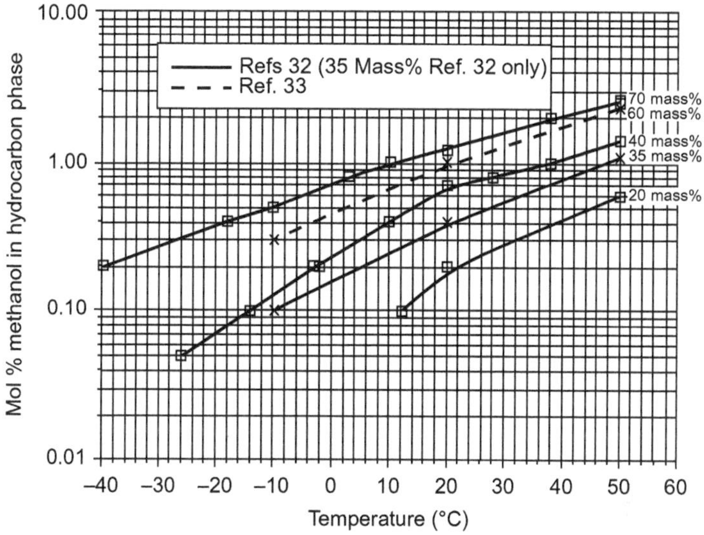

Figure 8.5 Methanol solubility in paraffinic hydrocarbons as a function of pressure, temperature, and aqueous phase composition (GPSA Engineering Data Book, 11th edition).

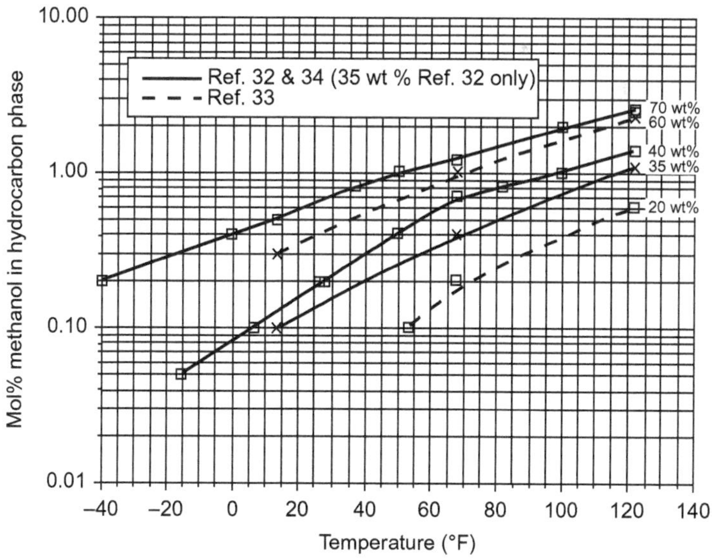

Figure 8.6 Methanol solubility in paraffinic hydrocarbons as a function of pressure, temperature, and aqueous phase composition in American engineering units (GPSA Engineering Data Book, 11th edition).

losses increase. This is proved by the data of Chen et al. (1988). In an paraffin-rich condensate, the methanol losses could be as low as 0.2 times those in an aromatic condensate. It should be noted that the graphs depict no consequence of pressure on the methanol distribution between the two liquid phases (Carroll, 2014).

8.5.3 Inhibitor Injection Rates

Injection rates of methanol in ranges of 0.15–1.5 m^3/day (1–10 bbl/day) are customary in the natural gas process. Sometimes they can be more than 0.15–1.5 m^3/day, but injection rates of more than 1.5 m^3/day become rather costly. Injection pressures greater than 1000 psia (7000 kPa) are usual (Carroll, 2014).

Consequently, the injection pump is intended to operate under the conditions of high pressures and flow rates. Two types of injection pumps are popular (1) a piston pump and (2) a diaphragm pump (Carroll, 2014).

Problems

8.1 Calculate the hydrate formation pressure of methane at 17°C by Makogon method. $\left(\text{Hint}: \ \gamma = \frac{\text{MW}}{28.96}\right)$

8.2 Consider the following gas mixture and calculate the hydrate formation pressure via iterative method (K-value approach) at 9°C. (Hint: You can use correlations for predicting K-value of hydrocarbons and nonhydrocarbons which are explained in chapter on gas condensate)

Component	Mole Fraction
N_2	0.0046
H_2S	0.0348
CO_2	0.0061
C_1	0.6864
C_2	0.139
C_3	0.0689
$i\text{-}C_4$	0.0066
$n\text{-}C_4$	0.0266
$i\text{-}C_5$	0.0062
$n\text{-}C_5$	0.0094
C_6	0.0114

8.3 Consider a gas processing plant in which feed from the gas reservoir operates at a rate of 90 × 10^5 m^3/day. The production also includes 0.08 m^3/day of water, which is to be transported in the same pipeline. The gas enters the pipeline at 39°C and 3887 kPa. The hydrate formation temperature of the gas is determined to be 33°C at 3487 kPa. In the transportation through the pipeline, the gas is expected to cool to 7°C. To prevent hydrate formation, calculate the amount of methanol that must be injected.

8.4 Consider the following gas mixture. Calculate the amount of required brine with 3 wt% salinity to surpass hydrate formation temperature by 5°C via McCain method.

Component	Mole Fraction
CO_2	0.08
H_2S	0.051
C_1	0.869

8.5 Consider the following mixture and calculate the hydrate formation pressure by Makogon method at 8°C.

Component	Mole Fraction
C_1	0.88
H_2O	0.02
CO_2	0.10

8.6 Natural gas flowing in a pipeline exits the line at 60°F and 950 psia and the flow rate of the gas is 8 million standard cubic feet per day (MMSCFD). To prevent hydrate formation, it is estimated that there should be 31 wt% methanol in the aqueous phase. Calculate the methanol losses to the vapor phase.

8.7 Consider the following gas mixture. Calculate the amount of required $CaCl_2$ concentration to surpass hydrate formation temperature by 2.5°C at $P = 5000$ kPa via Østergaard et al. (2005) method.

Component	Mole Fraction
CO_2	0.09
H_2S	0.051
C_1	0.859

The following table reports the values of constants in Østergaard et al. (2005) correlation.

Constant	$CaCl_2$
c_1	0.194
c_2	7.58×10^{-3}
c_3	1.953×10^{-4}
c_4	4.253×10^{-2}
c_5	1.023
c_6	2.8×10^{-5}

$R = 8.314\,\mathrm{J/mol\,K}$

$T_m = 273.15\mathrm{K}$

$h_{sl} = 6006\,\mathrm{J/mol}$

8.8 Consider a protein that inhibits the formation of ice is in the blood of a human. Further presume that this inhibition is simply a freezing-point depression. Estimate the concentration of the protein to achieve a 1.75°C depression. Typically, assume a value of 2682 g/mol. Consider the following information:

$R = 8.314$ J/mol K
$T_m = 273.15$K
$h_{sl} = 6006$ J/mol

8.9 Consider the following mixture and calculate the hydrate formation pressure by Kobayashi, and Bahadori and Vuthaluru approaches at 11°C.

Component	Mole Fraction
CO_2	0.08
H_2S	0.10
C_2	0.82

8.10 Consider the following gas mixture. Calculate the amount of hydrate formation depression when 4 wt% NaCl used at $P = 5000$ kPa via Østergaard et al. (2005) method.

Component	Mole Fraction
CO_2	0.09
H_2S	0.051
C_1	0.859

The following table reports the values of constants in Østergaard et al. (2005) correlation.

Constant	$CaCl_2$
c_1	0.3534
c_2	1.375×10^{-3}
c_3	2.433×10^{-4}
c_4	4.056×10^{-2}
c_5	0.7994
c_6	2.25×10^{-5}

8.11 Consider the following mixture and calculate the hydrate formation pressure by Motiee method at 6°C.

Component	Mole Fraction
CO_2	0.05
H_2S	0.06
C_1	0.77
C_2	0.12

8.12 Methane hydrate forms at 22.5°C and 32 MPa. Calculate the amount of methanol required to suppress this temperature by 14.5°C via

a. The Hammerschmidt equation

b. The Nielsen–Bucklin equation

8.13 Calculate the freezing point of a 23% solution of methanol in water. Consider the following information

$M_s = 18.015$ g/mol

$R = 8.314$ J/mol K

$T_m = 273.15$K

$h_{sl} = 6006$ J/mol

$M_i = 32.042$ g/mol

8.14 Consider the following gas mixture and calculate the hydrate formation pressure via iterative method (K-value approach) at 9°C. (Hint: You can use correlations for predicting K-value of hydrocarbons and nonhydrocarbons, which are explained in the chapter on gas condensate)

Component	Mole Fraction
N_2	0.0054
H_2S	0.0053
CO_2	0.035
C_1	0.689
C_2	0.1364
C_3	0.0689
n-C_4	0.0332
n-C_5	0.0156
C_6	0.0112

REFERENCES

Aspenes, G., Dieker, L.E., Aman, Z.M., Høiland, S., Sum, A.K., Koh, C.A., Sloan, E.D., 2010. Adhesion force between cyclopentane hydrates and solid surface materials. Journal of Colloid and Interface Science 343 (2), 529–536.

Bahadori, A., Vuthaluru, H.B., 2009. A novel correlation for estimation of hydrate forming condition of natural gases. Journal of Natural Gas Chemistry 18, 1–5.

Blas, F.J., Vega, L.F., 1998. Prediction of binary and ternary diagrams using the Statistical Associating Fluid Theory (SAFT) equation of state. Industrial and Engineering Chemistry Research 37, 660–674.

Blas, F.J., Vega, L.F., 1997. Thermodynamic behaviour of homonuclear and heteronuclear Lennard-Jones chains with association sites from simulation and theory. Molecular Physics 92, 135–150.

Bullin, K.A., Bullin, J.A. (Eds.), March 2004. Optimizing Methanol Usage for Hydrate Inhibition in a Gas Gathering System. 83rd Annual GPA Convention.

Camargo, R., Palermo, T., Sinquin, A., Glenat, P., 2000. Rheological characterization of hydrate suspensions in oil dominated systems. In: Gas Hydrates: Challenges for the Future, vol. 912. Wiley, New York, pp. 906–916.

Carroll, J., 2003. Natural Gas Hydrates: A Guide for Engineers. Gulf Professional Publishing, 225 Wyman Street, Waltham, MA, USA.
Carroll, J., 2009. Natural Gas Hydrates: A Guide for Engineers. Gulf Professional Publishing, 225 Wyman Street, Waltham, MA, USA.
Carroll, J., 2014. Natural Gas Hydrates: A Guide for Engineers, third ed. Gulf Professional Publishing. 225 Wyman Street, Waltham, MA 02451, USA.
Chapman, W., Gubbins, K., Jackson, G., Radosz, M., 1990. New reference equation of state for associating liquids. Industrial and Engineering Chemistry Research 29, 1709–1721.
Chapman, W.G., Gubbins, K., Jackson, G., Radosz, M., 1989. SAFT: equation-of-state solution model for associating fluids. Fluid Phase Equilibria 52, 31–38.
Chen, C.J., Ng, H.-J., Robinson, D.B., 1988. The Solubility of Methanol or Glycol in Water-Hydrocarbon Systems. GPA Research Report RR-117. Gas Processors Association, Tulsa, OK.
Chen, S., Kreglewski, A., 1977. Applications of the augmented van der Waals theory of fluids. I. Pure fluids. Berichte Der Bunsen-Gesekkschaft 81, 1048–1052.
Diamantonis, N., Boulougouris, G.C., Mansoor, E., Tsangaris, D., Economou, I.G., 2013. Evaluation of cubic, SAFT, and PC-SAFT equations of state for the vapor-liquid equilibrium modeling of CO2 mixtures with other gases. Industrial and Engineering Chemistry Research 52, 3933–3942.
Diamantonis, N.I., 2013. Mathematical Modeling of Thermophysical Properties and Phase Equilibria of Pure Carbon Dioxide and Multicomponent Mixtures. National Technical University of Athens (Ph.D. thesis).
Economou, I., 2002. Statistical associating fluid theory: a successful model for the calculation of thermodynamic and phase equilibrium properties of complex fluid mixtures. Industrial and Engineering Chemistry Research 41, 953–962.
Elgibaly, A.A., Elkamel, A.M., 1998. A new correlation for predicting hydrate formation conditions for various gas mixtures and inhibitors. Fluid Phase Equilibria 152 (1), 23–42.
Galindo, A., Davies, L., Gil-Villegas, A., Jackson, G., 1998. The thermodynamics of mixtures and the corresponding mixing rules in the SAFT-VR approach for potentials of variable range. Molecular Physics 106, 241–252.
Gil-Villegas, A., Galindo, A., Whitehead, P.J., Mills, S.J., Jackson, G., Burgess, A.N., 1997. Statistical associating fluid theory for chain molecules with attractive potentials of variable range. Journal of Chemical Physics 106, 4168–4186.
Gross, J., Sadowski, G., 2000. Application of perturbation theory to a hard-chain reference fluid: an equation of state for square-well chains. Fluid Phase Equilibria 168, 183–199.
Gross, J., Sadowski, G., 2002. Application of the perturbed-chain SAFT equation of state to associating systems. Industrial and Engineering Chemistry Research 41, 5510–5515.
Gross, J., Sadowski, G., 2001. Perturbed-chain SAFT: an equation of state based on a perturbation theory for chain molecules. Industrial and Engineering Chemistry Research 40, 1244–1260.
Hammerschmidt, E.G., 1934. Formation of gas hydrates in natural gas transmission lines. Industrial and Engineering Chemistry 26, 851.
Hernandez, O.C., Hensley, H., Sarica, C., Brill, J.P., Volk, M., Delle-Case, E., 2004. Improvements in single-phase paraffin deposition modeling. SPE Production and Facilities 19 (4), 237–244. http://www.pet.hw.ac.uk/research/hydrate/images/hydrates/structures_large.jpg.
Kane, M., Singh, A., Hanssen, R., USA, T. E. P, 2008. Hydrates blockage experience in a deepwater subsea dry gas pipeline: lessons learned. In: Proceedings of the Offshore Technology Conference, Houston, Texas, May 5–8, 2008. Society of Petroleum Engineers, Houston, TX.

Katz, D.L., 1945. Prediction of conditions for hydrate formation in natural gases. In: Petroleum Development and Technology 1945, vol. 160. Transactions of the American Institute of Mining and Metallurgical Engineers, AIME, New York. SPE-945140-G, 140.

Klauda, J.B., Sandler, S.I., 2000. A fugacity model for gas hydrate phase equilibria. Industrial and Engineering Chemistry Research 39 (9), 3377–3386.

Klauda, J.B., Sandler, S.I., 2002. Ab initio intermolecular potentials for gas hydrates and their predictions. Journal of Physical Chemistry B 106 (22), 5722–5732.

Klauda, J.B., Sandler, S.I., 2003. Phase behavior of clathrate hydrates: a model for single and multiple gas component hydrates. Chemical Engineering Science 58, 27–41.

Kobayashi, R., Song, K.Y., Sloan, E.D., 1987. In: Bradley, H.B. (Ed.), Petroleum Engineers Handbook. SPE, Richardson, TX, pp. 25-1–25-28.

Kontogeorgis, G.M., Folas, G.K., 2010. Thermodynamic Models for Industrial Applications. From Classical and Advanced Mixing Rules to Association Theories. John Wiley and Sons, United Kingdom.

Kontogeorgis, G.M., Michelsen, M.L., Folas, G.K., Derawi, S., von Solms, N., Stenby, E.H., 2006. Ten years with the CPA (Cubic-Plus-Association) equation of state. Part 1. Pure compounds and self-associating systems. Industrial and Engineering Chemistry Research 45, 4855–4868.

Kontogeorgis, G.M., Voutsas, E.C., Yakoumis, I.V., Tassios, D.P., 1996. An equation of state for associating fluids. Industrial and Engineering Chemistry Research 35, 4310–4318.

Kontogeorgis, G.M., Yakoumis, I.V., Meijer, H., Hendriks, E., Moorwood, T., 1999. Multicomponent phase equilibrium calculations for water-methanol-alkane mixtures. Fluid Phase Equilibria 158–160, 201–209.

Lachance, J.W., Talley, L.D., Shatto, D.P., Turner, D.J., Eaton, M.W., 2012. formation of hydrate slurries in a once-through operation. Energy Fuels 26 (7), 4059–4066.

Lee, J.D., Englezos, P., 2005. Enhancement of the performance of gas hydrate kinetic inhibitors with polyethylene oxide. Chemical Engineering Science 60 (19), 5323–5330.

Lemmon, E.W., McLinden, M.O., Friend, D.G., n.d. Thermophysical properties of fluid systems. In: Linstrom, P.J., Mallard, W.G. (Eds.), NIST Chem. WebBook, NIST Stand. Ref. Database Number 69. http://webbook.nist.gov.

Lingelem, M.N., Majeed, A.I., Stange, E., 1994. Industrial experience in evaluation of hydrate formation, inhibition, and dissociation in pipeline design and operation. In: Proceedings of International Conference on Natural Gas Hydrates. Wiley, New York.

Makogon, Y.F., 1981. Hydrates of Hydrocarbons. PennWell, Tulsa, OK.

Mansoori, G.A., Carnahan, N.F., Starling, K.E., Leland Jr., T.W., 1971. Equilibrium thermodynamic properties of the mixture of hard spheres. Journal of Chemical Physics 54, 1523–1525.

Matthews, P.N., Notz, P.K., Widener, M.W., Prukop, G., 2000. Flow loop experiments determine hydrate plugging tendencies in the field. In: Gas Hydrates: Challenges for the Future, vol. 912. Wiley, New York, pp. 330–338.

McCain, W.D., 1990. The Properties of Petroleum Fluids, second ed. PennWell Publishing Co., Tulsa, OK.

McIntyre, G., Hlavinka, M., Hernandez, V., Bryan, T. (Eds.), 2004. Hydrate Inhibition with Methanol—A Review and New Concerns over Experimental Data Presentation. 83rd Annual GPA Convention, New Orleans.

Michelsen, M.L., Hendriks, E.M., 2001. Physical Properties from association models. Fluid Phase Equilibria 180, 165–174.

Motiee, M., July 1991. Estimate possibility of hydrate. Hydrological Processes 70 (7), 98–99.

Müller, E., Gubbins, K., 2001. Molecular-based equations of state for associating fluids: a review of SAFT and related approaches. Industrial and Engineering Chemistry Research 40, 2193–2211.

Nicholas, J.W., Dieker, L.E., Sloan, E.D., Koh, C.A., 2009. Assessing the feasibility of hydrate deposition on pipeline walls-Adhesion force measurements of clathrate hydrate particles on carbon steel. Journal of Colloid and Interface Science 331 (2), 322–328.

Nielsen, R.B., Bucklin, R.W., April 1983. Why not use methanol for hydrate control? Hydrocarbon Processing 62 (4), 71–78.

Østergaard, K.K., Masoudi, R., Tohidi, B., Danesh, A., Todd, A.C., 2005. A general correlation for predicting the suppression of hydrate dissociation temperature in the presence of thermodynamic inhibitors. Journal of Petroleum Science and Engineering 48, 70–80.

Østergaard, K.K., Tohidi, B., Danesh, A., Todd, A.C., Burgass, R.W., 2000. A general correlation for predicting the hydrate-free zone of reservoir fluids. SPE Production and Facilities 15, 228–233.

Peng, D.Y., Robinson, D.B., 1976. A new two-constant equation of state. Industrial and Engineering Chemistry Fundamentals 15, 59–64.

Prausnitz, J.M., Lichtenthaler, R.N., de Azevedo, E.G., 1999. Molecular Thermodynamics of Fluid-Phase Equilibria, third ed.

Rao, I., Koh, C.A., Sloan, E.D., Sum, A.K., 2013. Gas hydrate deposition on a cold surface in water-saturated gas systems. Industrial and Engineering Chemistry Research 52 (18), 6262–6269.

Sloan, E.D., 2003. Fundamental principles and applications of natural gas hydrates. Nature 426 (6964), 353–363.

Sloan, E.D., Koh, C.A., Sum, A.K., Ballard, A.L., Creek, J.L., Eaton, M.W., Lachance, J.W., McMullen, N., Palermo, T., Shoup, G.J., Talley, L.D., 2011. Natural Gas Hydrates in Flow Assurance. Gulf Professional Pub./Elsevier, Burlington, MA p. xii, 200.

Sloan, E.D., 2005. Fluid Phase Equilibria 228–229, 67.

Sloan, E.D., 2000. Hydrate Engineering, SPE Inc. Texas, Richardson.

Sloan, E.D., Koh, C.A., 2008. Clathrate Hydrates of Natural Gases, third ed. CRC Press, Boca Raton, FL.

Sun, C.-Y., Chen, G.-J., Lin, W., Guo, T.-M., 2003. Hydrate formation conditions of sour natural gases. Journal of Chemical and Engineering Data 48, 600–603.

Towler, B.F., Mokhatab, S., April 2005. Quickly estimate hydrate formation conditions in natural gases. Hydrological Processes 61–62.

Turner, D.J., Miller, K.T., Sloan, E.D., 2009. Methane hydrate formation and an inward growing shell model in water-in-oil dispersions. Chemical Engineering Science 64 (18), 3996–4004.

van der Waals, J.H., Platteeuw, J.C., 1959. Advances in Chemical Physics, 2, pp. 1–57.

Walas, S.M., 1985. Phase Equilibria in Chemical Engineering. Butter-worth Publishers, Stoneham, MA.

Welling and Associates, 1999. Survey, cited by Macintosh, N., flow assurance still leading concern among producers. Offshore. October (2000).

Wertheim, M.S., 1984a. Fluids with highly directional attractive forces. I. Statistical thermodynamics. Journal of Statistical Physics 35, 19–34.

Wertheim, M.S., 1984b. Fluids with highly directional attractive forces. II. Thermodynamic perturbation theory and integral equations. Journal of Statistical Physics 35, 35–47.

Wertheim, M.S., 1986a. Fluids with highly directional attractive forces. III. Multiple attraction sites. Journal of Statistical Physics 42, 459–476.

Wertheim, M.S., 1986b. Fluids with highly directional attractive forces. IV. Equilibrium polymerization. Journal of Statistical Physics 42, 477–492.

CHAPTER NINE

Characterization of Shale Gas

M.A. Ahmadi[1], A. Bahadori[2,3]
[1]Petroleum University of Technology (PUT), Ahwaz, Iran
[2]Southern Cross University, Lismore, NSW, Australia
[3]Australian Oil and Gas Services, Lismore Pty Ltd, NSW, Australia

9.1 INTRODUCTION

During recent years the increasing global energy demand has engaged awareness concerning alternative sources for energy, comprising both unconventional petroleum resources and "renewable energy resources". Oil and gas production from unconventional petroleum reservoirs is possible just using the specific technologies (Zou et al., 2012). Such petroleum reservoirs comprise coal-bed methane (CBM), shale gas, basin-center gas, tight gas, gas hydrates, heavy oil, oil and gas in fractured shale and chalk, tar sands, and shallow biogenic gas [United States Geological Survey (USGS, 2005)].

Fundamental differences exist between unconventional and conventional reservoirs:

- Conventional reservoirs are completely separated from the source rock due to buoyancy-drive mechanism;
- Conventional reservoirs are found in stratigraphic or structural traps, which are defined as porous reservoir rocks sealed in place by impermeable caprocks or faults;
- The unconventional reservoirs consist of large volumes of rock formations laterally charged with hydrocarbons, and they do not depend on buoyancy and gravity of water, gas, and oil for production;
- The reservoirs in unconventional fields coexist with the source which usually encompasses only one formation;
- Conventional reservoirs are usually discrete fields, whereas unconventional reservoirs have large diffuse boundaries and spatial extension (see Fig. 9.1).

Unconventional petroleum accumulations are found in passive continental margin basins, foreland thrust zones, and in basins of the foreslope areas of foreland basins. They tend to occur in giant structures in regional

Fluid Phase Behavior for Conventional and Unconventional Oil and Gas Reservoirs
http://dx.doi.org/10.1016/B978-0-12-803437-8.00009-9

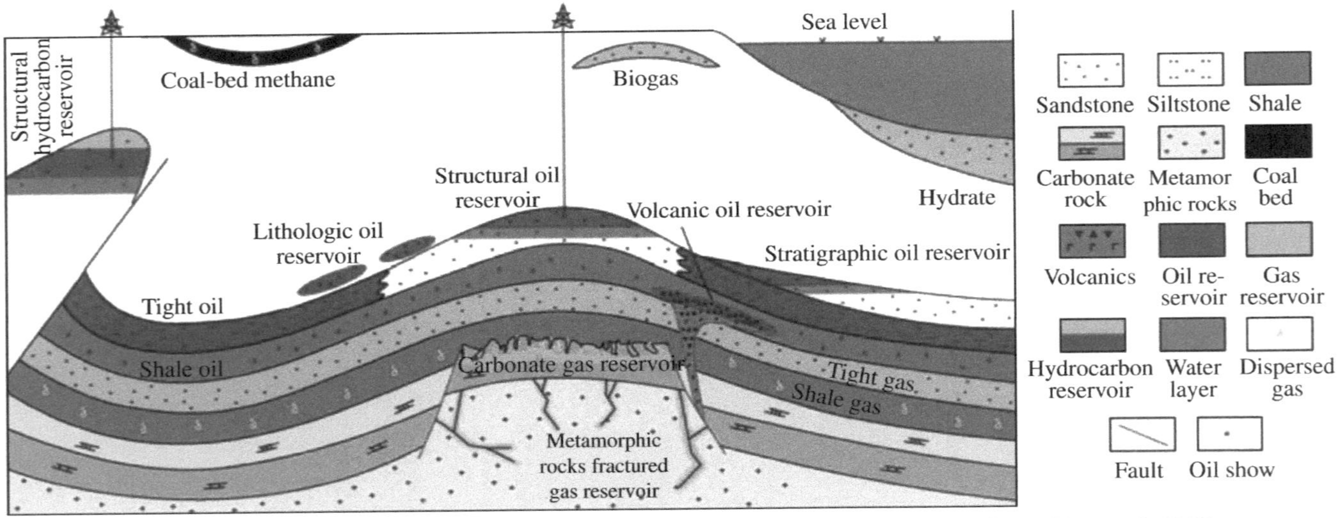

Figure 9.1 Distribution model of different unconventional and conventional hydrocarbons (Zou et al., 2012).

Table 9.1 Differences Between Unconventional Gas Reservoirs (Zou et al., 2012)

Characteristics	Shale Gas	Coal-Bed Methane (CBM)	Tight Gas
Location	Close to sedimentation center of the basin	Distribution area of continental higher plants	Basin center or slope
Porosity	<4–6%	Most less than 10%	Most less than 10%
Permeability (10^{-3} μm^2)	<0.001–2 × 10^3	Most less than 1	Most less than 1
Configuration of reservoir source rock	Source rocks, reservoirs, and seals are in one	Source rocks, reservoirs, and seals are in one	Reservoir contact source rocks directly or adjacent
Seepage	Desorption, diffusion	Non-Darcy flow dominates	Non-Darcy flow dominates
Fluid	Dry gas, adsorbed gas in kerogen and pores, free gas in fractures	Absorbed gas dominates, minor amount of free gas	Gas saturation varies greatly, most less than 60%
Occurrence	Diffused and gas enriched in fractures	Fracture or cleat area	Dissolution pores and fracture area

slope and basin centers—depressions where vast deposits of petroleum source rocks occur (Zou et al., 2012).

Global unconventional natural gas resources include shale gas, tight gas, coal-bed methane, and natural gas hydrates. According to recent research, the unconventional gas is approximately 8.3 times than that of the global conventional gas, pointing to a promising future [International Energy Agency (IEA, 2009); United States Geological Survey (USGS, 2001); Energy Information Administration (EIA, 2004)]. Table 9.1 demonstrated the major differences between unconventional gas reservoirs.

9.2 SHALE GAS RESERVOIR CHARACTERISTICS

Oil and gas exist in deep rock formations, where porosity is the bulk pore volume in which fluids and organic matter are found, and permeability is defined as the difficulty or ease of the movement of the fluid in the matrix.

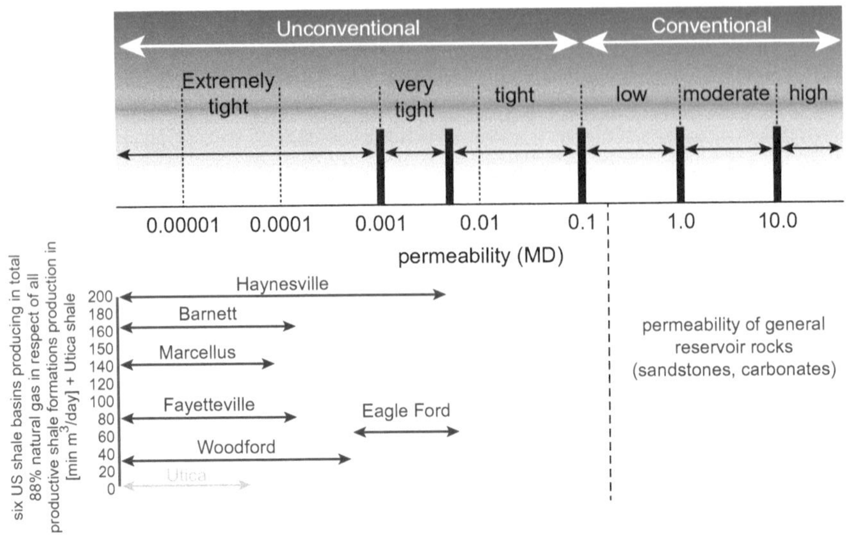

Figure 9.2 Ranges of permeability for both conventional and unconventional oil/gas reservoirs (http://www.ogj.com/articles/print/volume-111/issue-4/exploration—develo pment/economics-fiscal-competitiveness-eyed.html).

Typically, permeability and porosity indicate proportionality in their values. In shale formation, both parameters are typically reversely proportional, i.e., if porosity is high, there might not be adequate interconnectivity in the matrix and oil or gas resources are not recoverable. This is owing to the small pore sizes interconnected in the matrix. Fig. 9.2 depicts the ranges of permeability for both conventional and conventional oil/gas reservoirs along with several examples including US shale reservoirs.

9.3 BASIC SCIENCE BEHIND CONFINEMENT

Based on International Union of Pure and Applied Chemistry (IUPAC) pore-size distribution, Alharthy et al. (2013) proposed three different thermodynamic phase behavior' paths related to vapor—liquid equilibrium conditions in porous media:

- **Unconfined** pore phase behavior, which is mainly the unshifted phase behavior in macropores or what would be in a pressure—volume—temperature (PVT) cell with no pore-confinement effects. These occupy majority of the macro- and mesopores (60—80%).

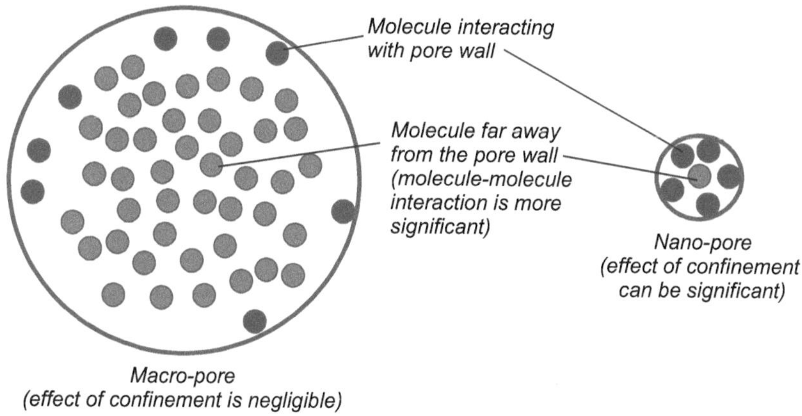

Figure 9.3 Schematic representation of confinement effect (Haider, 2015).

- **Mid-confined** pore phase behavior, which is the partially shifted phase behavior in mesopores and is considered between unconfined and confined pore phase behaviors. This category occupies about 10–15% of the pores.
- **Confined** pore phase behavior, which is the shifted phase behavior in nanopores in which the pore size is less than 3 nm; these occupy about 3–5% of pores and are dependent on the mineralogy (clay content) (Kuila, 2013).

In a porous solid with interconnected pathways, a molecule may collide with another molecule or with the pore walls. When the pore size is relatively larger, the number of interactions the molecules have with the pore walls is negligible as compared to the number of interactions molecules have with one another. This, however, does not hold true as the pore size gets smaller (Chandra, 2014; Kuila, 2013). In tight reservoirs, pore sizes become comparable to the size of the fluid molecules trying to flow through them (Nelson, 2009). Fig. 9.3 illustrates this effect schematically. The red particles indicate the molecules that get to interact with the pore wall at a given time. Because a nanopore can hold fewer molecules when compared to macropores, the interaction among molecules (van der Waals interactions) and between molecules and the pore wall increases. Fluid molecules in such a condition are termed to be "confined" or under "pore proximity" effect because the free path available to the molecules is restricted by the geometry of the void space of pore (Chandra, 2014). Roughly speaking, confinement effect is felt when the pore size to molecule size ratio is less than 20 (Devegowda, 2012). Pore throat diameter that is typical in shale gas formations has been shown to vary between 0.5 and 100 nm (Ambrose, 2010), whereas the chain diameter of straight hydrocarbons is in the range of 0.4–0.6 nm (Mitariten, 2005; Haider, 2015).

The properties of molecules in the surface layer (the ones close to the pore wall as indicated by molecules colored in red (light gray in print versions) in Fig. 9.3) would be affected by increased pore wall–fluid interactions, which leads to alteration in dynamics of molecules in the surface layer sticking to the pore wall (Chandra, 2014). Overall, the situation of molecules in confined geometry is both theoretically and experimentally very complicated and not fully understood (Haider, 2015).

9.3.1 Impact of Confinement on Critical Properties

When a phase envelope is crossed in gas condensate systems, there is a large gas–oil volume split in the nano-, meso-, and macropores (Alharthy, 2013). This is presumed to be responsible for economical production of liquids in such systems. Simulations and experimental data reveal that critical properties of many compounds change as pore size decreases (Singh, 2009; Devegowda, 2012). It was illustrated by Kuz (2002) that, to properly account for the behavior in confined fluids, the critical properties of components should be altered as a function of the ratio of molecule to pore size. They developed a correlation for the deviation of critical temperature and pressure from van der Waals equation of state (EOS) by studying confined fluids in square cross-section pores. Though they neglected the interaction between the fluid molecules and the wall, they did find good agreement between the predicted capillary condensation and critical temperature and experimental data (Haider, 2015).

Hamada (2007) used grand canonical Monte Carlo numerical simulations to study thermodynamic properties of confined Lennard–Jones (LJ) particles in silt and cylindrical pore systems and indicated changes in fluid-phase behavior as a function of pore radius (Zee Ma, 2016). Singh (2009) investigated the behavior of methane (C_1), n-butane (C_4), and n-octane (C_8) inside nanoscale slits with widths between 0.8 and 5 nm using grand canonical Monte Carlo simulations and found out that, whereas critical temperature decreased with reduction in pore radius, the critical pressures of *n*-butane and *n*-octane first increased and subsequently decreased. They also found that the critical property shift is dependent on pore-surface types and hence differed for mica and graphite. This work is of importance as shale rocks are characterized by organic and inorganic pore systems both of which vary in mineral composition and thus will cause different intensities of pore wall and molecule interaction (Devegowda, 2012; Haider, 2015).

Teklu (2014) extended the work of Singh (2009) to a Bakken fluid sample and found that shifts in critical properties led to the suppression of the Bakken fluid-phase envelope. Alharthy (2013) also used the correlations developed by Singh (2009) to investigate the impact of confinement on various variations of Eagle Ford composition. They found that a shift in critical properties led to an increase in condensate production, and this increase was a function of both pore size and composition (Haider, 2015).

Singh (2009) reported critical properties shift due to the pore-proximity effect for methane, *n*-butane, and *n*-octane. Ma et al. (2013) and Jin et al. (2013) developed a series of correlations to take into account the effect of confinement on hydrocarbon critical properties. These correlations are shown as follows (Sanaei et al., 2014):

$$\Delta T_c = \frac{T_c - T_{cz}}{T_c} = 1.1775\left(\frac{D}{\sigma}\right)^{-1.338} \quad \text{for} \quad \left(\frac{D}{\sigma}\right) \geq 1.5 \tag{9.1}$$

$$\Delta T_c = \frac{T_c - T_{cz}}{T_c} = 0.6 \quad \text{for} \quad \left(\frac{D}{\sigma}\right) \leq 1.5 \tag{9.2}$$

$$\Delta P_c = \frac{P_c - P_{cz}}{P_c} = 1.5686\left(\frac{D}{\sigma}\right)^{-0.783} \tag{9.3}$$

in which ΔT_c, ΔP_c are the critical temperature and pressure shift due to confinement, respectively. T_c and P_c are critical temperature (°F) and critical pressure (psi) for bulk state, respectively. T_{cz} and P_{cz} are critical temperature (°F) and critical pressure (psi) under confinement, respectively. D is the pore diameter (nm) and σ is the effective molecular diameter (nm), which is the diameter of the smallest cross section of a molecule (Ma et al., 2013; Jin et al., 2013). The effective molecular diameter can be calculated via the following equation (Haider, 2015):

$$\sigma_{LJ} = 0.244\sqrt[3]{\frac{T_{cb}}{P_{cb}}} \tag{9.4}$$

in which σ_{LJ} is Lennard–Jones size parameter (collision diameter in nm), T_{cb} is bulk critical temperature (K), and P_{cb} is pore critical pressure (atm) (Haider, 2015).

Example 9.1

Calculate the compressibility factor and viscosity of methane gas at a pressure range of 0–5000 pounds per square inch absolute (psia) and constant

(Continued)

temperature of 180°F when pore radius is equal to 1, 2, 5, 10, and 50 nm. Plot compressibility factor, ratio of gas viscosity of methane under confinement to its bulk state versus corresponding pressure for different pore radius.

Hint: To determine Z-factor use the Dranchuk and Abou-Kassem (1975) method and to calculate viscosity use Lee et al. (1966) equation.

Answer

To investigate the effect of confinement on gas properties, methane as the primary component of natural gas is considered, and gas compressibility factor and gas viscosity for different pore sizes are calculated (Sanaei et al., 2014).

First, methane critical properties were modified for each pore size using Eqs. (9.1) to (9.3). Second, using the modified critical pressure and temperature, gas compressibility factor (z), and viscosity of this component are calculated.

Fig. 9.4 demonstrates the calculated Z-factor for different pore radii. It can be seen that as pore size decreases, the Z-factor increases. This increase is negligible for a 50-nm diameter capillary and dramatic increase can be seen when the pore size is less than 5 nm. Fig. 9.5 shows the ratio of gas viscosity of methane under confinement to its bulk state for different pore sizes. This figure shows a decrease in gas viscosity with a decrease in pore size. Again, the similarity to the Z-factor, when pore size is less than 10 nm, gas viscosity deviates significantly from bulk value and the major change can be seen for pore sizes less than 5 nm (Sanaei et al., 2014).

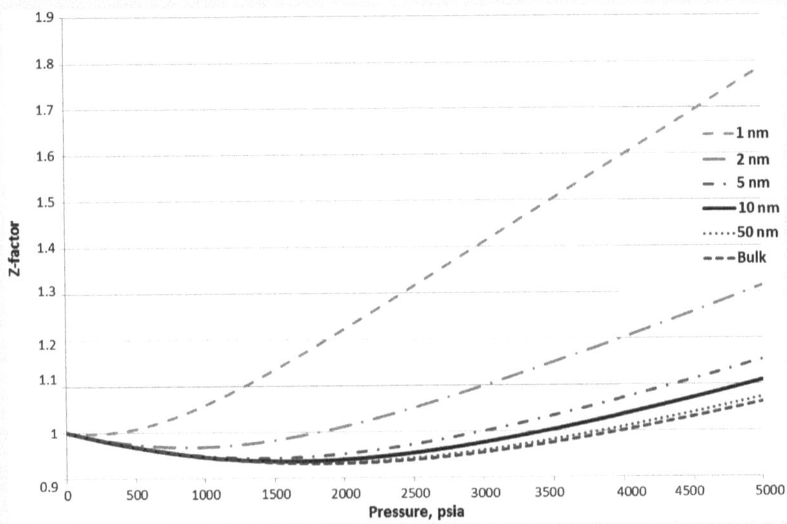

Figure 9.4 Effect of confinement on methane deviation factor at 180°F as a function of pressure (Sanaei et al., 2014).

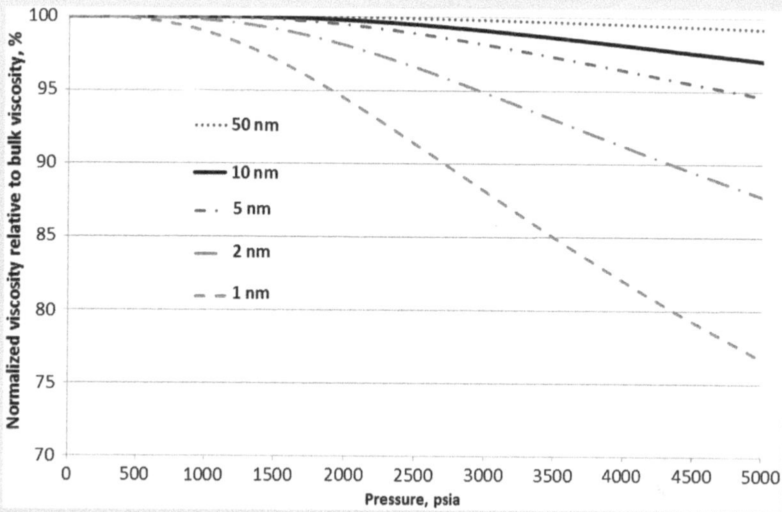

Figure 9.5 Normalized gas viscosity relative to bulk state viscosity for methane at 180°F (Sanaei et al., 2014).

Example 9.2

Consider the Eagle Ford sample fluid mixture with composition demonstrated in Fig. 9.6. Using Peng–Robinson EOS calculate the phase envelope of this fluid sample when pore radius is equal to 5, 10, 15, and 30 nm. Moreover, determine the effect of pore radius on dew-point pressure.

Answer

To see the pore-proximity effect on a two-phase diagram, first, critical pressure and temperature shift for each component of fluid mixture are calculated. Second, these updated critical properties are used in commercial PVT package software and modified phase envelope is calculated using the Peng–Robinson EOS. Fig. 9.7 shows different phase envelopes for 5, 10, 15, and 30 nm pore sizes and bulk state. As the pore size decreases, the phase envelope shrinks, critical pressure and temperature drop, and the critical point shifts to the left. The fluid behaves more like a dry gas as the pore size decreases. Additionally, by decreasing the pore size, dew-point pressure decreases between 5 and 24%.

(Continued)

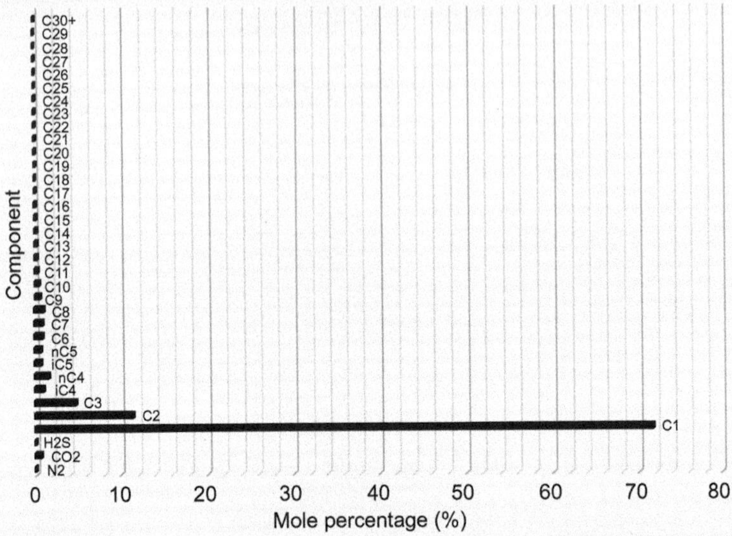

Figure 9.6 Reservoir fluid composition (Eagle Ford) (Sanaei et al., 2014).

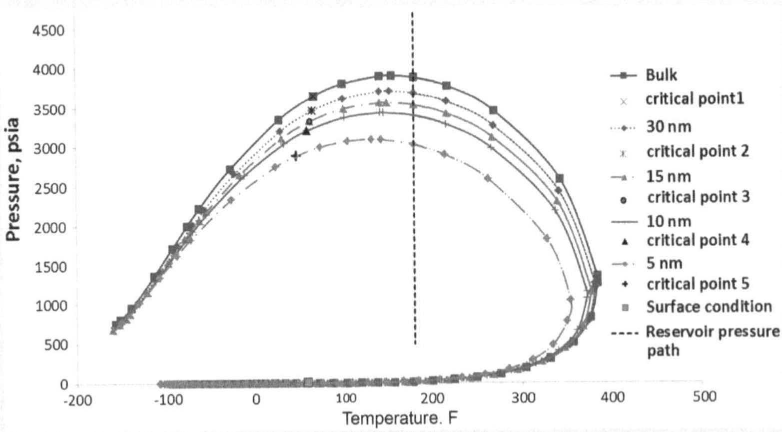

Figure 9.7 Two-phase envelope change for Eagle Ford gas condensate sample (Sanaei et al., 2014).

From this figure it can also be concluded that at a constant pressure and temperature significant decrease in liquid dropout is expected considering confinement. This result is very important because this indicates that less condensate drop out is expected for a reservoir with smaller pore sizes (Sanaei et al., 2014).

9.3.2 Diffusion Effect Due to Confinement

Confinement may give rise to Knudson diffusion. As discussed previously, in a porous solid, a molecule may collide with another molecule or with the pore walls. At high pressure, molecule–molecule collisions are dominant. At low pressure, collisions are dominantly between molecules and the walls, and the free path is restricted by the geometry of the void spaces (Rotelli, 2012). This regime is termed as Knudson diffusion. It combines both the geometry as well as the pressure information of the system. At low Knudson diffusion number the continuum flow regime is valid, but in the regime in which Knudson is approaching unity, the continuum validity possibly breaks down (Rotelli, 2012). The Knudson number is defined by:

$$K_n = \lambda/L \tag{9.5}$$

here, λ is the mean free path traveled by the fluid particle as shown in Fig. 9.8, L is the pore diameter, and K_n is Knudson number. The range of its values in different flow regimes is listed in Table 9.2.

Rotelli (2012) showed that, for gas condensates, diffusion can play an important role especially in small pore sizes and at lower pressures. In multiphase compositions, such as gas condensate reservoirs, the equilibrium gas composition at bubble point differs due to bubble-point suppression. This will be discussed later in this thesis. Having differing gas compositions

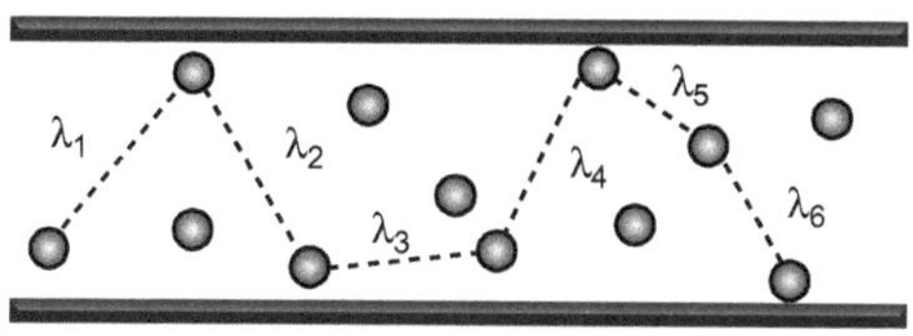

Figure 9.8 Schematic illustration of the mean free path taken by molecules in confinement (Blasingame, 2013).

Table 9.2 Flow Regimes Based on Knudson Diffusion Number

Knudsen Number (K_n)	Flow Regime
$K_n \leq 0.001$	Viscous flow
$0.001 < K_n < 0.1$	Slip flow
$0.1 < K_n < 10$	Transition flow
$K_n \geq 10$	Knudsen's (free molecular) flow

Kuila, U., 2012. Application of Knudsen flow in modelling gas-flow in shale reservoirs. Hyderabad, 9th Biennial International Conference and Exposition on Petroleum Geophysics.

(at the bubble point) in different-sized pores should impact the gas phase growth and may cause flow due to diffusion. Additionally, heterogeneity of the pore-size distribution may be one of the important reasons for concentration gradients causing diffusive flow in an unconventional liquid-rich reservoir (Firincioglu, 2014). Rotelli (2012) showed that, unlike gas condensates, oil is characterized by a more viscous flow, so it tends to move according to Darcy's equation. This is because in the case of oil, molecules tend to interact with each other before they are able to reach the pore wall (Haider, 2015).

9.3.3 Capillary Pressure

Nano size pores can affect the phase behavior of in situ oil and gas owing to increased capillary pressure (Alharthy, 2013; Nojabaei, 2014; Wang, 2013). Not accounting for increased capillarity in small pores can lead to inaccurate estimates of ultimate recovery and saturation pressures. It has been argued that in the presence of capillary forces, the classical thermodynamic behavior is not sufficient to explain gas bubble formation in porous medium (Alharthy, 2013). When capillary forces are considered, the classical thermodynamics approach requires very high super saturation values that are typically not observed in conventional hydrocarbon reservoirs (Firincioglu, 2014; Haider, 2015).

In tight-pore reservoirs, because a relatively significant number of molecules get to interact with the pore walls, the pressure difference between the wetting phase (the phase that sticks to the pore walls) and the nonwetting phase can no longer be ignored. This gives rise to capillary pressure which is (Haider, 2015):

$$P_{\text{cap}} = P_{\text{nw}} - P_{\text{w}} \tag{9.6}$$

in which P_{cap} stands for the capillary pressure, P_{nw} denotes the nonwetting phase pressure, and P_{w} represents the wetting phase pressure.

Investigating the impact of capillary pressure is the prime focus of this work. Based on previously published literature, the presence of capillarity leads to a reduction in oil density and viscosity but to an increase in gas density and viscosity (Nojabaei, 2012; Firincioglu, 2014). Fig. 9.9 shows the alteration of various fluid properties in the presence of capillary pressure. As shown, oil density reduces when capillary pressure becomes significant (Haider, 2015).

Reduction in oil density and viscosity can be attributed to suppression of bubble-point pressure, a phenomenon that arises under the influence of capillarity (Honarpour, 2013; Nojabaei, 2012; Alharthy, 2013; Wang,

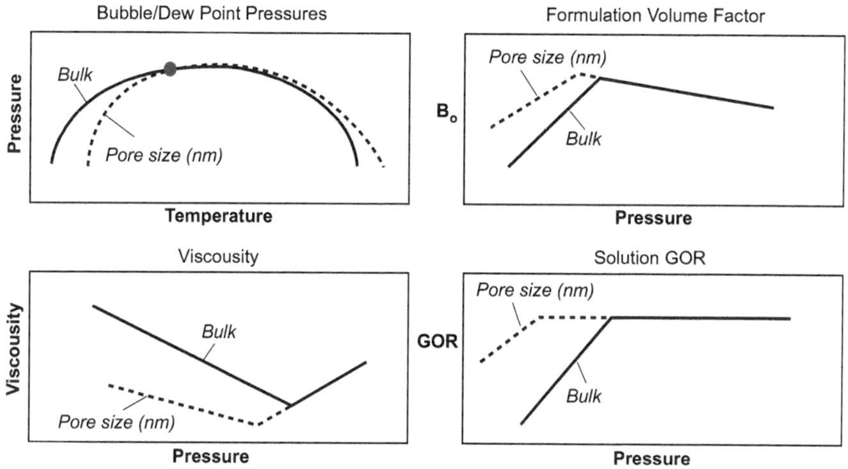

Figure 9.9 Alteration of fluid properties under the influence of capillarity (Honarpour, 2013; Haider, 2015).

2013; Teklu, 2014). Dew point, on the other hand, appears at relatively higher reservoir pressures. The suppression of bubble-point pressure causes gas to be in oil for a longer time as pressure is reduced. Fig. 9.10A shows a schematic representation of the unconfined scenario when gas starts evolving as soon as the reservoir pressure drops below the fluid's bubble-point pressure. Fig. 9.10B, on the other hand, illustrates what happens when confinement makes capillary pressure significant, which in turn causes suppression of the bubble point. Compared to an unconfined system, gas will stay dissolved in oil at lower pressures. This phenomenon is likely to cause an increase in oil production and recovery (Haider, 2015).

Nojabaei (2012) attempted to history match gas production data obtained from a well in the Bakken field using both suppressed bubble point and the original bubble point of the unconfined fluid. This is shown in Fig. 9.11. It was observed that predictions matched the field data well with suppressed bubble-point pressure by producing less gas compared to conventional unconfined systems. This indicates the strong need to further our understanding regarding the potential forces that alter the bubble-point pressure in tight pores (Nojabaei, 2012; Haider, 2015).

9.3.4 Adsorption Phenomenon in Shale Reservoirs

In dry-gas shale reservoirs, it is widely acknowledged that gas adsorption is one of the most important storage mechanisms, and that it accounts for

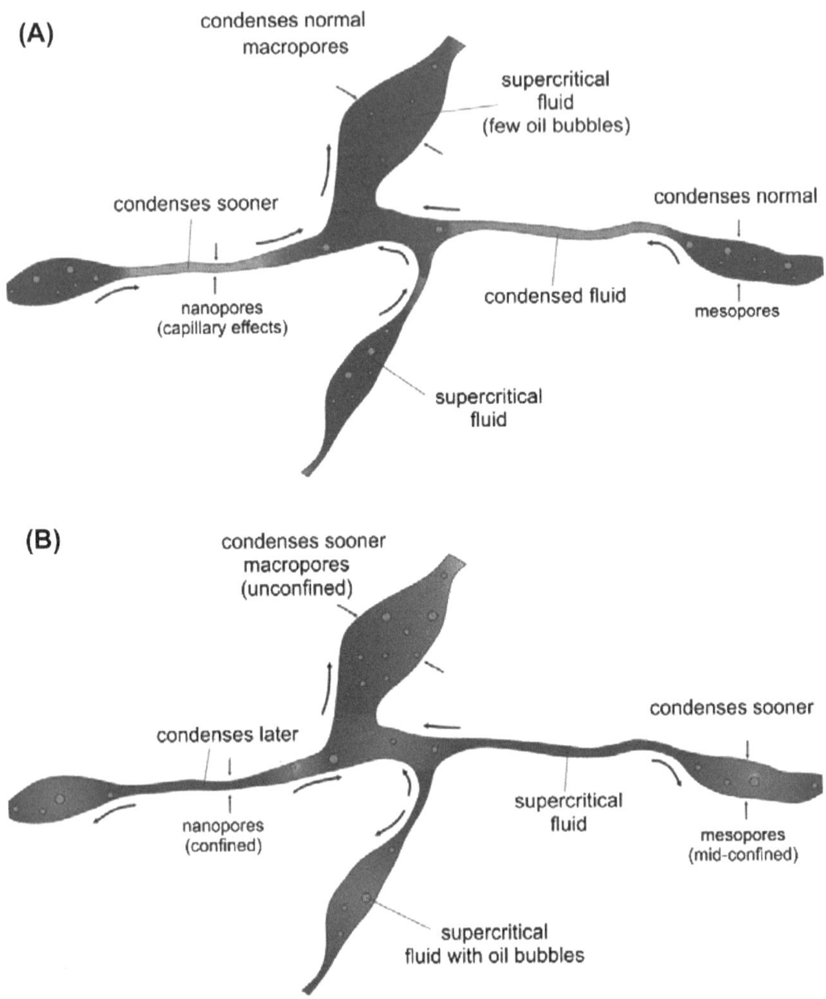

Figure 9.10 Conceptual pore network model showing different phase-behavior paths (A) with phase-behavior shift and (B) without phase-behavior shift (Alharthy, 2013; Haider, 2015).

close to 45% of initial gas storage (Rajput, 2014). In the case of liquid-rich shales, however, adsorption is not typically considered. Using adsorption modeling formalism based on thermodynamically Ideal Adsorbed Solution (IAS) theory, Rajput (2014) showed that 5−13% of the liquid fluid present in shale can be adsorbed onto shale and negligence of this additional storage mechanism can lead to considerable error in reserve estimation. The error values depend on the amount and adsorption parameters of adsorbent present, as well as the composition of liquid-rich shale. Fig. 9.12 shows the

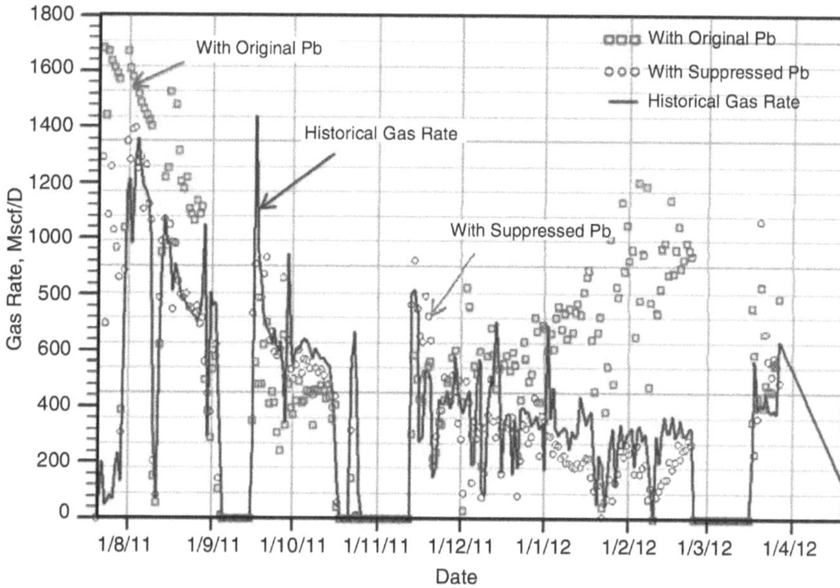

Figure 9.11 History match of gas rate for scenarios with or without PVT adjustments (Nojabaei, 2012).

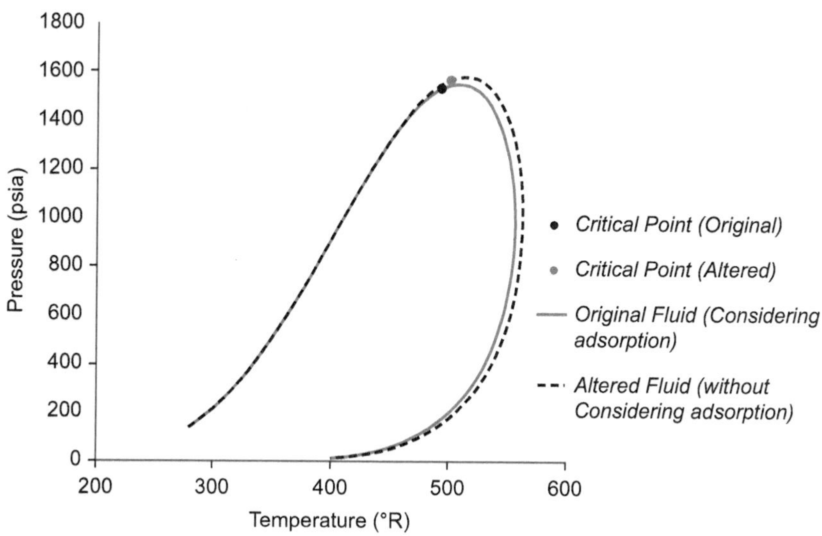

Figure 9.12 Comparison of phase envelopes of original and adsorption-altered reservoir fluid (Rajput, 2014).

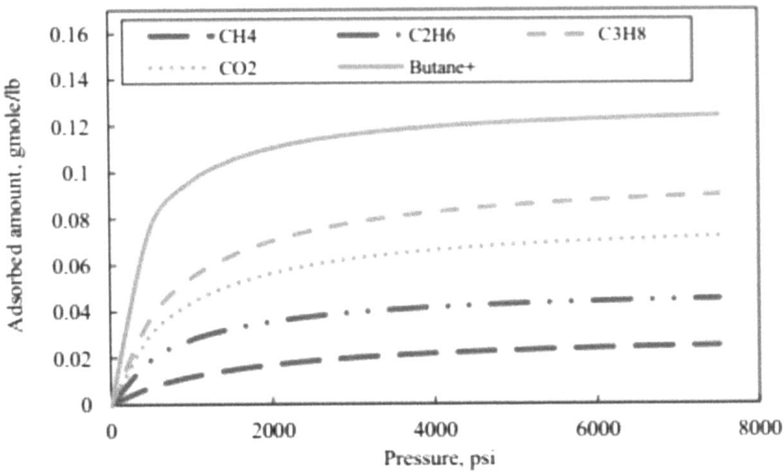

Figure 9.13 Adsorption isotherms for different components (Haghshenas, 2014).

differences in critical and dew-point line of the phase envelopes of fluid mixtures, with and without consideration of liquid-phase adsorption. There is, however, insignificant change in bubble-point line location, which could be attributed to the fact that heavier components are preferentially adsorbed (Haider, 2015).

Haghshenas (2014) modeled heavy hydrocarbon component adsorption using Langmiur isotherms. Fig. 9.13 shows that for hydrocarbon components, adsorption increases strongly with the molecular weight. This observation shows that in liquid-rich shales, adsorption on organic matter may be an important storage mechanism for the heavier fractions. Haghshenas (2014) also showed that the contribution of liquid desorption to the overall hydrocarbon recovery was dependent on fluid composition and pore connectivity/configuration (Haghshenas, 2014; Haider, 2015).

The adsorption phenomena in the porous media may have a significant impact on the reserve distribution of tight, shale, and coal-bed methane reservoirs. The adsorption process may largely distinguish from surface adsorption observed in the chemical labs. The main two differentiating reasons are: existence of capillary condensation phenomena in the narrow pores and possibility of flow access blocking in the porous network. The progress in the fundamentals of adsorption theory one may find in the Dabrowski (2001) research. The adsorption phenomena related to the porous media are discussed in many textbooks (i.e., Defay and Prigogine, 1966; Adamson, 1990; Dullien, 1992). The

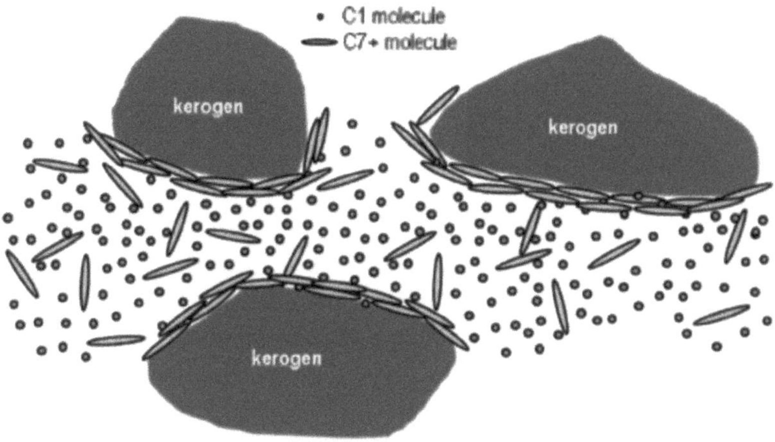

Figure 9.14 Example of selective molecule adsorption to the kerogen in a shale gas condensate system (Altman et al., 2014).

advances in the adsorption process in the high-pressure porous media may be found in Shapiro and Stenby (1996a,b, 2000, 2001), Guo et al. (1996), and Satik et al. (1995), Kang (2011) Altman et al. (2014), and Travalloni et al. (2010) works. The possible selective molecule adsorption to the kerogen in a shale gas condensate system is presented in Fig. 9.14.

9.4 EFFECT OF CONFINEMENT ON PHASE ENVELOPE

Phase behavior and fluid properties are governed by molecule—molecule and molecule—pore-wall interactions. In conventional reservoirs, the effect of molecule—pore-wall interactions is negligible because pore sizes are much larger than molecular mean free paths. However, this effect is very important in shale formations because the matrix is dominated by micro- to mesosized pores (pore size below 50 nm) (Kuila and Prasad, 2010). In general, fluids under confinement within pores of nanometer-scale size exhibit significant deviation from bulk thermophysical properties, such as critical properties, density, orientation profiles, and structural properties of chemical compounds (Singh and Singh, 2011; Singh, 2009; Thommes and Findenegg, 1994; Travalloni et al., 2010; Zarragoicoechea and Kuz, 2004). This is a consequence of finite size and increasingly significant effects of the interactions between molecule and molecule (van der Waals interactions) and interactions between the molecules and pore surface in such systems.

Sigmund et al. (1973) studied theoretically and experimentally the effect of pore size on phase behavior by including capillary pressure in flash calculations. He found that the decrease in bubble-point pressures and changes in vapor compositions for a $C_1-n\text{-}C_5$ binary mixture system are very small for pore radii more than 100 nm, but are significant for pore size less than 10 nm, because the difference between oil and gas pressures (capillary pressure) increased significantly. On the other hand, theoretical analyses have shown that when the pore radius decreased to the order of about 1 μm, the capillarity effect would be appreciable (Lee, 1989). Theoretical conclusions include: (1) the capillary pressure could influence the hydrocarbon distribution in mesopores (Shapiro and Stenby, 1996a,b); (2) with the decrease of pore radius, the bubble point would decrease or increase depending on fluid composition (Brusilovsky, 1992; Nojabaei et al., 2012; Pang et al., 2012); (3) the dew-point pressure would increase (Brusilovsky, 1992; Lee, 1989); and (4) the change of dew-point pressure depends on the value between pressure and cricondentherm pressure (Nojabaei et al., 2012).

Method 1: Modifying flash calculations. Flash calculation is a common approach used for phase equilibria calculations. In principle, flash calculations involve combining the vapor–liquid equilibrium (VLE) equations with the component mass balances and, in some cases, the energy balance. The influence of difference between oil and gas pressures (i.e., capillary pressure) is neglected in the flash calculations for conventional reservoirs. However, the capillary pressure is very high and cannot be ignored in phase-behavior calculations of shale formations. Lee (1989) developed an equation describing the influences of capillarity on phase equilibrium, which is shown in Eqs. (9.7) through (9.10). The chemical potential of each component is defined as:

$$\mu_i = \mu_i(P, T, z_i) \tag{9.7}$$

in which P, T, z_i are pressure, temperature, and composition, respectively.

Eq. (9.7) can be used to express equilibrium between vapor and liquid phases as follows:

$$\mu_{oi}(P_o = P_g - P_{cap}, T, x_i) = \mu_{gi}(P_g, T, y_i) \tag{9.8}$$

The difference between oil and gas pressures is capillary pressure P_{cap}, which has a relationship with interfacial curvature by invoking the Laplace equation.

$$P_{cap} = P_g - P_o = \frac{2 \times \gamma}{r_p} \tag{9.9}$$

in which γ is interfacial tension, r_p is pore radius, and P_{cap} is capillary pressure. So, modifying Eq. (9.8) by accounting for capillary pressure, one yields:

$$\mu_{oi}(P_o, T, x_i) - \left(\frac{\partial \mu_{oi}}{\partial P}\right) P_{cap} + \cdots = \mu_{gi}(P_g, T, y_i) \tag{9.10}$$

in which "…"refers to composition and temperature variables. By only considering the effect of capillary pressure, Eq. (9.10) reduces to:

$$\mu_{oi}(P_o, T, x_i) - \left(\frac{\partial \mu_{oi}}{\partial P}\right) P_{cap} = \mu_{gi}(P_g, T, y_i) \tag{9.11}$$

in which

$$\mu_{oi}(P_o, T, x_i) = RT \ln(f_{oi}) + \mu_{ref} \tag{9.12}$$

$$\mu_{gi}(P_o, T, y_i) = RT \ln(f_{gi}) + \mu_{ref} \tag{9.13}$$

in which f_{oi} and f_{gi} are the fugacities of component i in oil and gas phases, respectively; μ_{ref} is the chemical potential of the reference state. Substituting Eqs. (9.12) and (9.13) into Eq. (9.11) results in:

$$f_{gi} = f_{oi} \exp\left(-\frac{\frac{df_{oi}}{dp} P_{cap}}{f_{oi}} \right) \tag{9.14}$$

From Eq. (9.14), it can be seen that the effect of capillary pressure on phase equilibrium is expressed in the exponential term. Considering the effect of capillary pressure, the flash calculation is modified as demonstrated through Fig. 9.15. When the capillary pressure closes to 0, the exponential term in Eq. (9.14) goes to 1, and the modified flash calculation returns to a regular flash calculation. Interfacial tension can be calculated by the Parachor method and is estimated by:

$$\gamma = \left[\sum\nolimits_i \chi_i \left(x_i \rho^L - y_i \rho^V\right)\right]^4 \tag{9.15}$$

Method 2: Modifying critical properties. Results from molecular dynamic simulation studies have shown that critical properties of fluids under confinement deviate from their bulk values. This paper summarized the confined-fluid critical properties shift from molecular simulation studies, which include the critical properties shift for single components (C_1, n-C_4, and n-C_8) at different pore shape (slit and cylinder), and different

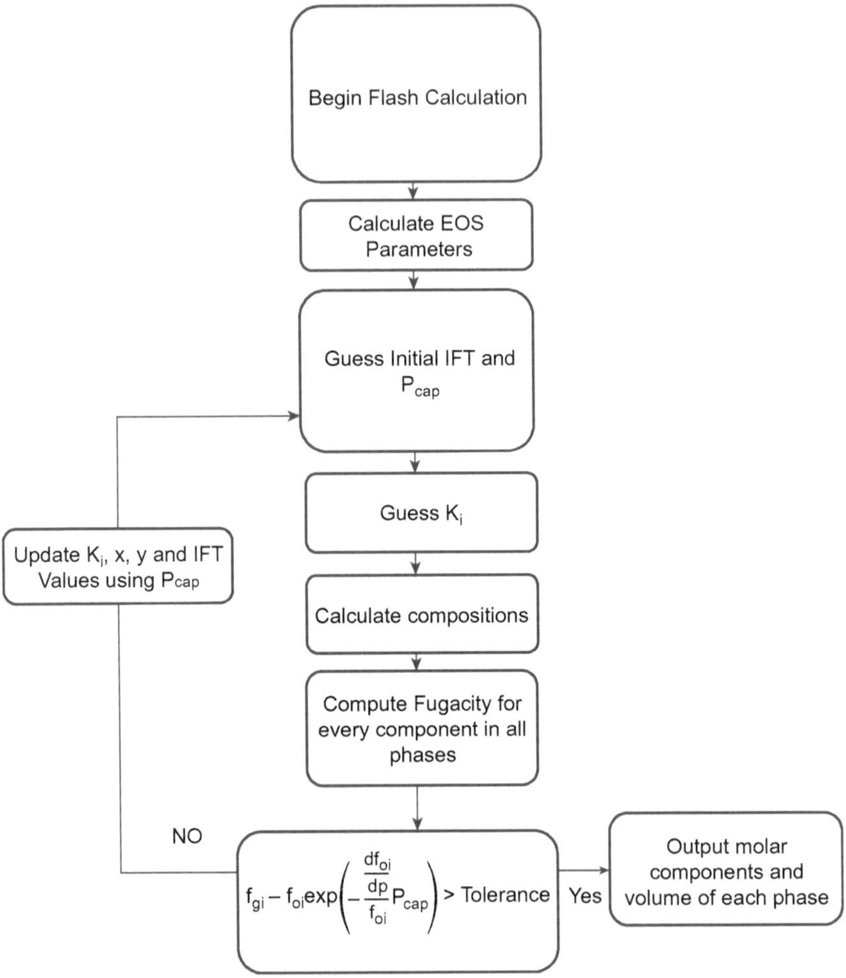

Figure 9.15 Flowchart of modified flash calculation (Jin et al., 2013).

pore surface (mica and graphite) (Singh and Singh, 2011; Singh, 2009; Vishnyakov et al., 2001). In the molecular simulation studies of Singh (Singh and Singh, 2011; Singh, 2009), molecule–molecule interactions are described with the Errington and Panagiotopoulos (1999) intermolecular potential, and molecule–wall interactions are described with the (9,3) Steele potential (Steele, 1973; Jin et al., 2013). Based on these data, this paper proposed new correlations between the shift of critical temperature and critical pressure of single component vs. the ratio of pore diameter to the molecule size. The effect of confinement on mixing rules was not included in the correlations, which are reported through (Jin et al., 2013).

Example 9.3

Consider a mixture of $C_1-n\text{-}C_5$; calculate the bubble-point pressures of a mixture at 100°F with different compositions using methods 1 and 2 when pore radius is equal to 10, 100, and infinity. Then, compare the calculated bubble-point pressures with experimental ones reported by Sigmund et al. (1973). Tables 9.4 and 9.5 present properties of each component and binary interaction coefficients. The binary interaction coefficient for $C_1-n\text{-}C_5$ was found by matching the experimental at $xC_1 = 0.0288$ and $xC_5 = 0.9712$ (Sigmund et al., 1973; Jin et al., 2013), and the binary interaction coefficients between C_1, $n\text{-}C_4$, and $n\text{-}C_8$ were calculated by correlations (Mehra et al., 1982; Jin et al., 2013).

Table 9.3 Compositional Data of Hydrocarbon Mixture for Flash Calculations (Jin et al., 2013)

Component	P_c (MPa)	T_c (K)	MW	ω	Parachor	σA
C_1	4.64	190.6	16.043	0.008	77	3.565
$n\text{-}C_4$	3.7997	425.12	58.123	0.200	189.9	4.687
$n\text{-}C_5$	3.3741	469.9	72.150	0.251	231.5	5.029
$n\text{-}C_8$	2.4825	568.70	114.23	0.399	309.022	7.098

Table 9.4 Binary Interaction Coefficients of Hydrocarbons (Jin et al., 2013)

Component	C_1	$n\text{-}C_4$	$n\text{-}C_5$	$n\text{-}C_8$
C_1	0	0.0035	0.029	0.0033
$n\text{-}C_4$	0.0035	0	0	0
$n\text{-}C_5$	0.029	0	0	0
$n\text{-}C_8$	0.0033	0	0	0

Answer

For method 1, flash calculations with different capillary pressure were performed to match the experimental data. Comparisons of bubble-point pressures, at different mole fractions of CH_4 in the mixture from the two methods and the experimental data of Sigmund et al. (1973) at 100°F, are presented in Tables 9.5 and 9.6. Moreover, the relative error of the simulated data and experimental data is shown in Fig. 9.16 (Jin et al., 2013). For method 2, critical properties of C_1 and $n\text{-}C_5$ were replaced by modified critical properties

(Continued)

Table 9.5 Comparison of Bubble Point From Method 1 (100°F) (Jin et al., 2013)

	Bulk Fluid Bubble Point Pressure (psi)		*r* = 100 nm Bubble Point Pressure (psi)		*r* = 10 nm Bubble Point Pressure (psi)	
X_{CH_4}	Sigmund	Method 1	Sigmund	Method 1	Sigmund	Method 1
0.0288	99.53	97.71	98.62	97.71	90.89	91.62
0.0628	200.75	200.05	199.09	197.61	184.83	185.43
0.0957	301.05	299.95	298.7	297.51	278.45	278.02
0.1282	402.4	402.28	399.41	397.41	373.66	370.61
0.1911	604.93	606.95	600.8	599.64	565.22	558.22
0.2508	804.65	811.62	799.62	799.43	756.35	745.83
0.3077	1001.14	1013.85	995.45	999.23	946.49	931.01
0.3748	1238.58	1259.94	1232.26	1242.90	1178.04	1157.60
0.439	1468.08	1506.03	1461.97	1485.30	1409.52	1381.80
0.5041	1697.94	1759.42	1691.69	1735.10	1638.18	1613.20
0.5788	1960.32	2049.37	1955.06	2021.40	1913.11	1878.80

Table 9.6 Comparison of Bubble Point From Method 2 (100°F) (Jin et al., 2013)

	Bulk Fluid Bubble Point Pressure (psi)		*r* = 100 nm Bubble Point Pressure (psi)		*r* = 10 nm Bubble Point Pressure (psi)	
X_{CH_4}	Sigmund	Method 2	Sigmund	Method 2	Sigmund	Method 2
0.0288	99.53	97.71	98.62	97.71	90.89	90.41
0.0628	200.75	200.05	199.09	197.61	184.83	185.43
0.0957	301.05	299.95	298.7	297.51	278.45	280.45
0.1282	402.4	402.28	399.41	399.84	373.66	345.48
0.1911	604.93	606.95	600.8	604.51	565.22	570.40
0.2508	804.65	811.62	799.62	806.74	756.35	765.32
0.3077	1001.14	1013.85	995.45	1006.54	946.49	960.24
0.3748	1238.58	1259.94	1232.26	1255.06	1178.04	1201.46
0.439	1468.08	1506.03	1461.97	1498.72	1409.52	1442.68
0.5041	1697.94	1759.42	1691.69	1753.33	1638.18	1697.29
0.5788	1960.32	2049.37	1955.06	2043.28	1913.11	1992.11

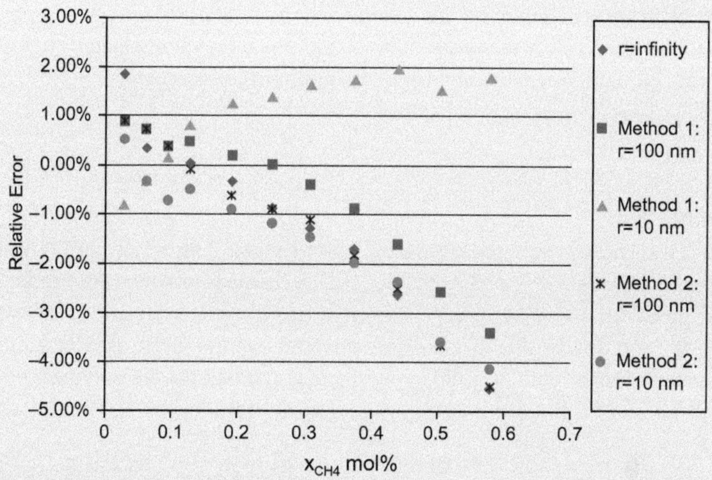

Figure 9.16 Relative error between this work (numerical) and Sigmund's experimental results (Jin et al., 2013).

according to Eqs. (9.1) to (9.4). The same mixtures 1, 2, and 3 consist of C_1, n-C_4, and n-C_8 with different compositions as mentioned before were used as sample fluids to investigate the influence of pore size on phase behavior. The critical properties of each component of the mixtures were modified by using the Eqs. (9.1) to (9.4) (Jin et al., 2013).

Fig. 9.16 indicates that the relative error increases with the mole fraction of methane in the mixture. This is because the C_1−n-C_5 binary interaction coefficient was obtained by matching the experimental data of one mixture at $xC_1 = 0.0288$ and $xC_5 = 0.9712$, and was used to predict the bubble-point pressures of other mixtures. Basically the binary interaction parameter was assumed constant and does not depend on composition. In the experimental and simulation results of method 1, the capillary pressure (P_{cap}) is more than gas pressure (P_g) at xC_1 less than 0.0957 mol fraction in a 10-nm pore (Sigmund et al., 1973), indicating negative liquid pressure, which leads to the transition from capillary condensation to thin-film adsorption (Udell, 1982). The maximum relative error is less than 5%, which means both methods are appropriate to study the effect of pore proximity on phase behavior and fluid properties (Jin et al., 2013).

The matrix in shale formations is characterized by micropores less than 2 nm in diameter to mesopores with diameters in the range of 2−50 nm (Kuila and Prasad, 2011). Therefore, the effect of capillary pressure is significant and cannot be ignored. To investigate the effect of capillarity on phase behavior in shale formations, the following examples plot the phase envelopes of

(Continued)

hydrocarbon mixtures with different compositions at different pore sizes ranging from infinity to 5 nm. Methane, *n*-butane, and *n*-octane were selected to represent the light, intermediate, and heavy components, respectively.

Example 9.4

Consider mixtures with the following compositions (Jin et al., 2013) using method 1; calculate the phase envelope of each mixture when pore radius is equal to 5, 10, 100, and infinity. Compare the phase envelopes of different pore radii and discuss the effect of pore proximity on the phase envelope shifts. The compositional data and binary interaction coefficients are summarized in Tables 9.3 and 9.4 (Jin et al., 2013).

Component	Mixture 1 (Mol%)	Mixture 2 (Mol%)	Mixture 3 (Mol%)
C_1	75	30	10
n-C_4	20	35	25
n-C_8	5	35	65

Answer

Fig. 9.17 shows the two-phase envelopes for mixture 1 at different pore sizes. It is seen that the two-phase region slightly shrinks when the pore size decreases.

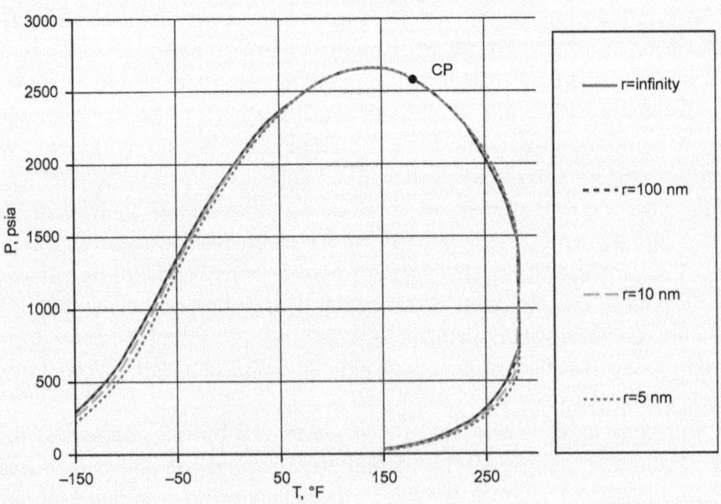

Figure 9.17 Phase envelopes for C_1 (75 mol%)—n-C_4 (20 mol%)—n-C_8 (5 mol%) mixtures at different pore radii (Method 1) (Jin et al., 2013).

Critical point does not change because capillary pressure goes to zero in the region close to the critical region. So capillary pressure cannot have any influence on the critical point. At 0°F, the capillary pressure for this mixture is only 12.3 psi, which cannot much influence the phase behavior. Nevertheless, the capillary pressure generally reduces the bubble-point pressure for every temperature. At lower temperatures, the pore size has much greater effect on the bubble-point pressures. At regions close to the critical point, there is no significant change in saturation pressures because the interfacial tension goes to zero (Jin et al., 2013). The upper dew-point pressures in the retrograde region slightly increases with the capillary pressure and the lower dew-point pressures slightly decrease, which has similar results with Nojabaei et al. (2012).

Figs. 9.18 and 9.19 present two-phase envelopes for mixtures 2 and 3, respectively. The two figures indicate that for these two mixtures, the two-phase envelopes do not show much difference when pore size decreases from infinity to 100 nm. This observation is in agreement with Sigmund's experimental results that the pore size has insignificant effect on phase behavior when it is more than 100 nm (Sigmund et al., 1973; Jin et al., 2013). With the decrease of pore radius, the saturation pressure of mixture 2 and mixture 3 decreases and has the same trend as mixture 1. When pore radius decreases from infinity to 5 nm, the bubble-point pressures of these three mixtures at 50°F are decreased by 0.7, 9.6, and 10.6%, respectively (Jin et al., 2013).

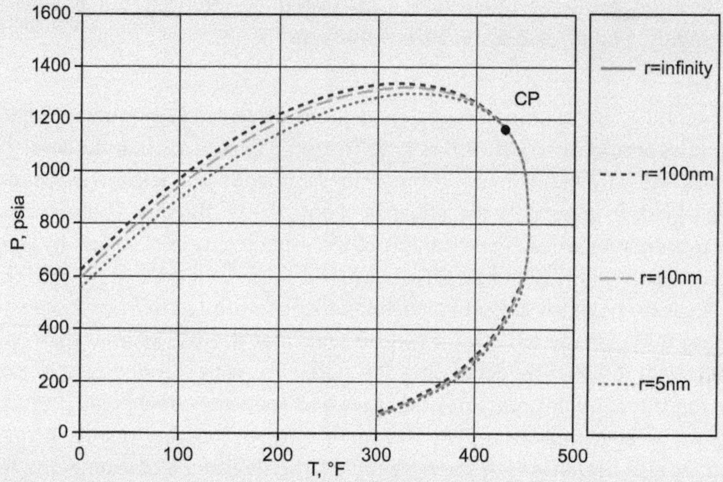

Figure 9.18 Phase envelopes for C_1 (30 mol%)—n-C_4 (35 mol%)—n-C_8 (35 mol%) mixtures at different pore radii (Method 1) (Jin et al., 2013).

(Continued)

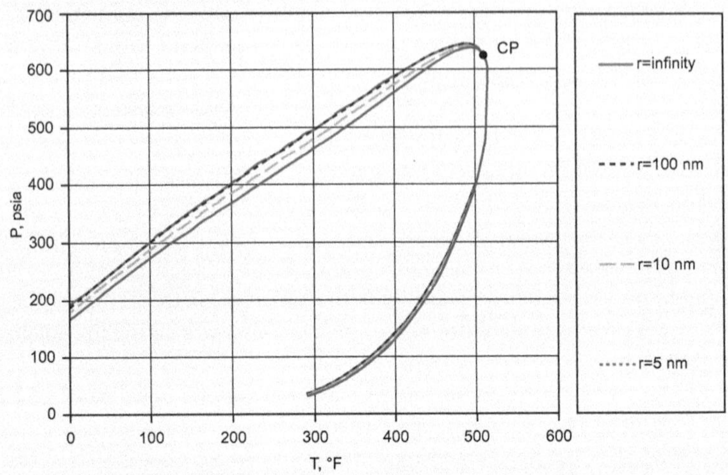

Figure 9.19 Phase envelopes for C_1 (10 mol%)—n-C_4 (25 mol%)—n-C_8 (65 mol%) mixtures at different pore radii (Method 1) (Jin et al., 2013).

Example 9.5

Consider the mixtures in example 9.4 using method 2; calculate the phase envelope of each mixture when pore radius is equal to 5, 10, 100, and infinity. Compare the phase envelope of different pore radii and discuss the effect of pore proximity on the phase envelope shifts. Moreover, compare the bubble-point pressures of mixture 2 calculated by methods 1 and 2 when pore radius is equal to 2, 5, 10, 50, and 100 nm.

Answer

For method 2, critical properties of C_1 and n-C_5 were replaced by modified critical properties according to Eqs. (9.1) to (9.4). The same mixtures 1, 2, and 3 consist of C_1, n-C_4, and n-C_8 with different compositions as mentioned before were used as sample fluids to investigate the influence of pore size on the phase behavior. The critical properties of each component of the mixtures were modified by using the Eqs. (9.1) to (9.4) (Jin et al., 2013).

Fig. 9.20 shows the two-phase envelopes of mixture 1 at different pore sizes ranging from infinity to 2 nm. It can be seen that the two-phase region was significantly reduced by decreasing the pore size. With the decrease of pore size, the bubble-point pressures decrease and the lower dew-point pressures increase at all temperatures. At the same time, the critical points of these mixtures also decrease with the pore radius. The deviations of saturation pressures under confinement are higher for temperatures and pressures closer to

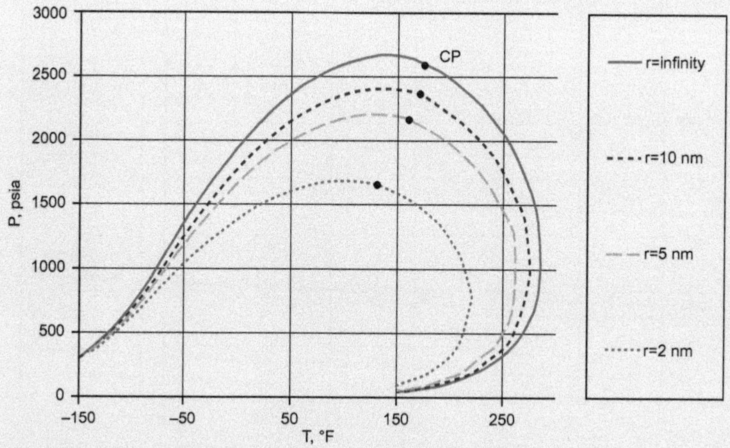

Figure 9.20 Phase envelope for C_1 (75 mol%)—*n*-C_4 (20 mol%)—*n*-C_8 (5 mol%) mixtures at different pore radii (Method 2) (Jin et al., 2013).

the critical point, which is the opposite of method 1. Figs. 9.21 and 9.22 plot the phase envelopes for mixture 2 and mixture 3, which have similar trends to mixture 1 (Jin et al., 2013).

In general, the two-phase region shrinks by decreasing the pore size in both methods. Fig. 9.23 compares the bubble-point pressures of mixture 2 vs. pore radius at 100°F from both methods. It can be seen that the bubble-point pressures decrease with decreasing the pore radius for both methods. Moreover, when pore radius is more than 10 nm, the difference between bubble-point

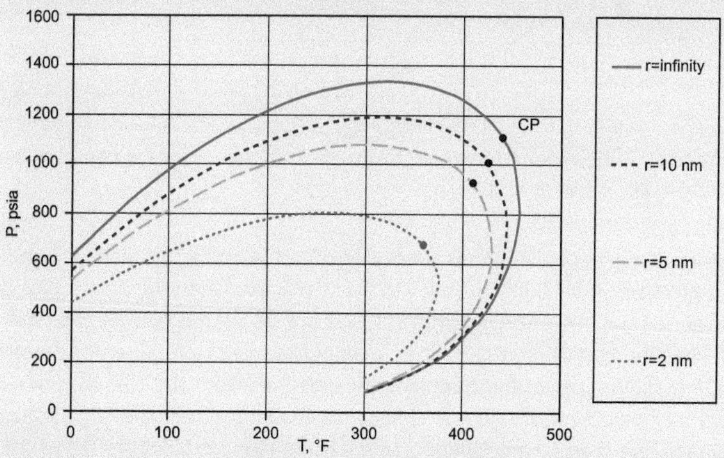

Figure 9.21 Phase envelope for C_1 (30 mol%)—*n*-C_4 (35 mol%)—*n*-C_8 (35 mol%) mixtures at different pore radii (Method 2) (Jin et al., 2013).

(Continued)

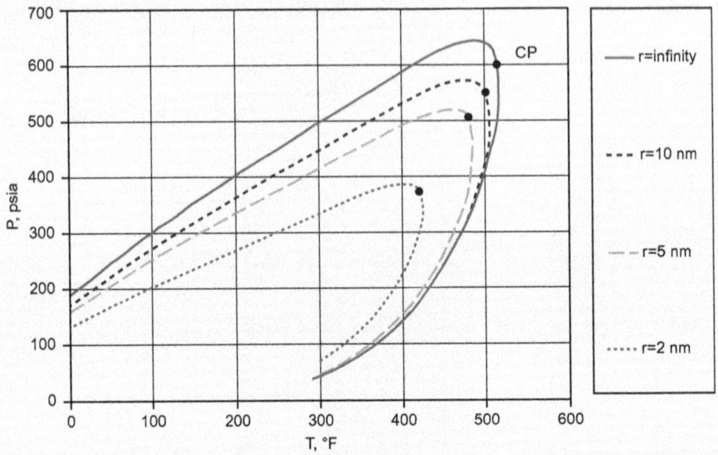

Figure 9.22 Phase envelope for C_1 (10 mol%)—n-C_4 (25 mol%)—n-C_8 (65 mol%) mixtures at different pore radii (Method 2) (Jin et al., 2013).

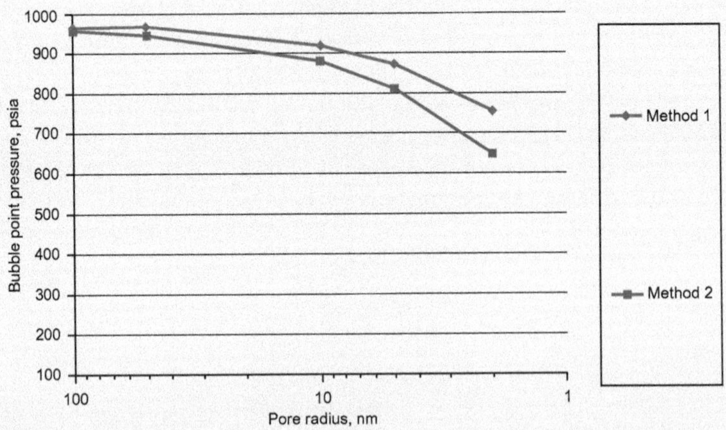

Figure 9.23 Bubble-point pressure of mixture 2 vs. pore radius at constant temperature (100°F) (Jin et al., 2013).

pressures from the two methods is less than 50 psi, and the deviations of bubble-point pressures from bulk flow are less than 10%. But when the pore radius decreases to 2 nm, the bubble-point pressures are decreased by 21% (method 1) and 32% (method 2) relative to their bulk bubble-point pressure. At the same time, the differences of the bubble-point pressures from the two methods increase by decreasing the pore size, because adsorption becomes significant in pore radii less than 10 nm (Shapiro and Stenby, 1996a,b; Udell, 1982), which is not taken into account in method 1 (Jin et al., 2013).

Example 9.6

Consider mixture 2 in example 13-4; calculate *k*-values of this mixture at 100°F and 400 psi using methods 1 and 2 when pore radius is equal to 2, 5, 10, 50, and 100 nm. Compare the *k*-values calculated by the two methods and discuss the effect of pore proximity on the *k*-value.

Answer

Fig. 9.24 presents the relationship between *k*-values (at 100°F, 400 psia) vs. pore radius for each component in mixture 2 obtained from both methods. The influence of pore size on *k*-value can be ignored when the pore radius is more than 10 nm, but is significant when pore radius is less than 10 nm. When the pore radius is more than 10 nm, the *k*-value for each component in mixture 2 obtained from the two methods are close to each other. But when pore radius is less than 10 nm, the differences between the *k*-value increase with the decrease of pore radius. The *k*-value of C_1 decreases with decreasing the pore radius for both methods. However, the *k*-value of n-C_4 and n-C_8 vs. decreasing the pore radius has the opposite trend from these two methods. Because method 2 changes critical temperature and pressure for each component, method 1 takes into account the effect of porous media on phase behavior by capillary pressure. Therefore, method 1 may have less sensitivity to composition than method 2. It is worth mentioning that experimental data are required to verify which method provides the right trends in *k*-values (Jin et al., 2013).

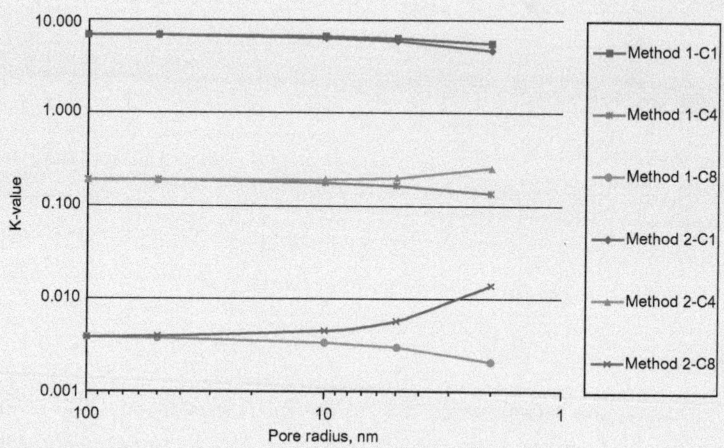

Figure 9.24 Phase equilibrium constant of mixture 2 vs. pore radius at constant temperature and pressure (100°F, 400 psia) (Jin et al., 2013).

Example 9.7

Consider mixture 2 in example 9.4; calculate interfacial tension (IFT) of this mixture at 100°F and 400 psi using methods 1 and 2 when pore radius is equal to 2, 5, 10, 50, and 100 nm. Compare the IFT calculated by the two methods and discuss the effect of pore proximity on the IFT values.

Answer

Fig. 9.25 presents the relationship between interfacial tension (IFT) of mixture 2 vs. decreasing the pore radius at 100°F and 400 psi from both methods. It can be observed that the IFT decreases in both methods with decrease in the pore radius. IFT from method 1 does not change significantly, but decreases sharply from method 2. This could be because: (1) two-phase region is smaller in method 2 than method 1, and (2) critical point does not change in method 1, but it changes significantly in method 2 (Jin et al., 2013).

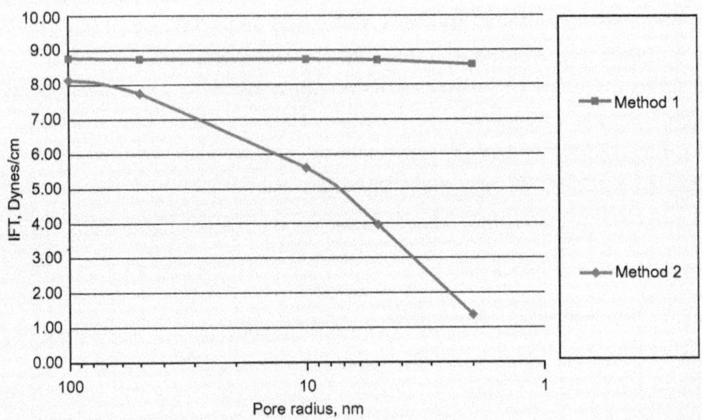

Figure 9.25 IFT of mixture 2 vs. pore radius at and constant temperature and pressure (100°F and 400 psia) (Jin et al., 2013).

Problems

9.1 Consider a gas sample with the following composition. The reservoir pressure and temperature are 3500 psi and 180°F, respectively. Determine the density of this fluid when the pore radius is equal to 8 nm.

Component	Mole Fraction	P_c (psi)	T_c (R)	ω_i
C_1	0.90	666.4	343.33	0.0104
C_2	0.065	706.5	549.92	0.0979
C_3	0.035	616.0	666.06	0.1522

Hint: you can use the following equations along with the Peng–Robinson equation of state for determining the density of gas mixtures.

$$\frac{\rho}{\rho_C} = 1 + d_1\varphi^{\frac{1}{3}} + d_2\varphi^{\frac{2}{3}} + d_3\varphi + d_4\varphi^{\frac{4}{3}}$$

in which

$$\varphi = 1 - \frac{T_r}{\alpha(T_r)}$$

in which $d_1 = 1.1688$, $d_2 = 1.8177$, $d_3 = -2.6581$, $d_4 = 2.1613$. The parameter T_r is the reduced temperature.

$$T_C = \sum_j^n x_j T_{C,j}$$

$$\alpha = \sum_{i=1}^n \sum_{j=1}^n x_i x_j \sqrt{\alpha_i \alpha_j}$$

$$\alpha = \left[1 + m\left(1 - \sqrt{T_R}\right)\right]^2$$

$$m = 0.3796 + 1.54226\omega - 0.2699\omega^2$$

$$\rho_C = \left[\sum_j^n x_j \rho_{C,j}^{-\frac{3}{4}}\right]^{-\frac{4}{3}}$$

9.2 Consider an ethane gas; calculate and compare the density of ethane at 195°F when pore radius is equal to 15 and 25 nm using the equation presented in problem 9.1. Reservoir pressure is equal to 4300 psi.

9.3 Consider a gas mixture with the following composition. The reservoir pressure and temperature are 4520 psi and 245°F, respectively. Determine the gas compressibility factor when the pore radius varies from 2 to 5 nm.

Component	Mole Fraction	P_c (psi)	T_c (R)
C_1	0.92	666.4	343.33
C_2	0.06	706.5	549.92
C_4	0.02	527.9	765.62

$$Z = A + BP_{\mathrm{pr}} + (1 - A)\exp(-C) - D\left(\frac{P_{\mathrm{pr}}}{10}\right)^4$$

$$A = -0.101 - 0.36T_{\mathrm{pr}} + 1.3868\sqrt{T_{\mathrm{pr}} - 0.919}$$

$$B = 0.021 + \frac{0.04275}{T_{\mathrm{pr}} - 0.65}$$

$$C = P_{\mathrm{pr}}\left(E + FP_{\mathrm{pr}} + GP_{\mathrm{pr}}^4\right)$$

$$D = 0.122\exp(-11.3(T_{\mathrm{pr}} - 1))$$

$$E = 0.6222 - 0.224T_{\mathrm{pr}}$$

$$F = \frac{0.0657}{T_{\mathrm{pr}} - 0.85} - 0.037$$

$$G = 0.32\exp(-19.53(T_{\mathrm{pr}} - 1))$$

9.4 Consider a gas mixture with the following composition. The reservoir pressure and temperature are 4250 psi and 210°F, respectively. Determine the viscosity of this gas when the pore radius is equal to 5 nm.

Component	Mole Fraction	MW	P_c (psi)	T_c (R)	ω_i
C_1	0.82	16.04	666.4	343.33	0.0104
C_2	0.13	30.07	706.5	549.92	0.0979
C_3	0.04	44.11	616.0	666.06	0.1522
C_4	0.01	58	527.9	765.62	0.1852

Hint: you can use the following equations for determining the viscosity of gas mixtures.

$$\mu_g = 10^{-4}K\exp\left[X\left(\frac{\rho_g}{62.4}\right)^Y\right]$$

$$K = \frac{(9.4 + 0.02\mathrm{MW})T^{1.5}}{209 + 19\mathrm{MW} + T}$$

$$X = 3.5 + \left(\frac{986}{T}\right) + 0.01\mathrm{MW}$$

$$Y = 2.4 - 0.2X$$

in which ρ stands for the density in g/cc, T denotes temperature in °R, and MW represents molecular weight of the gas.

9.5 Consider a methane gas and plot the Z-factor of methane at 180°F when pore radius is equal to 1, 2, 5, 10, and 50 nm. Consider maximum bulk pressure to 5000 psi.

9.6 Consider a gas mixture with the following composition. The reservoir pressure and temperature are 3880 psi and 188°F, respectively. Determine the gas compressibility factor when the pore radius is equal to 7 and 70 nm.

Component	Mole Fraction	P_c (psi)	T_c (R)
C_1	0.82	666.4	343.33
C_2	0.13	706.5	549.92
C_3	0.04	616.0	666.06
C_4	0.01	527.9	765.62

Hint: you can use the following equation for calculating the compressibility factor of gas mixtures.

$$Z = 1 - \frac{P_{\mathrm{pr}}}{T_{\mathrm{pr}}}\left[0.3648758 - 0.04188423\left(\frac{P_{\mathrm{pr}}}{T_{\mathrm{pr}}}\right)\right]$$

9.7 Consider a methane gas and plot the normalized gas viscosity relative to bulk state viscosity for methane at 200°F when pore radius is equal to 1, 2, 5, 10, and 50 nm. Consider maximum bulk pressure to 4800 psi.

9.8 Consider a gas mixture with the following composition. The reservoir pressure and temperature are 3900 psi and 180°F, respectively. Using the following equations to determine the viscosity of this gas when the pore radius is equal to 10 nm.

Component	Mole Fraction	MW	P_c (psi)	T_c (R)	ω_i
C_1	0.82	16.04	666.4	343.33	0.0104
C_2	0.13	30.07	706.5	549.92	0.0979
C_3	0.04	44.11	616.0	666.06	0.1522
C_4	0.01	58	527.9	765.62	0.1852

$$\mu_g = A_1 + A_2 + A_3$$

$$A_1 = -0.003338\left((\mathrm{MW} - P_{\mathrm{pr}})\rho_g\right)$$

$$-0.745356\left(\rho_g\left(\frac{\rho_g}{\left(T_{\mathrm{Pr}} - \left(\frac{T_{\mathrm{Pr}}}{\rho_g}\right)\right) - \left(\frac{\rho_g}{P_{\mathrm{Pr}} - \mathrm{MW}}\right)}\right)\right)$$

$$A_2 = -0.590871\left(\rho_g\left(\frac{\frac{MW}{T_{Pr}}}{\frac{T_{Pr}}{MW}}\right)\right) + 0.004602(T_{Pr}P_{Pr})$$
$$-0.007935P_{Pr} + 1.063654\rho_g$$

$$A_3 = -0.392638(\rho_g T_{Pr}) - 0.004755\left(\frac{P_{Pr}}{T_{Pr}}\right) + 0.000463MW$$
$$+0.011707T_{Pr} - 0.017994$$

μ stands for the viscosity of the hydrocarbon gas mixtures, P_{Pr} represents the pseudoreduced pressure, T_{Pr} denotes the pseudoreduced temperature, ρ_g stands for the density of the hydrocarbon gas mixtures, and MW stands for the molecular weight of the hydrocarbon gas mixtures.

REFERENCES

Adamson, A.W., 1990. Physical Chemistry of Surfaces. J. Wiley & Sons Inc.

Alharthy, N.S., 2013. Multiphase compositional modelling in small scale pores of unconventional shale reservoirs. In: SPE Annual Technical Conference and Exhibition, New Orleans, SPE 166306.

Alharthy, N., Nguyen, T.N., Teklu, T.W., et al., 2013. Multiphase compositional modeling in small-scale pores of unconventional shale reservoirs. In: Presented at SPE Annual Technical Conference and Exhibition. SPE-166306-MS, New Orleans, Louisiana, 30 September–2 October. http://dx.doi.org/10.2118/166306-MS.

Altman, R.M., et al., 2014. Understanding Mechanisms for Liquid Dropout from Horizontal Shale Gas Condensate Wells. Society of Petroleum Engineers. http://dx.doi.org/10.2118/170983-MS.

Ambrose, R.J., 2010. New pore scale considerations for shale gas in place calculations. In: Pittsburg, Pennsylvania, SPE 131772, SPE Unconvnetional Gas Conference.

Blasingame, T., 2013. Reservoir engineering aspects of unconventional reservoirs. In: Texas, USA, Analysis of Reservoir Performance for Unconventional Gas Condensate Reservoirs, Schlumberger CONDENFEST Webinar.

Brusilovsky, A.I., 1992. Mathematical Simulation of Phase Behavior of Natural Multicomponent Systems at High.

Chandra, S., 2014. Properties of Materials Confined in Nano-pores, vol. 221-005. Department of Physics, Banaras Hindu University, Varanasi, India.

Dabrowski, A., October 8, 2001. Adsorption-from theory to practice. Advances in Colloid and Interface Science 93 (1–3), 135–224.

Defay, R., Prigogine, I., 1966. Surface Tension and Adsorption. Longmans, London.

Devegowda, D., 2012. Phase behvaiour of gas condensates in shale due to pore proximity effects: implications for transport, reserves and well productivity. In: San Antonio, Texas, SPE 160099, SPE Annual Technical Conference and Exhibition.

Dranchuk, P.M., Abou-Kassem, H., 1975. Calculation of Z factors for natural gases using equations of state. Journal of Canadian Petroleum Technology 14 (3), 34. PETSOC-75-03-03. http://dx.doi.org/10.2118/75-03-03.

Dullien, F.A.L., 1992. Porous Media: Fluid Transport and Pore Structure, second ed. Academic Press, San Diego.
EIA, 2004. Annual Energy Review 2004, US Energy Information Administration, Office of Energy Markets and End Use U.S. Department of Energy Washington, DC 20585.
Errington, J.R., Panagiotopoulos, A.Z., 1999. New intermolecular potential models for benzene and cyclohexane. Journal of Chemical Physics 111, 9731–9738.
Firincioglu, T., 2014. Bubble Point Suppression in Unconventional Liquid Rich Reservoirs and its Impact on Oil Production (Ph.D. thesis). Colorado School of Mine, Colorado.
Guo, P., et al., 1996. A Theoretical Study of the Effect of Porous Media on the Dew Point Pressure of a Gas Condensate. SPE 25644.
Haghshenas, B., 2014. Simulation of liquid-rich shale gas reservoirs with heavy hydrocarbon fraction desorption. In: Woodlands, USA, SPE 168968, SPE Uncoventional Resources Conference.
Haider, B.A., 2015. Impact of Capillary Pressure and Critical Properties Shifts Due to Confinement on Hydrocarbon Production from Shale Reservoirs (M.Sc. thesis). Stanford University.
Hamada, 2007. Phase equilibria and interfacial tension of fluids confined in narrow pores. Journal of Chemical Physics 127 (8), 084908-1–084908-9.
Honarpour, 2013. The critical role of rock and fluids in unconventional shale reservoir performance. In: Stanford, Presentation Delivered at Stanford University, Energy Resources Engineering Department Weekly Meetings.
IEA, 2009. World Energy Outlook 2009. International Energy Agency, Paris, France.
Jin, L., Ma, Y., Jamili, A., 2013. Modifying van der Waals equation of state to consider influence of confinement on phase behavior. In: SPE Annual Technical Conference and Exhibition.
Kang, S.M., 2011. Carbon Dioxide Storage Capacity of Barnett Shale (M.Sc. thesis). Oklahoma University.
Kuila, U., 2013. Specific surface area and pore-size distribution in clays and shales. Geophysical Prospecting 61 (2), 341–362.
Kuila, U., 2012. Application of Knudsen flow in modelling gas-flow in shale reservoirs. In: Hyderabad, 9th Biennial International Conference and Exposition on Petroleum Geophysics.
Kuila, U., Prasad, M., 2010. Pore size distribution and ultrasonic velocities of compacted Na-montmorillonite clays. In: SEG Annual Meeing. Society of Exploration Geophysicists, Denver, Colorado, USA.
Kuila, U., Prasad, M., 2011. Surface area and pore-size distribution in clays and shales. In: SPE Annual Technical Conference and Exhibition held in Denver. Colorado, USA, 30 October–2 November.
Kuz, Z.A., 2002. Van der Waals equation of state for a fluid in a nanopor. Physical Review, Issue E 65, 021110(1-4).
Lee, A., González, M., Eakin, B., 1966. The viscosity of natural gases. Journal of Petroleum Technology 18 (8), 997–1000.
Lee, S.T., 1989. Capillary-gravity equilibria for hydrocarbon fluids in porous media. In: SPE Annual Technical.
Ma, Y., Jamili, A., Jin, L., 2013. Investigating the effect of pore proximity on phase behavior and fluid properties in shale formations. In: SPE Annual Technical Conference and Exhibition.
Mehra, R.K., Heidemann, R.A., Aziz, K., 1982. Computation of multiphase equilibrium for compositional simulation. Society of Petroleum Engineers Journal 22 (01), 61–68.
Mitariten, G., 2005. Molecular Gate Adsorption System for the Removal of Carbon Dioxide and/or Nitrogen from Coal Bed Abd Coal Mine Methane. Two Rivers Convention Centre, CO, Western States Coal Mine Methane Recovery and Use Workshop.

Nelson, March 2009. Pore-throat sizes in sandstones, tight sandstones, and shales. AAPG Bulletin 93 (3), 329–340.
Nojabaei, B., 2012. Effect of capillary pressure on fluid density and phase behavior in tight pore rocks and shales. In: San Antonio, Texas, SPE 159258, SPE Annual Technical Conference and Exhibition.
Nojabaei, B., 2014. Effect of saturation dependant capillary pressure on production in tight rocks and shales: a compositionally-extented black oil formulation. In: Charleston, WV, SUA, SPE 171028, SPE Eastern Regional Meeting.
Nojabaei, B., Johns, R.T., Chu, L., 2012. Effect of capillary pressure on fluid density and phase behavior in tight rocks and shales. In: SPE Annual Technical Conference and Exhibition. Society of Petroleum Engineers, San Antonio, Texas, USA.
Pang, J., Zuo, J.Y., Zhang, D., Du, L., 2012. Impact of porous media on saturation pressures of gas and oil in tight reservoirs. In: SPE Canadian Unconventional Resources Conference. Society of Petroleum Engineers, Calgary, Alberta, Canada.
Rajput, V., 2014. Thermodynamically consistent modelling of adsorption in liquid rich shales. In: Denver, Colorado, USA, SPE 169589 ,SPE Western North American and Rocky Mountain Joint Regional Meeting.
Rotelli, F., 2012. Shale Reservoirs: Pore-scale Flow Behaviour and its Effect on Production (Ph.D. dissertation). Politecnico Di Milano, Milano, Italy.
Sanaei, A., Jamili, A., Callard, J., Mathur, A., 2014. Production modeling in the Eagle Ford gas condensate window: integrating new relationships between core permeability, pore size, and confined PVT properties. In: SPE Western North American and Rocky Mountain Joint Regional Meeting Held in Denver, Colorado, USA, 16–18 April 2014.
Satik, C., Li, X., Yorsos, Y.C., 1995. Scaling of single-bubble growth in a porous medium. Physical Review E 51, 3286–3295.
Shapiro, A.A., Stanby, E.H., 1996a. Effect of Capillary Forces & Adsorption on Reserves Distributions. SPE 36922.
Shapiro, A.A., Stenby, E.H., 1996b. Effects of capillary forces and adsorption on reserves distribution. In: European Petroleum Conference. Society of Petroleum Engineers, Inc., Milan, Italy. Copyright 1996.
Shapiro, A., Stenby, E.H., 2000. Factorization of transport coefficients in macroporous media. Transport in Porous Media 41 (3), 305–323.
Shapiro, A., Stenby, E.H., 2001. Thermodynamics of the multicomponent vapor-liquid equilibrium under capillary pressure difference. Fluid Phase Equilibria 178 (1–2), 17–32.
Sigmund, P.M., Dranchuk, P.M., Morrow, N.R., Purvis, R.A., 1973. Retrograde condensation in porous media. Society of Petroleum Engineers Journal 13 (2), 93–104.
Singh, S.K., Singh, J.K., 2011. Effect of pore morphology on vapor– liquid phase transition and crossover behavior of critical properties from 3D to 2D. Fluid Phase Equilibria 300 (1–2), 182–187.
Steele, W.A., 1973. The physical interaction of gases with crystalline solids I. Gas-solid energies and properties of isolated adsorbed atoms. Surface Science 36, 317–352.
Singh, S., 2009. Vapor-liquid phase coexistence, critical properties and surface tension of confined alkanes. Journal of Physical Chemistry C 113, 7170–7180.
Teklu, T., 2014. Phase behaviour and minimum miscibility pressure in nanopores. In: Denver, SPE 168865, Unconventional Resources Technology Conference.
Thommes, M., Findenegg, G.H., 1994. Pore condensation and critical-point shift of a fluid in controlled-pore glass. Langmuir 10 (11), 4270–4277.
Travalloni, L., Castier, M., Tavaresa, F.W., Sandler, S.I., 2010. Critical behavior of pure confined fluids from an extension of the van der Waals equation of state. Journal of Supercritical Fluids 55 (2), 455–461.

Udell, K.S., March 1982. The thermodynamics of evaporation and condensation in a porous media. In: Society of Petroleum Engineers California Regional Meeting, Paper SPE-10779, San Francisco, CA.
USGS Fact Sheet FS-113-01, 2001. National Assessment of Oil and Gas Fact Sheet, Natural Gas Production in the United States.
U.S. Geological Survey Open-File Report 2005−1268, Published, 2005. Online Only Version 1.0, Assessment of Undiscovered Natural Gas Resources in Devonian Black Shales, Appalachian Basin, Eastern U.S.A. by Robert C. Milic.
Vishnyakov, A., Piotrovskaya, E.M., Brodskaya, E.N., Votyakov, E.V., Tovbin, Y.K., 2001. Critical properties of Lennard-Jones fluids in narrow slit-shaped pores. Langmuir 17 (14), 4451−4458.
Wang, Y., 2013. Compositional modelling of tight oil using dynamic nanopore properties. In: New Orleans, Loisiana, USA, SPE 166267, SPE Annual Technical Conference and Exhibition.
Zarragoicoechea, G.J., Kuz, V.A., 2004. Critical shift of a confined fluid in a nanopore. Fluid Phase Equilibria 220 (1), 7−9.
Zee Ma, Y., 2016. Stephen Holditch and Jean−Jacques Royer, Unconventional Oil and Gas Resources Handbook Evaluation and Development. Gulf Professional Publishing, 225 Wyman Street, Waltham, MA 02451, USA.
Zou, C., Zhu, R., Tao, S., Hou, L., Yuan, X., Song, Y., Niu, J., Dong, D., Liu, S., Jiang, L., Wang, S., Zhang, G., et al., 2012. Unconventional Petroleum Geology. Elsevier, 225 Wyman Street, Waltham, MA 02451, USA.

CHAPTER TEN

Characterization of Shale Oil

M.A. Ahmadi[1], A. Bahadori[2,3]
[1]Petroleum University of Technology (PUT), Ahwaz, Iran
[2]Southern Cross University, Lismore, NSW, Australia
[3]Australian Oil and Gas Services Pty Ltd, Lismore, NSW, Australia

10.1 INTRODUCTION

Shale resources have been changing the world's energy equation (Dyni, 2010). According to Energy Information Administration (EIA, 2013), estimated shale oil and shale gas resources in the United States and in 137 shale formations in 41 other countries represent 10% of the world's crude oil and 32% of the world's natural gas that is technically recoverable. This accounts for 345 billion barrels of technically recoverable shale oil and 7299 trillion cubic feet of recoverable shale gas in the world (EIA, 2013). Fig. 10.1 shows the

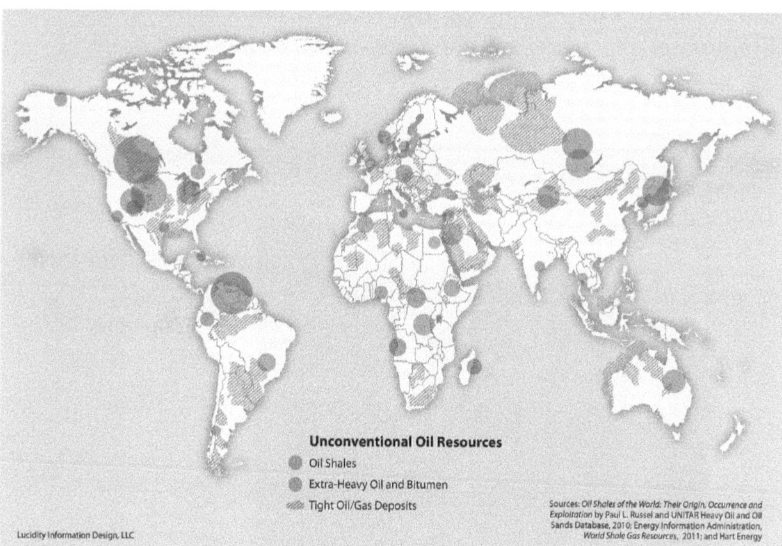

Figure 10.1 Unconventional oil resources across the globe (Gordon, 2012). *Oil Shales of the World: Their Origin, Occurrence and Exploitation by Paul L. Russel and UNITAR Heavy Oil and Oil Sands Database, 2010; Energy Information Administration, World Shale Gas Resources, 2011; and Hart Energy.*

Fluid Phase Behavior for Conventional and Unconventional Oil and Gas Reservoirs
http://dx.doi.org/10.1016/B978-0-12-803437-8.00010-5

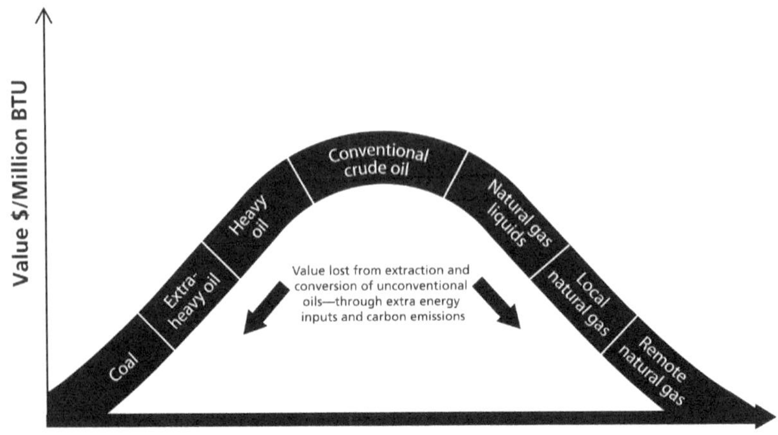

Figure 10.2 Hydrocarbon value hierarchy (Gordon, 2012).

unconventional oil deposits worldwide, which include oil shale, extraheavy oil and bitumen, and tight oil and gas.

Each of the fossil fuel energy resources comes with its own value. Fig. 10.2 shows the American Petroleum Institute (API) values and values in British Thermal Units (BTUs) of different oil and gas types. Fig. 10.3, on the other hand, illustrates the projected contribution of various hydrocarbon types.

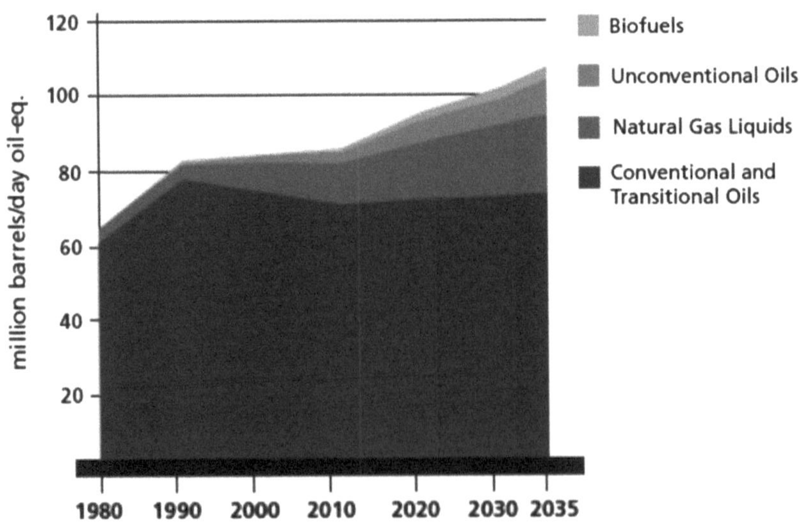

Figure 10.3 Projected new oil scenario (Gordon, 2012).

Figure 10.4 Selected light tight-oil plays worldwide (Ashraf and Satapathy, 2013).

Unconventional oil includes heavy and extra heavy oil, whereas conventional and tight oil have been lumped into "Conventional and Transitional Oil."

Fig. 10.4 shows some of the important light tight-oil basins across the world and Fig. 10.5 describes global shale resource exploration and extraction momentum in selected countries. Although shale resources exploration has been banned by many countries in Europe, North America has progressed a lot in this direction. On the other hand, Russia, China, and Australia have begun exploration of their shale resources (Brendow, 2009; Dyni, 2010).

Table 10.1 ranks countries by the amount of technically recoverable shale oil they bear. With 75 Billion Barrels of shale oil, Russian tops the list of shale oil reserves. It is interesting to see how small countries like Pakistan, too, have emerged as lands rich in shale oil reserves (Brendow, 2009; EIA, 2013; Dyni, 2010).

Fig. 10.6 shows the ranking of continents in terms of shale oil and gas reserves and in-place resources, respectively. The charts have been constructed using information extracted from EIA (2013). Europe tops the list of shale oil reserves and resource in-place, followed by Asia and South America. On the other hand, Africa and South America top the list of shale gas in-place and technically recoverable shale reserves, respectively (EIA, 2013).

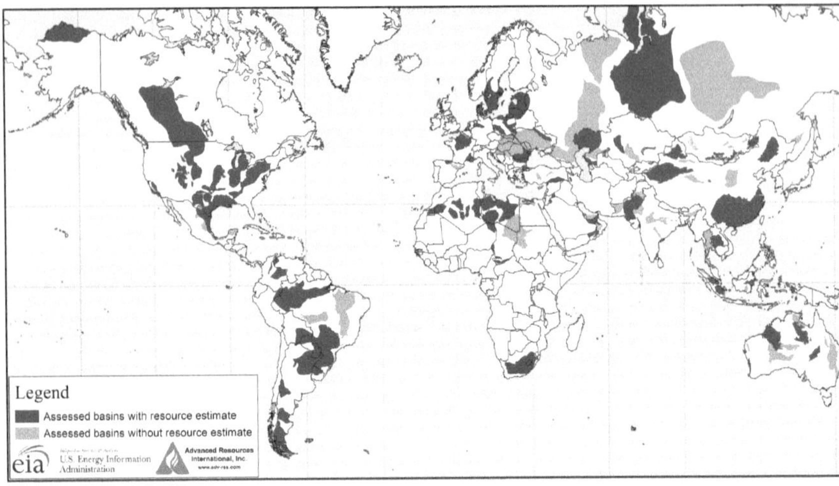

Figure 10.5 Approach of selected countries toward shale resource exploitation. http://www.eia.gov/analysis/studies/worldshalegas/.

Table 10.1 Top 10 Countries With Technically Recoverable Shale Gas and Oil Reserves

Rank	Country	Shale Oil (Billion Barrels)
1	Russia	75
2	United States	58
3	China	32
4	Argentina	27
5	Libya	26
6	Australia	18
7	Venezuela	13
8	Mexico	13
9	Pakistan	9
10	Canada	9

Adopted from EIA, 2013. Technically Recoverable Shale Oil and Shale Gas Resources: An Assessment of 137 Shale Formations in 41 Countries Outside the United States. US Energy Information Administration - Independent Statistics and Analysis, Washington, DC.

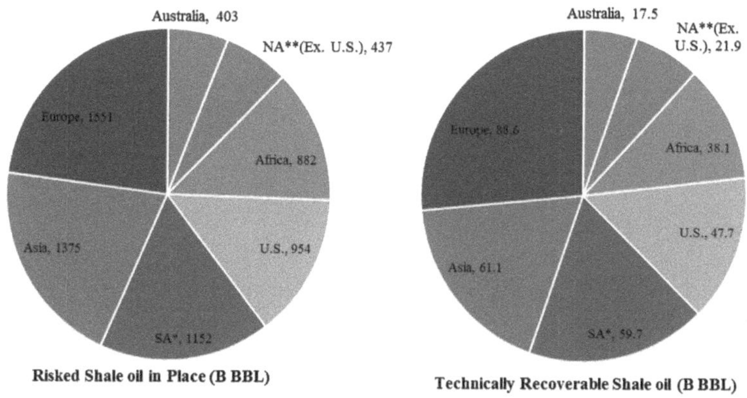

Figure 10.6 Continent-wise breakdown of risked in-place and technically recoverable shale oil; *SA is South America, **NA is North America (here excludes USA). *EIA, 2013. Technically Recoverable Shale Oil and Shale Gas Resources: An Assessment of 137 Shale Formations in 41 Countries Outside the United States. US Energy Information Administration - Independent Statistics and Analysis, Washington, DC.*

10.2 TYPES OF FLUIDS IN SHALE RESERVOIRS AND GENESIS OF LIQUID IN SHALE PORES

Shale resources can be broadly classified into three categories including oil shale, shale oil and gas condensate, and shale gas. Each of these differs greatly in flow characteristics. According to Colorado Oil and Gas Association (COGA, 2013), oil shale contains remains of "algae and plankton deposited millions of years ago that have not been buried deep enough to become sufficiently hot in order to break down into the hydrocarbons targeted in conventional oil projects." Shale oil and gas, on the other hand, are formed when the rock is buried deep enough to convert part of its kerogen into oil and gas. Horizontal drilling and fracturing is often required to produce them commercially, because these hydrocarbons are locked in place very tightly (COGA, 2013).

Several aspects determine whether shales are capable of generating hydrocarbons and whether they will generate oil or gas. During the process of hydrocarbon formulation, first, oxygen evolves as kerogen gives off CO_2 and H_2O, and later hydrogen evolves as hydrocarbons are formed (McCarthy, 2011). The general trend in the thermal transformation of kerogen to hydrocarbon starts with the generation of nonhydrocarbon gases and then progresses to oil, wet gas, and dry gas (McCarthy, 2011). Fig. 10.7 illustrates this progression for different types of kerogen. During

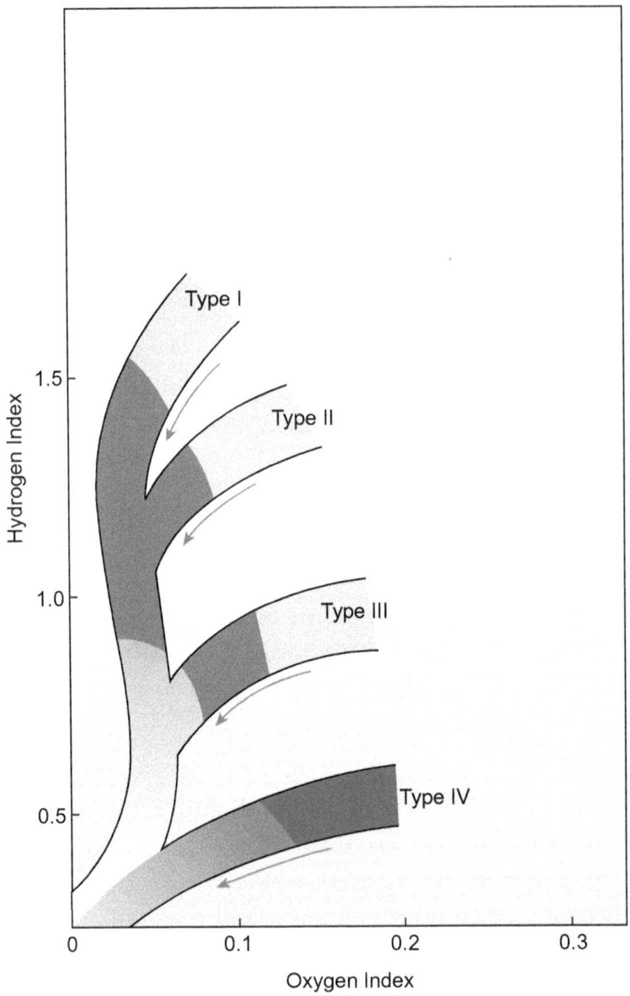

Figure 10.7 Kerogen type and oil and gas formulation (McCarthy, 2011; Haider, 2015).

thermal maturity of type I kerogen, liquid hydrocarbons tend to be generated. Type II, on the other hand, generates gas and oil, whereas type III generates gas, coal (often coal-bed methane), and oil in extreme conditions. It is generally considered that type IV kerogen is not capable of generating hydrocarbons (Rotelli, 2012; Synthetic Fuels Summary. Report No. FE-2468-82, March 1981). Fig. 10.8 highlights the conditions required to generate liquid hydrocarbons. Physical and chemical alteration of sediments and pore fluids take place at temperatures of 50−150°C (Pederson, 2010). This process is called "catagenesis." At these temperatures, chemical bonds

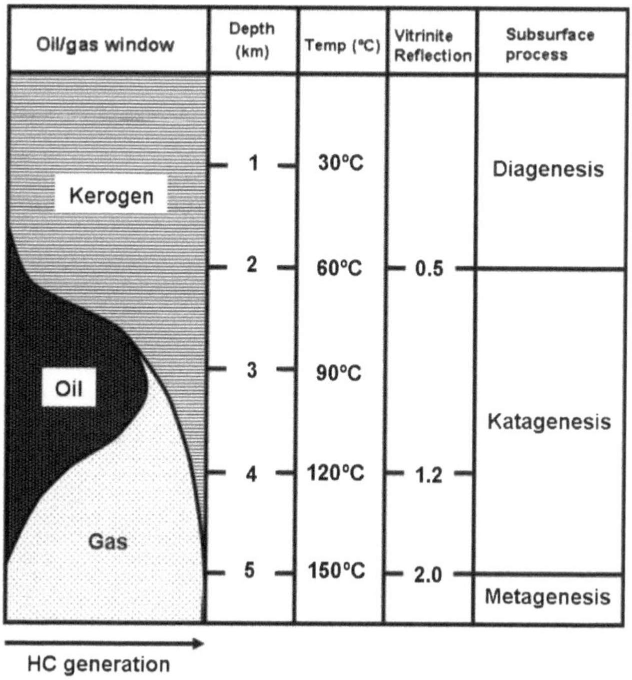

Figure 10.8 Depth and temperature condition for oil and gas formulation (Pederson, 2010; Haider, 2015).

break down in kerogen and clays within shale, generating liquid hydrocarbons (Pederson, 2010; Haider, 2015).

Liquid-rich shale (LRS) fluids can be divided into two categories—shale oil and shale gas condensate. At the original reservoir conditions, a gas condensate is a single-phase fluid (Fan, 2005). According to the work of Ismail (2010), shale condensate systems consist predominantly of "methane (C_1) and other short-chain hydrocarbons. The fluid also contains small amounts of long-chain hydrocarbons (heavy ends). The methane content in gas-condensate systems ranges from 65 to 90 mol%, whereas in crude oil systems, methane content ranges from 40 to 55 mol%." Fig. 10.9 shows a ternary diagram of these classifications (Haider, 2015).

10.3 SHALE PORE STRUCTURE AND HETEROGENEITY

The average size in currently producing liquid-rich reservoirs is estimated to be less than 100 nm (Firincioglu, 2013). According to Rotelli

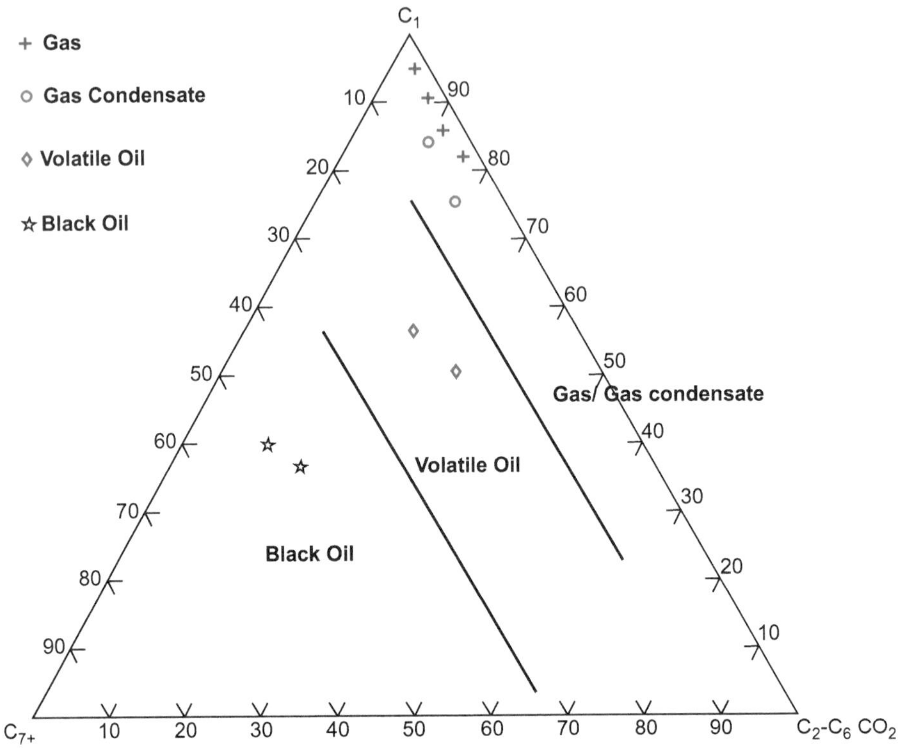

Figure 10.9 Ternary visualization of hydrocarbon classification (Ismail, 2010; Haider, 2015; Ismail and Horne, 2014). *Whitson, C.H., Brule, M.R., 2000. Phase behavior. In: SPE Monograph, vol. 20.*

(2012), to properly characterize shale, it is important to understand the following:

- Volume of pore network
- Characteristic dimension of pore network
- Pore type predominantly present
- Complexity of pore network

Based on the work of Kuila (2013), small pores in a shale matrix are associated with clay and kerogen. Bustin (2008) reported a bimodal pore-size distribution with modes around 10 nm and 10,000 nm for Barnett and Antrim formations. Loucks (2012) showed that scanning electron microscope (SEM) images of nanometer-scale pores associated with clays and kerogen in Barnett Shale revealed pores as small as 4 nm. Sondergeld (2013) stated that shale reservoirs exhibit hydrocarbon storage and flow characteristics that are "uniquely tied to nano-scale pore throat and pore

Pore Type	Image	Distinctive Features
Porous floccules		Clumps of electrostatically charged clay flakes arranged in edge-face or edge-to-edge cardhouse structure. Pores up to tens of micrometers in diameter. Pores may be connected.
Organopores		Pores in smooth surfaces of organic flakes or kerogen. Pore diameters are at nanometer scale. Pores are generally isolated. Porous organic coatings can also be adsorbed on clays.
Fecal pellets	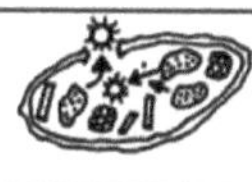	Spheres and ellipsoids with randomly oriented internal particles, giving rise to intrapellet pores. Pellets are sand size and may be aligned into laminae.
Fossil fragments	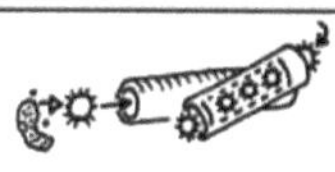	Porous fossil particles, including sponge spicules, radiolaria, and spores (*Tasmanites?*). Interior chamber may be open or filled with detrital or authigenic minerals.
Intraparticle grains/pores		Porous grains, such as pyrite framboids that have internal pores between microcrystals. Grains are of secondary origin and are usually dispersed within the shale matrix.
Microchannels and microfractures		Linear nanometer to micrometer size openings that often crosscut bedding planes. Occur at nanometer and larger scales.

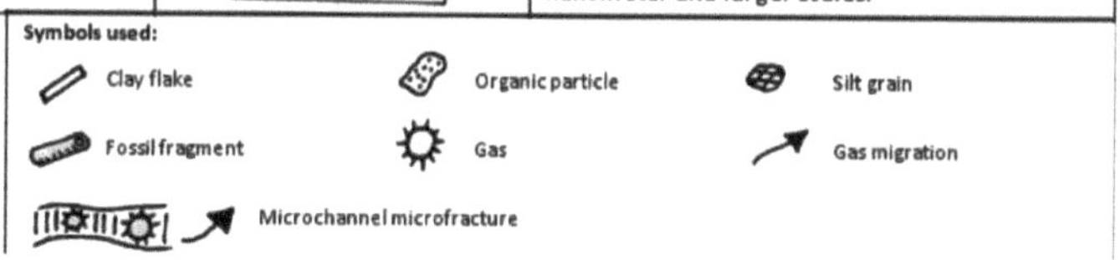

Figure 10.10 Pore types in the Barnett and Woodford gas shales (Slatt, 2011; Haider, 2015).

size distribution." Fig. 10.10 shows the different types of pores present in shale reservoirs. Each of them may alter fluid flow in a different manner (Haider, 2015).

10.4 SHALE OIL EXTRACTION

10.4.1 History

Three people who had "found a method to extract and make great quantities of tarr, pitch, and oyle out of a stone" were the inventors of the first shale oil extraction method which was granted by the British Crown in 1684 (Louw and Addison, 1985; Moody, 2007; Cane, 1976). The fundamentals of the modern industrial shale oil extraction referred to the methods invented firstly by Alexander Selligue in 1838, in France, and

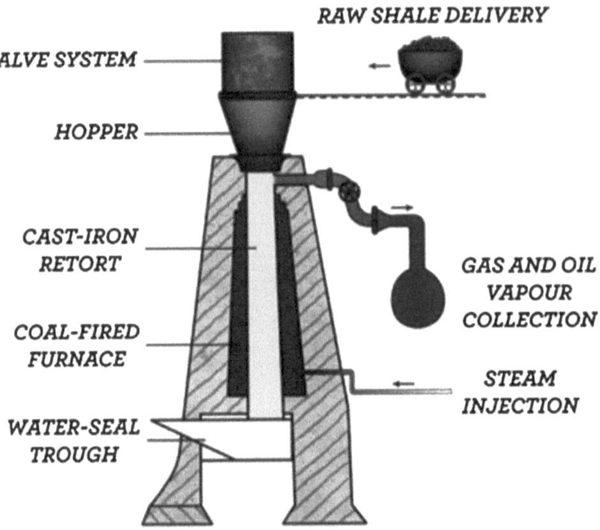

Figure 10.11 Schematic of Alexander C. Kirk's retort (Louw and Addison, 1985; www.en.wikipedia.org/wiki/Shale_oil_extraction).

modified by James Young in Scotland (Louw and Addison, 1985; Runnels et al., 1952). Alexander C. Kirk's retort was one of the first vertical oil shale retorts (Louw and Addison, 1985). Fig. 10.11 depicts the schematic of the Alexander C. Kirk's retort.

10.4.2 Processing Principles

Shale oil extraction defined as the decomposition process of the oil shale and converts its kerogen into synthetic crude oil. The extraction process is carried out via hydrogenation, pyrolysis, and/or thermal dissolution (Koel, 1999; Luik, 2009; Gorlov, 2007; Prien, 1976). The effectiveness of extraction process is assessed via contrasting their products to the products of a Fischer Analyze implemented on the shale sample (Speight, 2008; Baldwin et al., 1984; Smith et al., 2007; Francu et al., 2007; Prien, 1976; Synthetic Fuels Summary. Report No. FE-2468-82, March 1981).

Pyrolysis is the first and most common extraction technique for extracting the shale oils. In this process, oil shale is heated in the absence of oxygen until its kerogen decomposes into noncondensable flammable oil shale gas and condensable shale oil vapors. Oil shale gas and oil vapors are then accumulated and cooled, producing the shale oil to condense (Koel, 1999; Qian et al., 2007). The oil shale composition may provide added value to the process of extraction via the recovery of spin-offs, comprising ammonia, sulfur,

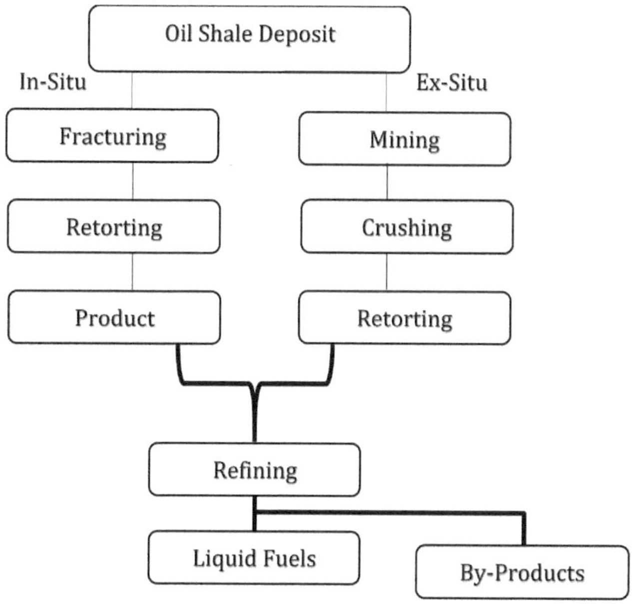

Figure 10.12 Overview of shale oil extraction techniques.

aromatic compounds, pitch, waxes, and asphalt (Johnson et al., 2004; Baldwin et al., 1984; Smith et al., 2007; Francu et al., 2007; Prien, 1976; Synthetic Fuels Summary. Report No. FE-2468-82, March 1981).

A source of energy is required for heating the oil shale to the temperature of pyrolysis and carrying out the endothermic reactions of the kerogen decomposition (Burnham and McConaghy, 2006). Two strategies are employed to reduce, and even eliminate, external heat energy requirements: the oil shale gas and char byproducts generated by pyrolysis may be burned as a source of energy, and the heat contained in hot spent oil shale and oil shale ash may be employed to preheat the raw oil shale (Koel, 1999). Fig. 10.12 depicts the overview of shale oil extraction techniques (Francu et al., 2007).

10.4.3 Extraction Technologies

Industry experts have generated different categorizations of the technologies employed to extract shale oil from oil shale.

By process principles: Based on the treatment of raw oil shale by heat and solvents, the techniques are categorized as thermal dissolution, hydrogenation, or pyrolysis (Luik, 2009; An Assessment of Oil Shale Technologies, June 1980; Baldwin et al., 1984; Smith et al., 2007; Forbes, 1970; Francu et al., 2007; Gorlov, 2007; Koel et al., 2001).

By location: Based on the location the methods are classified as in situ or *ex situ*. In *ex situ* processing, the oil shale is excavated either at the surface or underground and then delivered to a processing facility. On the other hand, in situ processing transforms the kerogen whereas it is still in the form of the deposit of an oil shale, after which it is then extracted through oil wells, where it rises in the same way as conventional oils (Burnham and McConaghy, 2006). Dissimilar ex situ method, it does not include mining or spent oil shale disposal aboveground as spent oil shale stays underground (Bartis et al., 2005; An Assessment of Oil Shale Technologies, June 1980; Baldwin et al., 1984; Smith et al., 2007; Forbes, 1970; Francu et al., 2007).

By heating method: The method of transferring heat from combustion products to the oil shale may be classified as direct or indirect. Although techniques that burn materials external to the retort, to heat another material that contacts the oil shale, are described as *indirect*, techniques that allow combustion products to contact the oil shale within the retort are classified as *direct* (Jialin and Jianqiu, 2006).

10.5 INCLUDING CONFINEMENT IN THERMODYNAMICS

This section discusses the methodology used to modify conventional thermodynamics to incorporate capillary pressure when modeling flow in liquid-rich shale reservoirs. The methodology described can be readily implemented in a modern reservoir simulator. The section begins by offering a detailed account on some of the key thermodynamic concepts and later extends these same concepts toward incorporating capillary pressure in vapor–liquid equilibrium computations, to offer a better representation of fluid flow in confinement (Nojabaei, 2012; Haider, 2015).

10.5.1 Classical Thermodynamics

Thermodynamics is a branch of physics concerned with heat and temperature and their relation to energy and work in near-equilibrium systems. Classical thermodynamics describes the bulk behavior of the body and not the microscopic behaviors of the very large numbers of its microscopic constituents, such as molecules (Vidal, 1997). These general constraints are expressed in the four laws of thermodynamics. In the petroleum industry, thermodynamics of phase equilibrium attempts to answer "Under given temperature and pressure and mass of components, what are the amounts and composition of phases that result?" (Kovscek, 1996; Nojabaei, 2012; Haider, 2015).

10.5.1.1 Equation of State

At the heart of thermodynamics lies the equation of state, which in simplest terms is a formula describing the interconnection between various macroscopically measurable properties of a system. More specifically, an equation of state is a thermodynamic equation describing the "state of matter under a given set of physical conditions. It is a constitutive equation which provides a mathematical relationship between two or more state functions associated with the matter, such as its temperature, pressure, volume, or internal energy" (Thijssen, 2013; Haider, 2015).

Equations of state are instrumental in the calculation of Pressure–Volume–Temperature (PVT) behavior of petroleum gas–liquid systems at equilibrium. Reservoir fluids contain a variety of substances of diverse chemical nature that include hydrocarbons and nonhydrocarbons (Ashour, 2011; Nojabaei, 2012; Haider, 2015). Hydrocarbons range from methane to substances that may contain 100 carbon atoms. Despite the complexity of hydrocarbon fluids found in underground reservoirs, equations of state have shown surprising performance in the phase-behavior calculations of these complex fluids (Ashour, 2011; Haider, 2015).

Although to date, no single equation of state (EOS) accurately predicts the properties of all substances under all conditions, a number of equations of state (EOSs) have been developed for gases and liquids over the course of thermodynamics history. Among the various categories of EOS, the Cubic EOS have been used in this work, as they have been widely used and tested for predicting the behavior of hydrocarbon systems (Ashour, 2011; Kovscek, 1996; Nojabaei, 2012; Haider, 2015).

The generalized form of cubic EOS is shown in Eq. (10.1) (Kovscek, 1996; Gmehling, 2012; Nojabaei, 2012; Haider, 2015), in which each of the four parameters a, b, u, and w, depend on the actual EOS as shown in Table 10.2.

$$P = \frac{RT}{V - b} - \frac{a}{V^2 + ubV + wb^2} \tag{10.1}$$

in which R stands for the ideal gas constant, T represents the temperature, and V denotes the molar volume.

In this table, T_c and P_c are the critical temperature and pressure, respectively, ω is the acentric factor, and f_ω is the acentric factor function.

Table 10.2 Parameters of the Conventional Equations of State (Nojabaei, 2012; Haider, 2015; Firincioglu, 2013)

EOS	u	w	a	b
Van der Waals	0	0	$\frac{27R^2T_c^2}{64P_c}$	$\frac{RT_c}{8P_c}$
Redlich–Kwong	1	0	$\frac{0.42748R^2T_c^{\frac{5}{2}}}{P_cT^{\frac{1}{2}}}$	$\frac{0.08664RT_c}{P_c}$
Soave–Redlich–Kwong	1	0	$\frac{0.42748R^2T_c^{\frac{5}{2}}}{P_cT^{\frac{1}{2}}}\left[1+f_w\left(1-T_r^{\frac{1}{2}}\right)\right]^2$ $f_w = 0.48 + 1.574\omega - 0.176\omega^2$	$\frac{0.08664RT_c}{P_c}$
Peng–Robinson	2	−1	$\frac{0.45724R^2T_c^2}{P_c}\left[1+f_w\left(1-T_r^{\frac{1}{2}}\right)\right]^2$ $f_w = 0.37464 + 1.54226\omega - 0.2699\omega^2$	$\frac{0.07780RT_c}{P_c}$

When dealing with mixtures, mixing rules (Kwak, 1986) are applied to parameters a and b:

$$a_v = \sum_{i=1}^{n}\sum_{j=1}^{m} y_i y_j \sqrt{\alpha_{ii}\alpha_{jj}}(1 - K_{ij}) \tag{10.2}$$

$$b_v = \sum_i y_i b_i \tag{10.3}$$

here, v represents the vapor phase. To calculate a_l and b_l (in which subscript l represents the liquid phase), y in Eqs. (10.2) and (10.3) will have to be replaced by liquid-phase molar compositions of each component, often denoted by x. Each equation requires an independent determination of $\overline{k_{ij}}$ or binary interaction coefficients, which are set to zero for Van der Waals and Redlich–Kwong (RK) EOS by definition.

More commonly, Eq. (10.1) is written in terms of the compressibility factor Z (Kovscek, 1996; Grguri, 2003; Nojabaei, 2012; Haider, 2015):

$$Z^3 - (1 + B^* - uB^*)Z^2 + \left(A^* + wB^{*2} - uB^* - uB^{*2}\right)Z - A^*B^* - wB^{*2} - wB^{*3} = 0 \tag{10.4}$$

in which,

$$A^* = \frac{aP}{R^2T^2} \tag{10.5}$$

$$B^* = \frac{bP}{RT} \tag{10.6}$$

10.5.1.2 Condition of Equilibrium

One of the most fundamental relationships in thermodynamics is given by Eq. (10.7) (Firincioglu, 2013; Haider, 2015):

$$\Delta U^T = \Delta Q - \Delta W + \sum_i^{N_c} \mu_i N_i \tag{10.7}$$

Substituting the expression for ΔQ and ΔW, and rearranging Eq. (10.7), we get a fundamental thermodynamic relationship and the definition of Gibbs free energy:

$$dG = VdP - Sdt + \sum_i^{N_c} \mu_i dN_i \tag{10.8}$$

here, G is Gibbs free energy, V is Volume is m^3, dP is change in pressure in bars, S is Entropy in joule/K, dt is the change in temperature, and K is the chemical potential, N is the number of moles, N_c is the total number of components, and i is the component index (Firoozabadi, 1999; Haider, 2015).

For a closed system to be in equilibrium, the chemical potential of a component, at a given temperature and pressure condition, must be the same in each phase. The equilibrium condition is thus given by (Firoozabadi, 1999; Haider, 2015):

$$\mu_i^\alpha = \mu_i^\beta = \cdots = \mu_i^{N_p} \quad i = 1, 2, 3, \ldots, N_c \tag{10.9}$$

in which α and β stand for the phases N_p and N_c and are the total number of phases and components, respectively. Lewis (1923) proposed the following expression for Gibbs free energy (Firoozabadi, 1999; Haider, 2015):

$$dG_i = RTd \ln f_i \tag{10.10}$$

in which f_i represents the fugacity of component i.

Fugacity is often computed from a relationship comprising a dimensionless variable called "fugacity coefficient" (Matar, 2009; Nojabaei, 2012; Haider, 2015). This is given by:

$$\varnothing_i = \frac{f_i}{P} \tag{10.11}$$

here, $\varnothing_i$ is the fugacity coefficient of component i and is computed using the following expression, which has been derived using the general form of cubic equation (Kovscek, 1996; Matar, 2009; Nojabaei, 2012; Haider, 2015):

$$\ln\widehat{\varnothing_l} = \frac{b_i}{b}(Z-1) - \ln(Z - B^*) + \frac{A^*}{B^*\sqrt{u^2 - 4w}}\left(\frac{b_i}{b} - \delta_i\right) \ln\frac{2Z + B^*\left(u + \sqrt{u^2 - 4w}\right)}{2Z + B^*\left(u - \sqrt{u^2 - 4w}\right)} \tag{10.12}$$

$$\frac{b_i}{b} = \frac{T_{ci}/P_{ci}}{\sum x_j T_{cj}/P_{cj}} \tag{10.13}$$

$$\delta_i = \frac{2a_i^{\frac{1}{2}}}{a} \sum_{j=1}^{N_c} x_j \sqrt{\alpha_j}(1 - K_{ij}) \tag{10.14}$$

here, A^* and B^* are given by Eqs. (10.5) and (10.6).

At equilibrium condition at constant temperature and pressure, we know $dG_i = 0$ and $dP = dt = 0$. Substituting this in Eq. (10.8) we have:

$$G_i = \mu_i \tag{10.15}$$

For the fugacity of component i then, it must hold that

$$dG_i = d\mu_i = RTd\ln f_i \tag{10.16}$$

and the equality of the chemical potential translates into an equality of fugacity (Kovscek, 1996). Thus, at equilibrium, we have:

$$\mu_i^{\alpha} = \mu_i^{\beta} = \cdots = \mu_i^{N_p} \quad \rightarrow f_i^{\alpha} = f_i^{\beta} = \cdots = f_i^{N_p} \quad i = 1, 2, 3, \ldots, N_c \tag{10.17}$$

10.5.1.3 Vapor–Liquid Equilibrium/Flash Computation

Using flash one can obtain the equilibrium composition of two coexisting phases and solve for bubble- and dew-point pressures. The general flash routine that Automatic Differentiation General Purpose Research Simulator (AD-GPRS) follows is outlined as follows. This method also closely follows the algorithm illustrated by Kovscek (1996). A simplified representation of flash is illustrated in the following flowchart (Nojabaei, 2012; Haider, 2015).

The first step to make an initial guess for *K-values*, in which K is the equilibrium ratio given by $K_i = y_i/x_i$ and x_i and y_i are the liquid and gaseous molar fractions of component. This initial guess can be computed using Wilson's Equation (Wilson, 1969; Nojabaei, 2012; Haider, 2015):

$$K_i = \frac{P_{ci}}{P} \exp\left[5.37(1 + \omega_i)\left(1 - \frac{T_{ci}}{T}\right)\right] \tag{10.18}$$

here, P_{ci} and T_{ci} are the critical pressure and temperature of a component with index i. ω is the acentric factor of the component.

In flash process, a liquid mixture is partially separated and the gas is allowed to come to equilibrium with the liquid. For two phases, a mass balance on 1 mol of mixture yields the following:

$$Z_i = x_i l + y_i(1 - l) \tag{10.19}$$

here, Z_i is the overall composition of a component in the system and l is the mole fraction of the mixture that is present in liquid phase. Plugging $K_i = y_i/x_i$ into Eq. (10.19), we get expressions for the liquid and gaseous molar fractions for each component as follows:

$$x_i = \frac{Z_i}{1 + (1 - l)K_i} \tag{10.20}$$

$$y_i = \frac{K_i Z_i}{1 + (1 - l)K_i} \tag{10.21}$$

Using the fact that the sum of all mole fractions in each phase must be 1, we can combine Eqs. (10.20) and (10.21) to yield:

$$f(l) = \sum_i^{N_c} \frac{Z_i(1 - K_i)}{K_i + (1 - K_i)l} = 0 \tag{10.22}$$

Eq. (10.22) is called the Rachford–Rice Equation (Rachford, 1952) and can be iteratively solved to obtain l (the unknown) liquid fraction. The converged value of l tells whether the system is in single vapor phase ($l < 0$), two phases ($0 < l < 1$) or single liquid phase ($l > 1$). Additionally, once l is known, Eqs. (10.20) and (10.21) can be used to obtain the liquid and vapor compositions of each component in the system. Mixed Newton/Bisection method is often used to solve for l.

Phase molar compositions thus obtained can be substituted in Eqs. (10.2) and (10.3) to obtain respective EOS parameter for each phase, that is a_v, a_l, b_v, and b_l.

If there are two phases present in the system, the EOS will be solved twice (one for each phase) using its respective phase EOS parameters. Each solution gives the volume of its respective phase. At given P and T, the compressibility factor Z is computed for each phase (that is Z_v and Z_l) using Eq. (10.4). Note that to do that, A^* and B^* in Eqs. (10.5) and (10.6) too are separately computed for each phase. For instance, A_v^* uses a_v and A_l^* uses a_l.

Once liquid and vapor volumes are computed, we use Eq. (10.12) to compute fugacity coefficients for every component i and Eq. (10.11) to compute the corresponding fugacities. The system is in equilibrium when the following is true for all components:

$$\widehat{f}_i^{\,\mathrm{l}} = \widehat{f}_i^{\,\mathrm{v}}, \quad i = 1, 2, \ldots, N_\mathrm{c} \tag{10.23}$$

Numerically, this is equivalent to

$$\left| \frac{\widehat{f}_i^{\,\mathrm{l}}}{\widehat{f}_i^{\,\mathrm{v}}} - 1 \right| < \varepsilon \tag{10.24}$$

here, ε is a small number, usually in the range of 10^{-4} to 10^{-6}.

Each time a new K value is calculated, the system is checked for equilibrium. This can be done using Successive Substitution (SSI) method. Thus, K can be computed as (Nojabaei, 2012; Haider, 2015):

$$(K_i)^{K+1} = \left(\frac{\widehat{f}_i^{\,\mathrm{l}}}{\widehat{f}_i^{\,\mathrm{v}}} K_i \right)^K \tag{10.25}$$

New values of l can thus be generated by computed K values and by solving Eq. (10.22).

10.5.2 Modification of Flash to Incorporate Capillary Pressure in Tight Pores

Conventional flash involves the computation of all EOS parameters and fugacity for each phase at a single pressure (Nojabaei, 2012; Haider, 2015); that is:

$$\widehat{f}_i^{\,\mathrm{l}} = \widehat{f}_i^{\,\mathrm{l}}(P, V^\mathrm{l}, T, x_1, x_2, \ldots) \tag{10.26}$$

$$\widehat{f}_i^{\,\mathrm{v}} = \widehat{f}_i^{\,\mathrm{v}}(P, V^\mathrm{v}, T, x_1, x_2, \ldots) \tag{10.27}$$

This works well for conventional reservoirs, but for tight reservoirs each phase has to be treated against its own respective phase pressure. Therefore, fugacity is now defined as following as (Nojabaei, 2012; Haider, 2015):

$$\widehat{f}_i^{\,\mathrm{l}} = \widehat{f}_i^{\,\mathrm{l}}(P^\mathrm{l}, V^\mathrm{l}, T, x_1, x_2, \ldots) \tag{10.28}$$

$$\widehat{f}_i^{\,\mathrm{v}} = \widehat{f}_i^{\,\mathrm{v}}(P^\mathrm{v}, V^\mathrm{v}, T, x_1, x_2, \ldots) \tag{10.29}$$

When the capillary forces are considered, the phase pressures are no longer equal and the difference is given by Laplace equation which is as follows (Nojabaei, 2012; Haider, 2015):

$$P_{\text{cap}} = P_{\text{g}} - P_{\text{l}} = \frac{2\sigma \cos\theta}{r} \tag{10.30}$$

here, P_{cap} is the capillary pressure, P_{g} and P_{l} are the gas (vapor) phase and liquid (oil) phase pressures, r is the pore radius, θ is the wettability angle and σ is the interfacial tension. Considering an oil wet system (the wettability angle to be 180°), which, in many cases, such as Bakken Formation shale reservoir, is a valid assumption (Fine, 2009). Therefore, Eq. (10.30) gets simplified to the following (Nojabaei, 2012; Haider, 2015):

$$P_{\text{cap}} = \frac{2\sigma}{r} \tag{10.31}$$

There are several correlations and methods to calculate the interfacial tension (IFT). According to Ayirala (2006), the most important among these models are the Parachor model (Macleod, 1923; Sugden, 1924), the corresponding states theory (Brock and Bird, 1955), thermodynamic correlations (Clever, 1963), and the gradient theory (Carey, 1979). In this work, Macleod–Sugden formulation has been used to calculate IFT because it is most widely used in the petroleum industry due to its simplicity (Ayirala, 2006; Nojabaei, 2012; Haider, 2015). Eq. (10.32) presents the Macleod–Sugden formulation:

$$\sigma = \left[\sum_i \gamma_i \left(x_i \rho^{\text{l}} - y_i \rho^{\text{v}} \right) \right]^4 \tag{10.32}$$

Here, γ_i is components' Parachor value and ρ^{l} and ρ^{v} are liquid and vapor densities, respectively. Thus, IFT is a function of changes in densities, compositions and Parachor, and becomes zero at the critical point in which phase properties start approaching each other (Haider, 2015).

The flash flowchart presented in Fig. 10.13 can thus be modified as illustrated in Fig. 10.14 to incorporate capillary pressure. Here, the red-boxed parameters get influenced by capillary pressure, which in turn influences the whole flash. Similar modifications in the vapor–liquid equilibrium (VLE) to accommodate capillary pressure have been done by Firincioglu (2013) and Nojabaei (2012), previously Haider (2015).

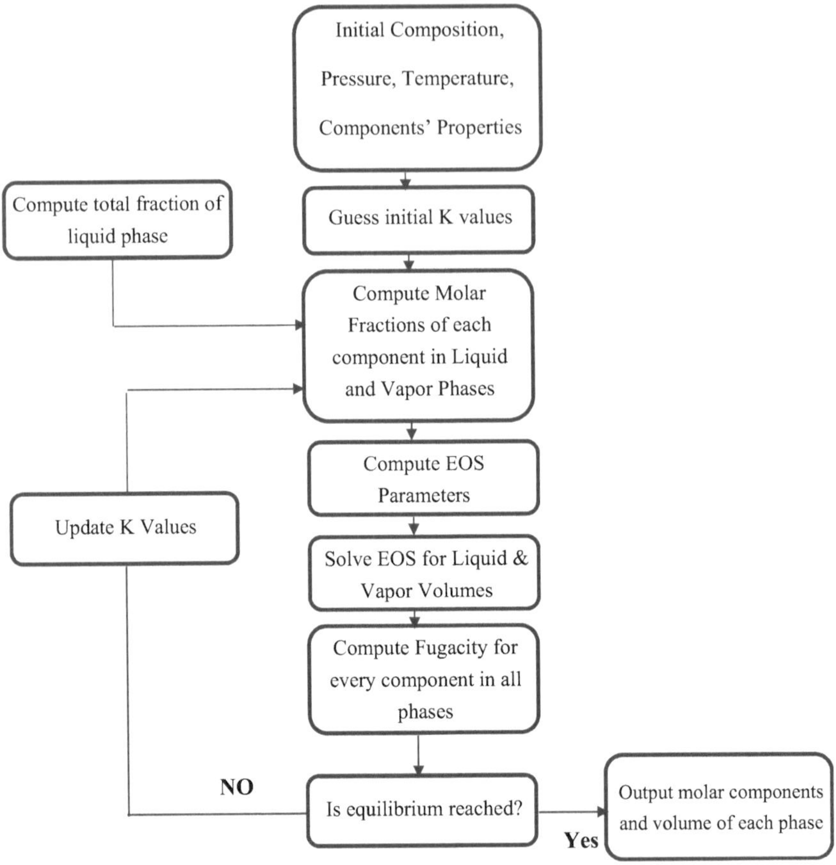

Figure 10.13 Flow chart of conventional flash calculation (Nojabaei, 2012; Haider, 2015).

10.5.3 Stability Test Using Gibbs Free Energy Approach

It has been shown by Wang (2013) that standard stability test based on tangent plane distance analysis can be extended to consider capillarity effect. Michelsen (1982) showed that Eq. (10.33) holds true if the original system is stable:

$$\sum_{i}^{M} y_i[\ln f_i(y) - \ln(z)] \geq 0 \tag{10.33}$$

in which $f_i(y)$ is the fugacity of the incipient phase of component i, whereas $f_i(z)$ is its fugacity in the original system. Here, the original phase is liquid and

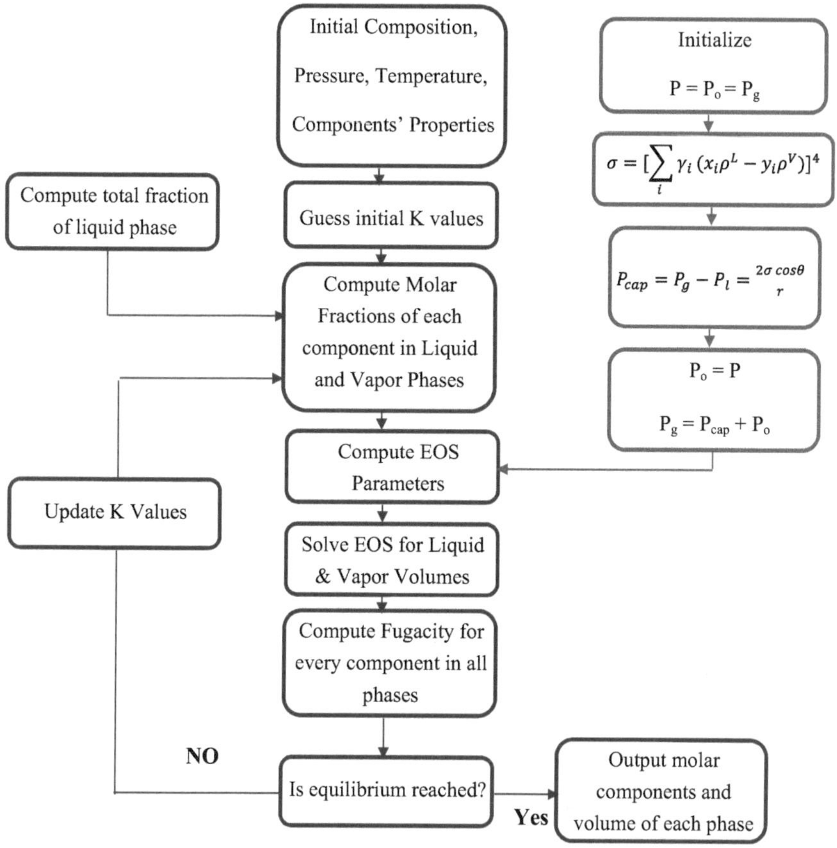

Figure 10.14 Modified workflow for incorporating capillary pressure in vapor–liquid flash calculation (Firincioglu, 2013; Nojabaei, 2012; Haider, 2015).

the incipient phase is vapor. If the original phase were vapor, the incipient must have been liquid. Considering the definition of fugacity (Haider, 2015):

$$f_i(z) = z_i \varnothing_i(z) P^{l} \tag{10.34}$$

$$f_i(y) = z_i \varnothing_i(z) P^{v} \tag{10.35}$$

in which $\varnothing_i$ is the fugacity coefficient of component i, we can substitute Eqs. (10.34) and (10.35) in Eq. (10.33) to get the following expression:

$$\sum_{i}^{M} y_i \left[\ln\left(y_i \varnothing_i(y) P^{v}\right) - \ln\left(z_i \varnothing(z) P^{l}\right)\right] \geq 0 \tag{10.36}$$

Some rearrangement of Eqs. (10.36) and (10.33) can yield the following:

$$\sum_{i}^{M} y_i\left[\ln y_i + \ln\varnothing(y) - \ln z_i - \ln\varnothing(z) + \left(\ln P^{\mathrm{v}} - \ln P^{\mathrm{l}}\right)\right] \geq 0 \quad (10.37)$$

Thus, using this approach, Wang (2013) showed that the capillary term ($\ln P^{\mathrm{v}} - \ln P^{\mathrm{l}}$) is naturally incorporated in the equilibrium test.

10.5.4 Impact of Critical Property Shifts Due to Confinement on Hydrocarbon Production

Confined space or pore proximity effect alters the fluid properties and phase behavior (Singh, 2009; Sapmanee, 2011; Teklu, 2014; Haider, 2015). This section presents works done by different scientists in the area of critical property shifts due to confinement. Moreover, this section contains applied examples for including confinement in PVT calculations of tight reservoirs.

10.5.4.1 Impact of Critical Properties Shift Due to Confinement Within Fluid-phase Envelope

Singh (2009) used Grand Monte Carlo simulation to study the impact of confinement on critical properties. He developed the following correlations for the shifts in critical temperature and pressure (Haider, 2015):

$$\Delta T_{\mathrm{c}}^{*} = \frac{T_{\mathrm{cb}} - T_{\mathrm{cp}}}{T_{\mathrm{cb}}} = 0.9409\frac{\sigma_{\mathrm{LJ}}}{r_{\mathrm{p}}} - 0.2415\left(\frac{\sigma_{\mathrm{LJ}}}{r_{\mathrm{p}}}\right)^{2} \quad (10.38)$$

$$\Delta P_{\mathrm{c}}^{*} = \frac{P_{\mathrm{cb}} - P_{\mathrm{cp}}}{P_{\mathrm{cb}}} = 0.9409\frac{\sigma_{\mathrm{LJ}}}{r_{\mathrm{p}}} - 0.2415\left(\frac{\sigma_{\mathrm{LJ}}}{r_{\mathrm{p}}}\right)^{2} \quad (10.39)$$

$$\sigma_{\mathrm{LJ}} = 0.244\sqrt[3]{\frac{T_{\mathrm{cb}}}{P_{\mathrm{cb}}}} \quad (10.40)$$

in which σ_{LJ} is Lennard–Jones size parameter (collision diameter in nm), r_{p} is pore radius (nm), $\Delta T_{\mathrm{c}}^{*}$ is relative critical temperature shift (dimensionless), T_{cb} is bulk critical temperature (K), T_{cp} is pore critical temperature (K), $\Delta P_{\mathrm{c}}^{*}$ relative critical pressure shift (dimensionless), P_{cb} is pore critical pressure (atm), and P_{cp} is pore critical pressure (atm).

The pressure at which a large quantity of gas is in equilibrium with an inconsiderable quantity of liquid is named the dew-point pressure (p_{d}) of a hydrocarbon system. For a total of 1 lb-mole of a hydrocarbon mixture, i.e., $n = 1$, at the dew-point pressure we have following conditions:

$$n_l \approx 0$$

$$n_v \approx 1$$

At aforementioned circumstances, the overall composition, z_i, and the composition of the vapor phase, y_i, are the same. Performing these limitations, Eq. (10.41) becomes:

$$\sum_i \frac{z_i}{K_i} = 1 \tag{10.41}$$

The pressure at which a large quantity of liquid is in equilibrium with an inconsiderable quantity of gas is named the bubble-point pressure (p_d), of a hydrocarbon system. For a total of 1 lb-mole of a hydrocarbon mixture, i.e., $n = 1$, at the bubble-point pressure we have following conditions:

$$n_v \approx 0$$

$$n_l \approx 1$$

At aforementioned conditions, the overall composition, z_i, and the composition of the liquid phase, x_i, are the same. Performing these limitations, Eq. (10.42) results in

$$\sum_i (z_i K_i) = 1 \tag{10.42}$$

Example 10.1

Using Table 10.3, determine the critical temperature and pressure shifts of a Bakken fluid sample. Moreover, calculate the pore critical temperature and pressure when pore radius is equal to 10 nm.

Table 10.3 Critical Properties of Bakken Fluid Components (Haider, 2015)

	C_1	C_2	C_3	C_4	CA	CB	CC	CD
T_{cb} (K)	186.2978	305.5384	369.9834	421.7823	486.3773	585.1389	740.0528	1024.717
P_{cb} (atm)	44.57146	49.1285	41.89997	37.18439	31.3889	24.72382	16.98495	12.9369

Answer

To study the impact of pore size on the critical property shifts of various components in a Bakken fluid system, the σ_{LJ}/r_p ratio is computed for every pore radius and reported in the following table. Moreover, the critical temperature and

(Continued)

pressure shifts are also listed in this table along with critical pressure and temperature when pore radius is equal to 10 nm.

	C_1	C_2	C_3	C_4	CA	CB	CC	CD
σ_{LJ} (nm)	0.393043	0.448712	0.504328	0.548234	0.608306	0.700548	0.858597	1.047887
T_{cb} (K)	186.2978	305.5384	369.9834	421.7823	486.3773	585.1389	740.0528	1024.717
P_{cb} (atm)	44.57146	49.1285	41.89997	37.18439	31.3889	24.72382	16.98495	12.9369
ΔT_c^*	0.036608	0.041733	0.046838	0.050857	0.056342	0.064729	0.079005	0.095944
ΔP_c^*	0.036608	0.041733	0.046838	0.050857	0.056342	0.064729	0.079005	0.095944
T_{cp} (K)	179.4777	292.7873	352.6541	400.3315	458.9738	547.2633	681.5849	926.402
P_{cp} (atm)	42.93977	47.07822	39.93746	35.29329	29.26039	23.12346	15.64305	11.69568
σ_{LJ}/r_P	0.039304	0.044871	0.050433	0.054823	0.060831	0.070055	0.08586	0.104789

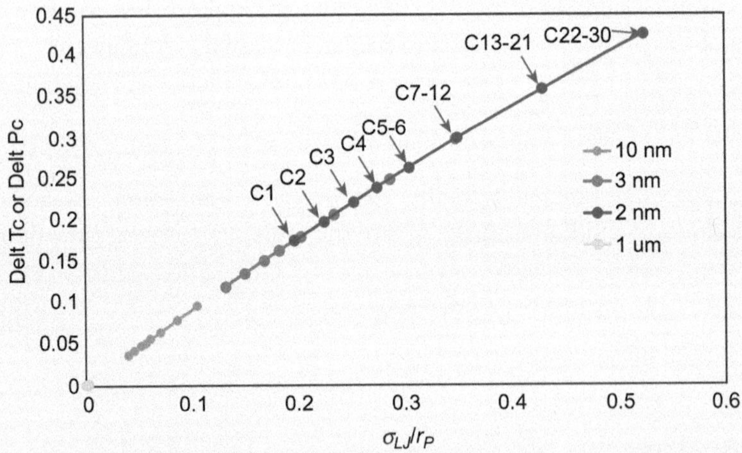

Figure 10.15 Impact of critical property shift as a function for Bakken fluid component (Haider, 2015).

Fig. 10.15 shows the plot of these critical shifts for Bakken fluid components at four different pore radii. As can be seen, the critical property shift increases as the molecules become heavier and the pore radius becomes smaller. Based on these findings, Teklu (2014) postulated that these shifts can be ignored for pore size >30 nm (Haider, 2015).

Using the new critical properties, Bakken fluid-phase envelope was constructed using Automatic Differentiation General Purpose Research Simulator (AD-GPRS). Fig. 10.16 shows significant suppression of the entire phase envelope and a pore radius of 10 nm (Haider, 2015).

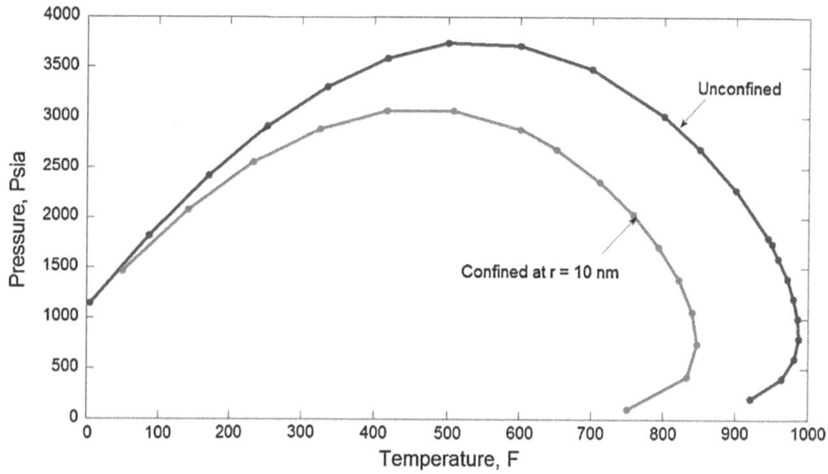

Figure 10.16 Suppression of Bakken fluid-phase envelope due to shifted critical properties at pore radius of 10 nm (Haider, 2015).

Example 10.2

Consider a binary mixture containing 30% C_1 and 70% C_6, determine the effect of capillary pressure on bubble-point pressure and phase envelope by varying the pore radius from 5 nm up to 100 nm.

Answer

At first, the impact of capillary pressure on bubble point by varying the pore size from 5 to 100 nm, using a binary mixture composed of 30% C_1 and 70% C_6, should be investigated. Then, the pore radius is assumed 10 nm and the effect of different composition on the bubble-point pressure is determined. For this mixture, the influence of capillary pressure fades away as the pore size approaches 100 nm. The phase envelope calculations use Peng–Robinson EOS through AD-GPRS. Fig. 10.17 illustrates that small pore radii can cause significant reduction in the bubble-point pressure. Based on published literature (Nojabaei, 2012; Teklu, 2014; Haider, 2015), dew points also get shifted but often at magnitudes that are less than the bubble-point shifts. Capillary pressure makes dew point appear sooner or at relatively higher reservoir pressures (Alharthy et al., 2013; Haider, 2015).

(Continued)

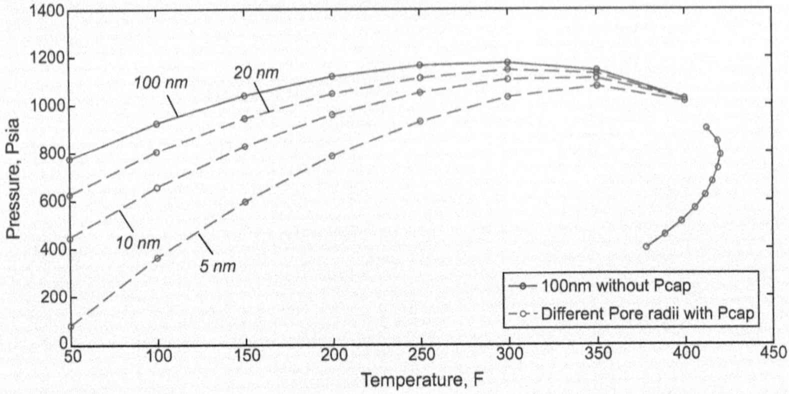

Figure 10.17 Impact of pore radius on bubble point suppression (Haider, 2015).

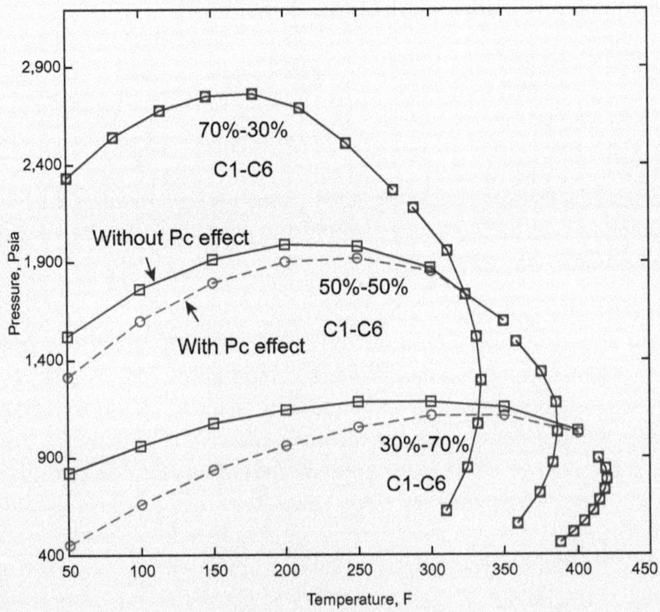

Figure 10.18 Influence of varying composition of C_1-C_6 on capillary pressure's influence using Automatic Differentiation General Purpose Research Simulator (AD-GPRS) (Haider, 2015).

Next, the pore size was fixed to 10 nm and the influence of varying compositions of the binary mixture on bubble point suppression was studied. The amount of methane was varied in the binary mixture comprising C_1-C_6. Fig. 10.18 shows the phase envelopes both with and without capillary pressure.

With the increase in the percentage of the heavier component C_6 in the system, the critical pressure positions shift. This influences the strength of capillary pressure, which in turn gets translated into higher bubble-point suppression. This is because higher bubble-point pressures reduce the density differences between liquid and vapor (Nojabaei, 2012) and thus suppress the impact of capillarity (Haider, 2015).

Example 10.3

Consider Bakken fluid sample with the following composition (see Table 10.4). Determine the effect of confinement on the phase envelope using Peng–Robinson EOS when pore radius is equal to 10 and 100 nm. Moreover, phase envelope of the Bakken fluid sample is depicted through Fig. 10.19.

Table 10.4 Composition of Bakken Fluid Sample (Nojabaei, 2012)

Component	C_1	C_2	C_3	C_4	C_{5-6}	C_{7-12}	C_{13-21}	C_{22-80}
Molar fraction	0.36736	0.14885	0.09334	0.05751	0.06406	0.15854	0.0733	0.03704

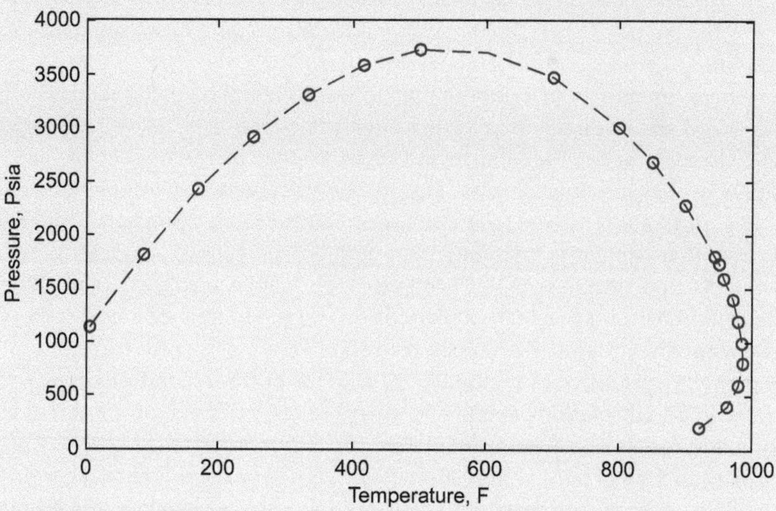

Figure 10.19 Phase envelope of Bakken fluid sample (Haider, 2015).

Answer

A typical tight reservoir such as Bakken has a pore radius ranging from 10 to 50 nm (Wang, 2013). Fig. 10.20 illustrates the bubble-point pressure suppression

(Continued)

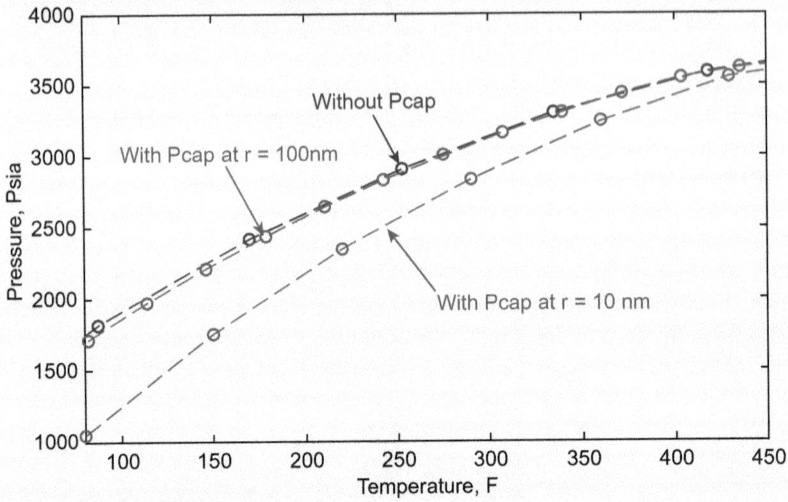

Figure 10.20 Bakken fluid phase envelope bubble-point suppression under the influence of confinement (Haider, 2015).

of Bakken fluid-phase envelope when the pore radius is equal to 10 nm. This suppression can lead to considerable deviation of fluid-flow properties from its respective unconfined state. Additionally, suppression in Bakken fluid's bubble point starts fading away quickly as the pore radius approaches 100 nm.

The suppression of bubble-point pressure leads to the retention of gas in oil for a longer time as the pressure is reduced, as will density and viscosity of oil. This led to the investigation of the impact of pore radius on Bakken fluid density. Fig. 10.21 depicts reduction in oil density as the pore radius increases and the reservoir pressure decreases. This indicates that capillarity becomes more influential at lower reservoir pressures. As depicted in Fig. 10.22, oil viscosity follows a similar trend. The results only display densities and viscosities till the bubble-point pressures as the capillary pressure becomes zero above that. It is worth mentioning that the IFT values were multiplied by a factor of three due to the uncertainty in IFT calculations and Macleod and Sugden correlation's tendency to underpredict IFT (Ayirala, 2006). Fig. 10.23 demonstrates the density of oil as a function of pore radius when the reference pressure is changed from oil-phase pressure to gas-phase pressure, that is, $P_o = P_g - P_{cap}$ instead of $P_g = P_o + P_{cap}$ (Haider, 2015).

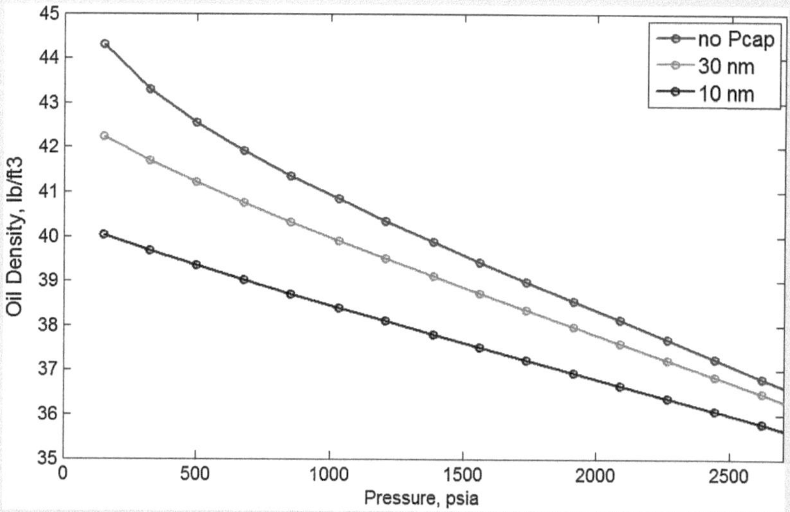

Figure 10.21 Bakken oil density as a function of pore radius and oil-phase (reservoir) pressure (Haider, 2015).

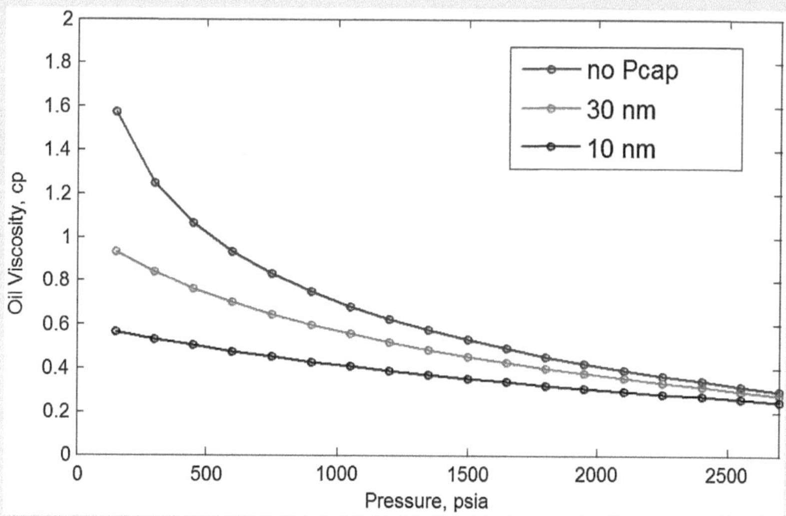

Figure 10.22 Bakken oil viscosity as a function of pore radius and oil-phase (reservoir) pressure (Haider, 2015).

(Continued)

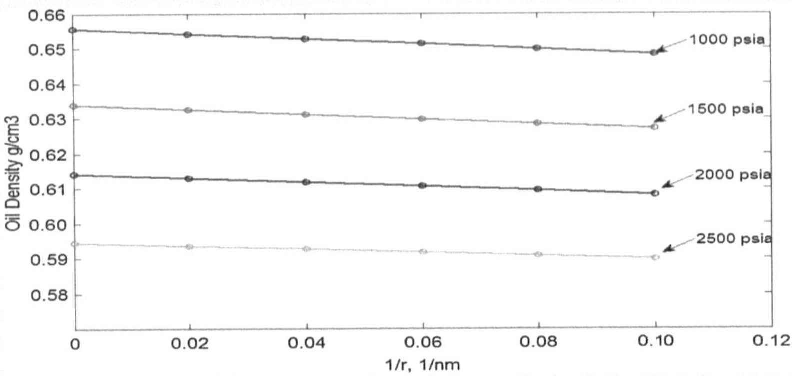

Figure 10.23 Bakken oil density as a function of pore radius and gas-phase pressure (Haider, 2015).

Problems

10.1 List the methods of oil shale extraction and explain two of them.

10.2 Consider a volatile oil with the following composition. The reservoir pressure and temperature are 3500 Psi and 180°F, respectively. Using Peng–Robinson equation of state, determine the density of this fluid when the pore radius is equal to 8 nm.

Component	Mol%
N_2	0.015
H_2S	0.015
CO_2	0.02
C_1	0.55
C_2	0.07
C_3	0.02
n-C_4	0.03
n-C_5	0.01
C_6	0.02
C_{7+}	0.25
$MW_{C7+} = 231$	
$\gamma_{C7+} = 0.887$	

10.3 Consider a volatile oil with composition given below. Calculate the dew point pressure of this fluid by Peng–Robinson equation of state when pore radius is equal to 5 nm. The reservoir temperature is 220°F.

Component	Mol%
N_2	0.012
H_2S	0.013
CO_2	0.025
C_1	0.55
C_2	0.07
C_3	0.03
i-C_4	0.01
n-C_4	0.03
i-C_5	0.01
n-C_5	0.01
C_6	0.02
C_{7+}	0.22
$MW_{C7+} = 231$	
$\gamma_{C7+} = 0.887$	

10.4 Using the blow table to determine the critical temperature and pressure shifts when the pore radius is equal to 12 nm.

	C_1	C_2	C_3	C_4
T_{cb} (K)	184.2978	307.5384	371.9834	418.7823
P_{cb} (atm)	42.5114	47.1085	40.12997	36.18439

10.5 Consider a volatile oil with composition given in the following table. Calculate the dew-point pressure of this fluid by Soave–Redlich–Kwong equation of state when pore radius is equal to 10 nm and compare the dew-point pressure when confinement is included and ignored. The reservoir temperature is 190°F.

Component	Mol%
N_2	0.01
H_2S	0.01
CO_2	0.03
C_1	0.59
C_2	0.09
C_3	0.01
i-C_4	0.005
n-C_4	0.005
i-C_5	0.005
n-C_5	0.005
C_6	0.01
C_{7+}	0.23
$MW_{C7+} = 229$	
$\gamma_{C7+} = 0.875$	

10.6 Consider a volatile oil with composition given in the following table. Calculate the bubble-point pressure of this fluid by Soave–Redlich–Kwong equation of state when pore radius is equal to 8 nm and compare the bubble-point pressure when confinement is included and ignored. The reservoir temperature is 212°F.

Component	Mol%
C_1	0.39
C_2	0.06
C_3	0.015
i-C_4	0.015
n-C_4	0.005
i-C_5	0.025
n-C_5	0.025
C_6	0.01
C_{7+}	0.455
$MW_{C7+} = 278$	
$\gamma_{C7+} = 0.919$	

10.7 Using following table plot the critical pressure and temperature shifts versus corresponding pore radius when pore diameter varies from 2 nm up to 30 nm.

Component	P_c (Psi)	T_c (R)
C_1	666.4	343.33
C_2	706.5	549.92
C_3	616.0	666.06
C_4	527.9	765.62
C_5	488.6	845.8
C_6	453	923
C_{7+}	285	1210

10.8 Consider a volatile oil with composition given in the following table. Calculate the dew-point pressure of this fluid by Soave–Redlich–Kwong equation of state when pore radius is equal to 10 nm and compare the dew-point pressure when confinement is included and ignored. The reservoir temperature is 190°F.

Component	Mol%
CO_2	0.05
C_1	0.61
C_2	0.07
C_3	0.01
i-C_4	0.005
n-C_4	0.005
i-C_5	0.005
n-C_5	0.005
C_6	0.01
C_{7+}	0.23
$MW_{C7+} = 234$	
$\Upsilon_{C7+} = 0.8915$	

10.9 Consider a crude oil with composition given in the following table. Plot phase envelop using Peng–Robinson equation of state when pore radius is equal to 5 nm and compare the phase envelop when confinement is included and ignored. The reservoir temperature is 235°F.

Component	Mol%
H_2S	0.025
CO_2	0.025
C_1	0.41
C_2	0.075
C_3	0.005
i-C_4	0.01
i-C_5	0.01
C_6	0.01
C_{7+}	0.43
$MW_{C7+} = 267$	
$\Upsilon_{C7+} = 0.9105$	

10.10 Consider a crude oil with composition given in the following table. Calculate oil formation volume factor when pore radius is equal to 11 nm. The reservoir pressure and temperature are 4225 Psi and 222°F, respectively. The stock tank pressure and temperature are 10.7 Psi and 77°F, respectively.

Component	Mol%
H_2S	0.05
C_1	0.36
C_2	0.055
C_3	0.025
i-C_4	0.01
i-C_5	0.01
C_6	0.01
C_{7+}	0.48
$MW_{C7+} = 279$	
$\Upsilon_{C7+} = 0.9212$	

REFERENCES

Alharthy, N., Nguyen, T.N., Teklu, T.W., et al., 2013. Multiphase Compositional Modeling in Small-Scale Pores of Unconventional Shale Reservoirs. Presented at SPE Annual Technical Conference and Exhibition. New Orleans, Louisiana, 30 September–2 October. SPE-166306-MS. http://dx.doi.org/10.2118/166306-MS.

An Assessment of Oil Shale Technologies, June 1980. United States Office of Technology Assessment (DIANE Publishing), ISBN 978-1-4289-2463-5, pp. 108–110, 133; 138–139; 148–150; NTIS order #PB80-210115.

Ashour, I., 2011. Applications of Equations of State in the Oil and Gas Industry. Oman: Department of Petroleum and Chemical Engineering College of Engineering, Sultan Qaboos University.

Ashraf, M., Satapathy, M., 2013. The Global Quest for Light Tight Oil: Myth or Reality? Schlumberger Newsroom. Available online: https://www.sbc.slb.com/Our_Ideas/Energy_Perspectives/1st%20Semester13_Content/1st%20Semester%202013_Global.aspx.

Ayirala, S., 2006. A new Parachor model to predict dynamic interfacial tension and miscibility in multicomponent hydrocarbon systems. Colloid and Interface Science (299), 321–331.

Baldwin, R.M., Bennett, D.P., Briley, R.A., 1984. Reactivity of Oil Shale Towards Solvent Hydrogenation. ISSN: 0569-3799, vol. 29 (1). American Chemical Society. Division of Petroleum Chemistry, pp. 148–153.

Bartis, J.T., LaTourrette, T., Dixon, L., Peterson, D.J., Cecchine, G., 2005. Oil Shale Development in the United States. Prospects and Policy Issues. Prepared for the National Energy Technology Laboratory of the United States Department of Energy. The RAND Corporation, ISBN 978-0-8330-3848-7, p. x, 15–18; 50.

Brendow, K., 2009. Oil shale – a local asset under global constraint. Oil Shale. A Scientific-Technical Journal (Estonian Academy Publishers). ISSN: 0208-189X 26 (3), 357–372. http://dx.doi.org/10.3176/oil.2009.3.02.

Brock, J.R., Bird, R.B., 1955. Surface tension and the principle of corresponding states. AIChE Journal 1, 174.

Burnham, A.K., McConaghy, J.R., October 16, 2006. Comparison of the acceptability of various oil shale processes. In: 26th Oil Shale Symposium, vol. 2 (17). Lawrence Livermore National Laboratory, Golden, Colorado. UCRL-CONF-226717.

Bustin, 2008. Impact of shale properties on pore structure and storage characteristics. In: Fort Worth, Texas, SPE Shale Gas Production Conference.

Cane, R.F., 1976. The origin and formation of oil shale. In: Yen, T.F., Chilingar, G.V. (Eds.), Oil Shale. Elsevier, Amsterdam, ISBN 978-0-444-41408-3, p. 56.
Carey, B.S., 1979. The Gradient Theory of Fluid Interfaces (Ph.D. dissertation). University of Minnesota, Minneapolis.
Clever, H.L., 1963. Journal of Chemical & Engineering Data 8, 291.
Colorado Oil and Gas Association, 2013. Oil Shale vs. Shale Oil. Industrial Commission, Oil and Gas Research Council, Denver, CO, USA. http://www.coga.org/pdf_Basics/Basics_OilShale.pdfDakota.
Dyni, J.R., 2010. Oil shale. In: Clarke, A.W., Trinnaman, J.A. (Eds.), Survey of Energy Resources, twenty second ed. World Energy Council, pp. 93–123. 978-0-946121-02-1.
EIA, 2013. Technically Recoverable Shale Oil and Shale Gas Resources: An Assessment of 137 Shale Formations in 41 Countries Outside the United States. US Energy Information Administration - Independent Statistics and Analysis, Washington, DC.
Fan, L., 2005. Understanding Gas-Condensate Reservoirs. Schlumberger: Oil Field Review, Texas, USA.
Fine, K., 2009. Bakken Water Opportunites Assesment - Phase 1, North Dakota.
Firincioglu, T., 2013. Bubble Point Suppression in Unconventional Liquids Rich Reservoirs and Its Impact on Oil Production (Ph.D. dissertation). Colorado School of Mines.
Firoozabadi, A., 1999. Thermodynamics of Hydrocarbon Reservoirs. McGraw-Hill, NY.
Forbes, R.J., 1970. A Short History of the Art of Distillation from the Beginnings Up to the Death of Cellier Blumenthal. Brill Publishers, ISBN 978-90-04-00617-1, pp. 41–42.
Francu, J., Harvie, B., Laenen, B., Siirde, A., Veiderma, M., May 2007. A Study on the EU Oil Shale Industry Viewed in the Light of the Estonian Experience. A Report by EASAC to the Committee on Industry, Research and Energy of the European Parliament. European Academies Science Advisory Council, pp. 12–13, 18–19; 23–24; 28.
Gmehling, J., 2012. Chemical Thermodynamics for Process Simulation, Germany.
Gordon, D., 2012. Understanding Unconventional Oil Resources, s.l.: Energy and Environment, Carnegie Papers.
Gorlov, E.G., October 2007. Thermal dissolution of solid fossil fuels. Solid Fuel Chemistry (Allerton Press, Inc.). ISSN: 1934-8029 41 (5), 290–298. http://dx.doi.org/10.3103/S0361521907050047 (subscription required).
Grguri, 2003. Excess molar volume of acetonitrile + alcohol systems at 298.15K. Part II: correlation by cubic equation of state. Journal of the Serbian Chemical Society 68 (1), 47–56. JSCS–3019.
Haider, B.A., 2015. Impact of Capillary Pressure and Critical Properties Shifts Due to Confinement on Hydrocarbon Production from Shale Reservoirs (M.Sc. thesis). Stanford University.
Ismail, M.A., 2010. Field Observation of Gas Condensate Well Testing. Energy Resources Engineering Department, Stanford University, Stanford, California. MS report.
Ismail, M.A., Horne, R.N., 2014. An investigation of gas-condensate flow in liquid-rich shales. In: SPE 169021. SPE Unconventional Resources Conference, 1–3 April, The Woodlands, Texas, USA. Journal of Chemical Society 25 (1924), 32.
Jialin, Q., Jianqiu, W., November 07, 2006. World oil shale retorting technologies. In: International Oil Shale Conference. China University of Petroleum, Amman, Jordan. Jordanian Natural Resources Authority.
Johnson, H.R., Crawford, P.M., Bunger, J.W., 2004. Strategic Significance of America's Oil Shale Resource vol. II: Oil Shale Resources, Technology and Economics. Office of Deputy Assistant Secretary for Petroleum Reserves; Office of Naval Petroleum and Oil Shale Reserves; United States Department of Energy, pp. 13–16. A2; B3–B5.
Koel, M., 1999. Estonian Oil Shale. ISSN: 0208-189X. Oil Shale. A Scientific-Technical Journal (Estonian Academy Publishers).

Koel, M., Ljovin, S., Hollis, K., Rubin, J., 2001. Using neoteric solvents in oil shale studies. Pure and Applied Chemistry (Blackwell Science). ISSN: 0033-4545 73 (1), 153–159. http://dx.doi.org/10.1351/pac200173010153.

Kovscek, A., 1996. Thermodynamics of Fluid Phase Equilibria, Class Notes. Stanford University, Stanford, California.

Kuila, U., 2013. Specific Surface Area and Pore-size Distribution in Clays and Shales.

Kwak, 1986. Van Der Waals Mixing Rules for Cubic Equation of State. Department of Chemical Engineering, University of Illinois, Chicago.

Lewis, 1923. Thermodynamics and the Free Energy of Chemical Substances. McGraw-Hill Book Co., Inc., New York.

Loucks, R., 2012. Spectrum of Pore Types and Networks in Mudrocks and a Descriptive Classification for Matrix-related Mudrock Pores. The American Association of Petroleum Geologists. http://aapgbull.geoscienceworld.org/content/96/6/1071.figures-only.

Louw, S.J., Addison, J., 1985. In: Seaton, A. (Ed.), Studies of the Scottish Oil Shale Industry. vol. 1 History of the Industry, Working Conditions, and Mineralogy of Scottish and Green River Formation Shales. Final Report on US Department of Energy. Institute of Occupational Medicine, p. 35, 38; 56–57. DE-ACO2 – 82ER60036.

Luik, H., June 08, 2009. Alternative technologies for oil shale liquefaction and upgrading. In: International Oil Shale Symposium. Tallinn University of Technology, Tallinn, Estonia.

Macleod, D.B., 1923. On a relation between surface tension and density. Transactions of Faraday Society 19, 38.

Matar, A., 2009. Chemical Engineering Thermodynamics II (0905323) 02-The Molar Gibbs Free Energy & Fugacity of a Pure Component. Chemical Engineering Department. University of Jordan.

McCarthy, K., 2011. Basic petroleum geochemistry for source rock evaluation. In: Oilfield Review Summer. Schlumberger, Houston, Texas.

Michelsen, L., 1982. The isothermal flash problem. Part I. Phase-split alculation. Fluid Phase Equilibria 9, 21–40.

Moody, R., April 20, 2007. Oil & gas shales, definitions & distribution in time & space. In: The History of On-shore Hydrocarbon Use in the UK. Geological Society of London, p. 1.

Nojabaei, B., 2012. Effect of capillary pressure on fluid density and phase behavior in tight pore rocks and shales. In: San Antonio, Texas, SPE 159258, SPE Annual Technical Conference and Exhibition.

Pederson, J., 2010. The Oil and Gas Window, Oil and Gas Geology.

Prien, C.H., 1976. Survey of oil-shale research in last three decades. In: Yen, T.F., Chilingar, G.V. (Eds.), Oil Shale. Elsevier, Amsterdam, ISBN 978-0-444-41408-3, pp. 237–243.

Qian, J., Wang, J., Li, S., October 15, 2007. One year's progress in the Chinese Oil Shale Business. In: 27th Oil Shale Symposium. China University of Petroleum, Golden, Colorado.

Rachford, H.H., 1952. Procedure for use of electronic digital computers in calculating flash vaporization hydrocarbon equilibrium. Journal of Petroleum Technology SPE 952327 4 (10), 19, 3.

Rotelli, F., 2012. Shale Reservoirs: Pore-scale Flow Behaviour and its Effect on Production (Ph.D. dissertation). Politecnico Di Milano, Milano, Italy.

Runnels, R.T., Kulstad, R.O., McDuffee, C., Schleicher, J.A., 1952. Oil shale in Kansas. Kansas Geological Survey Bulletin (University of Kansas Publications) 96 (part 3). Retrieved 2009-05-30.

Sapmanee, K., 2011. Effects of Pore Proximity on Behavior and Production Prediction of Gas/Condensate, Oklahoma (M.S. thesis). Mewbourne School of Petroleum and Geological Engineering, University of Oklahoma.

Singh, S., 2009. Vapor-liquid phase coexistence, critical properties and surface tension of confined alkanes. Journal of Physical Chemistry C 113, 7170–7180.
Slatt, R., 2011. Pore types in the Barnett and Woodford gas shales: contribution to understanding gas storage and. AAPG Bulletin 95, 2017–2030.
Smith, M.W., Shadle, L.J., Hill, D., 2007. Oil Shale Development from the Perspective of NETL's Unconventional Oil Resource Repository. United States Department of Energy. DOE/NETL-IR-2007-022.
Sondergeld, C., 2013. Petrophysical considerations in evaluating and producing shale gas resources. In: Pittsburgh, Pennsylvania, SPE 131768, SPE Unconventional Gas Conference.
Speight, J.G., 2008. Synthetic Fuels Handbook: Properties, Process, and Performance, 13. McGraw-Hill, ISBN 978-0-07-149023-8, pp. 182–186.
Sugden, S., 1924. Journal of the Chemical Society 125, 1177.
Synthetic Fuels Summary. Report No. FE-2468-82, March 1981. The Engineering Societies Commission on Energy, Inc. (United States Department of Energy), vol. 80, pp. 83–84; 90.
Teklu, T., 2014. Phase behaviour and minimum miscibility pressure in nanopores. In: Denver, SPE 168865, Unconventional Resources Technology Conference.
Thijssen, J., 2013. e-Study guide for: Computational Physics, second ed.
Vidal, J., 1997. Thermodynamics - Applications in Chemical Engineering and Petroleum Industry.
Wang, Y., 2013. Compositional modelling of tight oil using dynamic nanopore properties. In: New Orleans, Louisiana, USA, SPE 166267, SPE Annual Technical Conference and Exhibition.
Whitson, C.H., Brule, M.R., 2000. Phase behavior. In: SPE Monograph, vol. 20.
Wilson, G., 1969. A Modified Redlich-Kwong Equation-of-State, Application to General Physical Data Calculations. AIChE Natl, Cleveland, Ohio.

INDEX

'*Note:* Page numbers followed by "f" indicate figures, "t" indicate tables, and "b" indicate boxes.'

F

S

FSC
www.fsc.org
MIX
Papier | Fördert gute Waldnutzung
FSC® C083411

Kolibri 360 GmbH
Ferdinand-Jühlke-Straße 7
99095 Erfurt, Deutschland
produktsicherheit@kolibri360.de